AF361675

Ubiquitination in Health and Diseases

Ubiquitination in Health and Diseases

Editor

Tomoaki Ishigami

MDPI • Basel • Beijing • Wuhan • Barcelona • Belgrade • Manchester • Tokyo • Cluj • Tianjin

Editor
Tomoaki Ishigami
Yokohama City University
Graduate School of Medicine
Japan

Editorial Office
MDPI
St. Alban-Anlage 66
4052 Basel, Switzerland

This is a reprint of articles from the Special Issue published online in the open access journal *International Journal of Molecular Sciences* (ISSN 1422-0067) (available at: https://www.mdpi.com/journal/ijms/special_issues/ubiquitination_health).

For citation purposes, cite each article independently as indicated on the article page online and as indicated below:

LastName, A.A.; LastName, B.B.; LastName, C.C. Article Title. *Journal Name* **Year**, *Volume Number*, Page Range.

ISBN 978-3-0365-0546-6 (Hbk)
ISBN 978-3-0365-0547-3 (PDF)

Contents

About the Editor

Tomoaki Ishigami, Ph.D.
Associate Professor
Department of Medical Science and Cardio-Renal Medicine
Yokohama City University Graduate School of Medicine

Dr. Tomoaki Ishigami was born in Tokyo, Japan. He completed his graduation and obtained his medical degree and Ph.D. from the Yokohama City University, Yokohama, Japan. Following this, he moved to the Eccles Institute of Human Genetics, University of Utah, USA, where he worked as a Post-Doctoral Fellow. His primary research interest is elucidating the molecular pathophysiology of salt-sensitive hypertension and atherosclerosis [1–3]. To elucidate the molecular basis of human hypertension, he has used molecular genetic approaches, including single nucleotide polymorphism (SNP) analyses and extensive direct sequencing of candidate genes such as human angiotensinogen [4–6] and human *NEDD4L* [7–9] . Dr. Ishigami and his colleagues revealed that human angiotensinogen gene variants are responsible for human hypertension and appeared according to the migration and selection of human beings globally after leaving Africa, providing substantial evidence supporting the thrifty gene hypothesis for salt-sensitive hypertension [10,11]. Various ion transporters and their accessory proteins expressed along the urinary tubules are biologically responsible for salt sensitivity and salt-sensitive hypertension. Angiotensinogen in proximal tubules and renin in connecting tubules are the two major components of the tubular renin–angiotensin system. The discovery of alternative renin transcripts in the connecting tubules of transgenic mice in 2014 was a landmark achievement for Dr. Ishigami and his team [12]. Furthermore, using extensive direct sequencing, Dr. Ishigami and colleagues reported the molecular diversity of the human *NEDD4L* gene, which is critically involved in epithelial sodium channel ubiquitination [7–9,13,14]. Additionally, using genetically engineered mouse models, direct monitoring of blood pressure via an implantable telemetry system, and detailed metabolic analyses by measurement of urine and food consumption, Dr. Ishigami and colleagues successfully provided direct evidence of the relationship between the lack of a single molecular variant of the ubiquitin ligase gene, *Nedd4-2*, and dietary salt-sensitive hypertension [15–17]. Finally, a review by Dr. Ishigami and colleagues, entitled Regulators of Epithelial Sodium Channels in Aldosterone-Sensitive Distal Nephrons (ASDN): Critical Roles of Nedd4L/Nedd4-2 and Salt-Sensitive Hypertension, was selected to feature in the current Special Issue of *IJMS*, titled Ubiquitination in Health and Disease [18].

1.   Chen L, Ishigami T, Nakashima-Sasaki R, Kino T, Doi H, Minegishi S, Umemura S (2016) Commensal Microbe-specific Activation of B2 Cell Subsets Contributes to Atherosclerosis Development Independently of Lipid Metabolism. EBioMedicine 13: 237–247. DOI:10.1016/j.ebiom.2016.10.030.

2.   Chen L, Ishigami T (2016) Intestinal Microbiome and Atherosclerosis—Authors' Reply. EBioMedicine 13: 19–20. DOI:10.1016/j.ebiom.2016.10.039.

3. Arakawa K, Ishigami T, Nakai-Sugiyama M, Chen L, Doi H, Kino T, Minegishi S, Saigoh-Teranaka S, Sasaki-Nakashima R, Hibi K, et al. (2019) Lubiprostone as a potential therapeutic agent to improve intestinal permeability and prevent the development of atherosclerosis in apolipoprotein E-deficient mice. PLoS One 14: e0218096. DOI:10.1371/journal.pone.0218096.

4. Ishigami T, Umemura S, Iwamoto T, Tamura K, Hibi K, Yamaguchi S, Nyuui N, Kimura K, Miyazaki N, Ishii M (1995) Molecular variant of angiotensinogen gene is associated with coronary atherosclerosis. Circulation 91: 951–954.

5. Ishigami T, Umemura S, Tamura K, Hibi K, Nyui N, Kihara M, Yabana M, Watanabe Y, Sumida Y, Nagahara T, et al. (1997) Essential hypertension and 5′ upstream core promoter region of human angiotensinogen gene. Hypertension 30: 1325–1330.

6. Ishigami T, Tamura K, Fujita T, Kobayashi I, Hibi K, Kihara M, Toya Y, Ochiai H, Umemura S (1999) Angiotensinogen gene polymorphism near transcription start site and blood pressure: role of a T-to-C transition at intron I. Hypertension 34: 430–434.

7. Ishigami T, Araki N, Umemura S (2010) Human Nedd4L rs4149601 G allele generates evolutionary new isoform I with C2 domain. Hypertension 55: e10; author reply e11. DOI:10.1161/HYPERTENSIONAHA.109.146738.

8. Ishigami T, Araki N, Minegishi S, Umemura M, Umemura S (2014) Genetic variation in NEDD4L, salt sensitivity, and hypertension: human NEDD4L rs4149601 G allele generates evolutionary new isoform I with C2 domain. J Hypertens 32: 1905. DOI:10.1097/HJH.0000000000000293.

9. Dunn DM, Ishigami T, Pankow J, von Niederhausern A, Alder J, Hunt SC, Leppert MF, Lalouel JM, Weiss RB (2002) Common variant of human NEDD4L activates a cryptic splice site to form a frameshifted transcript. J Hum Genet 47: 665–676.

10. Nakajima T, Jorde LB, Ishigami T, Umemura S, Emi M, Lalouel JM, Inoue I (2002) Nucleotide diversity and haplotype structure of the human angiotensinogen gene in two populations. Am J Hum Genet 70: 108–123.

11. Nakajima T, Wooding S, Sakagami T, Emi M, Tokunaga K, Tamiya G, Ishigami T, Umemura S, Munkhbat B, Jin F, et al. (2004) Natural selection and population history in the human angiotensinogen gene (AGT): 736 complete AGT sequences in chromosomes from around the world. Am J Hum Genet 74: 898–916

12. Ishigami T, Kino T, Chen L, Minegishi S, Araki N, Umemura M, Abe K, Sasaki R, Yamana H, Umemura S (2014) Identification of bona fide alternative renin transcripts expressed along cortical tubules and potential roles in promoting insulin resistance in vivo without significant plasma renin activity elevation. Hypertension 64: 125–133. DOI:10.1161/HYPERTENSIONAHA.114.03394

13. Araki N, Umemura M, Miyagi Y, Yabana M, Miki Y, Tamura K, Uchino K, Aoki R, Goshima Y, Umemura S, et al. (2008) Expression, transcription, and possible antagonistic interaction of the human Nedd4L gene variant: implications for essential hypertension. Hypertension 51: 773–777

14. Umemura M, Ishigami T, Tamura K, Sakai M, Miyagi Y, Nagahama K, Aoki I, Uchino K, Rohrwasser A, Lalouel JM, et al. (2006) Transcriptional diversity and expression of NEDD4L gene in distal nephron. Biochem Biophys Res Commun 339: 1129–1137

15. Minegishi S, Ishigami T, Kino T, Chen L, Nakashima-Sasaki R, Araki N, Yatsu K, Fujita M, Umemura S (2016) An isoform of Nedd4-2 is critically involved in the renal adaptation to high salt intake in mice. Sci Rep 6: 27137. DOI:10.1038/srep27137

16. Minegishi S, Ishigami T, Kawamura H, Kino T, Chen L, Nakashima-Sasaki R, Doi H, Azushima K, Wakui H, Chiba Y, et al. (2017) An Isoform of Nedd4-2 Plays a Pivotal Role in Electrophysiological Cardiac Abnormalities. Int J Mol Sci 18. DOI:10.3390/ijms18061268

17. Kino T, Ishigami T, Murata T, Doi H, Nakashima-Sasaki R, Chen L, Sugiyama M, Azushima K, Wakui H, Minegishi S, et al. (2017) Eplerenone-Resistant Salt-Sensitive Hypertension in Nedd4-2 C2 KO Mice. Int J Mol Sci 18. DOI:10.3390/ijms18061250

18. Ishigami T, Kino T, Minegishi S, Araki N, Umemura M, Ushio H, Saigoh S, Sugiyama M (2020) Regulators of Epithelial Sodium Channels in Aldosterone-Sensitive Distal Nephrons (ASDN): Critical Roles of Nedd4L/Nedd4-2 and Salt-Sensitive Hypertension. Int J Mol Sci 21. DOI:10.3390/ijms21113871

Preface to "Ubiquitination in Health and Diseases"

Dear Colleagues,

Ubiquitination is a representative, reversible biological process for the post-translational modification of various proteins with multiple catalytic sequences, including ubiquitin itself, the E1 ubiquitin-activating enzymes, the E2 ubiquitin-conjugating enzymes, and the E3 ubiquitin ligase and deubiquitinating enzymes. Since the ubiquitin–proteasome system plays a pivotal role in various molecular life phenomena such as cell cycle control, protein quality control, and cell surface expression of ion transporters, its failure causes various diseases such as cancer, neurodegenerative diseases, cardiovascular diseases, and hypertension. Various genetic diseases derived from abnormalities in genes involved in ubiquitination have been reported, such as Parkinsonism, Cushing disease, and Liddle syndrome. Ubiquitination is a post-translational modification of proteins subsequent to phosphorylation, and approximately 40% of the proteins encoded by human genes undergo this modification. Although clinical applications targeting ubiquitination are still limited compared with those directed to kinase systems such as tyrosine kinases, for which an inventory of tyrosine kinase inhibitors is already available in clinical settings, many compounds affecting ubiquitination and presenting high pharmacological activity have been identified at the basic research level; therefore, future developments can be expected. Abnormalities of E3 ubiquitin ligase affect the phenotypes specific to each target substrate, which, thus, are also attractive targets for selective drug discovery. In this Special Issue of the *International Journal of Molecular Science*, we would like to invite your contributions in the form of either original research articles or reviews, in the expanding field of mechanic, functional, and pharmacological dissections, concerning the physiological and pathological implications of specific ubiquitination reactions.

Tomoaki Ishigami
Editor

Review

Role of Deubiquitinases in Human Cancers: Potential Targeted Therapy

Keng Po Lai [1], Jian Chen [1,*] and William Ka Fai Tse [2,*]

[1]   Guangxi Key Laboratory of Tumor Immunology and Microenvironmental Regulation, Guilin Medical University, Guilin 541004, China; kengplai@cityu.edu.hk

[2]   Center for Promotion of International Education and Research, Faculty of Agriculture, Kyushu University, Fukuoka 819-0395, Japan

*   Correspondence: chenjian@glmc.edu.cn (J.C.); kftse@agr.kyushu-u.ac.jp (W.K.F.T.);
    Tel.: +86-773-5895810 (J.C.); +81-92-802-4767 (W.K.F.T.)

Received: 25 February 2020; Accepted: 1 April 2020; Published: 6 April 2020

Abstract: Deubiquitinases (DUBs) are involved in various cellular functions. They deconjugate ubiquitin (UBQ) from ubiquitylated substrates to regulate their activity and stability. Studies on the roles of deubiquitylation have been conducted in various cancers to identify the carcinogenic roles of DUBs. In this review, we evaluate the biological roles of DUBs in cancer, including proliferation, cell cycle control, apoptosis, the DNA damage response, tumor suppression, oncogenesis, and metastasis. This review mainly focuses on the regulation of different downstream effectors and pathways via biochemical regulation and posttranslational modifications. We summarize the relationship between DUBs and human cancers and discuss the potential of DUBs as therapeutic targets for cancer treatment. This review also provides basic knowledge of DUBs in the development of cancers and highlights the importance of DUBs in cancer biology.

Keywords: deubiquitinase; degradation; therapeutic target; cancer

---

1. Introduction

Deubiquitinases (DUBs) deconjugate ubiquitin (UBQ) from ubiquitylated substrates to regulate their activities and stability. They are a heterogeneous group of cysteine proteases and metalloproteases [1] that cleave the isopeptide bond between a lysine and the C-terminus of UBQ. DUBs can also edit UBQ chains and process UBQ precursors. In addition, some DUBs can edit UBQ-like proteins and their conjugates. DUBs in the human genome can be classified into subclasses based on their UBQ-protease domains [1]: UBQ-specific proteases (USPs), which represent the largest class, otubain proteases (OTUs), UBQ C-terminal hydrolases (UCHs), Machado–Joseph disease proteases (MJDs), Jab1/Mov34/Mpr1 Pad1 N-terminal+ (MPN+) (JAMM) motif proteases, and motif interacting with ubiquitin-containing novel DUB family (MINDY) [2]. In addition, some new potential DUBs without the above typical domains were currently identified, such as the monocyte chemotactic protein-induced protein (MCPIP) [3] and Zn-finger and UFSP domain protein (ZUFSP) [4]. Approximately 100 DUBs have been identified in humans. They are expressed and located in various organelles in the cell [5]: USP1 and USP7 are found in the nucleus, USP30 in the mitochondria, and USP21 and USP33 in microtubules. More examples are shown in Table 1 [5–8]. Some DUBs have higher expressions in specific tissues, such as USP3 and UCHL3 in the pancreas and lung and USP14 in the brain [5].

Table 1. The sub-cellular localizations of DUBs in mammalian cells.

Organelle	DUBs
Nucleolus	USP36, USP39
Nucleus	BAP1, MYSM1, USP1, USP11, USP22, USP26, USP28, USP29, USP3, USP42, USP44, USP49, USP51, USP7, USPL1, ZUP1
Golgi	USP32, USP33
Endoplasmic reticulum	ATXN3, USP13, USP19, USP33, YOD1
Microtubules	CYLD, USP21
Centriole	USP21, USP33, USP9X
Early endosome and multivesicular body	AMSH, AMSH-LP, USP2a, USP8
Lipid droplet	USP35
Peroxisome and mitochondrion	USP30
Cajal body	USPL1
Stress granule	USP10, USP13, USP5
Plasma membrane	JOSD1, USP6
Cytoplasm	A20, CYLD, PSMD14, UCHL5, USP14

There has been extensive research on ubiquitination [9,10] and how DUBs regulate the deubiquitylation process and their relative functions [11]. Moreover, an increasing number of studies have uncovered the role of DUBs in cancer development [12]. Numerous informative reviews on DUBs have been published [13–18] and research on DUBs has been increasing in recent years. In this review, we aim to provide enriched content that summarizes the classical discoveries, and includes the current findings on DUBs that are related to different aspects of human cancer, including proliferation, cell cycle control, apoptosis, the DNA damage response (DDR), tumor suppression, oncogenesis, and metastasis. Summarized information is shown in Table 2. Lastly, we discuss the potential of DUBs as chemotherapeutic targets for cancer treatment.

2. DUBs and Cell Cycle Control

The cell cycle refers to a series of processes, including DNA synthesis, S phase; cell growth, G1 phase; evaluation of the accuracy of the genomic material, G2 phase; and cell division, M phase. The cycle is completed by duplicating the genetic information and equally segregating it into two daughter cells. Many cell cycle checkpoints are controlled by cyclins and cyclin-dependent kinases (CDKs) [19]. The E3 ligases participate at almost every phase, indicating the importance of ubiquitination and deubiquitination in regulating the cell cycle [20,21].

Table 2. Functional roles of DUBs in cancer properties.

Functions	DUBs	Targets	References
Cell cycle control	BAP1	KLF5	[22]
	DUB3	cyclin A	[23]
	OTUD6B-2	cyclin D1 and c-Myc	[24]
	OTUD7B	APC/C, GβL, HIF2α and E2F1	[25–28]
	USP10	SKP2, Bcr-Abl	[29]
	USP14	AR	[30]
	USP17	p21, ELK-1, Su(var)3-9, Enhancer-of-zeste, and Trithorax domain-containing protein 8	[31–33]
	USP21	FOXM1	[34]
	USP3	KLF5	[35]
	USP7	PHF8	[36]

Table 2. *Cont.*

Functions	DUBs	Targets	References
Cell proliferation	OTUB1	p53	[37]
	OTUD1	p53, SMAD7	[38,39]
	USP10	p53	[40]
	USP14	AR	[41]
	USP15	MDM2, TGF-β receptor	[42,43]
	USP2	MDM2	[44]
	USP28	p53, p21, and p16INK4a	[45,46]
	USP29	p53	[47]
	USP4	β-catenin, p53 and NF-κB	[48–50]
	USP42	P53	[45]
	USP49	FKBP51	[51]
	USP5	P53	[52]
	USP6NL	β-catenin	[53]
	USP7	MDM2	[54–58]
	USP9X	β-catenin, p53	[59,60]
Cell apoptosis	ATXN3	p53	[61]
	JOSD1	MCL1	[62]
	USP5	p53, MAF bZIP	[63,64]
DNA damage repair	BAP1	PR-DUB	[65]
	CYLD	p53	[66]
	OTUD5	SPT16	[67]
	OTUD7A	Rap80/BRCA1-A complex	[68]
	OTUD7B	Rap80/BRCA1-A complex	[68]
	UBP12	PCNA	[69]
	UBP2	PCNA	[69]
	UCHL5	NFRKB	[70]
	USP1	PCNA	[71–74]
	USP11	BRCA2	[75]
	UBP15	PCNA	[69]
	USP3	γH2AX and H2A	[76]
	USP48	BRCA1	[77]
	USP7	PHF8, pBmi1, Bmi1, RNF168, and BRCA1	[36,78]
	USP9X	claspin	[79]
Tumor suppression	CYLD	tumor necrosis factor receptor-associated factor 2, IKKγ	[80–82]
	USP11	PML	[83]
	USP13	PTEN	[84]
	USP46	PHLPP	[85]
Oncogene	BAP1	ASXL1	[86]
	USP22	c-Myc	[87]
	USP28	MYC	[88]
	USP9X	FBW7	[89]
Metastasis	DUB3	Snail, Slug and Twist	[90,91]
	OTUB1	Snail	[92]
	PSMD14	GRB2	[93]
	USP17	SMAD4	[94]
	USP3	SUZ12	[95]
	USP32	RAB7	[96]
	USP37	14-3-3γ	[97]

The ability to advance through different stages of the cell cycle regardless of inhibitory signals is one of the hallmarks of cancer. A large number of DUBs have been found to play roles in cell cycle control of cancers via the regulation of different cell cycle checkpoints. OTUD6B-2 and USP17 were reported to control the G1 phase; USP3, USP10, USP14, USP17, USP20, and BAP1 played roles in the G1/S transition. In addition, S/G2 transition was controlled by OTUD7B and DUB3. USP7 and OTUD7B were necessary for the regulation of mitotic phase (Figure 1).

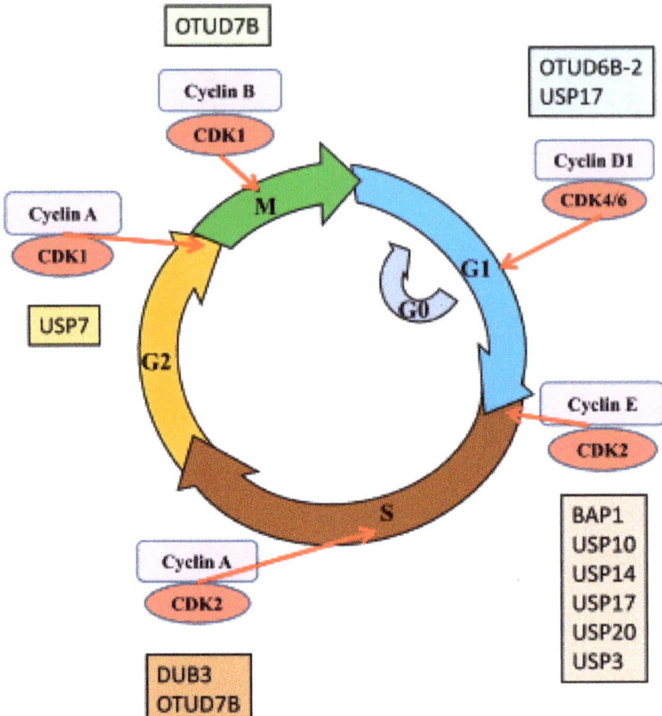

Figure 1. Roles of DUBs in cell cycle control in cancers. The eukaryotic cell cycle consists of the G1 phase (blue), the S-phase (brown), the G2 phase (yellow), and the M (mitosis) phase (green). Cells can enter a quiescent state, the G0 phase (grey). Cell cycle phases are indicated by different colored arrows. The cell cycle is regulated by complexes that are composed of cyclins (light purple), and its relative cyclin-dependent protein kinases (CDKs) (pink). The cyclin-CDK complex plays regulatory roles in the cell cycle. The red arrows indicate their targets, either within the designated cell cycle phase or in the transition state. Various DUBs have been shown to interact with the cyclin–cdk complex. DUBs that participate in G1 phase are labeled in light blue; S phase in light brown; G2 phase in light yellow; and M phase in light green. The detailed interaction partner of each individual DUB can be found in the main text and the Table 2.

For the G1 phase regulation, OTUD6B operates downstream of mTORC1 signaling in non-small cell lung cancer (NSCLC), and its isoform OTUD6B-2, was reported to control the stability of cyclin D1 and c-Myc [24]. USP17 is another cell cycle-regulating DUB. It was found to be highly expressed in colon, esophageal, and cervical cancers. The depletion of USP17 increases the levels of the CDK inhibitor p21 and impairs the G1-S transition, leading to cell cycle arrest [31]. In addition, USP17 deubiquitinates the transcription factor ELK-1. The stabilization of ELK-1 increases the expression of cyclin D1 [32]. USP17 further decreases Su(var)3-9, enhancer-of-zeste, and trithorax domain-containing protein 8 ubiquitination to trigger cellular senescence [33].

For the G1/S phase, USP20 deubiquitinates and stabilizes the DNA checkpoint protein claspin, and thus activates the ATR-Chk1 signaling in the DNA damage response pathway [98]. USP10 deubiquitinates SKP2 and augments the activation of Bcr-Abl by mediating deubiquitination and stabilization of SKP2 in chronic myelogenous leukemia cells [29]. An RNAi-based screening study discovered that USP21 binds and deubiquitinates FOXM1, leading to its increased stability, which induces cell cycle progression in basal-like breast cancer [34]. In addition, DUBs could regulate transcription factors for cell cycle control. The transcription factor Krüppel-like factor 5 (KLF5), which promotes cell proliferation by inhibiting the expression of the cell cycle inhibitor p27 [22], is highly expressed in breast cancer. A genome-wide siRNA library screen identified BAP1 and USP3 as KLF5 DUBs. Both BAP1 and USP3 bind to and stabilize KLF5 via deubiquitination [22,35], indicating the possible regulatory role of DUBs in cancer proliferation. Another example is the androgen receptor (AR), a key transcription factor in the development of breast cancer [99]. It has been reported that AR can be stabilized by USP14, and depletion of USP14 reduces cell proliferation by blocking the G0/G1–S phase transition in AR-responsive breast cancer cells [30].

For the S/G2/M phase, OTUD7B, also called cezanne, is frequently overexpressed in different cancer types, such as breast and lung cancer [100,101]. It is reported to be a cell cycle-dependent DUB because it deubiquitylates substrates of the mitotic cyclin anaphase-promoting complex/cyclosome (APC/C) and prevents their degradation during mitosis [25]. The APC/C is a key regulator of cell cycle progression through the regulation of CDK activity [26]. OTUD7B controls the cell cycle through HIF2α and E2F1 in response to oncogenic signaling [27]. In addition, it removes UBQ from GβL in the mTOR complex to regulate mTORC2 signaling in response to growth signals [28]. Besides, DUB3 can directly deubiquitinate cyclin A in NSCLC. The depletion of DUB3 decreases cyclin A levels, leading to cell cycle arrest at the G0/G1-S phase checkpoint in NSCLC cells [23]. Lastly, it is known that histone demethylases can regulate the cell cycle through transcriptional regulation [102]. The histone demethylase PHF8 is stabilized by USP7, leading to the upregulation of cyclin A2, which is critical for cell growth and proliferation in breast carcinomas [36].

3. DUBs and Cell Proliferation

In addition to their role in regulating the cell cycle, DUBs have been reported to regulate cell proliferation through different cell signaling pathways, such as Wnt/β-catenin signaling, p53-mouse double minute 2 (MDM2) signaling, PI3K-Akt signaling, AR signaling, and transforming growth factor beta (TGF-β) signaling. Aberrant canonical Wnt/β-catenin signaling is tightly associated with many solid and liquid tumors [103]. Furthermore, alteration or loss of differentiation control could facilitate the development of metastatic traits during tumorigenesis [104,105]. Numerous studies have demonstrated the control of Wnt/β-catenin signaling by DUBs in cancer [48,106]. USP6NL is elevated in colorectal cancer (CRC) and regulates β-catenin accumulation. Knockdown of USP6NL results in inhibition of cell proliferation and G0/G1 cell cycle arrest in human CRC cell lines [53]. In addition, USP4 is a candidate for a β-catenin-specific DUB. There is a positive correlation between the levels of USP4 and β-catenin in human colon cancer tissues. Further, knockdown of USP4 reduces invasiveness and migration in colon cancer cells [48]. β-catenin is also stabilized by USP9X, leading to high-grade glioma cell growth. USP9X removes the Lys48-linked polyubiquitin chains from β-catenin to prevent its proteasomal degradation. Depletion of USP9X induces G1-S cell cycle arrest and inhibits cell proliferation in glioblastoma cells [59].

The tumor suppressor p53 is a transcription factor able to control important cellular pathways. It prevents genome mutation and plays protective roles in tumor onset and progression. It is mainly regulated by ubiquitylation, indicating the importance of DUBs in monitoring its ubiquitin cycle [107]. Both MDM2 and p53 are targeted by different DUBs (Figure 2). Suppression of USP2 leads to MDM2 destabilization and results in p53 activation [44]. USP7 plays a key role in the p53 pathway by stabilizing both MDM2 and p53 (Figure 2) [54–58]. Under normal conditions, USP7 has a higher binding affinity to MDM2, the major E3 ligase of p53 [56], and thus deubiquitylates MDM2 more efficiently to prevent its self-degradation and maintain stable protein levels for controlling p53 via the UBQ-proteasome pathway [108]. USP10 regulates p53 localization and stability by deubiquitinating p53. It reverses MDM2-induced p53 nuclear export and degradation [40]. Moreover, USP29 is reported to cleave poly-ubiquitin chains from p53 and thus stabilize it [47], while USP15 stabilizes the E3 UBQ ligase MDM2 in cancer cells and regulates p53 function and cancer cell survival. Inhibition of USP15 induces apoptosis and boosts antitumor T cell responses in tumor cells [42]. Furthermore, a large number of DUBs have been found to target p53 or p53-associated proteins directly, leading to proliferation. USP5 regulates p53 levels and alters cell growth and cell cycle distribution associated with p21 induction in melanoma cells [52]. OTUD1 is required for p53 stabilization, and OTUD1 overexpression increases the cleavage of caspase-3 and PARP and subsequently increases apoptosis [38]. Another p53-associated DUB, otubain 1 (OTUB1), is expressed in high-grade tumor types, such as lung, breast, and ovarian tumors. OTUB1 regulates p53 to promote tumor cell survival and proliferation [37]. USP42 controls the level of p53 ubiquitination during the early phase of the DDR to promote DNA repair, resulting in the activation of p53-dependent transcription and cell-cycle arrest in response to stress [45]. In addition, USP28 depletion leads to increased ubiquitinated H2A-K119 and decreased expression of p53, p21, and p16INK4a, suggesting a role for USP28 in cell proliferation via the control of p53 and p53-associated proteins [109]. Additionally, USP28 deubiquitinates TP53-binding protein 1 to promote p53-mediated transcription [46]. USP4 is a potential oncogene that inhibits p53 and NF-κB via histone deacetylases 2 (HDAC2) [49,50]. USP9X-dependent p53 degradation was observed in hepatocellular carcinoma (HCC) cells treated with the small molecule DUB inhibitor WP1130 [60].

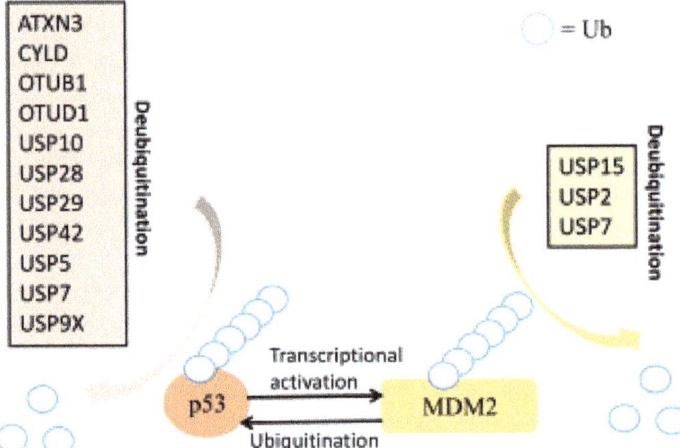

Figure 2. DUBs in MDM2-p53 pathways. Ubiquitination is found on both p53 and MDM2 molecules; various DUBs could revise that via deubiquitination to regulate the p53 pathway. DUBs' targets on p53 are labeled in light brown; those that interact with MDM2 are labeled in light yellow. Detailed descriptions can be referred to the main text.

DUBs are also involved in other signaling pathways that promote tumor proliferation. USP15, which stabilizes the type I TGF-β receptor and enhances the TGF-β pathway, is upregulated in various cancers [43]. In addition, OTUD1 mitigates TGF-β-induced pro-oncogenic responses via deubiquitination of SMAD7 at lysine 220 in breast cancer [39]. USP49 regulates the Akt pathway through the stabilization of FKBP51. FKBP51 activates PH domain leucine-rich-repeats protein phosphatase (PHLPP) to dephosphorylate Akt, which inhibits pancreatic cancer cell proliferation [51]. The AR pathway is commonly activated in prostate cancer (PCa), and it plays a critical role in PCa growth and progression. USP14 was reported to bind with and stabilize AR in androgen-responsive PCa cells. Overexpression of USP14 promotes the proliferation of LNCaP cells [41]. Furthermore, DUBs control different growth factors in tumor cells. For instance, USP8 prevents degradation of the epidermal growth factor receptor and thus promotes proliferation [110].

4. DUBs and Apoptosis

The ability to evade apoptosis is one of the essential changes in cancer cells that causes malignant transformation [111]. Apoptosis is a cellular self-destruction program in response to various cellular stresses. The two extrinsic and intrinsic pathways in apoptosis both involve the activation of caspase molecules. The activation of initiator caspase will further lead to the activation of executioner caspase in apoptosis [112]. DUBs were found to target different pro- and anti-apoptotic proteins in both the extrinsic and intrinsic pathways. ATXN3 stabilizes p53 by deubiquitination and promotes p53-mediated apoptosis [61]. USP5 targets p53-unanchored UBQ polymers and regulates p53-mediated transcription. Depletion of USP5 controls tumor necrosis factor alpha apoptosis-inducing ligand (TRAIL)-mediated apoptotic responsiveness in TRAIL-resistant tumor cells, and this function of USP5 ubiquitination can be blocked by caspase 8-specific inhibitors [63]. In addition, USP5 deubiquitinates the MAF bZIP transcription factor and prevents its degradation. Knockdown of USP5 leads to apoptosis in multiple myeloma cells [64]. In a chemoresistant xenograft model, JOSD1 was identified to be upregulated during the development of chemoresistance. Moreover, JOSD1 has been reported to deubiquitinate and stabilize MCL1, which plays a suppressive role in mitochondrial apoptotic signaling. Therefore, depletion of JOSD1 leads to severe apoptosis in gynecological cancer cells through the degradation of MCL1 [62]. There are several DUBs that regulate the apoptotic pathways via BCL-2 family, an inhibitor of apoptotic proteins (IAPs) and caspases. DUB3/USP17 induces apoptosis through caspase 3 activation [113], whereas USP15 plays a role in stabilizing procaspse 3 [114]. Besides, A20, a DUB belongs to the OTU subclass, interacts with caspase 8 to reverse the ubiquitination of a cullin 3-based E3 ligase [115]. As for the IAPs, they are a class of proteins that inhibit apoptosis. They contain the baculovirus IAP repeat domain and the RING domain that provides the E3 ligase property [116]. USP19 stabilizes the cellular IAP1 and cellular IAP2 during caspase activation and apoptosis [117]; OTUD1 was found to regulate the TNF-dependent cell death by modulating the cellular IAP1 stability [118]. Furthermore, USP9X was reported to interact with an E3 ligase X-linked IAP for mitotic cell fate decision [119]. In addition, USP27X was found to interact with the BIM. BIM is a pro-apoptotic BH3-only protein that regulates the cell death proteins such as BAX. Overexpression of Usp27x reduces BIM ubiquitination, and induces apoptosis in tumor cells. On the other hand, suppression of USP27X could reduce apoptosis [120].

5. DUBs and the DDR

Cells undergo DDR to sense and repair unique lesion structures in the damaged DNA. Efficient DDR protects cells from genomic instability [121,122]. Ubiquitination regulates DDR by controlling DDR protein localization, activity, and stability [123]. DUBs play critical roles in different stages of the DDR through the regulation of many molecules involved in DNA repair (Figure 3). DNA repair is important for preventing tumor formation [124]. Proliferating cell nuclear antigen (PCNA) is a key molecule that mediates the tolerance to DNA damage and allows the growth of tumors. PCNA is monoubiquitinated in response to DNA damage. A fission yeast study showed the importance of UBP2, UBP12, and UBP15 in the stabilization of mono, di, and polyubiquitylated forms of PCNA, which sensitize cells to DNA damage [69]. In addition, PCNA can be deubiquitinated by USP1 in the crosslink repair pathway in Fanconi anemia [71–73]. In a complex with its cofactor UAF1, USP1 reverses PCNA ubiquitination [74]. UCHL5 regulates double-strand break (DSB) resection and repair by homologous recombination through protecting its interactor, NFRKB, from degradation [70]. In addition, USP20 plays role in genome maintenance and DNA repair by enhancing recombinational repair of collapsed replication forks [125]. Furthermore, USP9X regulates the DNA checkpoint protein claspin during S phase, suggesting a role in DNA repair [79]. USP7-promoted PHF8 stabilization confers cellular resistance to genotoxic insults and is required for the recruitment of BLM and KU70, which are both essential for DNA DSB repair [36].

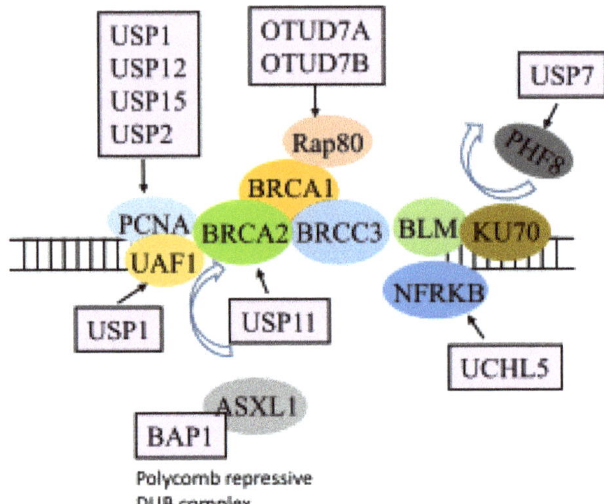

Figure 3. Roles of DUBs in DNA damage response. Various DUBs (light purple) have been shown to interact with molecules (various colors) that play roles in DNA repair and chromosomal stability during DNA damage. Proliferating cell nuclear antigen (PCNA) plays important roles during DNA replication and repair, while BRCA members are the key players in repairing the DNA lesions such as DNA double-strand breaks. In addition, BLM repairs DNA double-strand breaks to maintain genome stability. Detailed information can be found in the main text.

Breast-cancer susceptibility gene (BRCA) 1 contributes to DNA repair and the maintenance of chromosomal stability in response to DNA damage [126]. BRCA1 appears to play roles in two distinct pathways of DSB repair, non-homologous end joining and homology-directed repair, through the regulation of different effectors. It has been reported that several DUBs can regulate BRCA1. The BRCA1-associated DUB BAP1 is mutated in mesothelioma and melanoma [65]. BAP1 is a phosphorylation target for the DDR kinase ATM, and BAP1 mediates rapid poly(ADP-ribose)-dependent recruitment of the polycomb DUB complex PR-DUB to repair DNA DSBs [65]. In addition, both cezanne (OTUD7B) and cezanne2 (OTUD7A) promote the recruitment of the Rap80/BRCA1-A complex by binding to Lys63-polyubiquitin and targeting Lys11-polyubiquitin in response to DNA repair [68]. Another DUB, USP11, forms a complex with BRCA2. It deubiquitylates the partner and localizer of BRCA2 to enhance DNA repair [75]. BRCA1/BRCA2-containing complex 3 (BRCC3) is a Lys63-specific DUB involved in the DDR. BRCC3 inactivation increases the release of several cytokines, including G-CSF, which enhances proliferation in AML cell lines [127]. Further, OTUD5, a specific stabilizer of the UBR5 E3 ligase, is reported to localize at DNA DSBs. OTUD5 plays two roles at DSBs. First, OTUD5 interacts with UBR5 and represses RNA Pol II-mediated elongation and RNA synthesis. In addition, OTUD5 interacts with the FACT component SPT16 and antagonizes histone H2A deposition at DSBs [67].

Histone ubiquitination at DNA breaks is required for activation of the DDR and DNA repair. BRCA1-BARD1-catalyzed ubiquitination of histone H2A primes chromatin for repair by homologous recombination during the DDR. Ubiquitination of histone H2A and γH2AX by the UBQ ligases RNF168 and RNF8 generates a cascade of ubiquitination. USP3 deubiquitinates ubiquitinated γH2AX and H2A [76]. USP48 is another H2A DUB that is specific for the C-terminal BRCA1 ubiquitination site. USP48 promotes genomic stability by antagonizing the BRCA1 E3 ligase function. Depletion of USP48 increases the distance between p53-binding protein 1 (53BP1) from the DNA break point [77]. It should be noted that histone ubiquitination by RNF168 is a critical event for the recruitment of BRCA1 and 53BP1, and the stability of RNF168 can be regulated by USP7. Depletion of USP7 impairs H2A and γH2AX monoubiquitination, leading to decreases in the levels of pBmi1, Bmi1, RNF168, and BRCA1 under ultraviolet radiation-induced DNA damage [78]. Moreover, USP3, a histone H2A DUB, negatively regulates UBQ-dependent DDR signaling through regulation of chromatin ubiquitination in response to genotoxic stress [128]. Lastly, CYLD deubiquitinates p53 and facilitates its stabilization in response to genotoxic stress. Loss of CYLD catalytic activity causes impaired DNA damage-induced p53 stabilization and activation of skin tumorigenesis [66].

6. DUBs and Tumor Suppressors/Oncogenes

DUBs play an important role in cancer development by controlling various different tumor suppressors and oncogenes. CYLD was first identified as the tumor suppressor gene for cylindromatosis [129]. Its protein expression level is downregulated in various tumor types [130,131]. CYLD plays an essential role in NF-κB [82] and c-Jun N-terminal kinase pathways [132]. Briefly, it inhibits NF-κB activation by promoting deubiquitylation of several UBQ-dependent NF-κB positive regulators, such as tumor necrosis factor receptor-associated factor 2 and the NF-κB essential modulator/IKKγ subunit [80–82]. Enhanced and/or prolonged NF-κB signaling due to reduced CYLD activity increases cellular apoptosis resistance and the chances of tumor formation [133]. USP13 also acts as a tumor suppressor through its regulation of the phosphatase and tensin homolog deleted on chromosome 10 (PTEN)/AKT pathway in oral squamous cell carcinoma. Overexpression of USP13 induces PTEN expression and represses the activation of AKT, glucose transporter-1, and hexokinase-2, leading to growth inhibition [84]. In an RNAi screen, USP11 was identified as a promyelocytic leukemia (PML) regulator to deubiquitinate and stabilize PML, counteracting the functions of PML. UBQ ligases RNF4 and the KLHL20-Cullin 3-Roc1 complex [83]. This complex causes suppression of PML in many cancer types [83]. PHLPP is a family of Ser/Thr protein phosphatases that serve as tumor suppressors by negatively regulating AKT. In CRC, USP46 is reported to bind to PHLPP and directly remove its polyubiquitin chains, resulting in the stabilization of PHLPP. USP46-mediated stabilization of PHLPP subsequently inhibits AKT, blocking proliferation and tumorigenesis in colon cancer cells [85].

A large number of DUBs have been reported to bind with and stabilize oncogenes, such as c-MYC. USP22 promotes deubiquitination of c-MYC in breast cancer cells, resulting in increased levels of c-MYC. Overexpression of USP22 stimulates tumorigenic activity in breast cancer cells and is closely correlated with breast cancer progression [87]. USP9X acts as an FBW7 interactor, and the loss of FBW7 has been observed in many types of human cancer [134]. USP9X antagonizes FBW7-mediated ubiquitylation and causes FBW7 stabilization. USP9X suppresses tumor formation by regulating FBW7 protein stability, which reduces c-MYC levels [89]. The degradation of the oncogene product MYC is also enhanced by USP28 [88]. An integrated genomic analysis of malignant pleural mesotheliomas uncovered somatic inactivating mutations in the tumor-suppressive nuclear DUB BAP1. BAP1 targets histones with the polycomb repressor subunit ASXL1 [86].

7. DUBs and Metastasis

Metastasis, which is the ability of cancer cells spread to different tissues or organs, is regulated by many mechanisms. It is a series of biological processes including various invasion-metastasis cascades. Multiple reports have suggested the role of DUBs in controlling these mechanisms. The epithelial–mesenchymal transition (EMT) represents one of the most important invasive events in cancer metastasis. It refers to a change of a subset of adhesion molecules in cells: adopting a migratory and invasive behavior [135]. Numerous DUBs are involved in cancer cell invasiveness through the regulation of different EMT transcription factors (Figure 4).

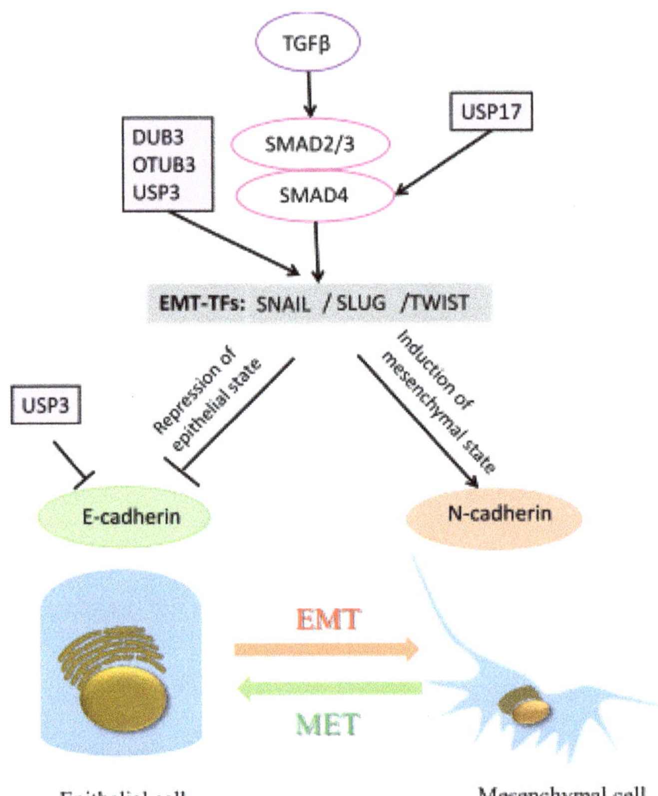

Figure 4. Roles of DUBs in epithelial–mesenchymal transition (EMT) in cancer metastasis. Epithelial cells are held together by numerous proteins, including tight junctions, adherens junctions, and desmosomes. These cells express molecules that are associated with the epithelial state, such as E-cadherin in epithelial state, and N-cadherin in mesenchymal state. Induction of EMT induces different EMT-inducing transcription factors (EMT-TFs) such as SNAIL, SLUG, and TWIST. These factors can then inhibit the epithelial state-related genes, such as E-cadherin, and activate the mesenchymal state related genes, such as N-cadherin. Various DUBs have been shown to interact with different EMT regulators. EMT is a reversible process, and mesenchymal cells can revert to the epithelial state by undergoing mesenchymal–epithelial transition (MET). A detailed description can be found in the main text.

SNAIL is a key regulator of EMT and plays an important role in tumor progression and metastasis. A group of DUBs, including OTUB1, DUB3, and USP3, are reported to stabilize Snail through preventing its ubiquitination and proteasomal degradation. OTUB1 promotes metastasis of esophageal squamous cell carcinoma through the stabilization of Snail [92]. DUB3 is found to be overexpressed in breast cancer, and depletion of DUB3 leads to Snail1 destabilization, which suppresses EMT, tumor invasiveness, and metastasis [90]. In addition, DUB3 also interacts with SLUG and TWIST and prevents their degradation, thereby promoting migration, invasion, and cancer stem cell-like properties in breast cancer cells [91]. Moreover, USP3 is significantly upregulated in glioblastomas and gastric cancer (GC). Clinicopathological data demonstrate that USP3 correlates with a shorter overall and relapse-free survival in glioblastomas [136]. It has also been reported that USP3 interacts with and stabilizes SUZ12 via deubiquitination. Expression of SUZ12 is negatively correlated with E-cadherin, which promotes migration and EMT in GC cells [95]. SMAD4 has been found to regulate EMT. USP17 is upregulated in osteosarcoma tissues and stabilizes SMAD4 through its DUB activity, leading to enhanced osteosarcoma cell invasion [94].

In addition to EMT mediators, DUBs target other molecules involved in cancer invasiveness. High expression of 14-3-3γ is found in various cancers, such as breast cancer and NSCLC [137,138]. Overexpression of 14-3-3γ promotes cell migration and invasion and correlates with the invasiveness of cancer cells. USP37 regulates the stability of 14-3-3γ through its DUB activity [97]. Another DUB, 26S proteasome non-ATPase regulatory subunit 14 (PSMD14), is a posttranslational regulator of growth factor receptor bound protein 2 (GRB2). PSMD14 is significantly upregulated in HCC tissues, and it inhibits the degradation of GRB2 via deubiquitination. Overexpression of PSMD14 correlates with vascular invasion, tumor recurrence, and poor tumor-free and overall survival in patients with HCC [93]. The small GTPase Ras-related protein RAB7 is an early-induced melanoma driver and endocytosis protein that favors tumor invasion [139]. It is suggested to play roles in modulating endosomal maturation and autophagosome resolution in various cell types [140,141]. It was recently shown to be regulated by USP32 [96].

8. DUBs as Therapeutic Targets for Cancer Treatment

As mentioned above, DUBs have been shown to deubiquitinate many targets involved in different characteristics of cancer (Table 2), suggesting that DUBs may be potential therapeutic targets in cancer treatment. Indeed, many studies have been conducted to examine the potential of DUBs in cancer therapeutics. As DUBs are part of the proteasome system, proteasome inhibitors target them, which has shown promising successes for cancer treatment. Several examples are given below. Bortezomib, the first proteasome inhibitor, has entered clinical practice to treat relapsed multiple myeloma and showed outstanding antimyeloma activity [142,143]. In addition, combination of bortezomib and epirubicin significantly increases the sensitivity of colorectal carcinoma cells to apoptosis [144]. Due to the resistance to bortezomib, next-generation proteasome inhibitors carfilzomib and ixazomib have been approved. Carfilzomib irreversibly binds to the β-5 subunit of the proteasome [145]. A preclinical study has demonstrated that carfilzomib increased efficacy against bortezomib-resistant multiple myeloma [146]. In the Phase 2 and Phase 3 clinical trials, single-agent carfilzomib provided durable anticancer activity in patients with relapsed and/or refractory multiple myeloma [147]. Ixazomib, the first oral proteasome inhibitor to enter the clinic, is now commonly used for multiple myeloma treatment. It is an efficacious and long-term therapy for patients with advanced stage multiple myeloma [148]. In a double-blind Phase 3 trial, the use of ixazomib significantly improved progression-free survival in patients with relapsed and/or refractory multiple myeloma [149].

In addition to proteasome inhibitors, numerous DUB therapeutic targets have been developed. One excellent and classical example is USP7. Activating p53 by inhibiting MDM2 is a major direction of cancer treatment [150,151]. Nutlin-3 from Roche and RITA (2,5-bis(5-hydroxymethyl-2-thienyl)furan (NSC652287)) from the National Cancer Institute have been developed for interfering with the MDM2/p53 interaction to induce p53 and therefore cell death in human tumor cells [152–154]. They represent an important class of small molecules that has significant antitumor effects without obvious toxicity in mice [153,155], which further suggests that promoting MDM2 degradation will provide a therapeutic benefit when treating p53-related cancers. Additionally, USP7 silencing promotes the degradation of MDM2 and thus abrogates p53 degradation. Targeting DUBs might provide a new direction for cancer treatment, as it has the advantage of a simpler mechanism than targeting UBQ ligases or the 26S proteasome [150,156]. A small molecule lead-like inhibitor of USP7, HBX41108, which stabilizes and activates p53, was identified using high-throughput screening [156]. This inhibitor symbolizes a milestone in DUB drug development and sheds light on new potential cancer therapies using DUB inhibitors.

In addition, many cancer studies have focused on the apoptotic role of DUBs and exploited this role for chemotherapy. A drug screening study demonstrated that the small molecule DUB inhibitor b-AP15 inhibits two DUBs, USP14 and UCHL5. Treatment with b-AP15 results in apoptosis of human

Waldenström macroglobulinemia (WM) cell lines and primary WM tumor cells [157]. In another chemotherapeutic study, pharmacological targeting of USP14 with the FDA-approved small-molecule inhibitor VLX1570 decreased viability in endometrial cancer cells through cell cycle arrest and caspase 3-mediated apoptosis [158]. The oncogenic transcription factor pre-B cell leukemia homeobox-1 (PBX1) promotes advanced PCa cell proliferation. USP9X interacts with and stabilizes the PBX1 protein by attenuating its Lys48-linked polyubiquitination. The USP9X inhibitor WP1130 markedly induces PBX1 degradation and promotes PCa cell apoptosis [159]. The selected DUB inhibitors that target on cancer cells are summarized in Table 3. To conclude, DUBs play multiple roles in cellular functions. The aberrant expression and regulation of these enzymes have been shown to contribute to promote tumorigenesis, making them promising therapeutic targets for cancer therapy.

Table 3. Summary of known DUB inhibitors that are targeted in cancer cells.

DUBs	DUBs Inhibitors	Therapeutic Targets	Functional Effects	References
USP8	9-Ethyloxyimino-9H-indeno [1,2-b]pyrazine-2,3-dicarbonitrile	Non-small cell lung cancer	Downregulation of receptor tyrosine kinases including EGFR, ERBB2, ERBB3, and MET	[160]
UCHL1	LDN-57444	Lung cancer cell line	Inhibit proliferation	[161]
UCHL1, UCHL3	TCID	Multiple myeloma	Induce apoptosis	[162]
USP1	Pimozide	Leukemic cell lines	Promoted the degradation of ID1	[163]
USP1-UAF1	ML323	Non-small cell lung cancer and osteosarcoma cells	Induced DNA damage	[164]
USP1-UAF1	Pimozide and GW7647	Non–small cell lung cancer	Inhibit cell proliferation	[165]
USP2	ML346	Colorectal cancer nad mantle cell lymphoma	Accelerate cyclin D1 degradation, cell cycle arrest	[166]
USP2a/USP2b/USP5/USP8	AM146, RA-9 and RA-14	Breast, ovarian and cervical cancer cell lines	Downregulation cell-cycle promoter, and upregulation of tumor suppressor	[167]
USP5/IsoT, USP4	Vialinin A	Basophilic leukemia cells	Inhibit the release of TNFα	[168]
USP7	HBX 41,108	Colorectal carcinoma	Induced p53-dependent apoptosis	[156]
USP7/USP47	P5091 and Compound 1	Multiple myeloma	Induce apoptosis, inhibit tumor growth	[169,170]
USP9X/USP5/USP24	WP1130	Mantle cell lymphoma	Downregulation of antiapoptotic and upregulation of proapoptotic proteins, such as MCL-1 and p53	[171,172]
USP14/ UCHL5	AC17	Human lung cancer cells	Inhibit NFκB pathway and reactive p53	[173]
USP14/UCHL5	b-AP15 (WO2013058691)	Multiple myeloma/ colorectal carcinoma	Downregulation of CDC25C, CDC2, and cyclin B1/ overexpression of the anti-apoptotic mediator Bcl-2 and anti-tumor activity	[162,174]
USP14/UCHL5	VLX1570	Colon carcinoma cell	Inhibit proteasome DUB activity	[175]

Author Contributions: Conceptualization, J.C. and W.K.F.T.; writing the review and editing, K.P.L., J.C. and W.K.F.T. All authors have read and agreed to the published version of the manuscript.

Funding: This work is supported by the funding of National Natural Science Foundation of China (number 81660610), and the National Natural Science Foundation of Guangxi Province (number 2019GXNSFFA245001, 2017GXNSFDA198019, 2018GXNSFAA281159) for Jian Chen. The pathology works in WKF Tse Laboratory were partially supported by the Japan Society for the Promotion of Science Bilateral Open Partnership Joint Research Project AJ179064, the joint research program of the Institute for Molecular and Cellular Regulation, Gunma University, Japan (18003); and the National Institute of Basic Biology Collaborative Research Program, Japan (19-307).

Acknowledgments: K.P.L. is supported by Hong Kong SAR, Macao SAR, and Taiwan Province Talent Young Scientist Program of Guangxi.

Conflicts of Interest: The authors declare no conflict of interest.

Abbreviations

53BP1	p53-binding protein 1
Akt	protein kinase B
APC/C	anaphase-promoting complex/cyclosome
AR	androgen receptor
BAP1	BRCA1 associated protein 1
BRCA	breast-cancer susceptibility gene
BRCC3	BRCA1/BRCA2-containing complex 3
CDK	cyclin-dependent kinase
CRC	colorectal cancer
DDR	DNA damage response
DSB	double-strand break
DUB	Deubiquitinase
ELK-1	ETS like-1 protein
EMT	epithelial-mesenchymal transition
FBW7	F-box and WD repeat domain-containing 7
FKBP51	FK506-binding protein 51
GC	gastric cancer
GRB2	growth factor receptor bound protein 2
HCC	hepatocellular carcinoma
JOSD1	Josephin domain containing 1
KLF5	Krüppel-like factor 5
MDM2	mouse double minute 2
mTORC1	mammalian target of rapamycin complex 1
NK-κB	nuclear factor kappa-light-chain-enhancer of activated B cells
NSCLC	non-small cell lung cancer
OTUB1	otubain 1
OTU	otubain protease
PBX1	pre-B cell leukemia homeobox-1
PCa	prostate cancer
PCNA	proliferating-cell nuclear antigen
PHF8	PHD finger protein 8
PHLPP	PH domain leucine-rich-repeats protein phosphatase
PML	promyelocytic leukemia
PSMD14	26S proteasome non-ATPase regulatory subunit 14
PTEN	phosphatase and tensin homolog deleted on chromosome 10
RNF	ring finger proteins
SKP2	S-phase kinase associated protein 2.
TGF-β	transforming growth factor beta
TRAIL	tumor necrosis factor alpha apoptosis-inducing ligand
UBQ	ubiquitin
UBR5	ubiquitin protein ligase E3 component N-recognin 5
UCH	ubiquitin C-terminal hydrolases
UCHL	ubiquitin C-terminal hydrolases like
USP	ubiquitin-specific protease
WM	Waldenström macroglobulinemia

References

1. Nijman, S.M.B.; Luna-Vargas, M.P.A.; Velds, A.; Brummelkamp, T.R.; Dirac, A.M.G.; Sixma, T.K.; Bernards, R. A genomic and functional inventory of deubiquitinating enzymes. *Cell* **2005**, *123*, 773–786. [CrossRef] [PubMed]
2. Abdul Rehman, S.A.; Kristariyanto, Y.A.; Choi, S.Y.; Nkosi, P.J.; Weidlich, S.; Labib, K.; Hofmann, K.; Kulathu, Y. Mindy-1 is a member of an evolutionarily conserved and structurally distinct new family of deubiquitinating enzymes. *Mol. Cell* **2016**, *63*, 146–155. [CrossRef] [PubMed]
3. Liang, J.; Saad, Y.; Lei, T.; Wang, J.; Qi, D.; Yang, Q.; Kolattukudy, P.E.; Fu, M. Mcp-induced protein 1 deubiquitinates traf proteins and negatively regulates jnk and nf-kappab signaling. *J. Exp. Med.* **2010**, *207*, 2959–2973. [CrossRef] [PubMed]
4. Hermanns, T.; Pichlo, C.; Woiwode, I.; Klopffleisch, K.; Witting, K.F.; Ovaa, H.; Baumann, U.; Hofmann, K. A family of unconventional deubiquitinases with modular chain specificity determinants. *Nat. Commun.* **2018**, *9*, 1–13. [CrossRef]
5. Clague, M.J.; Barsukov, I.; Coulson, J.M.; Liu, H.; Rigden, D.J.; Urbe, S. Deubiquitylases from genes to organism. *Physiol. Rev.* **2013**, *93*, 1289–1315. [CrossRef]
6. Sowa, M.E.; Bennett, E.J.; Gygi, S.P.; Harper, J.W. Defining the human deubiquitinating enzyme interaction landscape. *Cell* **2009**, *138*, 389–403. [CrossRef]
7. Komander, D.; Clague, M.J.; Urbe, S. Breaking the chains: Structure and function of the deubiquitinases. *Nat. Rev. Mol. Cell Biol.* **2009**, *10*, 550–563. [CrossRef]
8. Urbe, S.; Liu, H.; Hayes, S.D.; Heride, C.; Rigden, D.J.; Clague, M.J. Systematic survey of deubiquitinase localization identifies usp21 as a regulator of centrosome- and microtubule-associated functions. *Mol. Biol Cell* **2012**, *23*, 1095–1103. [CrossRef]
9. Hershko, A.; Ciechanover, A. The ubiquitin system. *Annu. Rev. Biochem.* **1998**, *67*, 425–479. [CrossRef]
10. Pickart, C.M. Ubiquitin enters the new millennium. *Mol. Cell* **2001**, *8*, 499–504. [CrossRef]
11. Ventii, K.H.; Wilkinson, K.D. Protein partners of deubiquitinating enzymes. *Biochem. J.* **2008**, *414*, 161–175. [CrossRef] [PubMed]
12. Mennerich, D.; Kubaichuk, K.; Kietzmann, T. Dubs, hypoxia, and cancer. *Trends Cancer* **2019**, *5*, 632–653. [CrossRef] [PubMed]
13. He, M.; Zhou, Z.; Wu, G.; Chen, Q.; Wan, Y. Emerging role of dubs in tumor metastasis and apoptosis: Therapeutic implication. *Pharmacol. Ther.* **2017**, *177*, 96–107. [CrossRef] [PubMed]
14. Hussain, S.; Zhang, Y.; Galardy, P.J. Dubs and cancer: The role of deubiquitinating enzymes as oncogenes, non-oncogenes and tumor suppressors. *Cell Cycle* **2009**, *8*, 1688–1697. [CrossRef]
15. Bednash, J.S.; Mallampalli, R.K. Targeting deubiquitinases in cancer. *Methods Mol. Biol.* **2018**, *1731*, 295–305.
16. Fraile, J.M.; Quesada, V.; Rodriguez, D.; Freije, J.M.; Lopez-Otin, C. Deubiquitinases in cancer: New functions and therapeutic options. *Oncogene* **2012**, *31*, 2373–2388. [CrossRef]
17. Cheng, J.; Guo, J.; North, B.J.; Wang, B.; Cui, C.P.; Li, H.; Tao, K.; Zhang, L.; Wei, W. Functional analysis of deubiquitylating enzymes in tumorigenesis and development. *Biochim. Et Biophys. Acta. Rev. Cancer* **2019**, *1872*, 188312. [CrossRef]
18. Clague, M.J.; Urbe, S.; Komander, D. Breaking the chains: Deubiquitylating enzyme specificity begets function. *Nat. Rev. Mol. Cell Biol.* **2019**, *20*, 338–352. [CrossRef]
19. Venuto, S.; Merla, G. E3 ubiquitin ligase trim proteins, cell cycle and mitosis. *Cells* **2019**, *8*, 510. [CrossRef]
20. Vodermaier, H.C. Apc/c and scf: Controlling each other and the cell cycle. *Curr. Biol.* **2004**, *14*, R787–R796. [CrossRef]
21. Sivakumar, S.; Gorbsky, G.J. Spatiotemporal regulation of the anaphase-promoting complex in mitosis. *Nat. Rev. Mol. Cell Biol.* **2015**, *16*, 82–94. [CrossRef] [PubMed]
22. Qin, J.; Zhou, Z.; Chen, W.; Wang, C.; Zhang, H.; Ge, G.; Shao, M.; You, D.; Fan, Z.; Xia, H.; et al. Bap1 promotes breast cancer cell proliferation and metastasis by deubiquitinating klf5. *Nat. Commun.* **2015**, *6*, 1–12. [CrossRef] [PubMed]
23. Hu, B.; Deng, T.; Ma, H.; Liu, Y.; Feng, P.; Wei, D.; Ling, N.; Li, L.; Qiu, S.; Zhang, L.; et al. Deubiquitinase dub3 regulates cell cycle progression via stabilizing cyclin a for proliferation of non-small cell lung cancer cells. *Cells* **2019**, *8*, 297. [CrossRef] [PubMed]

24. Sobol, A.; Askonas, C.; Alani, S.; Weber, M.J.; Ananthanarayanan, V.; Osipo, C.; Bocchetta, M. Deubiquitinase otud6b isoforms are important regulators of growth and proliferation. *Mol. Cancer Res.* **2017**, *15*, 117–127. [CrossRef]

25. Bonacci, T.; Suzuki, A.; Grant, G.D.; Stanley, N.; Cook, J.G.; Brown, N.G.; Emanuele, M.J. Cezanne/otud7b is a cell cycle-regulated deubiquitinase that antagonizes the degradation of apc/c substrates. *EMBO J.* **2018**, *37*. [CrossRef]

26. Bonacci, T.; Emanuele, M.J. Impressionist portraits of mitotic exit: Apc/c, k11-linked ubiquitin chains and cezanne. *Cell Cycle* **2019**, *18*, 652–660. [CrossRef]

27. Moniz, S.; Bandarra, D.; Biddlestone, J.; Campbell, K.J.; Komander, D.; Bremm, A.; Rocha, S. Cezanne regulates e2f1-dependent hif2alpha expression. *J. Cell Sci.* **2015**, *128*, 3082–3093. [CrossRef]

28. Wang, B.; Jie, Z.; Joo, D.; Ordureau, A.; Liu, P.; Gan, W.; Guo, J.; Zhang, J.; North, B.J.; Dai, X.; et al. Traf2 and otud7b govern a ubiquitin-dependent switch that regulates mtorc2 signalling. *Nature* **2017**, *545*, 365–369. [CrossRef]

29. Liao, Y.; Liu, N.; Xia, X.; Guo, Z.; Li, Y.; Jiang, L.; Zhou, R.; Tang, D.; Huang, H.; Liu, J. Usp10 modulates the skp2/bcr-abl axis via stabilizing skp2 in chronic myeloid leukemia. *Cell Discov.* **2019**, *5*, 1–15. [CrossRef]

30. Liao, Y.; Xia, X.; Liu, N.; Cai, J.; Guo, Z.; Li, Y.; Jiang, L.; Dou, Q.P.; Tang, D.; Huang, H.; et al. Growth arrest and apoptosis induction in androgen receptor-positive human breast cancer cells by inhibition of usp14-mediated androgen receptor deubiquitination. *Oncogene* **2018**, *37*, 1896–1910. [CrossRef]

31. McFarlane, C.; Kelvin, A.A.; de la Vega, M.; Govender, U.; Scott, C.J.; Burrows, J.F.; Johnston, J.A. The deubiquitinating enzyme usp17 is highly expressed in tumor biopsies, is cell cycle regulated, and is required for g1-s progression. *Cancer Res.* **2010**, *70*, 3329–3339. [CrossRef] [PubMed]

32. Ducker, C.; Chow, L.K.Y.; Saxton, J.; Handwerger, J.; McGregor, A.; Strahl, T.; Layfield, R.; Shaw, P.E. De-ubiquitination of elk-1 by usp17 potentiates mitogenic gene expression and cell proliferation. *Nucleic Acids Res.* **2019**, *47*, 4495–4508. [CrossRef] [PubMed]

33. Fukuura, K.; Inoue, Y.; Miyajima, C.; Watanabe, S.; Tokugawa, M.; Morishita, D.; Ohoka, N.; Komada, M.; Hayashi, H. The ubiquitin-specific protease usp17 prevents cellular senescence by stabilizing the methyltransferase set8 and transcriptionally repressing p21. *J. Biol. Chem.* **2019**, *294*, 16429–16439. [CrossRef] [PubMed]

34. Arceci, A.; Bonacci, T.; Wang, X.; Stewart, K.; Damrauer, J.S.; Hoadley, K.A.; Emanuele, M.J. Foxm1 deubiquitination by usp21 regulates cell cycle progression and paclitaxel sensitivity in basal-like breast cancer. *Cell Rep.* **2019**, *26*, 3076–3086. [CrossRef]

35. Wu, Y.; Qin, J.; Li, F.; Yang, C.; Li, Z.; Zhou, Z.; Zhang, H.; Li, Y.; Wang, X.; Liu, R.; et al. Usp3 promotes breast cancer cell proliferation by deubiquitinating klf5. *J. Biol. Chem.* **2019**, *294*, 17837–17847. [CrossRef]

36. Wang, Q.; Ma, S.; Song, N.; Li, X.; Liu, L.; Yang, S.; Ding, X.; Shan, L.; Zhou, X.; Su, D.; et al. Stabilization of histone demethylase phf8 by usp7 promotes breast carcinogenesis. *J. Clin. Investig.* **2016**, *126*, 2205–2220. [CrossRef]

37. Saldana, M.; VanderVorst, K.; Berg, A.L.; Lee, H.; Carraway, K.L. Otubain 1: A non-canonical deubiquitinase with an emerging role in cancer. *Endocr. Relat. Cancer* **2019**, *26*, R1–R14. [CrossRef]

38. Piao, S.; Pei, H.Z.; Huang, B.; Baek, S.H. Ovarian tumor domain-containing protein 1 deubiquitinates and stabilizes p53. *Cell. Signal.* **2017**, *33*, 22–29. [CrossRef]

39. Zhang, Z.; Fan, Y.; Xie, F.; Zhou, H.; Jin, K.; Shao, L.; Shi, W.; Fang, P.; Yang, B.; van Dam, H.; et al. Breast cancer metastasis suppressor otud1 deubiquitinates smad7. *Nat. Commun.* **2017**, *8*, 1–16. [CrossRef]

40. Yuan, J.; Luo, K.; Zhang, L.; Cheville, J.C.; Lou, Z. Usp10 regulates p53 localization and stability by deubiquitinating p53. *Cell* **2010**, *140*, 384–396. [CrossRef]

41. Liao, Y.; Liu, N.; Hua, X.; Cai, J.; Xia, X.; Wang, X.; Huang, H.; Liu, J. Proteasome-associated deubiquitinase ubiquitin-specific protease 14 regulates prostate cancer proliferation by deubiquitinating and stabilizing androgen receptor. *Cell Death Dis.* **2017**, *8*, e2585. [CrossRef] [PubMed]

42. Zou, Q.; Jin, J.; Hu, H.; Li, H.S.; Romano, S.; Xiao, Y.; Nakaya, M.; Zhou, X.; Cheng, X.; Yang, P.; et al. Usp15 stabilizes mdm2 to mediate cancer-cell survival and inhibit antitumor t cell responses. *Nat. Immunol.* **2014**, *15*, 562–570. [CrossRef] [PubMed]

43. Eichhorn, P.J.; Rodon, L.; Gonzalez-Junca, A.; Dirac, A.; Gili, M.; Martinez-Saez, E.; Aura, C.; Barba, I.; Peg, V.; Prat, A.; et al. Usp15 stabilizes tgf-beta receptor i and promotes oncogenesis through the activation of tgf-beta signaling in glioblastoma. *Nat. Med.* **2012**, *18*, 429–435. [CrossRef] [PubMed]

44. Stevenson, L.F.; Sparks, A.; Allende-Vega, N.; Xirodimas, D.P.; Lane, D.P.; Saville, M.K. The deubiquitinating enzyme usp2a regulates the p53 pathway by targeting mdm2. *EMBO J.* **2007**, *26*, 976–986. [CrossRef] [PubMed]

45. Hock, A.K.; Vigneron, A.M.; Carter, S.; Ludwig, R.L.; Vousden, K.H. Regulation of p53 stability and function by the deubiquitinating enzyme usp42. *EMBO J.* **2011**, *30*, 4921–4930. [CrossRef]

46. Cuella-Martin, R.; Oliveira, C.; Lockstone, H.E.; Snellenberg, S.; Grolmusova, N.; Chapman, J.R. 53bp1 integrates DNA repair and p53-dependent cell fate decisions via distinct mechanisms. *Mol. Cell* **2016**, *64*, 51–64. [CrossRef]

47. Liu, J.; Chung, H.J.; Vogt, M.; Jin, Y.; Malide, D.; He, L.; Dundr, M.; Levens, D. Jtv1 co-activates fbp to induce usp29 transcription and stabilize p53 in response to oxidative stress. *EMBO J.* **2011**, *30*, 846–858. [CrossRef]

48. Yun, S.I.; Kim, H.H.; Yoon, J.H.; Park, W.S.; Hahn, M.J.; Kim, H.C.; Chung, C.H.; Kim, K.K. Ubiquitin specific protease 4 positively regulates the wnt/beta-catenin signaling in colorectal cancer. *Mol. Oncol.* **2015**, *9*, 1834–1851. [CrossRef]

49. Li, Z.; Hao, Q.; Luo, J.; Xiong, J.; Zhang, S.; Wang, T.; Bai, L.; Wang, W.; Chen, M.; Wang, W.; et al. Usp4 inhibits p53 and nf-κb through deubiquitinating and stabilizing hdac2. *Oncogene* **2016**, *35*, 2902–2912. [CrossRef]

50. Zhang, X.; Berger, F.G.; Yang, J.; Lu, X. Usp4 inhibits p53 through deubiquitinating and stabilizing arf-bp1. *EMBO J.* **2011**, *30*, 2177–2189. [CrossRef]

51. Luo, K.; Li, Y.; Yin, Y.; Li, L.; Wu, C.; Chen, Y.; Nowsheen, S.; Hu, Q.; Zhang, L.; Lou, Z.; et al. Usp49 negatively regulates tumorigenesis and chemoresistance through fkbp51-akt signaling. *EMBO J.* **2017**, *36*, 1434–1446. [CrossRef] [PubMed]

52. Potu, H.; Peterson, L.F.; Pal, A.; Verhaegen, M.; Cao, J.; Talpaz, M.; Donato, N.J. Usp5 links suppression of p53 and fas levels in melanoma to the braf pathway. *Oncotarget* **2014**, *5*, 5559–5569. [CrossRef] [PubMed]

53. Sun, K.; He, S.B.; Yao, Y.Z.; Qu, J.G.; Xie, R.; Ma, Y.Q.; Zong, M.H.; Chen, J.X. Tre2 (usp6nl) promotes colorectal cancer cell proliferation via wnt/beta-catenin pathway. *Cancer Cell Int.* **2019**, *19*, 102. [CrossRef] [PubMed]

54. Cummins, J.M.; Vogelstein, B. Hausp is required for p53 destabilization. *Cell Cycle* **2004**, *3*, 689–692. [CrossRef]

55. Li, M.; Brooks, C.L.; Kon, N.; Gu, W. A dynamic role of hausp in the p53-mdm2 pathway. *Mol. Cell* **2004**, *13*, 879–886. [CrossRef]

56. Hu, M.; Gu, L.; Li, M.; Jeffrey, P.D.; Gu, W.; Shi, Y. Structural basis of competitive recognition of p53 and mdm2 by hausp/usp7: Implications for the regulation of the p53-mdm2 pathway. *PLoS Biol.* **2006**, *4*, e27. [CrossRef]

57. Sheng, Y.; Saridakis, V.; Sarkari, F.; Duan, S.; Wu, T.; Arrowsmith, C.H.; Frappier, L. Molecular recognition of p53 and mdm2 by usp7/hausp. *Nat. Struct. Mol. Biol.* **2006**, *13*, 285–291. [CrossRef]

58. Li, M.; Chen, D.; Shiloh, A.; Luo, J.; Nikolaev, A.Y.; Qin, J.; Gu, W. Deubiquitination of p53 by hausp is an important pathway for p53 stabilization. *Nature* **2002**, *416*, 648–653. [CrossRef]

59. Yang, B.; Zhang, S.; Wang, Z.; Yang, C.; Ouyang, W.; Zhou, F.; Zhou, Y.; Xie, C. Deubiquitinase usp9x deubiquitinates beta-catenin and promotes high grade glioma cell growth. *Oncotarget* **2016**, *7*, 79515–79525. [CrossRef]

60. Liu, H.; Chen, W.; Liang, C.; Chen, B.W.; Zhi, X.; Zhang, S.; Zheng, X.; Bai, X.; Liang, T. Wp1130 increases doxorubicin sensitivity in hepatocellular carcinoma cells through usp9x-dependent p53 degradation. *Cancer Lett.* **2015**, *361*, 218–225. [CrossRef]

61. Liu, H.; Li, X.; Ning, G.; Zhu, S.; Ma, X.; Liu, X.; Liu, C.; Huang, M.; Schmitt, I.; Wullner, U.; et al. The machado-joseph disease deubiquitinase ataxin-3 regulates the stability and apoptotic function of p53. *PLoS Biol.* **2016**, *14*, e2000733. [CrossRef] [PubMed]

62. Wu, X.; Luo, Q.; Zhao, P.; Chang, W.; Wang, Y.; Shu, T.; Ding, F.; Li, B.; Liu, Z. Josd1 inhibits mitochondrial apoptotic signalling to drive acquired chemoresistance in gynaecological cancer by stabilizing mcl1. *Cell Death Differ.* **2020**, *27*, 55–70. [CrossRef]

63. Potu, H.; Kandarpa, M.; Peterson, L.F.; Donato, N.J.; Talpaz, M. Tumor necrosis factor related apoptosis inducing ligand (trail) regulates deubiquitinase usp5 in tumor cells. *Oncotarget* **2019**, *10*, 5745–5754. [CrossRef]

64. Wang, S.; Juan, J.; Zhang, Z.; Du, Y.; Xu, Y.; Tong, J.; Cao, B.; Moran, M.F.; Zeng, Y.; Mao, X. Inhibition of the deubiquitinase usp5 leads to c-maf protein degradation and myeloma cell apoptosis. *Cell Death Dis.* **2017**, *8*, e3058. [CrossRef] [PubMed]

65. Ismail, I.H.; Davidson, R.; Gagne, J.P.; Xu, Z.Z.; Poirier, G.G.; Hendzel, M.J. Germline mutations in bap1 impair its function in DNA double-strand break repair. *Cancer Res.* **2014**, *74*, 4282–4294. [CrossRef] [PubMed]

66. Fernandez-Majada, V.; Welz, P.S.; Ermolaeva, M.A.; Schell, M.; Adam, A.; Dietlein, F.; Komander, D.; Buttner, R.; Thomas, R.K.; Schumacher, B.; et al. The tumour suppressor cyld regulates the p53 DNA damage response. *Nat. Commun.* **2016**, *7*, 1–14. [CrossRef] [PubMed]

67. de Vivo, A.; Sanchez, A.; Yegres, J.; Kim, J.; Emly, S.; Kee, Y. The otud5-ubr5 complex regulates fact-mediated transcription at damaged chromatin. *Nucleic Acids Res.* **2019**, *47*, 729–746. [CrossRef] [PubMed]

68. Wu, X.; Liu, S.; Sagum, C.; Chen, J.; Singh, R.; Chaturvedi, A.; Horton, J.R.; Kashyap, T.R.; Fushman, D.; Cheng, X.; et al. Crosstalk between lys63- and lys11-polyubiquitin signaling at DNA damage sites is driven by cezanne. *Genes Dev.* **2019**, *33*, 1702–1717. [CrossRef] [PubMed]

69. Alvarez, V.; Vinas, L.; Gallego-Sanchez, A.; Andres, S.; Sacristan, M.P.; Bueno, A. Orderly progression through s-phase requires dynamic ubiquitylation and deubiquitylation of pcna. *Sci. Rep.* **2016**, *6*, 25513. [CrossRef]

70. Nishi, R.; Wijnhoven, P.; le Sage, C.; Tjeertes, J.; Galanty, Y.; Forment, J.V.; Clague, M.J.; Urbe, S.; Jackson, S.P. Systematic characterization of deubiquitylating enzymes for roles in maintaining genome integrity. *Nat. Cell Biol* **2014**, *16*, 1016–1026, 1011–1018. [CrossRef]

71. Whitehurst, C.B.; Vaziri, C.; Shackelford, J.; Pagano, J.S. Epstein-barr virus bplf1 deubiquitinates pcna and attenuates polymerase eta recruitment to DNA damage sites. *J. Virol.* **2012**, *86*, 8097–8106. [CrossRef] [PubMed]

72. Castella, M.; Jacquemont, C.; Thompson, E.L.; Yeo, J.E.; Cheung, R.S.; Huang, J.W.; Sobeck, A.; Hendrickson, E.A.; Taniguchi, T. Fanci regulates recruitment of the fa core complex at sites of DNA damage independently of fancd2. *PLoS Genet.* **2015**, *11*, e1005563. [CrossRef] [PubMed]

73. Nijman, S.M.; Huang, T.T.; Dirac, A.M.; Brummelkamp, T.R.; Kerkhoven, R.M.; D'Andrea, A.D.; Bernards, R. The deubiquitinating enzyme usp1 regulates the fanconi anemia pathway. *Mol. Cell* **2005**, *17*, 331–339. [CrossRef] [PubMed]

74. Olazabal-Herrero, A.; Garcia-Santisteban, I.; Rodriguez, J.A. Structure-function analysis of usp1: Insights into the role of ser313 phosphorylation site and the effect of cancer-associated mutations on autocleavage. *Mol. Cancer* **2015**, *14*, 33. [CrossRef] [PubMed]

75. Orthwein, A.; Noordermeer, S.M.; Wilson, M.D.; Landry, S.; Enchev, R.I.; Sherker, A.; Munro, M.; Pinder, J.; Salsman, J.; Dellaire, G.; et al. A mechanism for the suppression of homologous recombination in g1 cells. *Nature* **2015**, *528*, 422–426. [CrossRef] [PubMed]

76. Sharma, N.; Zhu, Q.; Wani, G.; He, J.; Wang, Q.E.; Wani, A.A. Usp3 counteracts rnf168 via deubiquitinating h2a and gammah2ax at lysine 13 and 15. *Cell Cycle* **2014**, *13*, 106–114. [CrossRef]

77. Uckelmann, M.; Densham, R.M.; Baas, R.; Winterwerp, H.H.K.; Fish, A.; Sixma, T.K.; Morris, J.R. Usp48 restrains resection by site-specific cleavage of the brca1 ubiquitin mark from h2a. *Nat. Commun.* **2018**, *9*, 1–16. [CrossRef]

78. Zhu, Q.; Sharma, N.; He, J.; Wani, G.; Wani, A.A. Usp7 deubiquitinase promotes ubiquitin-dependent DNA damage signaling by stabilizing rnf168. *Cell Cycle* **2015**, *14*, 1413–1425. [CrossRef]

79. McGarry, E.; Gaboriau, D.; Rainey, M.D.; Restuccia, U.; Bachi, A.; Santocanale, C. The deubiquitinase usp9x maintains DNA replication fork stability and DNA damage checkpoint responses by regulating claspin during s-phase. *Cancer Res.* **2016**, *76*, 2384–2393. [CrossRef]

80. Kovalenko, A.; Chable-Bessia, C.; Cantarella, G.; Israel, A.; Wallach, D.; Courtois, G. The tumour suppressor cyld negatively regulates nf-kappa b signalling by deubiquitination. *Nature* **2003**, *424*, 801–805. [CrossRef]

81. Trompouki, E.; Hatzivassiliou, E.; Tsichritzis, T.; Farmer, H.; Ashworth, A.; Mosialos, G. Cyld is a deubiquitinating enzyme that negatively regulates nf-kappa b activation by tnfr family members. *Nature* **2003**, *424*, 793–796. [CrossRef] [PubMed]

82. Brummelkamp, T.R.; Nijman, S.M.; Dirac, A.M.; Bernards, R. Loss of the cylindromatosis tumour suppressor inhibits apoptosis by activating nf-kappab. *Nature* **2003**, *424*, 797–801. [CrossRef] [PubMed]

83. Wu, H.C.; Lin, Y.C.; Liu, C.H.; Chung, H.C.; Wang, Y.T.; Lin, Y.W.; Ma, H.I.; Tu, P.H.; Lawler, S.E.; Chen, R.H. Usp11 regulates pml stability to control notch-induced malignancy in brain tumours. *Nat. Commun.* **2014**, *5*, 1–16. [CrossRef] [PubMed]

84. Qu, Z.; Zhang, R.; Su, M.; Liu, W. Usp13 serves as a tumor suppressor via the pten/akt pathway in oral squamous cell carcinoma. *Cancer Manag. Res.* **2019**, *11*, 9175–9183. [CrossRef] [PubMed]

85. Li, X.; Stevens, P.D.; Yang, H.; Gulhati, P.; Wang, W.; Evers, B.M.; Gao, T. The deubiquitination enzyme usp46 functions as a tumor suppressor by controlling phlpp-dependent attenuation of akt signaling in colon cancer. *Oncogene* **2013**, *32*, 471–478. [CrossRef]

86. Bott, M.; Brevet, M.; Taylor, B.S.; Shimizu, S.; Ito, T.; Wang, L.; Creaney, J.; Lake, R.A.; Zakowski, M.F.; Reva, B.; et al. The nuclear deubiquitinase bap1 is commonly inactivated by somatic mutations and 3p21.1 losses in malignant pleural mesothelioma. *Nat. Genet.* **2011**, *43*, 668–672. [CrossRef]

87. Kim, D.; Hong, A.; Park, H.I.; Shin, W.H.; Yoo, L.; Jeon, S.J.; Chung, K.C. Deubiquitinating enzyme usp22 positively regulates c-myc stability and tumorigenic activity in mammalian and breast cancer cells. *J. Cell Physiol.* **2017**, *232*, 3664–3676. [CrossRef]

88. Diefenbacher, M.E.; Popov, N.; Blake, S.M.; Schulein-Volk, C.; Nye, E.; Spencer-Dene, B.; Jaenicke, L.A.; Eilers, M.; Behrens, A. The deubiquitinase usp28 controls intestinal homeostasis and promotes colorectal cancer. *J. Clin. Inv.* **2014**, *124*, 3407–3418. [CrossRef]

89. Khan, O.M.; Carvalho, J.; Spencer-Dene, B.; Mitter, R.; Frith, D.; Snijders, A.P.; Wood, S.A.; Behrens, A. The deubiquitinase usp9x regulates fbw7 stability and suppresses colorectal cancer. *J. Clin. Inv.* **2018**, *128*, 1326–1337. [CrossRef]

90. Wu, Y.; Wang, Y.; Lin, Y.; Liu, Y.; Wang, Y.; Jia, J.; Singh, P.; Chi, Y.I.; Wang, C.; Dong, C.; et al. Dub3 inhibition suppresses breast cancer invasion and metastasis by promoting snail1 degradation. *Nat. Commun.* **2017**, *8*, 1–16. [CrossRef]

91. Lin, Y.; Wang, Y.; Shi, Q.; Yu, Q.; Liu, C.; Feng, J.; Deng, J.; Evers, B.M.; Zhou, B.P.; Wu, Y. Stabilization of the transcription factors slug and twist by the deubiquitinase dub3 is a key requirement for tumor metastasis. *Oncotarget* **2017**, *8*, 75127–75140. [CrossRef] [PubMed]

92. Zhou, H.; Liu, Y.; Zhu, R.; Ding, F.; Cao, X.; Lin, D.; Liu, Z. Otub1 promotes esophageal squamous cell carcinoma metastasis through modulating snail stability. *Oncogene* **2018**, *37*, 3356–3368. [CrossRef] [PubMed]

93. Lv, J.; Zhang, S.; Wu, H.; Lu, J.; Lu, Y.; Wang, F.; Zhao, W.; Zhan, P.; Lu, J.; Fang, Q.; et al. Deubiquitinase psmd14 enhances hepatocellular carcinoma growth and metastasis by stabilizing grb2. *Cancer lett.* **2020**, *469*, 22–34. [CrossRef] [PubMed]

94. Song, C.; Liu, W.; Li, J. Usp17 is upregulated in osteosarcoma and promotes cell proliferation, metastasis, and epithelial-mesenchymal transition through stabilizing smad4. *Tumour Biol.* **2017**, *39*, 1010428317717138. [CrossRef]

95. Wu, X.; Liu, M.; Zhu, H.; Wang, J.; Dai, W.; Li, J.; Zhu, D.; Tang, W.; Xiao, Y.; Lin, J.; et al. Ubiquitin-specific protease 3 promotes cell migration and invasion by interacting with and deubiquitinating suz12 in gastric cancer. *J. Exp. Clin. Cancer Res.* **2019**, *38*, 277. [CrossRef]

96. Sapmaz, A.; Berlin, I.; Bos, E.; Wijdeven, R.H.; Janssen, H.; Konietzny, R.; Akkermans, J.J.; Erson-Bensan, A.E.; Koning, R.I.; Kessler, B.M.; et al. Usp32 regulates late endosomal transport and recycling through deubiquitylation of rab7. *Nat. Commun.* **2019**, *10*, 1–18. [CrossRef]

97. Kim, J.O.; Kim, S.R.; Lim, K.H.; Kim, J.H.; Ajjappala, B.; Lee, H.J.; Choi, J.I.; Baek, K.H. Deubiquitinating enzyme usp37 regulating oncogenic function of 14-3-3gamma. *Oncotarget* **2015**, *6*, 36551–36576. [CrossRef]

98. Yuan, J.; Luo, K.; Deng, M.; Li, Y.; Yin, P.; Gao, B.; Fang, Y.; Wu, P.; Liu, T.; Lou, Z. Herc2-usp20 axis regulates DNA damage checkpoint through claspin. *Nucleic Acids Res.* **2014**, *42*, 13110–13121. [CrossRef]

99. Jiang, Y.Z.; Ma, D.; Suo, C.; Shi, J.; Xue, M.; Hu, X.; Xiao, Y.; Yu, K.D.; Liu, Y.R.; Yu, Y.; et al. Genomic and transcriptomic landscape of triple-negative breast cancers: Subtypes and treatment strategies. *Cancer Cell* **2019**, *35*, 428–440. [CrossRef]

100. Chiu, H.W.; Lin, H.Y.; Tseng, I.J.; Lin, Y.F. Otud7b upregulation predicts a poor response to paclitaxel in patients with triple-negative breast cancer. *Oncotarget* **2018**, *9*, 553–565. [CrossRef]

101. Lin, D.D.; Shen, Y.; Qiao, S.; Liu, W.W.; Zheng, L.; Wang, Y.N.; Cui, N.; Wang, Y.F.; Zhao, S.; Shi, J.H. Upregulation of otud7b (cezanne) promotes tumor progression via akt/vegf pathway in lung squamous carcinoma and adenocarcinoma. *Front. Oncol.* **2019**, *9*, 862. [CrossRef] [PubMed]

102. Anastas, J.N.; Zee, B.M.; Kalin, J.H.; Kim, M.; Guo, R.; Alexandrescu, S.; Blanco, M.A.; Giera, S.; Gillespie, S.M.; Das, J.; et al. Re-programing chromatin with a bifunctional lsd1/hdac inhibitor induces therapeutic differentiation in dipg. *Cancer Cell* **2019**, *36*, 528–544. [CrossRef] [PubMed]

103. Zhan, T.; Rindtorff, N.; Boutros, M. Wnt signaling in cancer. *Oncogene* **2017**, *36*, 1461–1473. [CrossRef] [PubMed]

104. Ben-Porath, I.; Thomson, M.W.; Carey, V.J.; Ge, R.; Bell, G.W.; Regev, A.; Weinberg, R.A. An embryonic stem cell-like gene expression signature in poorly differentiated aggressive human tumors. *Nat. Genet.* **2008**, *40*, 499–507. [CrossRef]

105. Reya, T.; Morrison, S.J.; Clarke, M.F.; Weissman, I.L. Stem cells, cancer, and cancer stem cells. *Nature* **2001**, *414*, 105–111. [CrossRef]

106. Nguyen, H.H.; Kim, T.; Nguyen, T.; Hahn, M.J.; Yun, S.I.; Kim, K.K. A selective inhibitor of ubiquitin-specific protease 4 suppresses colorectal cancer progression by regulating beta-catenin signaling. *Cell. Physiol. Biochem.: Int. J. Exp. Cell. Physiol. Biochem. Pharmacol.* **2019**, *53*, 157–171.

107. Brooks, C.L.; Gu, W. P53 regulation by ubiquitin. *FEBS Lett.* **2011**, *585*, 2803–2809. [CrossRef]

108. Brooks, C.L.; Li, M.; Hu, M.; Shi, Y.; Gu, W. The p53–mdm2–hausp complex is involved in p53 stabilization by hausp. *Oncogene* **2007**, *26*, 7262–7266. [CrossRef]

109. Li, F.; Han, H.; Sun, Q.; Liu, K.; Lin, N.; Xu, C.; Zhao, Z.; Zhao, W. Usp28 regulates deubiquitination of histone h2a and cell proliferation. *Exp. Cell Res.* **2019**, *379*, 11–18. [CrossRef]

110. Niendorf, S.; Oksche, A.; Kisser, A.; Lohler, J.; Prinz, M.; Schorle, H.; Feller, S.; Lewitzky, M.; Horak, I.; Knobeloch, K.P. Essential role of ubiquitin-specific protease 8 for receptor tyrosine kinase stability and endocytic trafficking in vivo. *Mol. Cell Biol.* **2007**, *27*, 5029–5039. [CrossRef]

111. Hanahan, D.; Weinberg, R.A. The hallmarks of cancer. *Cell* **2000**, *100*, 57–70. [CrossRef]

112. Fernald, K.; Kurokawa, M. Evading apoptosis in cancer. *Trends Cell Biol.* **2013**, *23*, 620–633. [CrossRef] [PubMed]

113. Burrows, J.F.; McGrattan, M.J.; Rascle, A.; Humbert, M.; Baek, K.H.; Johnston, J.A. Dub-3, a cytokine-inducible deubiquitinating enzyme that blocks proliferation. *J. Biol. Chem.* **2004**, *279*, 13993–14000. [CrossRef] [PubMed]

114. Xu, M.; Takanashi, M.; Oikawa, K.; Tanaka, M.; Nishi, H.; Isaka, K.; Kudo, M.; Kuroda, M. Usp15 plays an essential role for caspase-3 activation during paclitaxel-induced apoptosis. *Biochem Biophys Res. Commun* **2009**, *388*, 366–371. [CrossRef]

115. Jin, Z.; Li, Y.; Pitti, R.; Lawrence, D.; Pham, V.C.; Lill, J.R.; Ashkenazi, A. Cullin3-based polyubiquitination and p62-dependent aggregation of caspase-8 mediate extrinsic apoptosis signaling. *Cell* **2009**, *137*, 721–735. [CrossRef]

116. Crook, N.E.; Clem, R.J.; Miller, L.K. An apoptosis-inhibiting baculovirus gene with a zinc finger-like motif. *J. Virol.* **1993**, *67*, 2168–2174. [CrossRef]

117. Mei, Y.; Hahn, A.A.; Hu, S.; Yang, X. The usp19 deubiquitinase regulates the stability of c-iap1 and c-iap2. *J. Biol. Chem.* **2011**, *286*, 35380–35387. [CrossRef]

118. Goncharov, T.; Niessen, K.; de Almagro, M.C.; Izrael-Tomasevic, A.; Fedorova, A.V.; Varfolomeev, E.; Arnott, D.; Deshayes, K.; Kirkpatrick, D.S.; Vucic, D. Otub1 modulates c-iap1 stability to regulate signalling pathways. *EMBO J.* **2013**, *32*, 1103–1114. [CrossRef]

119. Engel, K.; Rudelius, M.; Slawska, J.; Jacobs, L.; Ahangarian Abhari, B.; Altmann, B.; Kurutz, J.; Rathakrishnan, A.; Fernandez-Saiz, V.; Brunner, A.; et al. Usp9x stabilizes xiap to regulate mitotic cell death and chemoresistance in aggressive b-cell lymphoma. *EMBO Mol. Med.* **2016**, *8*, 851–862. [CrossRef]

120. Weber, A.; Heinlein, M.; Dengjel, J.; Alber, C.; Singh, P.K.; Hacker, G. The deubiquitinase usp27x stabilizes the bh3-only protein bim and enhances apoptosis. *EMBO Rep.* **2016**, *17*, 724–738. [CrossRef]

121. Ghosal, G.; Chen, J. DNA damage tolerance: A double-edged sword guarding the genome. *Transl. Cancer Res.* **2013**, *2*, 107–129. [PubMed]

122. Hoeijmakers, J.H. Genome maintenance mechanisms for preventing cancer. *Nature* **2001**, *411*, 366–374. [CrossRef] [PubMed]

123. Le, J.; Perez, E.; Nemzow, L.; Gong, F. Role of deubiquitinases in DNA damage response. *DNA Repair* **2019**, *76*, 89–98. [CrossRef] [PubMed]

124. Jackson, S.P.; Bartek, J. The DNA-damage response in human biology and disease. *Nature* **2009**, *461*, 1071–1078. [CrossRef]

125. Shanmugam, I.; Abbas, M.; Ayoub, F.; Mirabal, S.; Bsaili, M.; Caulder, E.K.; Weinstock, D.M.; Tomkinson, A.E.; Hromas, R.; Shaheen, M. Ubiquitin-specific peptidase 20 regulates rad17 stability, checkpoint kinase 1 phosphorylation and DNA repair by homologous recombination. *J. Biol. Chem.* **2014**, *289*, 22739–22748. [CrossRef]

126. Yoshida, K.; Miki, Y. Role of brca1 and brca2 as regulators of DNA repair, transcription, and cell cycle in response to DNA damage. *Cancer Sci.* **2004**, *95*, 866–871. [CrossRef]

127. Meyer, T.; Jahn, N.; Lindner, S.; Rohner, L.; Dolnik, A.; Weber, D.; Scheffold, A.; Kopff, S.; Paschka, P.; Gaidzik, V.I.; et al. Functional characterization of brcc3 mutations in acute myeloid leukemia with t(8;21)(q22;q22.1). *Leukemia* **2020**, *34*, 404–415. [CrossRef]

128. Lancini, C.; van den Berk, P.C.; Vissers, J.H.; Gargiulo, G.; Song, J.Y.; Hulsman, D.; Serresi, M.; Tanger, E.; Blom, M.; Vens, C.; et al. Tight regulation of ubiquitin-mediated DNA damage response by usp3 preserves the functional integrity of hematopoietic stem cells. *J. Exp. Med.* **2014**, *211*, 1759–1777. [CrossRef]

129. Bignell, G.R.; Warren, W.; Seal, S.; Takahashi, M.; Rapley, E.; Barfoot, R.; Green, H.; Brown, C.; Biggs, P.J.; Lakhani, S.R.; et al. Identification of the familial cylindromatosis tumour-suppressor gene. *Nat. Genet.* **2000**, *25*, 160–165. [CrossRef]

130. Strobel, P.; Zettl, A.; Ren, Z.; Starostik, P.; Riedmiller, H.; Storkel, S.; Muller-Hermelink, H.K.; Marx, A. Spiradenocylindroma of the kidney: Clinical and genetic findings suggesting a role of somatic mutation of the cyld1 gene in the oncogenesis of an unusual renal neoplasm. *Am. J. Surg. Pathol.* **2002**, *26*, 119–124. [CrossRef]

131. Hirai, Y.; Kawamata, Y.; Takeshima, N.; Furuta, R.; Kitagawa, T.; Kawaguchi, T.; Hasumi, K.; Sugai, S.; Noda, T. Conventional and array-based comparative genomic hybridization analyses of novel cell lines harboring hpv18 from glassy cell carcinoma of the uterine cervix. *Int. J. Oncol.* **2004**, *24*, 977–986. [CrossRef] [PubMed]

132. Reiley, W.; Zhang, M.Y.; Sun, S.C. Negative regulation of jnk signaling by the tumor suppressor cyld. *J. Biol. Chem.* **2004**, *279*, 55161–55167. [CrossRef] [PubMed]

133. Ikeda, F.; Dikic, I. Cyld in ubiquitin signaling and tumor pathogenesis. *Cell* **2006**, *125*, 643–645. [CrossRef] [PubMed]

134. Wang, Z.; Inuzuka, H.; Fukushima, H.; Wan, L.; Gao, D.; Shaik, S.; Sarkar, F.H.; Wei, W. Emerging roles of the fbw7 tumour suppressor in stem cell differentiation. *EMBO Rep.* **2011**, *13*, 36–43. [CrossRef] [PubMed]

135. Nieto, M.A.; Huang, R.Y.; Jackson, R.A.; Thiery, J.P. Emt: 2016. *Cell* **2016**, *166*, 21–45. [CrossRef] [PubMed]

136. Fan, L.; Chen, Z.; Wu, X.; Cai, X.; Feng, S.; Lu, J.; Wang, H.; Liu, N. Ubiquitin-specific protease 3 promotes glioblastoma cell invasion and epithelial-mesenchymal transition via stabilizing snail. *Mol. Cancer Res. MCR* **2019**, *17*, 1975–1984. [CrossRef] [PubMed]

137. Hiraoka, E.; Mimae, T.; Ito, M.; Kadoya, T.; Miyata, Y.; Ito, A.; Okada, M. Correction to: Breast cancer cell motility is promoted by 14-3-3gamma. *Breast Cancer* **2019**, *26*, 594. [CrossRef]

138. Raungrut, P.; Wongkotsila, A.; Champoochana, N.; Lirdprapamongkol, K.; Svasti, J.; Thongsuksai, P. Knockdown of 14-3-3gamma suppresses epithelial-mesenchymal transition and reduces metastatic potential of human non-small cell lung cancer cells. *Anticancer Res.* **2018**, *38*, 3507–3514. [CrossRef]

139. Alonso-Curbelo, D.; Riveiro-Falkenbach, E.; Perez-Guijarro, E.; Cifdaloz, M.; Karras, P.; Osterloh, L.; Megias, D.; Canon, E.; Calvo, T.G.; Olmeda, D.; et al. Rab7 controls melanoma progression by exploiting a lineage-specific wiring of the endolysosomal pathway. *Cancer Cell* **2014**, *26*, 61–76. [CrossRef]

140. Wang, T.; Ming, Z.; Xiaochun, W.; Hong, W. Rab7: Role of its protein interaction cascades in endo-lysosomal traffic. *Cell. Sign.* **2011**, *23*, 516–521. [CrossRef]

141. Zhang, M.; Chen, L.; Wang, S.; Wang, T. Rab7: Roles in membrane trafficking and disease. *Biosci. Rep.* **2009**, *29*, 193–209. [CrossRef] [PubMed]

142. Richardson, P.G.; Barlogie, B.; Berenson, J.; Singhal, S.; Jagannath, S.; Irwin, D.; Rajkumar, S.V.; Srkalovic, G.; Alsina, M.; Alexanian, R.; et al. A phase 2 study of bortezomib in relapsed, refractory myeloma. *N. Engl. J. Med.* **2003**, *348*, 2609–2617. [CrossRef] [PubMed]

143. Richardson, P.G.; Sonneveld, P.; Schuster, M.W.; Irwin, D.; Stadtmauer, E.A.; Facon, T.; Harousseau, J.L.; Ben-Yehuda, D.; Lonial, S.; Goldschmidt, H.; et al. Bortezomib or high-dose dexamethasone for relapsed multiple myeloma. *N. Engl. J. Med.* **2005**, *352*, 2487–2498. [CrossRef] [PubMed]

144. Cacan, E.; Ozmen, Z.C. Regulation of fas in response to bortezomib and epirubicin in colorectal cancer cells. *J. Chemother.* **2020**, 1–9. [CrossRef] [PubMed]

145. Okazuka, K.; Ishida, T. Proteasome inhibitors for multiple myelomaJap. *J. Clin. Oncol.* **2018**, *48*, 785–793.

146. Kuhn, D.J.; Chen, Q.; Voorhees, P.M.; Strader, J.S.; Shenk, K.D.; Sun, C.M.; Demo, S.D.; Bennett, M.K.; van Leeuwen, F.W.; Chanan-Khan, A.A.; et al. Potent activity of carfilzomib, a novel, irreversible inhibitor of the ubiquitin-proteasome pathway, against preclinical models of multiple myeloma. *Blood* **2007**, *110*, 3281–3290. [CrossRef] [PubMed]

147. Jakubowiak, A.J.; Dytfeld, D.; Griffith, K.A.; Lebovic, D.; Vesole, D.H.; Jagannath, S.; Al-Zoubi, A.; Anderson, T.; Nordgren, B.; Detweiler-Short, K.; et al. A phase 1/2 study of carfilzomib in combination with lenalidomide and low-dose dexamethasone as a frontline treatment for multiple myeloma. *Blood* **2012**, *120*, 1801–1809. [CrossRef]

148. Richardson, P.G.; Zweegman, S.; O'Donnell, E.K.; Laubach, J.P.; Raje, N.; Voorhees, P.; Ferrari, R.H.; Skacel, T.; Kumar, S.K.; Lonial, S. Ixazomib for the treatment of multiple myeloma. *Expert Opin. Pharmacother.* **2018**, *19*, 1949–1968. [CrossRef]

149. Moreau, P.; Masszi, T.; Grzasko, N.; Bahlis, N.J.; Hansson, M.; Pour, L.; Sandhu, I.; Ganly, P.; Baker, B.W.; Jackson, S.R.; et al. Oral ixazomib, lenalidomide, and dexamethasone for multiple myeloma. *N. Engl. J. Med.* **2016**, *374*, 1621–1634. [CrossRef]

150. Petroski, M.D. The ubiquitin system, disease, and drug discovery. *BMC Biochem.* **2008**, *9*, S7. [CrossRef]

151. Cheon, K.W.; Baek, K.H. Hausp as a therapeutic target for hematopoietic tumors (review). *Int. J. Oncol.* **2006**, *28*, 1209–1215. [CrossRef] [PubMed]

152. Issaeva, N.; Bozko, P.; Enge, M.; Protopopova, M.; Verhoef, L.G.; Masucci, M.; Pramanik, A.; Selivanova, G. Small molecule rita binds to p53, blocks p53-hdm-2 interaction and activates p53 function in tumors. *Nature Med.* **2004**, *10*, 1321–1328. [CrossRef] [PubMed]

153. Vassilev, L.T.; Vu, B.T.; Graves, B.; Carvajal, D.; Podlaski, F.; Filipovic, Z.; Kong, N.; Kammlott, U.; Lukacs, C.; Klein, C.; et al. In vivo activation of the p53 pathway by small-molecule antagonists of mdm2. *Science* **2004**, *303*, 844–848. [CrossRef] [PubMed]

154. Tovar, C.; Rosinski, J.; Filipovic, Z.; Higgins, B.; Kolinsky, K.; Hilton, H.; Zhao, X.; Vu, B.T.; Qing, W.; Packman, K.; et al. Small-molecule mdm2 antagonists reveal aberrant p53 signaling in cancer: Implications for therapy. *Proc. Natl. Acad. Sci. USA* **2006**, *103*, 1888–1893. [CrossRef]

155. Stuhmer, T.; Chatterjee, M.; Hildebrandt, M.; Herrmann, P.; Gollasch, H.; Gerecke, C.; Theurich, S.; Cigliano, L.; Manz, R.A.; Daniel, P.T.; et al. Nongenotoxic activation of the p53 pathway as a therapeutic strategy for multiple myeloma. *Blood* **2005**, *106*, 3609–3617. [CrossRef]

156. Colland, F.; Formstecher, E.; Jacq, X.; Reverdy, C.; Planquette, C.; Conrath, S.; Trouplin, V.; Bianchi, J.; Aushev, V.N.; Camonis, J.; et al. Small-molecule inhibitor of usp7/hausp ubiquitin protease stabilizes and activates p53 in cells. *Mol. Cancer Ther.* **2009**, *8*, 2286–2295. [CrossRef]

157. Chitta, K.; Paulus, A.; Akhtar, S.; Blake, M.K.; Caulfield, T.R.; Novak, A.J.; Ansell, S.M.; Advani, P.; Ailawadhi, S.; Sher, T.; et al. Targeted inhibition of the deubiquitinating enzymes, usp14 and uchl5, induces proteotoxic stress and apoptosis in waldenstrom macroglobulinaemia tumour cells. *Br. J. Haematol.* **2015**, *169*, 377–390. [CrossRef]

158. Vogel, R.I.; Pulver, T.; Heilmann, W.; Mooneyham, A.; Mullany, S.; Zhao, X.; Shahi, M.; Richter, J.; Klein, M.; Chen, L.; et al. Usp14 is a predictor of recurrence in endometrial cancer and a molecular target for endometrial cancer treatment. *Oncotarget* **2016**, *7*, 30962–30976. [CrossRef]

159. Liu, Y.; Xu, X.; Lin, P.; He, Y.; Zhang, Y.; Cao, B.; Zhang, Z.; Sethi, G.; Liu, J.; Zhou, X.; et al. Inhibition of the deubiquitinase usp9x induces pre-b cell homeobox 1 (pbx1) degradation and thereby stimulates prostate cancer cell apoptosis. *J. Biol. Chem.* **2019**, *294*, 4572–4582. [CrossRef]

160. Byun, S.; Lee, S.Y.; Lee, J.; Jeong, C.H.; Farrand, L.; Lim, S.; Reddy, K.; Kim, J.Y.; Lee, M.H.; Lee, H.J.; et al. Usp8 is a novel target for overcoming gefitinib resistance in lung cancer. *Clin. Cancer Res.* **2013**, *19*, 3894–3904. [CrossRef]

161. Liu, Y.; Lashuel, H.A.; Choi, S.; Xing, X.; Case, A.; Ni, J.; Yeh, L.A.; Cuny, G.D.; Stein, R.L.; Lansbury, P.T., Jr. Discovery of inhibitors that elucidate the role of uch-l1 activity in the h1299 lung cancer cell line. *Chem. Biol.* **2003**, *10*, 837–846. [CrossRef] [PubMed]

162. Tian, Z.; D'Arcy, P.; Wang, X.; Ray, A.; Tai, Y.T.; Hu, Y.; Carrasco, R.D.; Richardson, P.; Linder, S.; Chauhan, D.; et al. A novel small molecule inhibitor of deubiquitylating enzyme usp14 and uchl5 induces apoptosis in multiple myeloma and overcomes bortezomib resistance. *Blood* **2014**, *123*, 706–716. [CrossRef] [PubMed]

163. Mistry, H.; Hsieh, G.; Buhrlage, S.J.; Huang, M.; Park, E.; Cuny, G.D.; Galinsky, I.; Stone, R.M.; Gray, N.S.; D'Andrea, A.D.; et al. Small-molecule inhibitors of usp1 target id1 degradation in leukemic cells. *Mol. Cancer Ther.* **2013**, *12*, 2651–2662. [CrossRef] [PubMed]

164. Liang, Q.; Dexheimer, T.S.; Zhang, P.; Rosenthal, A.S.; Villamil, M.A.; You, C.; Zhang, Q.; Chen, J.; Ott, C.A.; Sun, H.; et al. A selective usp1-uaf1 inhibitor links deubiquitination to DNA damage responses. *Nat. Chem. Biol.* **2014**, *10*, 298–304. [CrossRef]

165. Chen, J.; Dexheimer, T.S.; Ai, Y.; Liang, Q.; Villamil, M.A.; Inglese, J.; Maloney, D.J.; Jadhav, A.; Simeonov, A.; Zhuang, Z. Selective and cell-active inhibitors of the usp1/uaf1 deubiquitinase complex reverse cisplatin resistance in non-small cell lung cancer cells. *Chem. Biol.* **2011**, *18*, 1390–1400. [CrossRef]

166. Davis, M.I.; Pragani, R.; Fox, J.T.; Shen, M.; Parmar, K.; Gaudiano, E.F.; Liu, L.; Tanega, C.; McGee, L.; Hall, M.D.; et al. Small molecule inhibition of the ubiquitin-specific protease usp2 accelerates cyclin d1 degradation and leads to cell cycle arrest in colorectal cancer and mantle cell lymphoma models. *J. Biol. Chem.* **2016**, *291*, 24628–24640. [CrossRef]

167. Issaenko, O.A.; Amerik, A.Y. Chalcone-based small-molecule inhibitors attenuate malignant phenotype via targeting deubiquitinating enzymes. *Cell Cycle* **2012**, *11*, 1804–1817. [CrossRef]

168. Okada, K.; Ye, Y.Q.; Taniguchi, K.; Yoshida, A.; Akiyama, T.; Yoshioka, Y.; Onose, J.; Koshino, H.; Takahashi, S.; Yajima, A.; et al. Vialinin a is a ubiquitin-specific peptidase inhibitor. *Bioorg. Med. Chem. Lett.* **2013**, *23*, 4328–4331. [CrossRef]

169. Chauhan, D.; Tian, Z.; Nicholson, B.; Kumar, K.G.; Zhou, B.; Carrasco, R.; McDermott, J.L.; Leach, C.A.; Fulcinniti, M.; Kodrasov, M.P.; et al. A small molecule inhibitor of ubiquitin-specific protease-7 induces apoptosis in multiple myeloma cells and overcomes bortezomib resistance. *Cancer Cell* **2012**, *22*, 345–358. [CrossRef]

170. Weinstock, J.; Wu, J.; Cao, P.; Kingsbury, W.D.; McDermott, J.L.; Kodrasov, M.P.; McKelvey, D.M.; Suresh Kumar, K.G.; Goldenberg, S.J.; Mattern, M.R.; et al. Selective dual inhibitors of the cancer-related deubiquitylating proteases usp7 and usp47. *ACS Med. Chem. Lett.* **2012**, *3*, 789–792. [CrossRef]

171. Pal, A.; Young, M.A.; Donato, N.J. Emerging potential of therapeutic targeting of ubiquitin-specific proteases in the treatment of cancer. *Cancer Res.* **2014**, *74*, 4955–4966. [CrossRef] [PubMed]

172. Kapuria, V.; Peterson, L.F.; Fang, D.; Bornmann, W.G.; Talpaz, M.; Donato, N.J. Deubiquitinase inhibition by small-molecule wp1130 triggers aggresome formation and tumor cell apoptosis. *Cancer Res.* **2010**, *70*, 9265–9276. [CrossRef]

173. Zhou, B.; Zuo, Y.; Li, B.; Wang, H.; Liu, H.; Wang, X.; Qiu, X.; Hu, Y.; Wen, S.; Du, J.; et al. Deubiquitinase inhibition of 19s regulatory particles by 4-arylidene curcumin analog ac17 causes nf-kappab inhibition and p53 reactivation in human lung cancer cells. *Mol. Cancer Ther.* **2013**, *12*, 1381–1392. [CrossRef] [PubMed]

174. D'Arcy, P.; Linder, S. Proteasome deubiquitinases as novel targets for cancer therapy. *Int. J. Biochem. Cell Biol.* **2012**, *44*, 1729–1738. [CrossRef] [PubMed]

175. Wang, X.; D'Arcy, P.; Caulfield, T.R.; Paulus, A.; Chitta, K.; Mohanty, C.; Gullbo, J.; Chanan-Khan, A.; Linder, S. Synthesis and evaluation of derivatives of the proteasome deubiquitinase inhibitor b-ap15. *Chem. Biol. Drug Des.* **2015**, *86*, 1036–1048. [CrossRef] [PubMed]

International Journal of
Molecular Sciences

Article

Semisynthetic Modification of Tau Protein with Di-Ubiquitin Chains for Aggregation Studies

Francesca Munari [1], Carlo Giorgio Barracchia [1], Francesca Parolini [1], Roberto Tira [1], Luigi Bubacco [2], Michael Assfalg [1] and Mariapina D'Onofrio [1,*]

[1] Department of Biotechnology, University of Verona, 37128 Verona, Italy; francesca.munari@univr.it (F.M.); carlogiorgio.barracchia@univr.it (C.G.B.); francesca.parolini@univr.it (F.P.); roberto.tira@univr.it (R.T.); michael.assfalg@univr.it (M.A.)

[2] Department of Biology, University of Padova, 35121 Padova, Italy; luigi.bubacco@unipd.it

* Correspondence: mariapina.donofrio@univr.it; Tel.: +39-045-802-7801

Received: 12 May 2020; Accepted: 19 June 2020; Published: 20 June 2020

Abstract: Ubiquitin, a protein modifier that regulates diverse essential cellular processes, is also a component of the protein inclusions characteristic of many neurodegenerative disorders. In Alzheimer's disease, the microtubule associated tau protein accumulates within damaged neurons in the form of cross-beta structured filaments. Both mono- and polyubiquitin were found linked to several lysine residues belonging to the region of tau protein that forms the structured core of the filaments. Thus, besides priming the substrate protein for proteasomal degradation, ubiquitin could also contribute to the assembly and stabilization of tau protein filaments. To advance our understanding of the impact of ubiquitination on tau protein aggregation and function, we applied disulfide-coupling chemistry to modify tau protein at position 353 with Lys48- or Lys63-linked di-ubiquitin, two representative polyubiquitin chains that differ in topology and structure. Aggregation kinetics experiments performed on these conjugates reveal that di-ubiquitination retards filament formation and perturbs the fibril elongation rate more than mono-ubiquitination. We further show that di-ubiquitination modulates tau-mediated microtubule assembly. The effects on tau protein aggregation and microtubule polymerization are essentially independent from polyubiquitin chain topology. Altogether, our findings provide novel insight into the consequences of ubiquitination on the functional activity and disease-related behavior of tau protein.

Keywords: tau protein; ubiquitination; semisynthesis; disulfide-coupling; polyubiquitin; fibrils; aggregation; neurodegeneration

---

1. Introduction

Ubiquitination is a prevalent post-translational process that regulates essential eukaryotic cell functions such as protein homeostasis, signaling, and cellular localization [1–3]. It consists of the covalent addition of the small protein ubiquitin to the side chain of lysine residues in a target protein. The reaction is controlled by the coordinated action of a ubiquitin-activating enzyme (E1), a ubiquitin-conjugating enzyme (E2) and a ubiquitin ligase (E3), which catalyze the formation of an isopeptide bond between the carboxyl group of the C-terminal glycine of ubiquitin and the ε-amino group of the substrate's lysine [2–5]. Ubiquitination of a ubiquitin molecule through one (or more) of its seven lysine residues (Lys6, Lys11, Lys27, Lys29, Lys33, Lys48 and Lys63) or its amino terminus produces polymers named polyubiquitin chains [2,5]. Polyubiquitin chains have different biological functions depending on the type of linkage that connects the ubiquitin units. Lys48- and Lys11-linked chains regulate proteasome-mediated protein degradation, Lys63-linked chains are involved in endocytosis, cell signaling, and cellular response to DNA damage [2,6], while the functions of the remaining polyubiquitin chain types remain less clear [7]. Each chain type exhibits unique conformational properties that correlate with the diverse biological activities [7,8]. Prototypical polyubiquitin chains with Lys48- and Lys63- linkage display distinct structures: prevalently compact in the case of Lys48-linked chain [9] and extended in the case of Lys63-linked polyubiquitin [10]. Differences in polyubiquitin chain topology generate unique molecular interfaces that govern the selective recognition of cellular receptors and the fine modulation of specific cellular pathways [11].

Due to the central role of protein ubiquitination in fundamental cellular pathways, the dysfunction of the ubiquitin system is involved in the onset of many human pathologies [12]. In particular, the impairment of protein turnover mediated by the ubiquitin proteasome system (UPS) is implicated in neurodegenerative disorders, including Alzheimer's disease (AD), characterized by the accumulation of misfolded proteins [4]. The understanding of the molecular mechanisms of UPS malfunction in relation to protein aggregation has been the target of intense and recent research aimed at finding new therapeutic strategies for the treatment of these diseases. The major incidence of neurodegeneration in the elderly can be justified in part by the fact that a progressive worsening of clearance activity in aging brains may promote the accumulation of toxic and misfolded proteins within neurons [13,14]. Additionally, it is well established that ubiquitin is a key component of the intracellular deposits formed by misfolded proteins in damaged neurons [4,15] and that certain types of protein aggregates can directly inhibit or obstruct the proteasome machinery [12,16,17]. The connection between UPS and protein aggregation was extensively studied in the case of AD, a progressive brain degeneration that is still incurable and which has a major incidence worldwide.

One of the key players in AD pathogenesis is the microtubule associated protein tau, an axonal protein mainly expressed in the central and peripheral nervous system [18]. In the brain, tau protein occurs in six isoforms of different length, generated by alternative splicing [19,20]. Tau protein belongs to the class of intrinsically disordered proteins. The large conformational flexibility that characterizes disordered proteins and regions allows them to carry out a variety of biological activities via distinct recognition mechanisms unfeasible for rigid proteins [21–23]. Tau protein controls the stability and assembly of microtubules, a function that is finely regulated by phosphorylation [18]. The extensively studied, 441-residue isoform (Figure 1) includes an N-terminal half that projects from the microtubule surface and a C-terminal half that promotes the assembly of microtubules. The four pseudo-repeats R1-R4 spanning residues 244-369, together with their flanking regions, constitute the microtubule binding domain (MBD) [19]. The MBD includes two hexapeptide motifs (VQIVYK and VQIINK) that are critical in promoting nucleation of tau aggregates [24]. Under pathological conditions, tau protein becomes hyperphosphorylated, detaches from the microtubules and undergoes a complex aggregation process characterized by a conformational transition to β-sheet rich structures and formation of straight and paired helical filaments (PHFs) that accumulate in neurofibrillary tangles (NFT) [19,20]. PHFs isolated from AD-brain or obtained from recombinant tau protein have been shown to be able to inhibit the proteasome [17]. The solved structure of tau filaments purified from AD-brain revealed residues 306-378 as the ordered fibril core, with the N- and C termini forming the fuzzy coat [25].

In addition to being heavily phosphorylated, tau protein isolated from AD-PHFs was found to be ubiquitinated at several lysine sites within the MBD [26–28]. Specifically, mono-ubiquitin was found to be linked to Lys254, Lys257, Lys311 and Lys317 [27], while polyubiquitin chains were found to be conjugated to Lys254, Lys311 and Lys353 [28]. PHF-tau is modified by three polyubiquitin linkages, Lys6-, Lys11-, and Lys48-, with the latter one being the most prevalent [28]. Recent cryo-EM studies revealed that, besides a proteasome-targeting role, ubiquitination of tau filaments could be structurally involved in mediating specific inter-protofilament packing [26]. Thus, it is plausible that ubiquitination severely affects the assembly and stability of tau filaments and the understanding of the underlying mechanism has the potential to shed new light into the molecular basis of the disease.

To study the impact of ubiquitination on the mechanism of tau protein aggregation, detailed biophysical experiments necessitate highly homogenous and uniquely modified protein samples. However, homogenous ubiquitination of lysine side chains is rarely achieved with the use of ubiquitin ligase enzymes. Indeed, we recently showed that CHIP (Carboxy terminus of Hsp70-interacting protein), an E3 ligase of tau protein [29], ubiquitinates the protein at more than ten sites [30]. To overcome the inherent limitation of the enzymatic approach and to obtain site-specific ubiquitination of target proteins, chemists have developed a vast array of semisynthetic methods that are based on the chemical conjugation of protein precursors [31]. Chemical ubiquitination strategies based on non-native isopeptide bond formation are often easier to implement and give high yields of the product conjugates. Among these, disulfide-coupling chemistry has proven to be highly versatile, efficient and robust, and has already been applied in many studies [30–34]. As a replacement for isopeptide linkage, this semisynthetic method generates a disulfide bond between a Cys residue placed in a specific position of the target protein and the C-terminal aminoethanethiol of a ubiquitin derivative obtained by intein processing. Recently, we described the production and characterization of tau protein mono-ubiquitinated at three different positions using disulfide-directed methodology [30]. With this approach, we obtained novel insight into the diverse effects that lysine mono-ubiquitination at different sites exerts on tau protein aggregation.

Here, we introduce a method based on directed disulfide bond formation in combination with enzymatic synthesis of ubiquitin chains to obtain controlled polyubiquitination of tau protein, thus adding a further level of complexity in the study of ubiquitinated tau protein. We produced and characterized the aggregation process of tau protein species homogenously modified with di-ubiquitin molecules at position 353, one of the ubiquitinated sites found in AD-brain filaments. The versatility of the method allowed us to incorporate two types of di-ubiquitin chains (Lys48- or Lys63- chains) characterized by different structures, and to explore the impact of these prototypical ubiquitin polymers on tau protein aggregation.

2. Results

In the present work, we optimized a semisynthetic strategy based on disulfide-coupling chemistry to covalently attach a di-ubiquitin molecule to the tau protein. Ubiquitin-substrate conjugates obtained using this method are functional mimics of the native ubiquitinated counterparts, with the synthetic linkage being only one atom longer than the native isopeptide bond [32,34]. We selected the construct tau4RD, which represents a short form of the protein, spanning residues Q244-E372 plus an initial Met (Figure 1). Tau4RD has been widely used to study the aggregation behavior of tau protein as it includes the majority of the microtubule binding region and most of the residues involved in the assembly of the cross-β structure forming the core of the tau filaments [35]. To obtain site-specificity, we used a tau4RD mutant devoid of the endogenous cysteines and where a single cysteine was installed at the desired position. We chose to introduce the modification at position 353 because this site was found polyubiquitinated in PHF-tau [26,28], it is part of the fibril core found in different tauopathies [26], and mono-ubiquitination at this position had an observable influence on the aggregation kinetics without hindering the formation of short fibrils [30]. Thus, it was deemed suitable for evaluating the effect of an extended ubiquitin chain on tau filament assembly.

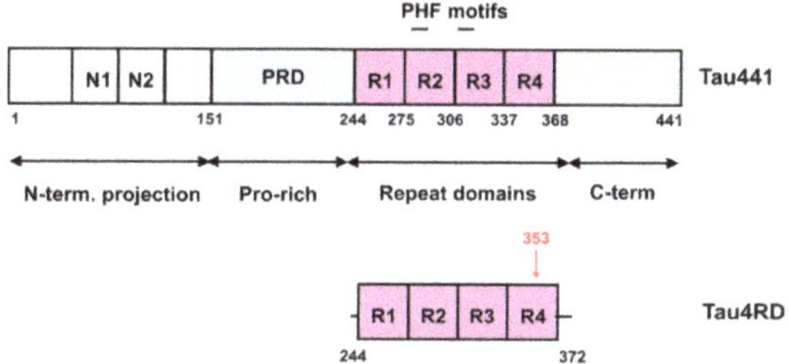

Figure 1. Schematic representation of Tau441 and Tau4RD proteins with their domain organization. The position of residue 353, that is, the conjugation site with di-ubiquitin molecules used in this work, is highlighted in red.

The method of preparation of di-ubiquitinated tau4RD is illustrated in Figure 2. First, using linkage-specific enzymes, we assembled di-ubiquitin molecules with Lys48- and Lys63-linkage (Figures 2b and 3a,b). We used the ubiquitin mutants Lys48Arg or Lys63Arg as distal units and Ub-SH, a ubiquitin derivative bearing a C-terminal aminoethanethiol linker obtained from intein cleavage with cysteamine, as proximal unit (Figure 2a). Homogenous di-ubiquitin chains ending with a C-terminal thiol (Ub$_2$(48/63)-SH) were then purified and allowed to react with 5,5′-Dithiobis(2-nitrobenzoic acid) (DTNB) to produce asymmetric activated disulfides (Figures 2c and 4a,b). Finally, the activated disulfides were incubated with tau4RD, modified with a unique cysteine at position 353, to produce the disulfide-linked conjugates Ub$_2$(48)tau4RD(353) and Ub$_2$(63)tau4RD(353) (Figure 2c). The protein conjugates were obtained at purity of >95% (Figure 3c) and their identity was verified by mass spectrometry (Figure 4c,d). These samples were then used to perform aggregation experiments and their behavior was compared with that of mono-ubiquitinated Ub-tau4RD(353) and of the unconjugated cysteine-free protein (tau4RDΔC).

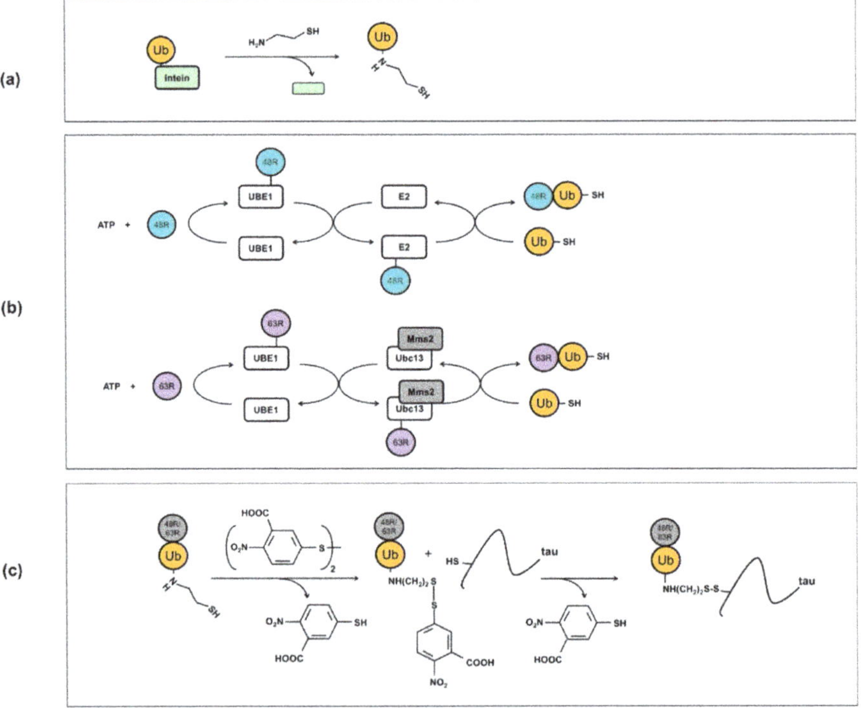

Figure 2. (**a**) Scheme of cysteamine-mediated cleavage of Ub-SH from a ubiquitin-intein fusion protein. We produced a chimeric protein where ubiquitin was cloned to the N-terminal of the GyrA intein. A ubiquitin with a C-terminal aminoethanethiol linker (Ub-SH) was obtained through a trans-thioesterification reaction between intein and cysteamine, followed by a S-to-N acyl shift. (**b**) Scheme of production of Ub$_2$(48)-SH and Ub$_2$(63)-SH di-ubiquitin molecules by enzymatic reaction. For the controlled assembly of K48-linked chains, the enzymes E1 and E2-25K were used. For the K63-linked chains, we used E1 and the complex Mms2/Ubc13. (**c**) Reaction of Ub$_2$(48/63)-SH with 5,5′-Dithiobis(2-nitrobenzoic acid) gave an activated asymmetric disulfide. This was then allowed to react with the unique cysteine placed in position 353 in tau4RD to obtain the disulfide-linked Ub$_2$(48)tau4RD(353) and Ub$_2$(63)tau4RD(353) conjugates.

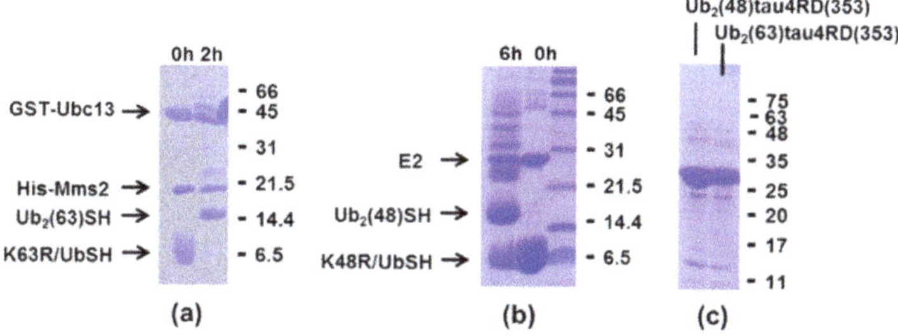

Figure 3. SDS-PAGE showing the enzymatic reaction to obtain the Ub_2(63)-SH (**a**) or Ub_2(48)-SH (**b**) di-ubiquitin molecules. In (**c**), the purified Ub_2(48)tau4RD(353) and Ub_2(63)tau4RD(353) are shown.

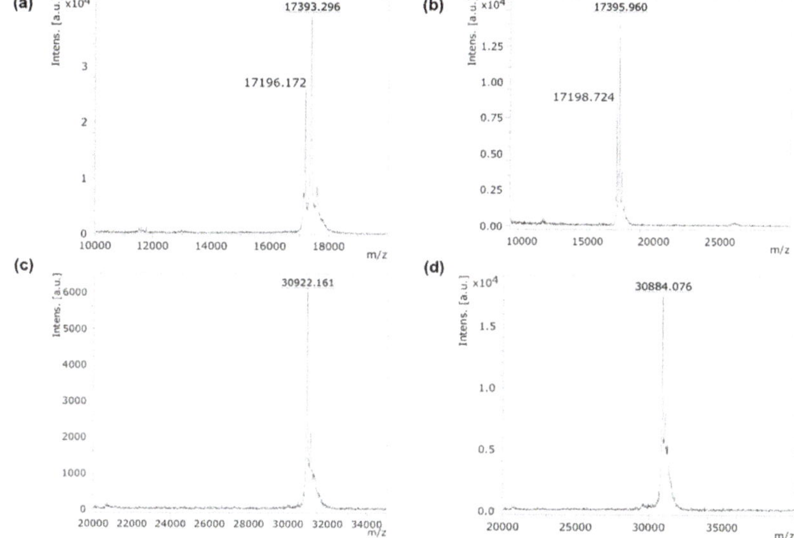

Figure 4. Maldi TOF MS analysis of protein samples. We obtained mass values of 17,393 and 17,396 for the Ub_2(48)-S-TNB (**a**) and Ub_2(63)-S-TNB (**b**) adducts, respectively (expected mass 17397), where TNB stands for (2-nitro-5-thiobenzoic acid). In (**a**) and (**b**), the MS peak of the unconjugated Ub_2(48/63)SH protein is also present (expected mass 17200). We obtained mass values of 30,922 and 30,884 for the Ub_2(48)tau4RD(353) (**c**) and Ub_2(63)tau4RD(353) (**d**), respectively (expected mass 30923).

The kinetics of filament formation of the prepared proteins was followed by monitoring changes in Thioflavin T (ThT) fluorescence (Figure 5a,b). In this assay, the fluorescence emission of ThT increases upon its specific binding to the β-sheet rich structure characteristic of tau filaments. The sigmoidal profile of the kinetic curves reflects the cooperative nature of the nucleation-dependent aggregation process. The initial flat portion of the curve, corresponding to the lag phase, is followed by a steep transition (the growth or elongation phase) and a flat terminal part (plateau phase). As shown previously [30], aggregation of tau4RDΔC in the presence of heparin is very rapid, characterized by an early transition midpoint at ~5 h, and a fast fibril growth, with an elongation time of 0.7 ± 0.2 h (Figure 5a,b, Table 1). The addition of unconjugated di-ubiquitin molecules (Ub_2(48/63)) at an equimolar ratio with tau4RDΔC did not affect the aggregation curves (Figure 5a), excluding the possibility that significant tau-ubiquitin inter-molecular contacts interfered with filament formation. Moreover, in the experimental conditions used, neither Lys48- nor Lys63-linked di-ubiquitin formed fibrils by themselves

(Figure 5a), thereby allowing us to interpret the experimental data in terms of ubiquitination-induced perturbations of the aggregation of tau protein.

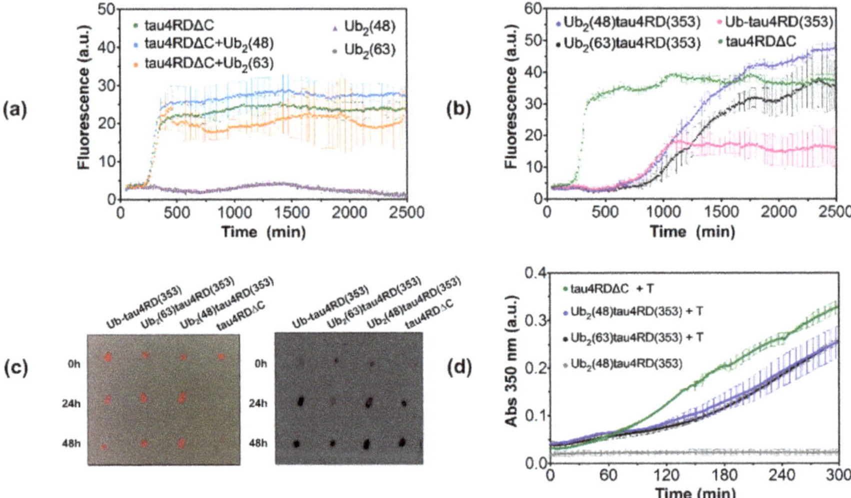

Figure 5. (**a**,**b**) Thioflavin T-based aggregation kinetics experiments. Error bars of fluorescence curves correspond to standard deviations of at least three independent experiments. In (**a**), aggregation data of tau4RDΔC are reported, in the absence or presence of equimolar Ub$_2$(48/63). Kinetics data for the Ub$_2$(48/63) alone are also reported. For these control experiments Ub$_2$(48/63) were produced with a distal D77 ubiquitin mutant, to avoid insertion of free thiol groups. In (**b**), aggregation data of di-ubiquitinated forms Ub$_2$(48)tau4RD(353) and Ub$_2$(63)tau4RD(353) are reported in comparison with tau4RDΔC, and mono-ubiquitinated Ub-tau4RD(353). (**c**) On the right, dot Blot of Ub$_2$(48)tau4RD(353) and Ub$_2$(63)tau4RD(353) at different times of aggregation induced with equimolar heparin, probed with the A11 antibody. On the left, deposited proteins were stained with Ponceau. Samples of tau4RDΔC and Ub-tau4RD(353) were used as controls. (**d**) Microtubule assembly in the presence of tau protein samples was measured monitoring the absorbance at 350 nm. Error bars of absorbance curves correspond to standard deviations of three independent experiments. Ub$_2$(48)tau4RD(353) sample in the absence of tubulin was used as control.

The sigmoidal aggregation profiles obtained with Ub$_2$(48)tau4RD(353) and Ub$_2$(63)tau4RD(353) indicate that di-ubiquitinated tau protein samples were capable of forming filamentous aggregates (Figure 5b). From visual inspection of the aggregation curves measured on all of the tau protein samples, we noted a significant difference in the maximum fluorescence at plateau for the mono-ubiquitinated tau protein in comparison with the other proteins. Because signal intensity is influenced by specific amyloid properties and the proteins under investigation have different structures and shapes, the reduction in ThT fluorescence observed for mono-ubiquitinated tau protein was likely due to a different affinity for ThT and not to the number of fibrils formed.

Based on a quantitative analysis of the aggregation kinetics, we found that filaments formation by di-ubiquitinated tau protein was significantly delayed compared to the unconjugated protein, as deduced by the longer lag phase and elongation time of the di-ubiquitinated proteins (Figure 5b, Table 1). Additionally, di-ubiquitination was found to specifically affect the transition midpoint and elongation time more than mono-ubiquitination. Indeed, Ub$_2$(48)tau4RD(353) and Ub$_2$(63)tau4RD(353) displayed a similar transition half time of ~21 h, which was significantly shifted with respect to the value determined for Ub-tau4RD(353) ($t_{0.5}$ ~15 h), and all the values of modified tau protein were larger compared to that of tau4RD ΔC ($t_{0.5}$ ~5 h). Likewise, the elongation time showed an analogous trend for the investigated samples (Table 1). By comparison of the aggregation curves, it emerged that the kinetics of the two di-ubiquitinated proteins was similar.

Ub$_2$(48)tau4RD(353) and, to a lesser extent Ub$_2$(63)tau4RD(353), produced aggregates that reacted with A11 (Figure 5c), an antibody capable of recognizing prefibrillar oligomers of diverse proteins [36]. Thus, the ability of the tau protein component to form intermediate amyloidogenic species was maintained after modification, as shown previously for mono-ubiquitinated tau4RD [30]. The ability of both Ub$_2$(48)tau4RD(353) and Ub$_2$(63)tau4RD(353) to form mature fibrils was confirmed by the TEM images (Figure 6a,b), which showed the presence of well-formed twisted filaments. The morphological analysis revealed that filaments were characterized by a large width of 20 ± 2 nm and a narrow width of 15–16 nm, and a twist crossover repeat of 57-58 nm (Table 2), indicating that the overall morphology of tau4RDΔC filaments was not heavily modified by di- or mono-ubiquitin conjugation (Figure 6a–d, Table 2).

Taken together, the obtained data clearly indicate that the incorporation of protein modifiers in the microtubule binding domain of tau protein at position 353 interferes with the aggregation mechanism but does not abrogate the formation of mature fibrils. The addition of one ubiquitin moiety to tau protein determines the inhibition of fibrils formation. The inhibitory effect is even stronger when a second ubiquitin moiety is attached to the proximal ubiquitin unit, as it resulted from the substantial increase of the duration of both midpoint transition and elongation time for the di-ubiquitinated species. However, despite their known structural differences, the topology of the investigated di-ubiquitin molecules (Lys48- or Lys63-linked) did not influence the process of fibrils formation. Thus, it appears that the inhibitory effect of di-ubiquitination at the 353 site is caused by the increased steric hindrance which impairs microscopic events that lead to fibrils formation.

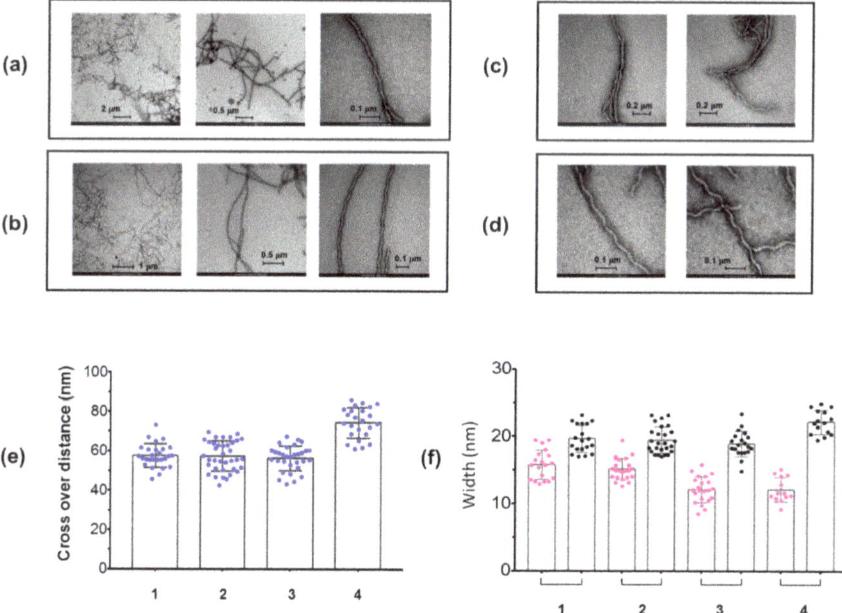

Figure 6. TEM representative images of (**a**) Ub$_2$(48)tau4RD(353), (**b**) Ub$_2$(63)tau4RD(353), (**c**) tau4RDΔC, and (**d**) Ub-tau4RD(353), after 48 h of incubation at 37 °C under static condition. 30 μL of sample at a concentration of 2.5 μM were deposited. Distributions of (**e**) cross-over distances and (**f**) widths of the twisted filaments measured from TEM images of Ub$_2$(48)tau4RD(353) (1), Ub$_2$(63)tau4RD(353) (2) and Ub-tau4RD(353) (3) conjugates and tau4RDΔC (4). In (**e**) and (**f**), dots indicate single measurements and bars the positions of means ± SD. In (**f**), dots in magenta refer to the measures of narrow widths and in black to large widths.

After having established how di-ubiquitination affects tau fibril formation, a process associated with disease, we set out to describe its consequence on tau protein functional activity. Specifically, we investigated the impact of Lys48- and Lys63-linked di-ubiquitin on tau-mediated tubulin polymerization. The assay was performed using $Ub_2(48)$tau4RD(353) or $Ub_2(63)$tau4RD(353), and in the presence of tau4RDΔC as a control (Figure 5d). Microtubule (MT) polymerization was monitored by following the increase in absorbance at 350 nm. The kinetics of MT assembly in the presence of tau4RDΔC resembled previous results reported on tau4RD [37]. The data acquired in the presence of $Ub_2(48)$tau4RD(353) or $Ub_2(63)$tau4RD(353) indicates that the presence of the di-ubiquitin chains at position 353 of tau protein, moderately but significantly inhibits MT polymerization. Indeed, after 300 min of incubation of tubulin with the conjugates, we observed ~77% of MT formation (referred to 100% for tau4RDΔC). This effect was independent from the topology of the di-ubiquitin chain, as the observed curves were almost superimposable.

Table 1. Kinetic parameters for the aggregation of tau protein samples, determined on the basis of ThT fluorescence assays.[1]

	$t_{0.5}$ (h)	t_{lag} (h)	τ (h)
tau4RDΔC	5.2 ± 0.1	3.8 ± 0.5	0.7 ± 0.2
Ub-tau4RD(353)	14.8 ± 1.0	12.4 ± 0.1	1.1 ± 0.6
$Ub_2(48)$-tau4RD(353)	20.7 ± 1.2	12.1 ± 1.1	4.3 ± 0.1
$Ub_2(63)$-tau4RD(353)	21.1 ± 3.3	15.1 ± 1.4	3.0 ± 0.9

[1] $t_{0.5}$: midpoint of the transition; τ: elongation time constant; $t_{lag} = t_{0.5} - 2\tau$.

Table 2. Morphological properties of the twisted filaments of tau protein samples obtained from the analysis of TEM images.

	Large Width	Narrow Width	Crossover Distance
tau4RDΔC	22 ± 2 nm (n = 14)	12 ± 2 nm (n = 13)	74 ± 8 nm (n = 24)
Ub-tau4RD(353)	19 ± 2 nm (n = 20)	12 ± 2 nm (n = 22)	56 ± 6 nm (n = 31)
$Ub_2(48)$-tau4RD(353)	20 ± 2 nm (n =19)	16 ± 2 nm (n = 19)	58 ± 6 nm (n = 27)
$Ub_2(63)$-tau4RD(353)	20 ± 2 nm (n = 26)	15 ± 2 nm (n = 24)	57 ± 8 nm (n = 37)

3. Discussion

The microtubule-associated protein tau plays a central role in the pathogenesis of AD. Besides being heavily phosphorylated [20], tau protein in PHFs from AD brains is found to be mono- or polyubiquitinated at multiple sites [26–28]. The observation that most of these sites belong to the microtubule binding region, a domain involved in the formation of the filaments core, suggests the attempt of the neurons to get rid of aggregated tau protein forms via the UPS. This hypothesis is supported by the fact that PHF-tau was found to be modified by Lys48- and Lys11- linked chains [28], both being signals for proteasomal degradation, and by mono-ubiquitin, recently recognized as an additional signal for proteasomal targeting [38]. These observations suggest a possible role of ubiquitination in the formation and clearance of pathological tau protein species.

In a bid to advance our understanding of the impact of ubiquitination on tau protein aggregation and function, we recently started developing methods to attain controlled ubiquitination of tau protein samples for molecular-level investigations. In a previous work, we produced mono-ubiquitinated tau4RD samples and found that the impact of the modification on tau protein fibrillogenesis was site-dependent [30]. While ubiquitination at Lys311, located within the PHF motif of the R3 domain, essentially abolished filament formation, modification at Lys254 and Lys353 changed the aggregation kinetics but did not arrest fibrillation completely. To follow up on this study, in the present work, we set out to investigate whether a minimal polyubiquitin chain, a di-ubiquitin, could affect the aggregation behavior of tau protein.

To obtain site-specific di-ubiquitination, we combined enzyme-mediated preparation of di-ubiquitin derivatives with disulfide-directed ligation to tau4RD, expanding our approach originally developed to produce mono-ubiquitination [30]. Based on the previous findings with mono-ubiquitination, di-ubiquitin was installed at position 353, a site that was deemed suitable for evaluating the effect of an extended ubiquitin chain on filament assembly. Because polyubiquitin chains exist in a variety of topologies and structures, here we chose to investigate the prototypical Lys48- and Lys63-linked chains as being representative of polymers that preferentially adopt compact or extended conformations, respectively [7]. Lys48-linked di-ubiquitin predominantly adopts a closed conformation in which the functional hydrophobic patches of mono-ubiquitin are sequestered at the ubiquitin/ubiquitin interface [9]. In contrast, Lys63-linked di-ubiquitin chains are extended and adopt an open conformation with no direct contact between the neighboring ubiquitin subunits, thereby exposing the hydrophobic patches of mono-ubiquitin [10]. Thus, we hypothesized that different architectures of polyubiquitin chains could differently affect substrate aggregation.

The performed aggregation experiments revealed that both di-ubiquitinated tau protein conjugates were less prone to form fibrils compared to the unconjugated protein (Figure 5b), although they retained the capability to form A11-positive prefibrillar oligomeric species (Figure 5c) and twisted filaments (Figure 6). Moreover, the dimeric ubiquitin modifier increased significantly both the half-time of the transition and the fibril elongation time with respect to mono-ubiquitin, with the time constants following the order: di-ubiquitinated >> mono-ubiquitinated > unmodified (Figure 5b, Table 1). However, the topological differences between the two di-ubiquitin chains (Lys48- or Lys63-linked) tested do not differentially interfere with the process of fibrils formation.

It is established that the duration of the lag phase depends on the rates of multiple parallel microscopic processes, such as the formation of primary nuclei and their amplification through elongation and secondary nucleation processes [39]. Because variations of the former process do not affect the growth phase, while changes in elongation and secondary nucleation modify both the lag phase and the growth phase, it seems possible that di-ubiquitination of tau protein could interfere with the growth and proliferation of primary nuclei rather than with their formation. This was more evident for di-ubiquitinated, rather than mono-ubiquitinated, tau protein. Since the inhibitory effect was dependent on the length of the ubiquitin chain, the impact on tau protein aggregation is likely be a result of a combination of structural motif at the tau-ubiquitin conjugating point and of the presence of the distal ubiquitin moiety which adds steric hindrance. Yet, we do not exclude the possibility that a further elongation of the polyubiquitin chain could elicit different results, particularly if considering that longer ubiquitin chains have lower thermodynamic stability and can form fibrils themselves under specific conditions [40].

To interpret our results in the wider context of post-translational modifications of tau protein, we reviewed recent studies which explored the impact of single or multiple phosphorylation of full-length tau protein or tau4RD [37,41]. Phosphorylation of both Ser262 and Ser356 was found to significantly alter the aggregation kinetics of tau4RD [37], and single-site phosphorylation affected tau protein aggregation in a sequence-specific manner [37,41]: phosphorylation of Ser305 had a significant impact on fibril formation; however modification at Ser356 did not significantly perturb the aggregation of tau protein. Here we have shown that di-ubiquitination at 353, close to residue 356, allows the formation of fibrils, although with a reduced aggregation rate, in line with the observation that this site is excluded from the ordered core (residues 272–330) in heparin-induced tau filaments [42]. Additionally, recent cryo-EM studies show that tau filaments from brain tissues [25,26] are characterized by a long core that includes Lys353 and suggest that bulky modifiers, such as mono- or di-ubiquitin, in that position could affect the mechanism of formation of tau protein aggregates.

Tau protein is known to play a central role in the assembly and stabilization of microtubules [18]. Therefore, we explored whether di-ubiquitination could regulate tau-mediated tubulin polymerization. Based on our observations, it emerges that the presence of either Lys48- or Lys63-linked di-ubiquitin at position 353 moderately but significantly inhibits MT polymerization. This effect was found to be independent of the di-ubiquitin linkage topology (Figure 5d). Our findings are consistent with previous studies reporting that multiple mono-ubiquitination of tau protein reduces MT binding affinity [43], and phosphorylation at Ser356 (close to Lys353) slightly impairs MT assembly [37]. Taken together, the evidence on site-specific phosphorylated or on multi mono-ubiquitinated tau protein, and our results obtained with the di-ubiquitinated proteins indicate that post-translational modifications around position 353 do not crucially affect tau-assisted MT polymerization and that the functional activity of tau protein is highly dependent on the sites of modification.

In conclusion, we were able to obtain, for the first time, samples of tau protein conjugated to Lys48- or Lys63- linked di-ubiquitin at a specific site and to study the effect of these modifications on tau protein aggregation and function in comparison with the mono-ubiquitinated species. We demonstrated that the conjugation of tau protein to di-ubiquitin chains at the 353 position significantly delayed, but did not completely inhibit, tau protein aggregation, in analogy with mono-ubiquitination. Quantitative analysis of the aggregation kinetics revealed a more pronounced effect of di-ubiquitin compared to mono-ubiquitin during elongation of filaments, rather than during primary nucleation events. However, linkage topology had a minor effect on the measured kinetic parameters. Finally, we demonstrated that di-ubiquitination modulated the tau-mediated microtubule assembly, thus highlighting the potential role of ubiquitination in the regulation of tau protein function. For a more comprehensive overview, further efforts will have to be directed to produce tau protein modified with longer polyubiquitin chains linked in different positions.

We believe that our findings also provide crucial information in light of recent structural studies on tau filaments from degenerated brain tissues. Ubiquitin is proposed to play a structural role by mediating specific inter-protofilament packing and promoting the formation of fibril subtypes, specific of the different tauopathies [26]. Moreover, future investigations are required to elucidate the synergic effects played by different post-translational modifications, such as ubiquitination, phosphorylation and acetylation, in the transition of tau protein to toxic species, with the aim to understand their implications in the different pathologies.

4. Materials and Methods

4.1. Materials

Cysteamine, Tris(2-carboxyethyl)phosphine, 5,5′-Dithiobis(2-nitrobenzoic acid), Thioflavin T, Heparin (H3393), dithiothreitol (DTT), acetonitrile, and Trifluoroacetic acid (TFA) were purchased from Sigma Aldrich (St Louis, MO, USA); Tubulin (T240) was purchased from Cytoskeleton (Denver, CO, USA), italian distributor Società italiana chimici (Rome, IT).

4.2. Methods

4.2.1. Protein Expression and Purification

In this work we used the following tau4RD (Q244-E372 plus initial Met) protein variants: tau4RD C291A, C322A (here named tau4RDΔC) and tau4RD C291A, C322A, K353C, here named tau4RD(353). The two proteins were expressed without affinity tag and purified as described in [30].

The ubiquitin mutants K48R, K63R, D77 were produced with the same protocol used for wild-type ubiquitin that is described in [44].

The recombinant enzymes human His-tagged E1, GST-tagged E2–25K, yeast His-tagged Mms2, yeast GST-tagged Ubc13 were produced as previously described [45]. In the present work, the GST tag was removed from the E2–25K protein by incubating the clean fusion protein attached to the GSH-resin with thrombin.

To obtain a ubiquitin bearing an aminoethanethiol C-terminal group (Ub-SH) required for the disulfide-coupling reaction, we first produced a chimeric protein where ubiquitin was cloned to the N-terminal of the GyrA intein in a pET22 vector. In this way, the chimeric protein has a C-terminal

His-tag. The Ub-intein-His protein was produced in BL21(DE3) cells at 37 °C overnight in auto-inducing medium and purified by immobilized nickel affinity chromatography (IMAC) according to standard protocols. Cleavage of Ub-SH was obtained by incubating the clean fusion protein in a buffer at pH 7.5 containing: Tris-HCl 20 mM, EDTA 1 mM, cysteamine 40 mM, and Tris(2-carboxyethyl)phosphine (TCEP) 3 mM for 48 h at 10 °C. Ub-SH was further purified by reverse -IMAC and a superdex-75 gel filtration column when required.

4.2.2. Production of Di-Ubiquitin Chains

Di-ubiquitin chains (Ub_2) were obtained from Ub variants through enzymatic reactions, with the strategy described in [46,47]. The reaction buffer contained 50 mM Tris HCl pH 8.0, 5 mM $MgCl_2$, 10 mM ATP, 3 mM TCEP, 0.02% NaN_3 and protease inhibitors. The ubiquitin variants K48R+D77 and K48R+UbSH were used as building blocks for the reconstitution of $Ub_2(48)$ and $Ub_2(48)$-SH, respectively. For the assembly of these K48-linked chains, we used 1 µM E1 and 20 µM E2-25K. The ubiquitin variants K63R+D77 and K63R+UbSH were used to assemble $Ub_2(63)$ and $Ub_2(63)$-SH, respectively. For these K63-linked chains, we used 1 µM E1, 25 µM Ubc13 and 25 µM Mms2. After reaction, we incubated the solutions with 0.2% perchloric acid to obtain precipitation of most of the enzymes. Then, SP cation-exchange and superdex-75 gel filtration were performed to obtain the pure di-ubiquitin chains.

4.2.3. Disulfide-Coupling Reaction

First, $Ub_2(48/63)$-SH chains were activated with 5,5′-Dithiobis(2-nitrobenzoic acid) (DTNB). Typically, 4-6 mg of DTNB were dissolved in 3 mL of 100 mM Hepes buffer pH 7.0 containing Ub_2-SH at a concentration of 1-2 mg/mL. After incubation at 10 °C overnight, the obtained $Ub_2(48/63)$-S-(2-nitro-5-thiobenzoic acid) (hereafter $Ub_2(48/63)$-S-TNB) disulfide adducts were purified by desalting. The activated proteins were verified by MALDI (Figure 4a,b).

Then, the $Ub_2(48/63)$-S-TNB disulfide adducts were incubated with tau4RD(353) in equimolar amount in 100 mM Hepes buffer pH 7.0, for 20′ at 25 °C. The $Ub_2(48/63)$tau4RD(353) disulfide conjugates were then purified by SP-ion exchange chromatography and verified by MALDI (Figure 4c,d).

It is important to note the thiol containing proteins $Ub_2(48/63)$-SH and tau4RD(353) were first incubated with a large excess of DTT that was then properly removed by size exclusion chromatography just before the disulfide coupling reaction.

The procedure to obtain mono-ubiquitinated tau protein at position 353 (Ub-tau4RD(353)) by a similar disulfide-coupling strategy is described in [30].

4.2.4. Thioflavin T Aggregation Assay

Thioflavin T aggregation assays were carried out in 96-well dark plates in a Tecan Infinite M200PRO Plex plate reader at 30 °C with cycles of 30 s of orbital shaking at 140 rpm and 10 min of rest throughout the incubation. ThT fluorescence measurements were taken every 11 min, using excitation wavelength of 450 nm and recording fluorescence emission at 480 nm.

Samples contained 0.01 mM proteins in 20 mM sodium phosphate buffer at pH 7.4 and 50 mM NaCl (with 0.02% NaN_3 and protease inhibitors with EDTA), incubated with equimolar amount of heparin and ThT. Samples containing tau protein variants were filtered through a 100 kDa cut-off filter (Sartorius) to remove pre-existing large oligomers. Each measurement was performed in three replicates. The aggregation curves were analyzed by fitting each individual experimental data set with the following sigmoidal function [48]:

$$y = y_i + \frac{y_f}{1 + e^{-[(t-t_{0.5})/\tau]}}$$

where y is the fluorescence intensity as a function of time t, y_i and y_f are the intercept of the initial and final baselines with the y-axis, $t_{0.5}$ is the time needed to reach halfway through the elongation phase and τ is the elongation time constant. The lag time is defined as $t_{lag} = t_{0.5} - 2\tau$. The values reported in the text correspond to the mean $\pm$ SD of the individual values computed separately on each curve.

Analysis and figures production were carried out with GraphPad Prism 7 (GraphPad Software Inc., La Jolla, CA, USA).

4.2.5. TEM Analysis

Ub_2(48)tau4RD(353) sample was further purified by HPLC-reverse phase C18 done with TFA 0.1%/acetonitrile gradient, before buffer exchange to the aggregation buffer.

Samples of Ub_2(48/63)tau4RD(353) conjugates, tau4RDΔC, and Ub-tau4RD(353) were incubated at concentration of 0.05 mM in 20 mM sodium phosphate buffer at pH 7.4 and 50 mM NaCl (with 0.02% NaN_3 and protease inhibitors with EDTA) at 37 °C without agitation, with the addition of equimolar amount of heparin as aggregation initiator. After 48 h of incubation, we used 100 kDa cut-off filters (Sartorius, Aubagne, FR) to exchange the buffer to H_2O mQ.

For TEM measurements, we used a Tecnai G^2 (FEI, Hillsboro, OR; USA) transmission electron microscope operating at 100 kV, with the procedure described in [30]. Images were analyzed with the ImageJ software.

4.2.6. Dot Blotting

Dot blotting of Ub_2(48)tau4RD(353) and Ub_2(63)tau4RD(353) conjugates, tau4RDΔC, and Ub-tau4RD(353) at different aggregation times, using the anti-oligomer antibody A11 (ThermoFisher, Waltham, MA, USA), was carried out according to the protocol described in [30].

4.2.7. Tubulin Polymerization Assay

The tubulin polymerization assay for tau4RDΔC, Ub_2(48)tau4RD(353) and Ub_2(63)tau4RD(353) was initiated by mixing in a 96-well plate 25 µM protein with tubulin (36 µM, in 100 µL total volume) in MT assembly buffer (80 mM PIPES pH 6.9, 2 mM $MgCl_2$ and 0.5 mM EGTA) supplemented with 1 mM GTP. The plate was incubated at 37 °C for 4 min. Then, the polymerization was monitored by measuring the absorbance at 350 nm every 30 sec, using a Tecan Infinite M200PRO Plex plate reader (Männendorf, CH) at 37 °C. Each experiment was performed in triplicate.

4.2.8. Mass Spectrometry

Maldi TOF MS analysis was performed on a Bruker Ultraflextreme MALDI-TOF/TOF instrument (Bruker Daltonics, Billerica, MA, USA) by the Centro Piattaforme Tecnologiche of the University of Verona as previously described [30].

Author Contributions: Conceptualization, M.D., F.M. and M.A.; investigation, F.M., C.G.B., F.P. and R.T.; formal analysis, L.B.; writing—original draft preparation, F.M.; writing—review and editing, M.D. and M.A.; supervision, M.D. and M.A.; funding acquisition, M.D. All authors have read and agreed to the published version of the manuscript.

Funding: This research was funded by the Alzheimer's Association, grant number AARG-17-529221 awarded to MD, and by the University of Verona, Progetto Ricerca di Base 2015 awarded to MD.

Acknowledgments: FM thanks "Fondazione Umberto Veronesi" for granting a postdoctoral fellowship. We thank Centro Piattaforme Tecnologiche of the University of Verona for providing access to the Mass Spectrometry Platform, and David Fushman who kindly provided the plasmids for expression of recombinant E2-25K, Ubc13 and Mms2.

Abbreviations

AD	Alzheimer's Disease
UPS	Ubiquitin Proteasome System
MBD	Microtubule Binding Domain
PHFs	Paired Helical Filaments
NFT	Neurofibrillary Tangles
DTNB	5,5'-Dithiobis(2-nitrobenzoic acid)
MT	Microtubules

References

1. Varshavsky, A. The Ubiquitin System, Autophagy, and Regulated Protein Degradation. *Annu. Rev. Biochem.* **2017**, *86*, 123–128. [CrossRef] [PubMed]
2. Komander, D. The emerging complexity of protein ubiquitination. *Biochem. Soc. Trans.* **2009**, *37*, 937–953. [CrossRef] [PubMed]
3. Varshavsky, A. The Ubiquitin System, an Immense Realm. *Annu. Rev. Biochem.* **2012**, *81*, 167–176. [CrossRef] [PubMed]
4. Dantuma, N.P.; Bott, L.C. The ubiquitin-proteasome system in neurodegenerative diseases: Precipitating factor, yet part of the solution. *Front. Mol. Neurosci.* **2014**, *7*. [CrossRef]
5. Pickart, C.M.; Fushman, D. Polyubiquitin chains: Polymeric protein signals. *Curr. Opin. Chem. Biol.* **2004**, *8*, 610–616. [CrossRef]
6. Chen, R.-H.; Chen, Y.-H.; Huang, T.-Y. Ubiquitin-mediated regulation of autophagy. *J. Biomed. Sci.* **2019**, *26*, 80. [CrossRef]
7. Kulathu, Y.; Komander, D. Atypical ubiquitylation — The unexplored world of polyubiquitin beyond Lys48 and Lys63 linkages. *Nat. Rev. Mol. Cell Biol.* **2012**, *13*, 508–523. [CrossRef]
8. Castañeda, C.A.; Chaturvedi, A.; Camara, C.M.; Curtis, J.E.; Krueger, S.; Fushman, D. Linkage-specific conformational ensembles of non-canonical polyubiquitin chains. *Phys. Chem. Chem. Phys.* **2016**, *18*, 5771–5788. [CrossRef]
9. Varadan, R.; Walker, O.; Pickart, C.; Fushman, D. Structural Properties of Polyubiquitin Chains in Solution. *J. Mol. Biol.* **2002**, *324*, 637–647. [CrossRef]
10. Varadan, R.; Assfalg, M.; Haririnia, A.; Raasi, S.; Pickart, C.; Fushman, D. Solution Conformation of Lys [63] -linked Di-ubiquitin Chain Provides Clues to Functional Diversity of Polyubiquitin Signaling. *J. Biol. Chem.* **2004**, *279*, 7055–7063. [CrossRef]
11. Ye, Y.; Blaser, G.; Horrocks, M.H.; Ruedas-Rama, M.J.; Ibrahim, S.; Zhukov, A.A.; Orte, A.; Klenerman, D.; Jackson, S.E.; Komander, D. Ubiquitin chain conformation regulates recognition and activity of interacting proteins. *Nature* **2012**, *492*, 266–270. [CrossRef] [PubMed]
12. Popovic, D.; Vucic, D.; Dikic, I. Ubiquitination in disease pathogenesis and treatment. *Nat. Med.* **2014**, *20*, 1242–1253. [CrossRef] [PubMed]
13. Sulistio, Y.A.; Heese, K. The Ubiquitin-Proteasome System and Molecular Chaperone Deregulation in Alzheimer's Disease. *Mol. Neurobiol.* **2016**, *53*, 905–931. [CrossRef] [PubMed]
14. Hegde, A.N.; Smith, S.G.; Duke, L.M.; Pourquoi, A.; Vaz, S. Perturbations of Ubiquitin-Proteasome-Mediated Proteolysis in Aging and Alzheimer's Disease. *Front. Aging Neurosci.* **2019**, *11*, 324. [CrossRef]
15. Mori, H.; Kondo, J.; Ihara, Y. Ubiquitin is a component of paired helical filaments in Alzheimer's disease. *Science* **1987**, *235*, 1641–1644. [CrossRef]
16. Oddo, S. The ubiquitin-proteasome system in Alzheimer's disease. *J. Cell. Mol. Med.* **2008**, *12*, 363–373. [CrossRef]
17. Keck, S.; Nitsch, R.; Grune, T.; Ullrich, O. Proteasome inhibition by paired helical filament-tau in brains of patients with Alzheimer's disease: Proteasome inhibition in Alzheimer's disease. *J. Neurochem.* **2003**, *85*, 115–122. [CrossRef]
18. Wang, Y.; Mandelkow, E. Tau in physiology and pathology. *Nat. Rev. Neurosci.* **2016**, *17*, 22–35. [CrossRef]
19. Mandelkow, E.-M.; Mandelkow, E. Biochemistry and Cell Biology of Tau Protein in Neurofibrillary Degeneration. *Cold Spring Harb. Perspect. Med.* **2012**, *2*, a006247. [CrossRef]

20. Martin, L.; Latypova, X.; Terro, F. Post-translational modifications of tau protein: Implications for Alzheimer's disease. *Neurochem. Int.* **2011**, *58*, 458–471. [CrossRef]

21. Uversky, V.N. Multitude of binding modes attainable by intrinsically disordered proteins: A portrait gallery of disorder-based complexes. *Chem. Soc. Rev.* **2011**, *40*, 1623–1634. [CrossRef] [PubMed]

22. Bajaj, R.; Munari, F.; Becker, S.; Zweckstetter, M. Interaction of the intermembrane space domain of Tim23 protein with mitochondrial membranes. *J. Biol. Chem.* **2014**, *289*, 34620–34626. [CrossRef] [PubMed]

23. Watson, M.; Stott, K. Disordered domains in chromatin-binding proteins. *Essays Biochem.* **2019**, *63*, 147–156. [CrossRef] [PubMed]

24. Von Bergen, M.; Friedhoff, P.; Biernat, J.; Heberle, J.; Mandelkow, E.-M.; Mandelkow, E. Assembly of tau protein into Alzheimer paired helical filaments depends on a local sequence motif (306VQIVYK311) forming beta structure. *Proc. Natl. Acad. Sci. USA* **2000**, *97*, 5129–5134. [CrossRef]

25. Fitzpatrick, A.W.P.; Falcon, B.; He, S.; Murzin, A.G.; Murshudov, G.; Garringer, H.J.; Crowther, R.A.; Ghetti, B.; Goedert, M.; Scheres, S.H.W. Cryo-EM structures of tau filaments from Alzheimer's disease. *Nature* **2017**, *547*, 185–190. [CrossRef]

26. Arakhamia, T.; Lee, C.E.; Carlomagno, Y.; Duong, D.M.; Kundinger, S.R.; Wang, K.; Williams, D.; DeTure, M.; Dickson, D.W.; Cook, C.N.; et al. Posttranslational Modifications Mediate the Structural Diversity of Tauopathy Strains. *Cell* **2020**, *180*, 633–644.e12. [CrossRef]

27. Morishima-Kawashima, M.; Hasegawa, M.; Takio, K.; Suzuki, M.; Titani, K.; Ihara, Y. Ubiquitin is conjugated with amino-terminally processed tau in paired helical filaments. *Neuron* **1993**, *10*, 1151–1160. [CrossRef]

28. Cripps, D.; Thomas, S.N.; Jeng, Y.; Yang, F.; Davies, P.; Yang, A.J. Alzheimer disease-specific conformation of hyperphosphorylated paired helical filament-Tau is polyubiquitinated through Lys-48, Lys-11, and Lys-6 ubiquitin conjugation. *J. Biol. Chem.* **2006**, *281*, 10825–10838. [CrossRef]

29. Petrucelli, L. CHIP and Hsp70 regulate tau ubiquitination, degradation and aggregation. *Hum. Mol. Genet.* **2004**, *13*, 703–714. [CrossRef]

30. Munari, F.; Barracchia, C.G.; Franchin, C.; Parolini, F.; Capaldi, S.; Romeo, A.; Bubacco, L.; Assfalg, M.; Arrigoni, G.; D'Onofrio, M. Semisynthetic and Enzyme-Mediated Conjugate Preparations Illuminate the Ubiquitination-Dependent Aggregation of Tau Protein. *Angew. Chem. Int. Ed.* **2020**. [CrossRef]

31. Mali, S.M.; Singh, S.K.; Eid, E.; Brik, A. Ubiquitin Signaling: Chemistry Comes to the Rescue. *J. Am. Chem. Soc.* **2017**, *139*, 4971–4986. [CrossRef] [PubMed]

32. Meier, F.; Abeywardana, T.; Dhall, A.; Marotta, N.P.; Varkey, J.; Langen, R.; Chatterjee, C.; Pratt, M.R. Semisynthetic, Site-Specific Ubiquitin Modification of α-Synuclein Reveals Differential Effects on Aggregation. *J. Am. Chem. Soc.* **2012**, *134*, 5468–5471. [CrossRef] [PubMed]

33. Chen, J.; Ai, Y.; Wang, J.; Haracska, L.; Zhuang, Z. Chemically ubiquitylated PCNA as a probe for eukaryotic translesion DNA synthesis. *Nat. Chem. Biol.* **2010**, *6*, 270–272. [CrossRef] [PubMed]

34. Chatterjee, C.; McGinty, R.K.; Fierz, B.; Muir, T.W. Disulfide-directed histone ubiquitylation reveals plasticity in hDot1L activation. *Nat. Chem. Biol.* **2010**, *6*, 267–269. [CrossRef] [PubMed]

35. Barghorn, S.; Biernat, J.; Mandelkow, E. Purification of Recombinant Tau Protein and Preparation of Alzheimer-Paired Helical Filaments In Vitro. In *Amyloid Proteins*; Humana Press: Totowa, NJ, USA, 2004; Volume 299, pp. 035–052. ISBN 978-1-59259-874-8.

36. Kayed, R. Common Structure of Soluble Amyloid Oligomers Implies Common Mechanism of Pathogenesis. *Science* **2003**, *300*, 486–489. [CrossRef] [PubMed]

37. Haj-Yahya, M.; Gopinath, P.; Rajasekhar, K.; Mirbaha, H.; Diamond, M.I.; Lashuel, H.A. Site-Specific Hyperphosphorylation Inhibits, Rather than Promotes, Tau Fibrillization, Seeding Capacity, and Its Microtubule Binding. *Angew. Chem. Int. Ed.* **2020**, *59*, 4059–4067. [CrossRef]

38. Livneh, I.; Kravtsova-Ivantsiv, Y.; Braten, O.; Kwon, Y.T.; Ciechanover, A. Monoubiquitination joins polyubiquitination as an esteemed proteasomal targeting signal. *BioEssays* **2017**, *39*, 1700027. [CrossRef]

39. Arosio, P.; Knowles, T.P.J.; Linse, S. On the lag phase in amyloid fibril formation. *Phys. Chem. Chem. Phys.* **2015**, *17*, 7606–7618. [CrossRef]

40. Morimoto, D.; Walinda, E.; Fukada, H.; Sou, Y.-S.; Kageyama, S.; Hoshino, M.; Fujii, T.; Tsuchiya, H.; Saeki, Y.; Arita, K.; et al. The unexpected role of polyubiquitin chains in the formation of fibrillar aggregates. *Nat. Commun.* **2015**, *6*, 6116. [CrossRef]

41. Ellmer, D.; Brehs, M.; Haj-Yahya, M.; Lashuel, H.A.; Becker, C.F.W. Single Posttranslational Modifications in the Central Repeat Domains of Tau4 Impact its Aggregation and Tubulin Binding. *Angew. Chem. Int. Ed.* **2019**, *58*, 1616–1620. [CrossRef]

42. Zhang, W.; Falcon, B.; Murzin, A.G.; Fan, J.; Crowther, R.A.; Goedert, M.; Scheres, S.H. Heparin-induced tau filaments are polymorphic and differ from those in Alzheimer's and Pick's diseases. *eLife* **2019**, *8*, e43584. [CrossRef] [PubMed]

43. Flach, K.; Ramminger, E.; Hilbrich, I.; Arsalan-Werner, A.; Albrecht, F.; Herrmann, L.; Goedert, M.; Arendt, T.; Holzer, M. Axotrophin/MARCH7 acts as an E3 ubiquitin ligase and ubiquitinates tau protein in vitro impairing microtubule binding. *Biochim. Biophys. Acta (BBA) - Mol. Basis Dis.* **2014**, *1842*, 1527–1538. [CrossRef] [PubMed]

44. Munari, F.; Bortot, A.; Zanzoni, S.; D'Onofrio, M.; Fushman, D.; Assfalg, M. Identification of primary and secondary UBA footprints on the surface of ubiquitin in cell-mimicking crowded solution. *FEBS Lett.* **2017**, *591*, 979–990. [CrossRef]

45. Munari, F.; Bortot, A.; Assfalg, M.; D'Onofrio, M. Alzheimer's disease-associated ubiquitin mutant Ubb+1: Properties of the carboxy-terminal domain and its influence on biomolecular interactions. *Int. J. Biol. Macromol.* **2018**, *108*, 24–31. [CrossRef]

46. Varadan, R.; Assfalg, M.; Fushman, D. Ranjani Varadan; Assfalg, M.; Fushman, D. Using NMR Spectroscopy to Monitor Ubiquitin Chain Conformation and Interactions with Ubiquitin-Binding Domains. *Methods Enzymol.* **2005**, *399*, 177–192.

47. Bortot, A.; Zanzoni, S.; D'Onofrio, M.; Assfalg, M. Specific Interaction Sites Determine Differential Adsorption of Protein Structural Isomers on Nanoparticle Surfaces. *Chem. Eur. J.* **2018**, *24*, 5911–5919. [CrossRef] [PubMed]

48. Nielsen, L.; Khurana, R.; Coats, A.; Frokjaer, S.; Brange, J.; Vyas, S.; Uversky, V.N.; Fink, A.L. Effect of Environmental Factors on the Kinetics of Insulin Fibril Formation: Elucidation of the Molecular Mechanism. *Biochemistry* **2001**, *40*, 6036–6046. [CrossRef] [PubMed]

International Journal of
Molecular Sciences

Review

Regulators of Epithelial Sodium Channels in Aldosterone-Sensitive Distal Nephrons (ASDN): Critical Roles of Nedd4L/Nedd4-2 and Salt-Sensitive Hypertension

Tomoaki Ishigami *, Tabito Kino, Shintaro Minegishi, Naomi Araki, Masanari Umemura, Hisako Ushio, Sae Saigoh and Michiko Sugiyama

Department of Medical Science and Cardio-Renal Medicine, Yokohama City University Graduate School of Medicine, 3-9 Fukuura, Yokohama, Kanazawa-Ku 236-0004, Japan; kino-tabio@umin.ac.jp (T.K.); shintaro.minegishi@gmail.com (S.M.); naomiaraki@gmail.com (N.A.); umemurma@yokohama-cu.ac.jp (M.U.); hisahisa5172@yahoo.co.jp (H.U.); saesaigo1019@googlemail.com (S.S.); vn_nv2525smile@yahoo.co.jp (M.S.)
* Correspondence: tommmish@yokohama-cu.ac.jp; Tel.: +81-45-787-2635 (ext. 6312)

Received: 19 April 2020; Accepted: 27 May 2020; Published: 29 May 2020

Abstract: Ubiquitination is a representative, reversible biological process of the post-translational modification of various proteins with multiple catalytic reaction sequences, including ubiquitin itself, in addition to E1 ubiquitin activating enzymes, E2 ubiquitin conjugating enzymes, E3 ubiquitin ligase, deubiquitinating enzymes, and proteasome degradation. The ubiquitin–proteasome system is known to play a pivotal role in various molecular life phenomena, including the cell cycle, protein quality, and cell surface expressions of ion-transporters. As such, the failure of this system can lead to cancer, neurodegenerative diseases, cardiovascular diseases, and hypertension. This review article discusses *Nedd4-2/NEDD4L*, an E3-ubiquitin ligase involved in salt-sensitive hypertension, drawing from detailed genetic dissection analysis and the development of genetically engineered mice model. Based on our analyses, targeting therapeutic regulations of ubiquitination in the fields of cardio-vascular medicine might be a promising strategy in future. Although the clinical applications of this strategy are limited, compared to those of kinase systems, many compounds with a high pharmacological activity were identified at the basic research level. Therefore, future development could be expected.

Keywords: salt-sensitive hypertension; ubiquitination; Nedd4L/Nedd4-2; epithelial sodium channel; aldosterone sensitive distal nephron; excitation-transcription coupling

1. Introduction

Hypertension is a multifactorial disease determined by both genetic and environmental factors, including dietary habit, and is a representative public health issue due to its high prevalence and risk for cardiovascular diseases. The relationship between oral salt intake and elevation of blood pressure were widely observed in clinical trials such as the INTERSALT Study [1], PURE Study [2], and others [3]. The salt sensitivity of blood pressure is defined as "a physiological trait present in rodents and other mammals, including humans, by which the blood pressure (BP) of some members of the population exhibits changes parallel to changes in salt intake" [4]. Genetic analyses for hereditary and familial hypertension by positional cloning in the 1990's revealed that ion transporters and their accessory proteins, located in the renal tubules are responsible for the development of genetic hypertension [5]. Based on data from our own [6–14] and other [15–21] studies, we previously reported that important factors in the pathogenesis of salt-sensitive hypertension include the mechanism of sodium reabsorption in the renal tubule and the abnormalities that enhance this reabsorption. Furthermore, we also reported that angiotensinogen gene single-nucleotide polymorphisms (SNPs) might explain the pathogenesis of this disease at the levels of molecular genetics, human evolution, developmental engineering, and kidney physiology [7–9,11,12,22,23]. Currently, this report describes the already enforced and ongoing investigations into epithelial sodium channels (ENaCs) in terminal nephrons and its regulatory ubiquitinating enzyme, Nedd4L/Nedd4-2, based on our findings, which might provide significant molecular insight into the onset and development of salt-sensitive hypertension.

2. Sodium Reabsorption in Terminal Nephrons

Lifton et al. determined that the genes involved in the development of hereditary human hypertension are restricted to the sodium transporter and its associated proteins in the renal tubule [5]. The most important message that can be derived from this finding might be the fact that the first hit in the development of naturally occurring hereditary hypertension in human is due to the enhanced sodium reabsorption mechanism in the renal tubule. We considered that the mechanism of sodium reabsorption in tubules need to be understood to determine the genetic factors involved in salt-sensitive hypertension, i.e., "hypothesis-driven approaches", which is the purpose of this review.

The function of the kidney is to produce urine by ultrafiltration, during which the primitive urine passes through the renal tubules, and the quality and quantity of the body fluids and electrolytes are maintained through concentration by water re-absorption, appropriate reabsorption of electrolytes, and appropriate secretion of the bicarbonate ions [24,25]. Most sodium ions in the serum are filtered into primitive urine and at the rate according to the nephron segments; >99% of the sodium that is reabsorbed and ultimately filtered in each segment of the renal tubules is excreted into urine. Considering the mechanism that regulates sodium reabsorption, it can be divided into two segments involving the glomerulus-proximal, tubule-loop of the Henle-distal tubule to the macula densa and the tubules distal to the macula densa (connecting tubules, cortical collecting tubules).

During the first part of the process, >90% of the ultrafiltered sodium is reabsorbed. A physiological regulatory mechanism of the renal microcirculation (tubulo-glomerular-feedback; TGF) between the macula densa and the afferent arterioles, detects the chloride ions in the renal tubular fluid flowing into the distal nephron and imports the anions, thus adjusting the diameter of the afferent arterioles. By adjusting the diameter of the afferent arterioles according to the difference between the amount of filtered and reabsorbed sodium chloride (NaCl), the glomerular filtration pressure and amount were adjusted, and the glomerular filtration rate was maintained according to the oral salt intake. A myogenic reaction specific to the afferent arterioles occurs, in which the smooth muscle around the arterioles reflexively contracts due to the blood flow brought to the afferent arterioles, which regulates the diameter of the afferent arterioles. Similar to TGF, this mechanism adjusts the diameter of the afferent arterioles and maintains a constant glomerular filtration pressure and volume. Moreover, the renal tubules distal to the macula densa are collectively called aldosterone-sensitive-distal nephrons (ASDN), and sodium reabsorption is fine-tuned by the regulation of the expression of sodium ion transporters on the apical side of renal tubular cells, mainly through the action of aldosterone, an endocrine factor. Sodium reabsorption at this site comprises <5% of the total reabsorption, but because of an absent or insufficient feedback/compensation mechanism as in TGF, abnormalities about sodium reabsorption at this site is thought to cause blood pressure abnormalities such as the Liddle syndrome or type I pseudo-hypoaldosteronism (type I PHA). Conversely, abnormalities in sodium reabsorption due to irregularities in the ion transporter from the proximal tubule to the macula densa can be canceled and corrected by powerful TGF and myogenic reactions. Even if such irregularities exist in nature, they might not be recognized due to these physiological compensations.

The onset of salt-sensitive hypertension caused by an enhanced renal tubular renin-angiotensin system (RAS) might be caused by abnormalities in the sodium reabsorption mechanism in the terminal nephron, i.e., the ASDN. Our findings in the C57Bl6/J mice showed that angiotensinogen secretion in the proximal tubular cells was enhanced as a result of the enhanced filtered sodium ion [9,11,12], due to excess oral salt intake. We then proposed a model of the onset of salt-sensitive hypertension by paradoxically enhancing angiotensin II stimulation on the apical side of ASDN, in which the likely ion transporter of the angiotensin II action was the epithelial sodium channels (ENaCs) [26–28].

The function of renal tubules, that comprise a highly differentiated and complex regulatory system, is difficult to describe in a few words. However, their function can be summarized to facilitate understanding, as follows [29–32]. The cellular effects of aldosterone are exerted through intracellular mineralocorticoid receptors (MR) that bind not only to aldosterone, but also to the steroid hormone, cortisol, to produce intracellular effects. As the plasma concentration of aldosterone is much lower than that of cortisol, when competing for binding with MR under the same conditions, most MR probably binds to cortisol. Thus, epithelial cells (such as renal tubule cells) that are sensitive to aldosterone express 11-βHSD2 (type 2 11β-hydroxy-steroid dehydrogenase) and cortisol is converted into cortisone, thereby losing its MR binding activity. As a result, aldosterone-specific cellular action can occur by reducing the competitive inhibitory effect of cortisol on aldosterone with regards to MR. MR is expressed in nephrons beyond the early distal convoluted tubule (DCT), but because 11-βHSD2 is expressed only in nephrons after late DCT, the effects of aldosterone are not evident in early DCT. Mineralocorticoid receptors bind cortisol, and the NaCl co-transporter (NCCT) that is sensitive to thiazide diuretics and that is expressed in early DCT, is controlled by aldosterone to a lesser degree. Regulators of NCCT were discovered in the Gordon syndrome (type II PHA) in which hereditary hypertension was thought to be due to abnormal sodium reabsorption through the NCCT as well as abnormalities in WNK1 and WNK4 [33], but details of the mechanism of hypertension onset remain unknown. Through consistent expression of MR and 11-βHSD2, because "cortisol" was inactivated into "cortisone", an aldosterone was able to exert effects by binding to the MR in the nephrons below the late DCT. Thus, aldosterone regulates the ENaC expression by binding to MR as ASDN, via the aldosterone–MR complex and through the expression of regulatory factors known as AIP (aldosterone inducible protein), such as SGK1 (serum and glucocorticoid-regulated kinase 1), in nephrons after late DCT.

The outline of sodium transport in the ASDN renal tubular epithelial cells is as shown [34] (Figure 1). Na-K-ATPase are expressed on the basolateral side of all ASDN and functions, by ATP-dependently pumping intracellular sodium out of cells and the uptake of potassium into the cell. As a result of this sodium pump and K-ATPase, a transmembrane Na gradient arises, and the sodium is reabsorbed through the ENaCs on the apical side. After intracellular uptake, potassium is secreted lumenally through the action of the potassium channel ROMK. This channel was also under the control of aldosterone, and aldosterone action in the renal tubule appears to act on potassium secretion as well as sodium reabsorption.

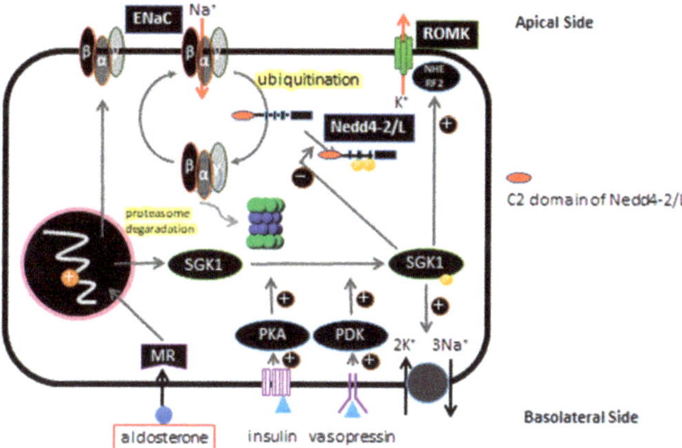

Figure 1. Schematic presentation of the ASDN epithelial cell is shown. ENaC is ubiquitinated by Nedd4-2/Nedd4L, and subsequently degraded by proteasome. Aldosterone, insulin, and vasopressin regulate ENaC gene expression via the basolateral side.

3. ENaC and Regulation of ENaC Expression in Renal Tubular Epithelial Cells

As ENaC is inhibited by the diuretic amiloride, it is referred to as an "amiloride-sensitive" epithelial sodium channel and is localized in epithelial cells with polarity. That is, it is expressed only in the apical side of the distal tubule, the alveolar epithelium, and colonic mucosa. It comprises α, β, and γ subunits, at a ratio of $1\alpha{:}1\beta{:}1\gamma$. The expression of its function requires all three subunits [29,35]. However, the α subunit alone or the combination of $\alpha\beta$ and $\alpha\gamma$ generates a sodium current, albeit incompletely. ENaC molecules have two transmembrane domains comprising an extracellular domain with multiple glycosylation sites, a short NH2 terminus, and COOH terminus, within cells. All three subunits have two proline-rich sequences (P1 and P2) called PY motifs, at their COOH terminal.

The endocrine factors principally regulate intracellular ENaC expression. In addition to aldosterone action through binding to MR, the action of vasopressin (ADH) secreted from the posterior pituitary gland appears to regulate the expression of ENaC via V2 receptor-adenylyl cyclase-cAMP, as well as insulin via the insulin receptor. As it acts on renal tubule cells through the blood stream, ADH acts via a receptor present on the basolateral side [26–28]. Abnormalities in the ENaC gene result in diseases such as cystic fibrosis, type I pseudohypoaldosteronism (type I PHA) [36], and Liddle syndrome [37].

4. Liddle Syndrome and ENaC

Liddle syndrome is a type of hereditary hypertension described by Liddle et al., in 1963, that has an autosomal-dominant inheritance pattern similar to that of primary aldosteronism, without elevated renin and aldosterone. Patients with Liddle syndrome exhibit hypokalemic alkalosis. The mechanism of this disease was discovered by Lifton et al. in 1995. [38,39] They found a gene mutation common to the P2 region of the proline-rich PY motif at the COOH terminus of β and γ ENaC. In 1996, Staub et al. [40] identified NEDD4 (neural precursor cell-expressed, developmentally, down-regulated 4) as a ubiquitinating enzyme (E3) protein that binds to this PY motif. Subsequent studies showed that NEDD4 ubiquitinates the Lys residue at the NH2 terminus by binding to the PY motif of ENaC, and the ubiquitinated ENaC was internalized into the cells where it was transported to the proteasome and was then degraded [41–43]. The post-translational modification and degradation of this protein is critical for the expression of ENaC on the cell membrane surface, and its physiological half-life is a maximum of one hour. A mutation of the PY motif results in impairments of ENaC ubiquitination via Nedd4 inhibition, sustained ENaC expression on the cell surface membrane, and enhanced sodium reabsorption; thus, a model of salt-sensitive hypertension was proposed. Snyder et al. concurrently revealed a similar mechanism [44].

Sodium reabsorption mediated by ENaC in ASDN does not possess sufficient physiological compensatory/feedback mechanisms such as TGF in the upstream nephron and the myogenic reaction in afferent arterioles. Therefore, the sodium reabsorption mechanism fails due to a dysfunctional ENaC-Nedd4 system, whether it is a gain or loss of function, and can eventually lead to abnormal blood pressure, without effective intra-renal correction.

5. Background on *NEDD4* and *NEDD4L/Nedd4-2*

The current paradigm of *NEDD4L* and salt-sensitive hypertension was confirmed, based on the following—the discovery and naming of *NEDD4L*, ENaC as a binding factor with respect to Liddle mutation, the discovery of KIAA0439 and *NEDD4L* (*Nedd4-2*), and our discovery of the human *NEDD4L* C2 domain and the V13 G/A mutation [45] (Figure 2). These findings provide a basis for our investigation, as described in the following.

Kumar et al. (1992) studied changes in gene expression during the development of neural precursor cells, using subtraction cloning [46]. A group of discovered genes was named the "neural precursor cell-expressed, developmentally, down-regulated" (*NEDD*) and the fourth among the molecules numbered 1 to 10, in order of identification, was *NEDD4*, which Staub et al. (1996) found to be associated with hypertension [40]. Using the yeast 2 hybrid method, they found that NEDD4 bound to the PY motif where the Liddle mutation of ENaC was concentrated, and that it was a specific ubiquitin ligase of ENaC. Thereafter, *NEDD4* was highlighted, but its significance shifted. With the discovery of the mechanism of ENaC degradation and the failure of retrieval from the cell membrane during the onset of Liddle syndrome, the gene encoding a protein related to this system is notable as a promising candidate associated with salt-sensitive hypertension. However, *NEDD4* was not a potential candidate gene because, new findings came to light, which led to other insights into *NEDD4/NEDD4L*.

Schematic presentation of exon usage of each human Nedd4L isoform (Chromosome 18)

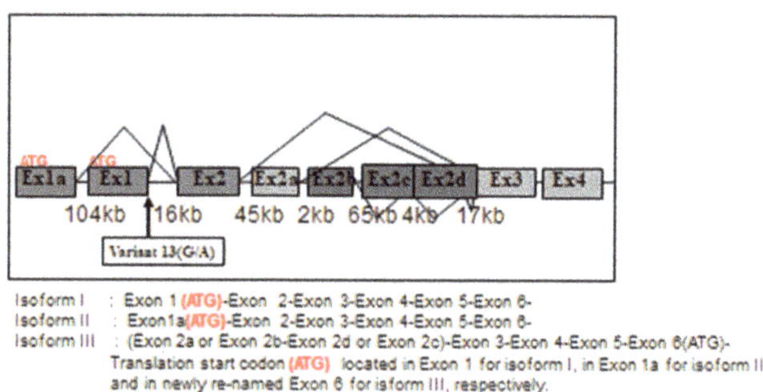

Figure 2. Schematic presentation of exon usage of each human Nedd4L isoform is shown. Exon in dark square shows the newly discovered exons through re-sequencing. Variant 13 (G→A) located at the end of the newly discovered exon 1. Translation start codon (ATG) located in Exon 1 for isoform I, in Exon 1a for isoform II, and newly re-named Exon 6 for isoform III, respectively. (modified from [45]).

Ishikawa et al. (1997) sequenced and published 78 unknown cDNA sequences from a human brain cDNA library [47]. Among these, Harvey and Kumar (2001) identified the KIAA0439 gene as having a Hect domain with ubiquitin ligase activity, while analyzing a Nedd4-like gene in a database [48] and found that it could ubiquitinate ENaC. Chen et al. also located a *NEDD4L* gene on human chromosome 18 with a 67% homology to human *NEDD4* [49]. Around the same time, both Staub and Kamynina et al. discovered *Nedd4-2*, which showed a high homology with *Nedd4* (*Nedd4-1*) in mouse cultured cells [31,50]. A comparison of binding between the WW domain and ENaC revealed that NEDD4-2 showed more ubiquitination activity, and that the endogenous ENaC negative regulator was likely to be *Nedd4-2* [50–52]. Thus, despite species differences between mice and humans, NEDD4-2 (NEDD4L) was essentially established as the main ubiquitinating enzyme in contrast to NEDD4, which was conventionally regarded as the first ubiquitinating enzyme of ENaC.

While the model of the ENaC-NEDD4 system continued to shift towards NEDD4L/NEDD4-2, we clarified significant differences that led to further findings. We and other authors, compared human/mouse *NEDD4/Nedd4* with *NEDD4L/Nedd4-2* and found that *NEDD4* had a C2 domain unlike *NEDD4L/Nedd4-2*, which was a critical difference between the two molecules. However, phylogenetic analyses of gene homology showed that human/mouse *NEDD4L/Nedd4-2* (KIAA0439) has a high homology with *NEDD4* in Xenopus, which has only one *NEDD4*, and the *NEDD4L* gene was thought to have appeared early during the evolutionary process. Based on this finding, the apparent difference between the two genes could not be explained.

6. Human NEDD4L Is a Causative Gene of Salt-Sensitive Hypertension

The ENaC-NEDD4L system plays an important role in post-translational modification through ubiquitination, which regulates the ENaC expression on the cell membrane of the terminal nephron. Genetic analyses of the Liddle syndrome and hereditary hypertension found that ENaC, which is restricted to the terminal nephron and regulates sodium reabsorption, and its regulatory protein,

NEDD4L, are promising candidate genes (*NEDD4L*) associated with salt-sensitive hypertension. We focused on the importance of the ENaC-NEDD4L system and selected *NEDD4L* as a target in a genetic analysis of salt-sensitive hypertension.

Based on a draft sequence of human chromosome 18 determined using genomic cDNA expression sequence tag (EST) data such as KIAA0439, Xenopus Nedd4 cDNA extracted from GenBank, Human and Mouse EST, and other databases, we performed Basic Local Alignment Search Tool (BLAST) and cross-match analyses. We searched for SNP by analyzing common polymorphisms determined by re-sequencing the genomic DNA of 48 normotensive persons whose DNA samples were registered in the HyperGEN network. We also analyzed human *NEDD4L* gene transcripts using RT-PCR, 5′RACE, and quantitative PCR targeting human RNA. In addition to known exon-introns, including the ATG codon, two exons (1a and 1) for which the splice site on the genomic DNA sequence exon-end was noted and an exon (exon 2) common to the transcription products from both, were discovered in chromosome 18. Exon 2 had a high homology with the C2 domain of the ancestral Nedd4, such as that in Xenopus and fugu and was thought to be a shared C2 domain that was conserved during molecular evolution [45] (Figure 3).

The start codon ATG of KIAA0439 was found in exon 6 of the conventional gene and in exon 7 of the gene that we discovered. Genomic DNA re-sequencing revealed 38 SNP. Among these, the 13th gene mutation (Variant 13 G/A) was found just before the exon 1 splice site. We spliced cDNA synthesized from human RNA through RT-PCR, immediately after the G base in those with a G allele, and the full-length C2 domain after exon 2 was encoded. In contrast, a frame shift occurred in those possessing an A allele because the splice site moved downstream by 10 bases, ending at the stop codon at exon 2. As a result, the gene product of exon 1 was knocked out (Figure 2). The results of the 5′RACE analyses of mRNA from human kidney and adrenal tissues revealed six transcripts. Among these, isoform I was a transcriptional product of exon 1, isoform II was a transcriptional product of exon 1a, and isoform III was located between exons 2 and 3. Exon 2a encoded the transcription initiation site of KIAA0439. Further analysis using quantitative PCR showed isoform I and II expression in the kidney and lung, respectively, suggesting that it might be under tissue-specific transcriptional control [45].

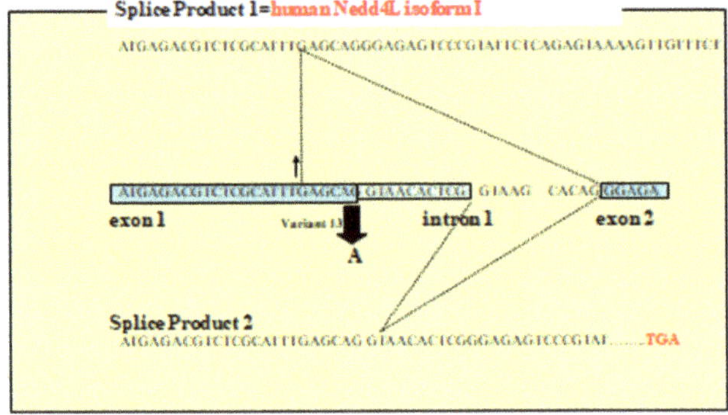

Figure 3. Detailed analyses of variant 13G/A (rs4149601) generating the splice variant of human. Nedd4L isoform I is shown. Splice site moves 10 base pairs and splice product 2 ends in the stop codon in exon 2.(modified from [45]).

These results showed that the human *NEDD4L* gene expresses two isoforms each with a C2 domain and another without a C2 domain. Among these, isoform I is a gene product resulting from a G/A mutation, a common variant, and isoform I was hypothesized to be associated with salt-sensitive hypertension. Therefore, we studied variant 13 (G/A) in 367 Japanese persons. An analysis based on genotype, alleles, and three inheritance patterns revealed that the GG and G alleles significantly correlated with essential hypertension. We then assessed the significance of the two types of C2 domains in the human Nedd4L gene and six types of transcriptional diversity. The results of a related study of Japanese patients revealed that a variant 13 (G/A) mutation correlated with essential hypertension [53–55]. In summary, the human NEDD4L gene is likely to be a causative gene of salt-sensitive hypertension in humans [56–58].

7. NEDD4L Is the Causative Protein of Salt-Sensitive Hypertension

We investigated the *Nedd4L* gene function in salt-sensitive hypertensive Dahl rats [59]. We initially analyzed the transcriptomes of rat RNA using the 5 'RACE method, to determine whether or not the *Nedd4L* gene in rats has a C2 domain. The results revealed an isoform with a C2 domain in rats and a novel exon on rat genomic DNA. The C2 domain of rat *Nedd4L* was 100% identical to the human C2 domain. Northern blotting using a novel "isoform A" possessing a C2 domain and a probe of the WW domain common to other isoforms, revealed the tissue-specific expression of *Nedd4L* gene transcription products, especially of isoform A, the expression of which was localized in the kidneys, lung, brain, and heart [59].

Changes in salt loading were examined by quantitative PCR and in situ hybridization. The results showed that salt intake increased *Nedd4L* expression in normal DR rats, but did not alter *Nedd4L* expression in salt-sensitive DS rats. The trend of *Nedd4L* expression determined by in situ hybridization was similar in kidney tissues [59]. A study of human kidney tissues obtained under written informed consent, showed that *NEDD4L* expression was restricted to late DCT and lower CNT, CCD, and collecting renal tubules, as in rats, and this distribution was consistent with that of ENaC expression, suggesting that *NEDD4L* is involved in the regulation of ENaC expression [60].

Subsequently, we performed in vitro functional experiments using heterologous gene expression systems using Xenopus oocytes. Initially, we successfully cloned three isoforms of human *NEDD4L* both with and without the C2 domain. A significant reduction in the amiloride-sensitive ENaC current by isoform II and III with ENaC cRNA injected was observed, when either isoform II or III cRNAs were injected into the Xenopus oocyte. The current was significantly restored when isoform I cRNAs were coinjected with other isoforms, indicating the dominant negative effects of the isoform I product against the downregulation of cell surface ENaC by isoforms II and III. Such interactions might abnormally increase sodium reabsorption in ASDN, suggesting that the human *NEDD4L* gene, especially the evolutionarily new isoform I, is a candidate gene for salt-sensitive hypertension in human [53,54,60,61].

8. Generation of Nedd4-2 C2 KO Mice and Discovery of Salt-Sensitive Hypertension with Potential Contributions to Cardio-Renal Involvements

Lastly, we decided to develop genetic engineered model of salt-sensitive hypertension such as *Nedd4-2* (*NEDD4L* in human) using knockout mice and examined the detailed phenotypic manifestations to determine the critical roles of *Nedd4-2* in salt-sensitive hypertension. First, we determined genetic variations of mice *Nedd4-2* using *in silico* exploration of the transcriptional start site of the gene. As reported previously, both human and rodent *NEDD4L/Nedd4-2* showed molecular diversity, with and without a C2 domain in their N-terminal. The *NEDD4L/Nedd4-2* isoforms with a C2 domain were hypothesized to be related closely to ubiquitination of ENaCs. *In silico* gene identification analysis of mice *Nedd4-2* C2 domain coding exon was performed using the BLAST and cross-match program. Then, the sequences of EST and cDNA in the GenBank database (nr) were aligned and compared with the sequence of mice *Nedd4-2* exon 4, which was already known as the transcription start site of mice *Nedd4-2* without a C2 domain. Cross-match analyses were performed repeatedly between the cDNA/EST sequences and genomic sequences, and the results were parsed with chromosome 18 of C57Bl6/J, to form a consistent assembly of EST, cDNA, and a genomic sequence. Thus, the expressed sequence tag and cDNA alignment and the results of the *in silico* bioinformatic database analysis of

Nedd4-2 on mouse chromosome 18q showed a newly identified exon2 coding the C2 domain of mice *Nedd4-2* [62].

Subsequently, we started to create a targeting vector for the newly discovered "exon 2", which codes the C2 domain of NEDD4-2 and generated genetically engineered mice without a C2 domain of NEDD4-2 [62]. Mice without a NEDD4-2 C2 domain did not show any growth retardation or infertility. We performed a detailed metabolic balance study and continuous blood pressure monitoring analyses, using metabolic cages for individual mice and unrestrained telemetry systems. Under normal oral salt intake, both the wild littermates and *Nedd4-2* C2 KO mice did not show any phenotypic differences, according to blood pressure, urinary sodium excretion, urinary osmotic pressure, urine volume, and water intake. However, under high oral salt intake, *Nedd4-2* C2 KO mice showed blood pressure elevation with reduced urinary sodium excretion [62]. Detailed quantitative analyses for the mRNA expressions levels along laser-captured urinary tubules showed a significant step-wise elevation of ENaC mRNA expressions, in accordance with both their genetic background and oral salt intake, which was abolished into the normal, through amiloride treatment [62]. ENaC mRNA expressions were paradoxically enhanced, despite a high oral salt intake, with a condition of single exon ablation for *Nedd4-2* gene. These results suggested that ENaC itself act as a "sensor" of intra-tubular salt with intra-tubular epithelial "excite-transcription coupling", which might regulate ENaC gene expressions paradoxically without the NEDD4-2 C2 domain [62]. This could be a pathological molecular mechanism underlying the salt-sensitive hypertension.

As NEDD4-2 with C2 domain is expressed not only in renal tubular cells but also in cardiomyocytes working as ubiquitinating enzyme for voltage-gated sodium-channel, SCN5a, we performed subsequent experiments examining the electrophysiological change of the heart after myocardial infarction, in mice. Additionally, to determine detailed characteristics of salt-sensitive hypertension of the mice, we tried a mineral corticoid receptor antagonist treatment for *Nedd4-2* C2 KO mice. Finally, us and other authors [63,64], have previously found that *Nedd4-2* C2 KO mice showed eplerenone-resistant salt-sensitive hypertension [65] and enhanced electrophysiological abnormalities, after myocardial infarction [66], suggesting a pivotal role of the *Nedd4-2* isoform with C2 domain for cardio-renal association, with regards to target-organ damages of the subjects with salt-sensitive hypertension.

9. Summary

In summary, from earlier detailed re-sequencing experiments for the human *NEDD4L* gene [45], we performed gene targeting experiments for the newly discovered exon, which encoded a C2 domain of mice Nedd4-2 [62]. Other experiments were performed including the discovery of a rodent *NEDD4L* gene C2 domain expressed along urinary tubules [59], heterologous gene expression in the Xenopus oocytes experiments with dominant negative effects of newly discovered human isoform I [60], the discovery of the NPC2 protein as a C2 domain binding protein in urinary tubules [61], and a genetic association study for human hypertension [53–55]. Ultimately, we found that the lack of single isoform of the gene caused a significant change *in vivo*, resulting in a higher oral salt intake suppressing sodium excretion in urine and elevated blood pressures, the pathophysiology of which is called "salt-sensitive" [62]. In accordance with human genetic studies, the impairment of tubular sodium transport might be pivotal in the onset and development of salt-sensitive hypertension [65,66].

The author's studies of *NEDD4L* were derived from a collaboration between 2000 and 2003 with Professor Robert Weiss at the Eccles Institute of Human Genetics at the University of Utah. Professor Robert Weiss is a bioinformatics expert who also participated in the Human Genome project, and some of the results described herein were derived from *in silico* analyses. This is an example of the ability to discover valuable information from vast databases using a desktop computer.

The Human Genome project was completed in 2003 which provided many insights, such as the fact that only 30,000 to 40,000 genes are needed to create diverse humans, instead of the original prediction of 100,000 to 120,000 genes. We found that one *NEDD4L* gene produces multiple transcripts. Thus, many genes might generate and maintain biological diversity via transcriptional diversity. The *NEDD4L* gene is also notable as an example of transcriptome analysis after genomic and proteomic analyses.

Human *NEDD4L* might be the next candidate gene for salt-sensitive hypertension, following angiotensinogen. Although there is currently insufficient evidence to fully support this hypothesis, when the amount of information available about the *NEDD4L* gene becomes similar to that of the angiotensinogen gene, the significance of the *NEDD4L* gene in the onset and progression of salt-sensitive hypertension will surely be further clarified by investigating its functions in vitro, in vivo, and clinically. Furthermore, therapeutic and diagnostic applications targeting the *NEDD4L* gene should become available in the foreseeable future, including de-ubiquitination enzyme activating agents for antagonizing human *NEDD4L*.

Funding: Tomoaki Ishigami was supported by a Grant-in-Aid for Scientific Research from the Ministry of Education, Culture, Sports, Science, and Technology (MEXT) No. 17K09730, and the Life Innovation Platform Yokohama (LIP.Yokohama).

Conflicts of Interest: The authors declare no conflict of interest.

Abbreviations

ASDN	Aldosterone Sensitive Distal Nephron
ENaC	Epithelial Sodium Channel
EPL	Eplerenone
DCT	Distal Convoluted Tubule
CNT	Conecting Tubule
CCD	Cortical Collecting Duct
ROMK	Renal outer medullary potassium channel
SGK1	Serum/glucocorticoid regulator kinase 1
MR	Mineral corticoid receptor
PKA	cAMP-dependent protein kinase A
PDK	Phosphoinositide-dependent kinase 1

References

1. Intersalt Cooperative Research Group. Intersalt: An international study of electrolyte excretion and blood pressure. Results for 24 hour urinary sodium and potassium excretion. *BMJ* **1988**, *297*, 319–328.
2. Mente, A.; O'Donnell, M.J.; Rangarajan, S.; McQueen, M.J.; Poirier, P.; Wielgosz, A.; Morrison, H.; Li, W.; Wang, X.; Di, C.; et al. Association of urinary sodium and potassium excretion with blood pressure. *N. Engl. J. Med.* **2014**, *371*, 601–611. [CrossRef] [PubMed]
3. O'Donnell, M.; Mente, A.; Rangarajan, S.; McQueen, M.J.; Wang, X.; Liu, L.; Yan, H.; Lee, S.F.; Mony, P.; Devanath, A.; et al. Urinary sodium and potassium excretion, mortality, and cardiovascular events. *N. Engl. J. Med.* **2014**, *371*, 612–623. [PubMed]
4. Elijovich, F.; Weinberger, M.H.; Anderson, C.A.; Appel, L.J.; Bursztyn, M.; Cook, N.R.; Dart, R.A.; Newton-Cheh, C.H.; Sacks, F.M.; Laffer, C.L.; et al. Salt Sensitivity of Blood Pressure: A Scientific Statement From the American Heart Association. *Hypertension* **2016**, *68*, e7–e46. [CrossRef] [PubMed]
5. Lifton, R.P.; Gharavi, A.G.; Geller, D.S. Molecular mechanisms of human hypertension. *Cell* **2001**, *104*, 545–556. [CrossRef]
6. Ishigami, T.; Umemura, S.; Iwamoto, T.; Tamura, K.; Hibi, K.; Yamaguchi, S.; Nyuui, N.; Kimura, K.; Miyazaki, N.; Ishii, M. Molecular variant of angiotensinogen gene is associated with coronary atherosclerosis. *Circulation* **1995**, *91*, 951–954. [CrossRef]
7. Ishigami, T.; Umemura, S.; Tamura, K.; Hibi, K.; Nyui, N.; Kihara, M.; Yabana, M.; Watanabe, Y.; Sumida, Y.; Nagahara, T.; et al. Essential hypertension and 5′ upstream core promoter region of human angiotensinogen gene. *Hypertension* **1997**, *30*, 1325–1330.
8. Ishigami, T.; Tamura, K.; Fujita, T.; Kobayashi, I.; Hibi, K.; Kihara, M.; Toya, Y.; Ochiai, H.; Umemura, S. Angiotensinogen gene polymorphism near transcription start site and blood pressure: Role of a T-to-C transition at intron I. *Hypertension* **1999**, *34*, 430–434. [CrossRef]

9. Lantelme, P.; Rohrwasser, A.; Gociman, B.; Hillas, E.; Cheng, T.; Petty, G.; Thomas, J.; Xiao, S.; Ishigami, T.; Herrmann, T.; et al. Effects of dietary sodium and genetic background on angiotensinogen and Renin in mouse. *Hypertension* **2002**, *39*, 1007–1014. [CrossRef]

10. Nakajima, T.; Jorde, L.B.; Ishigami, T.; Umemura, S.; Emi, M.; Lalouel, J.M.; Inoue, I. Nucleotide diversity and haplotype structure of the human angiotensinogen gene in two populations. *Am. J. Hum. Genet.* **2002**, *70*, 108–123.

11. Rohrwasser, A.; Ishigami, T.; Gociman, B.; Lantelme, P.; Morgan, T.; Cheng, T.; Hillas, E.; Zhang, S.; Ward, K.; Bloch-Faure, M.; et al. Renin and kallikrein in connecting tubule of mouse. *Kidney Int.* **2003**, *64*, 2155–2162. [CrossRef] [PubMed]

12. Gociman, B.; Rohrwasser, A.; Lantelme, P.; Cheng, T.; Hunter, G.; Monson, S.; Hunter, J.; Hillas, E.; Lott, P.; Ishigami, T.; et al. Expression of angiotensinogen in proximal tubule as a function of glomerular filtration rate. *Kidney Int.* **2004**, *65*, 2153–2160. [CrossRef] [PubMed]

13. Nakajima, T.; Wooding, S.; Sakagami, T.; Emi, M.; Tokunaga, K.; Tamiya, G.; Ishigami, T.; Umemura, S.; Munkhbat, B.; Jin, F.; et al. Natural selection and population history in the human angiotensinogen gene (AGT): 736 complete AGT sequences in chromosomes from around the world. *Am. J. Hum. Genet.* **2004**, *74*, 898–916. [PubMed]

14. Rohrwasser, A.; Morgan, T.; Dillon, H.F.; Zhao, L.; Callaway, C.W.; Hillas, E.; Zhang, S.; Cheng, T.; Inagami, T.; Ward, K.; et al. Elements of a paracrine tubular renin-angiotensin system along the entire nephron. *Hypertension* **1999**, *34*, 1265–1274. [CrossRef] [PubMed]

15. Jeunemaitre, X.; Soubrier, F.; Kotelevtsev, Y.V.; Lifton, R.P.; Williams, C.S.; Charru, A.; Hunt, S.C.; Hopkins, P.N.; Williams, R.R.; Lalouel, J.M.; et al. Molecular basis of human hypertension: Role of angiotensinogen. *Cell* **1992**, *71*, 169–180. [CrossRef]

16. Kimura, S.; Mullins, J.J.; Bunnemann, B.; Metzger, R.; Hilgenfeldt, U.; Zimmermann, F.; Jacob, H.; Fuxe, K.; Ganten, D.; Kaling, M. High blood pressure in transgenic mice carrying the rat angiotensinogen gene. *Embo. J.* **1992**, *11*, 821–827.

17. Tanimoto, K.; Sugiyama, F.; Goto, Y.; Ishida, J.; Takimoto, E.; Yagami, K.; Fukamizu, A.; Murakami, K. Angiotensinogen-deficient mice with hypotension. *J. Biol. Chem.* **1994**, *269*, 31334–31337.

18. Inoue, I.; Nakajima, T.; Williams, C.S.; Quackenbush, J.; Puryear, R.; Powers, M.; Cheng, T.; Ludwig, E.H.; Sharma, A.M.; Hata, A.; et al. A nucleotide substitution in the promoter of human angiotensinogen is associated with essential hypertension and affects basal transcription in vitro. *J. Clin. Invest.* **1997**, *99*, 1786–1797. [CrossRef]

19. Yanai, K.; Saito, T.; Hirota, K.; Kobayashi, H.; Murakami, K.; Fukamizu, A. Molecular variation of the human angiotensinogen core promoter element located between the TATA box and transcription initiation site affects its transcriptional activity. *J. Biol. Chem.* **1997**, *272*, 30558–30562.

20. Davisson, R.L.; Ding, Y.; Stec, D.E.; Catterall, J.F.; Sigmund, C.D. Novel mechanism of hypertension revealed by cell-specific targeting of human angiotensinogen in transgenic mice. *Physiol. Genomics* **1999**, *1*, 3–9. [CrossRef]

21. Jeunemaitre, X.; Inoue, I.; Williams, C.; Charru, A.; Tichet, J.; Powers, M.; Sharma, A.M.; Gimenez-Roqueplo, A.P.; Hata, A.; Corvol, P.; et al. Haplotypes of angiotensinogen in essential hypertension. *Am. J. Hum. Genet.* **1997**, *60*, 1448–1460. [PubMed]

22. Ishigami, T.; Iwamoto, T.; Tamura, K.; Yamaguchi, S.; Iwasawa, K.; Uchino, K.; Umemura, S.; Ishii, M. Angiotensin I converting enzyme (ACE) gene polymorphism and essential hypertension in Japan. Ethnic difference of ACE genotype. *Am. J. Hypertens* **1995**, *8*, 95–97. [CrossRef]

23. Ishigami, T.; Kino, T.; Chen, L.; Minegishi, S.; Araki, N.; Umemura, M.; Abe, K.; Sasaki, R.; Yamana, H.; Umemura, S. Identification of bona fide alternative renin transcripts expressed along cortical tubules and potential roles in promoting insulin resistance in vivo without significant plasma renin activity elevation. *Hypertension* **2014**, *64*, 125–133. [PubMed]

24. Guyton, A.C. Blood pressure control—Special role of the kidneys and body fluids. *Science* **1991**, *252*, 1813–1816. [CrossRef] [PubMed]

25. O'Shaughnessy, K.M.; Karet, F.E. Salt handling and hypertension. *J. Clin. Invest.* **2004**, *113*, 1075–1081. [CrossRef] [PubMed]

26. Peti-Peterdi, J.; Warnock, D.G.; Bell, P.D. Angiotensin II directly stimulates ENaC activity in the cortical collecting duct via AT(1) receptors. *J. Am. Soc. Nephrol.* **2002**, *13*, 1131–1135.

27. Komlosi, P.; Fuson, A.L.; Fintha, A.; Peti-Peterdi, J.; Rosivall, L.; Warnock, D.G.; Bell, P.D. Angiotensin I conversion to angiotensin II stimulates cortical collecting duct sodium transport. *Hypertension* **2003**, *42*, 195–199. [CrossRef] [PubMed]

28. Beutler, K.T.; Masilamani, S.; Turban, S.; Nielsen, J.; Brooks, H.L.; Ageloff, S.; Fenton, R.A.; Packer, R.K.; Knepper, M.A. Long-term regulation of ENaC expression in kidney by angiotensin II. *Hypertension* **2003**, *41*, 1143–1150.

29. Gormley, K.; Dong, Y.; Sagnella, G.A. Regulation of the epithelial sodium channel by accessory proteins. *Biochem. J.* **2003**, *371*, 1–14.

30. Booth, R.E.; Johnson, J.P.; Stockand, J.D. Aldosterone. *Adv. Physiol. Educ.* **2002**, *26*, 8–20.

31. Kamynina, E.; Staub, O. Concerted action of ENaC, Nedd4-2, and Sgk1 in transepithelial Na(+) transport. *Am. J. Physiol. Renal. Physiol.* **2002**, *283*, F377–F387. [CrossRef]

32. Meneton, P.; Loffing, J.; Warnock, D.G. Sodium and potassium handling by the aldosterone-sensitive distal nephron: The pivotal role of the distal and connecting tubule. *Am. J. Physiol. Renal. Physiol.* **2004**, *287*, F593–F601. [CrossRef] [PubMed]

33. Wilson, F.H.; Disse-Nicodeme, S.; Choate, K.A.; Ishikawa, K.; Nelson-Williams, C.; Desitter, I.; Gunel, M.; Milford, D.V.; Lipkin, G.W.; Achard, J.M.; et al. Human hypertension caused by mutations in WNK kinases. *Science* **2001**, *293*, 1107–1112. [CrossRef] [PubMed]

34. Vallon, V.; Wulff, P.; Huang, D.Y.; Loffing, J.; Volkl, H.; Kuhl, D.; Lang, F. Role of Sgk1 in salt and potassium homeostasis. *Am. J. Physiol. Regul. Integr. Comp. Physiol.* **2005**, *288*, R4–R10. [CrossRef] [PubMed]

35. Bhargava, A.; Wang, J.; Pearce, D. Regulation of epithelial ion transport by aldosterone through changes in gene expression. *Mol. Cell. Endocrinol.* **2004**, *217*, 189–196. [CrossRef] [PubMed]

36. Chang, S.S.; Grunder, S.; Hanukoglu, A.; Rosler, A.; Mathew, P.M.; Hanukoglu, I.; Schild, L.; Lu, Y.; Shimkets, R.A.; Nelson-Williams, C.; et al. Mutations in subunits of the epithelial sodium channel cause salt wasting with hyperkalaemic acidosis, pseudohypoaldosteronism type 1. *Nat. Genet.* **1996**, *12*, 248–253. [CrossRef] [PubMed]

37. Hansson, J.H.; Nelson-Williams, C.; Suzuki, H.; Schild, L.; Shimkets, R.; Lu, Y.; Canessa, C.; Iwasaki, T.; Rossier, B.; Lifton, R.P. Hypertension caused by a truncated epithelial sodium channel gamma subunit: Genetic heterogeneity of Liddle syndrome. *Nat. Genet.* **1995**, *11*, 76–82. [CrossRef]

38. Shimkets, R.A.; Warnock, D.G.; Bositis, C.M.; Nelson-Williams, C.; Hansson, J.H.; Schambelan, M.; Gill, J.R., Jr.; Ulick, S.; Milora, R.V.; Findling, J.W.; et al. Liddle's syndrome: Heritable human hypertension caused by mutations in the beta subunit of the epithelial sodium channel. *Cell* **1994**, *79*, 407–414. [CrossRef]

39. Schild, L.; Lu, Y.; Gautschi, I.; Schneeberger, E.; Lifton, R.P.; Rossier, B.C. Identification of a PY motif in the epithelial Na channel subunits as a target sequence for mutations causing channel activation found in Liddle syndrome. *EMBO J.* **1996**, *15*, 2381–2387. [CrossRef]

40. Staub, O.; Dho, S.; Henry, P.; Correa, J.; Ishikawa, T.; McGlade, J.; Rotin, D. WW domains of Nedd4 bind to the proline-rich PY motifs in the epithelial Na+ channel deleted in Liddle's syndrome. *EMBO J.* **1996**, *15*, 2371–2380. [CrossRef]

41. Plant, P.J.; Yeger, H.; Staub, O.; Howard, P.; Rotin, D. The C2 domain of the ubiquitin protein ligase Nedd4 mediates Ca2+-dependent plasma membrane localization. *J. Biol. Chem.* **1997**, *272*, 32329–32336. [CrossRef] [PubMed]

42. Staub, O.; Gautschi, I.; Ishikawa, T.; Breitschopf, K.; Ciechanover, A.; Schild, L.; Rotin, D. Regulation of stability and function of the epithelial Na+ channel (ENaC) by ubiquitination. *EMBO J.* **1997**, *16*, 6325–6336. [CrossRef] [PubMed]

43. Abriel, H.; Loffing, J.; Rebhun, J.F.; Pratt, J.H.; Schild, L.; Horisberger, J.D.; Rotin, D.; Staub, O. Defective regulation of the epithelial Na+ channel by Nedd4 in Liddle's syndrome. *J. Clin. Invest.* **1999**, *103*, 667–673. [CrossRef] [PubMed]

44. Goulet, C.C.; Volk, K.A.; Adams, C.M.; Prince, L.S.; Stokes, J.B.; Snyder, P.M. Inhibition of the epithelial Na+ channel by interaction of Nedd4 with a PY motif deleted in Liddle's syndrome. *J. Biol. Chem.* **1998**, *273*, 30012–30017. [CrossRef] [PubMed]

45. Dunn, D.M.; Ishigami, T.; Pankow, J.; von Niederhausern, A.; Alder, J.; Hunt, S.C.; Leppert, M.F.; Lalouel, J.M.; Weiss, R.B. Common variant of human NEDD4L activates a cryptic splice site to form a frameshifted transcript. *J. Hum. Genet.* **2002**, *47*, 665–676. [CrossRef] [PubMed]

46. Kumar, S.; Tomooka, Y.; Noda, M. Identification of a set of genes with developmentally down-regulated expression in the mouse brain. *Biochem. Biophys. Res. Commun.* **1992**, *185*, 1155–1161. [CrossRef]
47. Ishikawa, K.; Nagase, T.; Nakajima, D.; Seki, N.; Ohira, M.; Miyajima, N.; Tanaka, A.; Kotani, H.; Nomura, N.; Ohara, O. Prediction of the coding sequences of unidentified human genes. VIII. 78 new cDNA clones from brain which code for large proteins in vitro. *DNA Res.* **1997**, *4*, 307–313. [CrossRef]
48. Harvey, K.F.; Dinudom, A.; Cook, D.I.; Kumar, S. The Nedd4-like protein KIAA0439 is a potential regulator of the epithelial sodium channel. *J. Biol. Chem.* **2001**, *276*, 8597–8601. [CrossRef]
49. Chen, H.; Ross, C.A.; Wang, N.; Huo, Y.; MacKinnon, D.F.; Potash, J.B.; Simpson, S.G.; McMahon, F.J.; DePaulo, J.R., Jr.; McInnis, M.G. NEDD4L on human chromosome 18q21 has multiple forms of transcripts and is a homologue of the mouse Nedd4-2 gene. *Eur. J. Hum. Genet.* **2001**, *9*, 922–930. [CrossRef]
50. Kamynina, E.; Debonneville, C.; Hirt, R.P.; Staub, O. Liddle's syndrome: A novel mouse Nedd4 isoform regulates the activity of the epithelial Na(+) channel. *Kidney Int.* **2001**, *60*, 466–471. [CrossRef]
51. Kamynina, E.; Debonneville, C.; Bens, M.; Vandewalle, A.; Staub, O. A novel mouse Nedd4 protein suppresses the activity of the epithelial Na+ channel. *FASEB J* **2001**, *15*, 204–214. [CrossRef] [PubMed]
52. Kamynina, E.; Tauxe, C.; Staub, O. Distinct characteristics of two human Nedd4 proteins with respect to epithelial Na(+) channel regulation. *Am. J. Physiol. Renal Physiol.* **2001**, *281*, F469–F477. [CrossRef] [PubMed]
53. Ishigami, T.; Araki, N.; Umemura, S. Human Nedd4L rs4149601 G allele generates evolutionary new isoform I with C2 domain. *Hypertension* **2010**, *55*, e11. [CrossRef] [PubMed]
54. Ishigami, T.; Araki, N.; Minegishi, S.; Umemura, M.; Umemura, S. Genetic variation in NEDD4L, salt sensitivity, and hypertension: Human NEDD4L rs4149601 G allele generates evolutionary new isoform I with C2 domain. *J. Hypertens.* **2014**, *32*, 1905. [CrossRef] [PubMed]
55. Ishigami, T.; Umemura, M.; Araki, N.; Hirawa, N.; Tamura, K.; Uchino, K.; Umemura, S.; Rohrwasser, A.; Lalouel, J.M. NEDD4L protein truncating variant (v13(G/A):rs4149601) is associated with essential hypertension (EH) in a sample of the Japanese population. *Geriatrics Geront. Int.* **2007**, *7*, 114–118. [CrossRef]
56. Dahlberg, J.; Nilsson, L.O.; von Wowern, F.; Melander, O. Polymorphism in NEDD4L is associated with increased salt sensitivity, reduced levels of P-renin and increased levels of Nt-proANP. *PLoS ONE* **2007**, *2*, e432. [CrossRef]
57. Fava, C.; von Wowern, F.; Berglund, G.; Carlson, J.; Hedblad, B.; Rosberg, L.; Burri, P.; Almgren, P.; Melander, O. 24-h ambulatory blood pressure is linked to chromosome 18q21-22 and genetic variation of NEDD4L associates with cross-sectional and longitudinal blood pressure in Swedes. *Kidney Int.* **2006**, *70*, 562–569. [CrossRef]
58. Svensson-Farbom, P.; Wahlstrand, B.; Almgren, P.; Dahlberg, J.; Fava, C.; Kjeldsen, S.; Hedner, T.; Melander, O. A functional variant of the NEDD4L gene is associated with beneficial treatment response with beta-blockers and diuretics in hypertensive patients. *J. Hypertens* **2011**, *29*, 388–395. [CrossRef]
59. Umemura, M.; Ishigami, T.; Tamura, K.; Sakai, M.; Miyagi, Y.; Nagahama, K.; Aoki, I.; Uchino, K.; Rohrwasser, A.; Lalouel, J.M.; et al. Transcriptional diversity and expression of NEDD4L gene in distal nephron. *Biochem. Biophys. Res. Commun.* **2006**, *339*, 1129–1137. [CrossRef]
60. Araki, N.; Umemura, M.; Miyagi, Y.; Yabana, M.; Miki, Y.; Tamura, K.; Uchino, K.; Aoki, R.; Goshima, Y.; Umemura, S.; et al. Expression, transcription, and possible antagonistic interaction of the human Nedd4L gene variant: Implications for essential hypertension. *Hypertension* **2008**, *51*, 773–777. [CrossRef]
61. Araki, N.; Ishigami, T.; Ushio, H.; Minegishi, S.; Umemura, M.; Miyagi, Y.; Aoki, I.; Morinaga, H.; Tamura, K.; Toya, Y.; et al. Identification of NPC2 protein as interaction molecule with C2 domain of human Nedd4L. *Biochem. Biophys. Res. Commun.* **2009**, *388*, 290–296. [PubMed]
62. Minegishi, S.; Ishigami, T.; Kino, T.; Chen, L.; Nakashima-Sasaki, R.; Araki, N.; Yatsu, K.; Fujita, M.; Umemura, S. An isoform of Nedd4-2 is critically involved in the renal adaptation to high salt intake in mice. *Sci. Rep.* **2016**, *6*, 27137. [CrossRef] [PubMed]
63. Henshall, T.L.; Manning, J.A.; Alfassy, O.S.; Goel, P.; Boase, N.A.; Kawabe, H.; Kumar, S. Deletion of Nedd4-2 results in progressive kidney disease in mice. *Cell Death Differ.* **2017**, *24*, 2150–2160. [CrossRef] [PubMed]
64. Manning, J.A.; Shah, S.S.; Henshall, T.L.; Nikolic, A.; Finnie, J.; Kumar, S. Dietary sodium modulates nephropathy in Nedd4-2-deficient mice. *Cell Death Differ.* **2019**, *27*, 1832–1843. [CrossRef]

65. Kino, T.; Ishigami, T.; Murata, T.; Doi, H.; Nakashima-Sasaki, R.; Chen, L.; Sugiyama, M.; Azushima, K.; Wakui, H.; Minegishi, S.; et al. Eplerenone-Resistant Salt-Sensitive Hypertension in Nedd4-2 C2 KO Mice. *Int. J. Mol. Sci.* **2017**, *18*, 1250. [CrossRef]
66. Minegishi, S.; Ishigami, T.; Kawamura, H.; Kino, T.; Chen, L.; Nakashima-Sasaki, R.; Doi, H.; Azushima, K.; Wakui, H.; Chiba, Y.; et al. An Isoform of Nedd4-2 Plays a Pivotal Role in Electrophysiological Cardiac Abnormalities. *Int. J. Mol. Sci.* **2017**, *18*, 1268.

International Journal of
Molecular Sciences

Review

The Effect of Dysfunctional Ubiquitin Enzymes in the Pathogenesis of Most Common Diseases

Gizem Celebi [1], Hale Kesim [1], Ebru Ozer [1] and Ozlem Kutlu [2,3,*]

[1]   Faculty of Engineering and Natural Sciences, Molecular Biology, Genetics, and Bioengineering Program, Sabanci University, Istanbul 34956, Turkey; gizemcelebi@sabanciuniv.edu (G.C.); halekesim@sabanciuniv.edu (H.K.); ebruozer@sabanciuniv.edu (E.O.)
[2]   Sabanci University Nanotechnology Research and Application Center (SUNUM), Istanbul 34956, Turkey
[3]   Center of Excellence for Functional Surfaces and Interfaces for Nano Diagnostics (EFSUN), Sabanci University, Istanbul 34956, Turkey
*   Correspondence: ozlemkutlu@sabanciuniv.edu; Tel.: +90-216-483-9000 (ext. 2413)

Received: 16 June 2020; Accepted: 18 July 2020; Published: 1 September 2020

Abstract: Ubiquitination is a multi-step enzymatic process that involves the marking of a substrate protein by bonding a ubiquitin and protein for proteolytic degradation mainly via the ubiquitin–proteasome system (UPS). The process is regulated by three main types of enzymes, namely ubiquitin-activating enzymes (E1), ubiquitin-conjugating enzymes (E2), and ubiquitin ligases (E3). Under physiological conditions, ubiquitination is highly reversible reaction, and deubiquitinases or deubiquitinating enzymes (DUBs) can reverse the effect of E3 ligases by the removal of ubiquitin from substrate proteins, thus maintaining the protein quality control and homeostasis in the cell. The dysfunction or dysregulation of these multi-step reactions is closely related to pathogenic conditions; therefore, understanding the role of ubiquitination in diseases is highly valuable for therapeutic approaches. In this review, we first provide an overview of the molecular mechanism of ubiquitination and UPS; then, we attempt to summarize the most common diseases affecting the dysfunction or dysregulation of these mechanisms.

Keywords: ubiquitination; E3s; DUBs; UPS; cancer; neurodegenerative disease; immune-related diseases

---

1. Introduction

Cellular functions are highly dependent on the precise control of a single protein abundance within cells that is regulated by the equilibration of protein translation, folding, and degradation. The ubiquitination of proteins is one of the most important post-translational modifications, in which an ubiquitin, a small (8.6 kDa) regulatory protein, is attached to a substrate protein. This mechanism maintains protein homeostasis by regulating the degradation of cellular proteins, such as short-lived or long-lived regulatory proteins and damaged or misfolded proteins, mainly via the ubiquitin–proteasome or the autophagosome–lysosomal pathway (autophagy). Moreover, ubiquitination is involved in cellular processes by coordinating the cellular localization of proteins, activating or inactivating them, and modulating protein–protein interactions [1–3]. These effects are mediated by the addition of either a single ubiquitin (monoubiquitination) or a chain of ubiquitin proteins (polyubiquitination) [4].

In recent years, a considerable amount of research has been focused on the understanding of molecular action of ubiquitination in signaling pathways and how alterations in this mechanism ultimately lead to the development of human diseases. In this review, we summarize current knowledge of ubiquitination types and the sequential enzymatic cascade of ubiquitination as well as the involvement of this cascade with proteasomal degradation, which is known as the ubiquitin–proteasome system (UPS). Although protein homeostasis is regulated by the extensive crosstalk between the UPS and other degradation pathways (e.g., autophagy), the only focus of this review is providing insight into the failure of accurately regulating cellular enzymatic and proteolytic processes in the most common diseases, such as cancer, neurodegenerative, and immune-related diseases.

2. Ubiquitin and Ubiquitin Proteasome System (UPS)

Ubiquitin is a small, evolutionarily conserved 76 amino acid polypeptide encoded by four genes (*UBA52, RPS27A, UBB*, and *UBC*) in mammals. This protein was first identified in the 1970s as a protein of unknown function expressed in all eukaryotic cells; later on, the key features of this protein, including its C-terminal tail and the seven Lysine (Lys-K) residues, were revealed in the early 1980s [5–8]. The identification of ubiquitin–protein conjugates followed the discovery of the ubiquitination pathway, which was initially characterized as an ATP-dependent proteolytic system. ATP-dependent proteolysis factor 1 (APF-1) was found as a polypeptide that was capable of covalently binding to protein substrates in an ATP- and Mg^{2+}-dependent manner [9]. Afterwards, the APF-1 protein was named as ubiquitin.

Ubiquitination is initiated by the attachment of the last amino acid (Glycine-76) of a single ubiquitin molecule to the Lys residue of one substrate protein that is called monoubiquitination. In case of the addition of one ubiquitin molecule to multiple substrate residues, the mechanism is called multi-monoubiquitination (Figure 1A). Indeed, monoubiquitination is required for the formation of a ubiquitin chain on a single Lys residue of the substrate protein, which is known as polyubiquitination. Moreover, polyubiquitin chains can be modified and turned into more complex structures by the addition of ubiquitin-like proteins (e.g., SUMO "Small Ubiquitin-like Modifier ", NEDD "Neural Precursor Cell Expressed Developmentally Down-Regulated Protein") or some chemical modifications (e.g., acetylation or phosphorylation) [9] (Figure 1B). In polyubiquitination, the ubiquitin protein can be ubiquitinated on its seven Lys residues (Lys6, Lys11, Lys27, Lys29, Lys33, Lys48, and Lys63) or on its N-terminus, which leads to the formation of different ubiquitin chain topologies. In fact, the ubiquitinated Lys residues as well as the ubiquitination of either the same (homotypic) or the different Lys residues (heterotypic) determine the fate of the substrate protein (Figure 1B). In other words, Lys63-linked polyubiquitin chains are associated with non-proteolytic cellular functions such as inflammation, protein trafficking, or DNA repair, while Lys48-linked polyubiquitin chains target substrates that are mostly related to the proteasomal degradation, such that the Lys48-linked ubiquitin chain can be recognized by a specific subunit of the proteasome [10].

The enzymatic mechanism of ubiquitination occurs in a proteolytic and non-proteolytic pathway consisting of three main steps: activation (E1), conjugation (E2), and ligation (E3) (Figure 2A). Initially, the binding of an E1-activating enzyme with a thioester bond activates ubiquitin in an ATP-dependent manner. Then, ubiquitin is transferred from an E1-activating enzyme to the E2-conjugation enzyme; thereafter, the E2 enzyme binds to the substrate-bound E3 ligation enzyme, resulting in a covalent attachment between the C-terminal of ubiquitin and the target substrate. As the cycle repeats, a polyubiquitination chain is formed, and the polyubiquitinated substrate is transferred to the proteasome for proteolytic degradation [11]. Considering the essential role of this enzymatic pathway in the protein homeostasis in the cell, three different enzymes (E1, E2, and E3) have been the subject of detailed research in the last decade. According to current knowledge, there are two E1 and approximately 40 E2 genes that exist in the human genome. Besides, more than 600 E3 ligase genes that are critical for the balance between ubiquitination and deubiquitination are known in the human genome [12,13]. E3 ligases are devided in three distinct groups based on their catalytic domains. The first group of E3s contains two domains: Really Interesting New Gene (RING) and UFD2 homology (U-box). The RING E3s domain coordinates Zn^{2+} binding and recruits the ubiquitin-charged E2, while the U-box E3s domain acts in a similar manner without Zn2+ coordination. The second group of E3s includes the homologous to E6–AP carboxyl terminus (HETC) domain. This domain consists of C (contains a catalytic cysteine) and N (recruits ubiquitin-charged E2) lobes that can be modulated depending on the cellular pathway. The third and last group of E3s has a RING-between-RING (RBR) motif. RING1 recruits the ubiquitin-charged E2 and transfers ubiquitin to catalytic cysteine of RING2 that conjugates ubiquitin to the substrate [14]. It is worthy of note that E3 ligases are the well-studied enzyme in this enzymatic cascade, which is most probably due to its critical role in protein quality control and homeostasis in the cell.

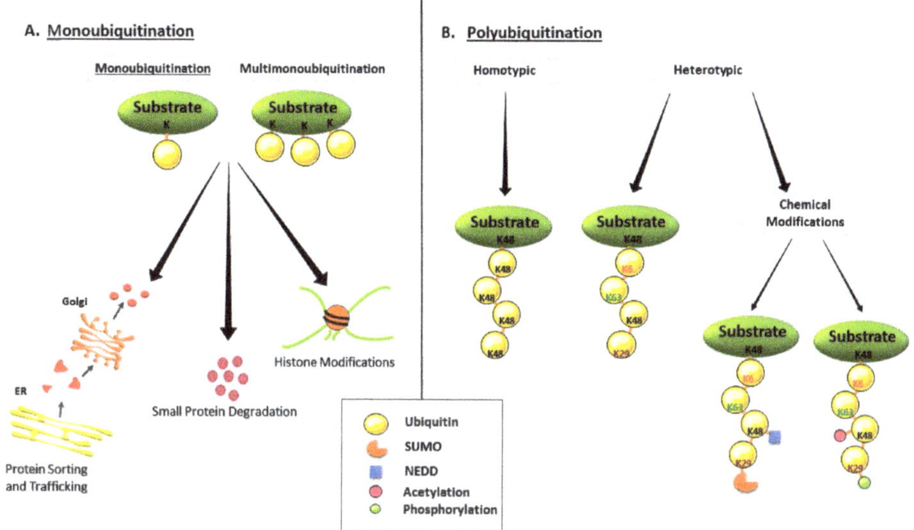

Figure 1. Types of Ubiquitination. (**A**) Monoubiquitination, (**B**) Polyubiquitination.

The proteasomes (26S proteasome) are large and multi-catalytic proteases that are composed of a 20S barrel-shaped core complex and a 19S regulatory complex. Protein degradation occurs in the 20S complex, which consists of four stacked hollow rings, each of which has coordinately functional distinct subunits: outer α rings and inner β rings (Figure 2B). Outer α rings contain seven α subunits, and these subunits are controlled by the binding of cap structures that recognize polyubiquitin tags on the substrate proteins. β subunits contain three to seven protease active sites involved in degradation processes. Notably, the 20S proteasome has also a critical role in the dissociation of ubiquitin proteins from the substrates via deubiquitinating enzymes (DUB) [15]. Once the ubiquitinated proteins reach the 19S regulatory complex, this complex delivers them to the 20S proteasome. The 19S proteasome has 18 different subunits, and its central part consists of six ATPases that bind to substrate proteins for degradation, unfolding, and translocation into the 20S complex [16]. As a final process, proteins are degraded into polypeptides, which are then chopped into small fragments and ultimately into the single amino acids by the activation of peptidases [17]. The overall system of ubiquitination and proteasomal degradation is known as the ubiquitin–proteasome system (UPS).

Ubiquitination is a reversible and dynamic phenomenon. Deubiquitinating enzymes (DUBs) are a large group of proteases that can partially or completely remove ubiquitin and ubiquitin-like proteins NEDD and SUMO from the target substrate. Currently, 100 DUBs are encoded by the human genome, and they are classified into six different groups: ubiquitin-specific proteases (USPs), ubiquitin COOH-terminal hydrolases (UCHs), Machado–Joseph Domain (Josephin domain)-containing proteins (MJD), the JAB1/MPN/MOV34 family (JAMMs), motif interacting with Ub-containing novel DUB family (MINDYs) [18], and ovarian tumor proteases (OTUs) [19]. DUBs are critical for the regulation of cell survival, immune response [20], and cell differentiation [21,22]; therefore, they are considered as potential drug targets in various diseases.

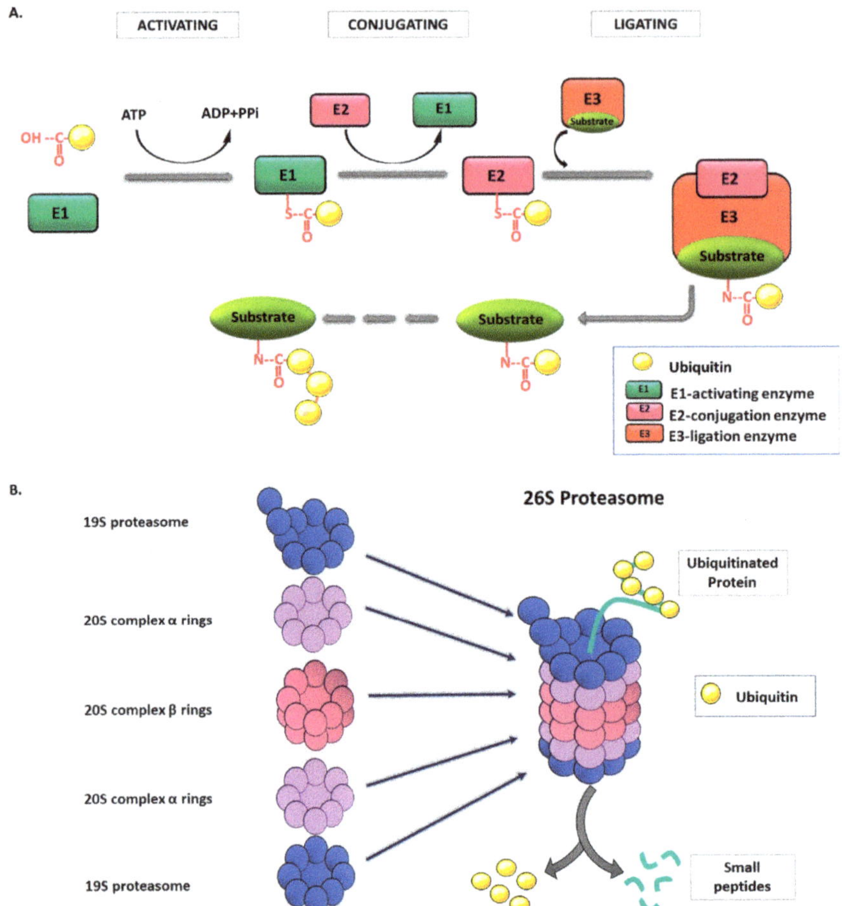

Figure 2. Ubiquitin proteasome system (UPS). (**A**) Enzymatic reaction of ubiquitination. For activation of the ubiquitination process, the E1 enzyme makes a thioester bond with the ubiquitin molecule, and the required energy is supplied by the hydrolysis of ATP. In the conjugation step, the ubiquitin on the E1 enzyme is transferred to the E2 enzyme. Finally, the substrate–E3 enzyme bond is involved, and the C-terminal of the ubiquitin molecule makes a covalent bond with the substrate, which is bound to the E3 enzyme. (**B**) Structure of proteasome and degradation of ubiquitinated substrates. The 26S proteasome has a multicomplex structure that is composed of a 19S regulatory units and a 20S catalytic core unit. In the 19S regulatory complex, the ubiquitinated proteins are unfolded, and ubiquitin tags are separated from the protein by deubiquitinase enzymes. Consequently, the unfolded polypeptide chain is delivered to the 20S catalytic core complex where they are degraded into small peptides.

In the following sections, we would like to emphasize the current understanding of the dysfunction or dysregulation of ubiquitin enzymes and proteasomal degradation in several common diseases. Since the subject is too broad, we try to narrow down the topics by focusing specifically on E1, E2, and E3 enzymes, DUBs, UPS, and their association regarding the progression of cancer, neurodegenerative diseases, immune regulation, and immune-related diseases.

3. Ubiquitination-Mediated Regulation in Cancer

Increasing evidence demonstrates the implication of ubiquitin enzymes in carcinogenesis. Although there are plenty of studies about E1 and E2 enzymes in cancer development, most of them focus on E3 ligases. In these studies, E1-activating enzymes are often used for targeting the inhibition of the UPS in cancer treatment [23,24]. On the other hand, E2-conjugated enzymes have been reported in cell cycle stimulation, DNA repair, and the induction of oncogenic signaling pathways during cancer progression via its distinct members: UBE2A, UBE2C, and UBE2D1. Although the overexpression of these E2 enzymes is highly correlated with a poor prognosis of pancreas, lung, breast, skin, and thyroid cancers, several E2 inhibitors have been developed [25]. To date, several compounds were reported to target E2 enzymes as a new class of potential treatment for cancer patients (Table 1). These compounds target the conjugation of E2 enzymes to their substrates [23,24].

Table 1. Summary of compounds targeting E2 enzymes.

Compounds	Targeted E2 Enzyme	References
CC0651	Ube2R1	[26]
Leucettamol A	Ubc13–Uev1A interaction	[27]
Manadosterols A and B	Ubc13–Uev1A interaction	[28]
BAY 11-7082	Ubc13	[29]
UBC12N26	UBC12–NAE binding	[30]
Sumoylation Inhibitors	Ubc9	[31]
Suramin	Cdc34–CRL interaction	[32]
NSC697923	Ube2N	[29]
Honokiol	UbcH8	[33]
Triazines Compounds	Rad6B	[34]

Ube2R1: Ubiquitin-Conjugating Enzyme E2, R1, or Cdc34, Ubc13: Ubiquitin-Conjugating Enzyme E2N, UBC12: NEDD8-Conjugating Enzyme, Uev1A: Ubiquitin-conjugating enzyme E2 variant 1A, NAE: NEDD8 activating enzyme, Ubc9: Ubiquitin-Conjugating Enzyme 9, CRL: Cullin-RING E3 ubiquitin ligases, Ube2N: Ubiquitin-Conjugating Enzyme E2N, UbcH8: Ubiquitin Conjugating Enzyme E2 L6, Rad6B: Ubiquitin-conjugating enzyme E2 B or Ube2B.

In contrast to E1 and E2 enzymes, accumulated data exist regarding the role of E3 ligases in cancer. FBW7 (or SCF^FBW7) is an E3–ubiquitin ligase and a substrate recognition component of SCF, regulating many pro-oncogenic proteins and pathways, such as c-Myc, Cyclin E, mTOR, and Notch [35–37]. The phosphorylation of Thr58 and Ser62 residues via FBW7 was shown to be an important regulation for the proteasomal degradation of c-Myc, and the mutation of Thr58 residue causes the tumor progression in Burkitt's lymphoma [38,39]. Relatively, in chronic myelogenous leukemia (CML), the deletion of FBW7 enhances the expression of c-Myc; it also leads p53-dependent apoptosis in human leukemia-initiating cell (LIC), and eventually, tumorigenesis is inhibited [40]. Moreover, FBW7 inhibited the activity of an oncogenic protein, enhancer of zeste homolog 2 (EZH2); hence, it restrained the migration and invasion of the pancreatic tumor by degrading EZH2 in a FBW7-dependent manner [41]. FBW7 was also associated with the mTOR pathway, in such a way that mTORC2 inhibition induces lipogenesis through the FBW7-mediated degradation of sterol regulatory element-binding protein 1 (SREBP1), which in turn decelerates tumor progression in lung, thyroid, melanoma, and cervical cancer [42]. Additionally, suppressing Notch signaling by FBW7 inhibits the improvement of cancer in adult T-cell leukemia lymphoma (ATLL) [43]. Consistent with the previous observations, the decreased expression level of FBW7 has been demonstrated in other types of cancer, including hepatocellular carcinoma [44], colorectal cancer [45], esophageal squamous cell carcinoma [46], etc. Although many studies strongly support the protective role of FBW7 in cancer, there is still controversy whether targeting FBW7 with a drug promotes or inhibits cancer, since an impairment of FBW7 function in cancer cells was shown to induce chemoresistance by stabilizing oncoproteins [47]. Thus, extensive research is necessary to clarify its drug-resistant role for therapeutic approaches.

MDM2 (Mouse double minute 2 homolog) is another E3 ligase that is involved in cancer. Under normal physiological conditions, the phosphorylation of tumor suppressor protein p53 restrains MDM2–p53 binding and further leads to the overexpression of p53 protein in the cells. In cancer cells, MDM2 polyubiquitinates p53 and cause its proteasomal degradation, which is one of the most frequently altered pathways in human cancers [48,49]. Accordingly, the overexpression of MDM2 has been observed in many cancer types, and thus it has been attracted as a drug target for cancer therapy [50,51].

APC/C, Cdc20, Cdh1, βTrCP, and Skp2 are several E3 ligases that have been suggested to be potential therapeutic molecules in breast cancer. Anaphase Promoting Complex/Cyclosome (APC/C) has two co-activators: Cdc20 and Cdh1. Cdc20 was shown as a prognostic candidate for breast cancer because of its elevated Cdc20 mRNA expression level and its correlation with increased tumor size in cancer patients [52]. Moreover, the inhibition of Cdc20 prevented the migration of breast cancer cell lines and consistently, the overexpression of Cdc20 accelerated the metastatic ability of cancer cells in in vitro conditions [53]. Cdh1 is also dysregulated in breast cancer and melanoma, causing an impairment in genome stability and DNA damage response [54]. The very well-known role of Cdh1 is delaying G1 to S transition in cell cycle regulation via targeting Cdc25A, Skp2, and USP37, which makes it a key protein for cancer development [55]. Similarly, the overexpression of βTrCP (Beta-Transducin Repeats-containing Proteins) was observed in breast and prostate cancer. The underlying molecular basis of βTrCP in breast and prostate cancer is that βTrCP targets the MTSS1 (metastasis suppressor 1) protein and impedes its degradation by UPS, thus promoting tumorigenesis. Therefore, in aggressive breast and prostate cancer, the prevention of MTSS1 degradation is considered as a potential treatment approach for these types of cancer [56]. Furthermore, βTrCP was found to be associated with lung cancer by ubiquitinating its one of the targets, FOXN2 (Forkhead box transcription factor), which plays a role in cell proliferation and radiosensitivity in lung cancer [57] Additionally, it was shown that βTRCP stimulates the ubiquitination and degradation of VEGF (Vascular Endothelial Growth Factor) receptor 2, thereby inhibiting angiogenesis and the migration of papillary thyroid cancer cells [58]. Considering the reported data about βTRCP, its effect in cancer regulation seems to be either positive or negative in a context-dependent manner. Lastly, Skp2 is an oncogene, functioning in the ubiquitination of programmed cell death protein 4 (PDCD4) and inhibiting apoptotic cell death. The enhanced expression of Skp2 was observed in breast and prostate cancer, thereby, therapeutic approaches combining radiotherapy and Skp2 targeting were suggested for breast cancer patients [59].

In fact, the data for prostate cancer obtained from various in vitro and in vivo studies imply a scenario in which several E3 ligases are involved in reciprocal interaction between distinct pathways that lead to carcinogenesis. For instance, SPOP (Speckle-type PO2 Protein) E3 ligase promotes ATF2 (Activating Transcription Factor 2) ubiquitination and degradation under normal conditions; however, defective SPOP function induced cell proliferation and invasion [60]. In aggressive prostate cancer, TRIM28 (tripartite motif28) E3 ligase is upregulated and promotes cell proliferation. SPOP was also shown to prevent the interaction of TRIM28–TRIM24 ligases via ubiquitination [61]. E6AP is another important E3 ligase, managing proteasomal degradation of tumor suppressor promyelocytic leukemia protein (PML). The suppression of E6AP results in the reduced growth of prostate cancer cell lines in in vitro conditions, whereas it promote cell senescence in in vivo models. In addition, the knockdown of E6AP was found to make cells more prone to radiation-inducing death [62]. Actually, the underlying mechanism of the tumor-promoting function of E6AP highly depends on its targets. For instance, E6AP targets a cell cycle regulator p27 and inhibits its expression via the E2F1-dependent pathway [63]. On the other hand, E6AP targets the metastasis suppressor NDRG1 (N-Myc Downstream Regulated 1) that was reported in mesenchymal phenotypes of prostate cancer. Pharmacological agents suppressed

E6AP-induced cell migration by increasing NDRG1 expression, indicating the E6AP–NDRG1 axis as an appealing target for prostate cancer therapy [64].

TRIM proteins are one of the largest subfamilies of E3 ligases directly involve in cancer by controlling cell cycle transition and regulating different oncogenic pathways [65]. Recently, the TRIM7 protein has gained attention in cancer studies, because of its association with Ras, Src, and NF-κB signaling pathways. In hepatocellular carcinoma models, cancer progression was suppressed by TRIM7, which was negatively regulating overactive Src [66]. Besides, a clinical study showed a decrease of Trim7 mRNA expression comparing adjacent normal cells in patients with lung cancer. Thus, the TRIM7 expression has been described as a negative regulation in lung tumors. At the molecular level, TRIM7 promoted apoptosis via the NF-κB signaling pathway and suppressed the proliferation and migration of cancer cells [67]. However, in the Ras-driven lung cancer model, transgenic overexpression of the *Trim7* gene increases the tumor size, while TRIM7 protein deficiency leads to a decrease in tumor growth. The reason was that Ras signaling promoted AP-1 transcription factor activation and stimulated TRIM7 protein that tagged ubiquitin to its co-activator RACO-1 and stabilized it [68]. HUWEI is another E3 ligase that is highly expressed in lung cancer. Inhibition of this E3 ligase activity gave rise to the prevention of cell proliferation and colony formation because of elevated p53 expression [69]. Very recently, the structure and function of HUWE1 and potential drug development targeting the HUWE1–p53 axis were reviewed in a large perspective by Gong et al. [70].

One of the most studied E3 ligases, Park2, has been studied in several types of cancer. In hepatocellular carcinoma, defective Park2 function causes abnormal hepatocyte proliferation and leads to avoiding cancer cells from apoptotic cell death [71]. Moreover, the deletion or insufficient expression of Park2 is often described in human glioma, which is correlated with poor survival rates in these patients [72].

Recently, deubiquitinases (DUBs) have gained substantial interest as anticancer agents due to their ability to target and catalyze ubiquitin on the substrate proteins. It was reported that the activation or inactivation of specific DUBs, particularly ubiquitin-specific proteases (USP), a major subfamily of DUBs, induced apoptosis in cultured tumor cells [73,74]. Relatively, the overexpression of DUBs was shown in various types of cancer [75]. However, several DUBs were reported as a tumor promoter, such as USP7, USP15, and USP32; some of them were classified as cancer-associated DUBs, such as DUB3, USP19, and USP25 [76–78]. Importantly, the coordinative action of DUBs and E3 ligases has also been reported in some cancer types. For instance, USP18 was shown to promote breast cancer growth by activating the Skp2/AKT pathway [79].

In breast cancer, the elevated levels of USP37 regulate stemness and cell migration via the hedgehog pathway [58]. Moreover, USP37 induces c-Myc activity by suppressing its degradation, while the inhibition of USP37 increases c-Myc recycling in lung cancer [80]. In contrast, c-Myc inactivation is controlled by the USP28 deubiquitinase [81]. USP7 is commonly found in aggressive brain tumors, leading to p53 stabilization and inhibiting cell death [82]. This mechanism is suppressed by a synthetic drug, 7-chloro-9-oxo-9H-indeno [1,2-b]pyrazine-2,3-dicarbonitrile (HBX 41108). In cancer cells, HBX 41108 treatment increases p53 transcription and recovers p53-mediated cell death. Moreover, USP7 inactivates the ubiquitinated form of Lysine-specific demethylase 1 (LSD1) and promotes cell proliferation through the suppression of cell cycle arrest in brain cancer cells [83]. On the other hand, USP39 influences the function of transcriptional co-activator with PDZ-binding motif (TAZ), which is one of the Hippo tumor suppressor pathway's proteins. Besides, USP39 inhibits TAZ mRNA expression and promotes tumor growth in glioma [84]. Furthermore, the inhibition of USP14 by a specific inhibitor b-AP15 results in the blockage of cell proliferation through the activation of cell cycle arrest in cancer cells [85,86]. USP14 inhibition also increases the effect of chemotherapeutic agent cisplatin in gastric cancer cells [87]. An in vivo study about the ubiquitin C-terminal hydrolase 5 (UCHL5) and USP14 inhibition by a nickel pyrithione complex displays the suppression of tumor growth in human acute myeloid leukemia [88].

Recent studies showed that OTU deubiquitinases can act as either positive or negative regulators in different types of cancer. For instance, OTU deubiquitinase 3 (OTUD3) expression level decreases directly via microRNA-32 targeting, which leads to enhanced proliferation in the HCT116 colorectal cell line [89]. On the other hand, the overexpression of ubiquitin aldehyde binding 1 (OTUB1) was detected in colorectal cancer tissues, which induced the metastasis both in vivo and in vitro studies via triggering the epithelial–mesenchymal transition [90]. In lung cancer, OTUD3 overexpression induced tumorigenesis by stabilizing glucose-regulated protein 78 (GRP78). In addition, Carboxyl terminus of Hsc70-Interacting Protein (CHIP) was found as a negative regulator of OTUD3, and it suppressed the metastasis of lung cancer [91]. Furthermore, the overexpression of OTU domain containing 4 (OTUD4) was revealed to suppress the migration, proliferation, and invasion of cancer cells in breast, liver, and lung via the stimulation of apoptosis by blocking the AKT signaling pathway [92].

DUBs are highly specific proteases that act against ubiquitin activity, and their function is crucial for UPS to maintain protein homeostasis. The UPS includes E3 ligases, DUBs, ubiquitin hydrolases, and the proteasome itself. In order to recover proteasomal function, several types of proteasome inhibitors have been investigated, and most of these mainly prevent the excessive degradation of tumor-suppressor proteins. Thiostrepton, dexamethasone, 2-methoxyestradiol, δ-tocotrienol, and quercetin are such inhibitors, which are sufficiently used for the treatment of cancer cells in liver, pancreas, prostate, breast, lung, and melanoma [93]. However, many types of cancer cells are still resistant to those proteasome inhibitors. In line with the current knowledge, there is a consensus on the concept that DUBs generally act as positive regulators in cancer progression. Thus, DUBs may be an ideal candidate for therapeutic approaches in cancer cells, which are resistant to proteasome inhibitors.

The discussed E3 ubiquitin–protein ligases and DUBs in related cancer types are summarized in Table 2.

Table 2. Summary of related gene expressions of discussed E3 ubiquitin–protein ligases and deubiquitinases or deubiquitinating enzymes (DUBs) in specific types of cancer.

E3s and DUBs	Disease Association	Related Gene Expression	Ref
FBW7	Hepatocellular carcinoma, colorectal cancer, esophageal squamous cell carcinoma	Underexpressed	[44,45]
MDM2	Breast cancer	Overexpressed	[50]
Cdc20	Breast, pancreatic cancer	Overexpressed	[94]
APC/C(Cdh1)	Breast cancer, melanoma	Downregulated	[95,96]
βTrCP	Breast, prostate, lung cancer	Expression depends on tissue type	[97,98]
Skp2	Breast cancer, prostate cancer	Overexpressed	[99,100]
SPOP	Prostate cancer	Downregulated	[101,102]
E6AP	Prostate cancer, cervical and lung cancer	Expression depends on tissue type	[103,104]
TRIM7	Lung cancer	Expression depends on tissue type	[67,68]
HUWE1	Lung, breast, and colorectal carcinoma	Expression depends on tissue type	[105–107]
Park2	Glioma, hepatocellular carcinoma, and lymphoma	Underexpressed	[71,108]
USP7	Prostate cancer, brain tumors	Overexpressed	[78,82]
USP14	Gastric cancer	Overexpressed	[87]
USP15	Breast Cancer	Overexpressed	[109]
USP18	Breast cancer	Overexpressed	[79]
USP32	Breast cancer	Overexpressed	[76]

Table 2. *Cont.*

E3s and DUBs	Disease Association	Related Gene Expression	Ref
USP37	Breast and lung cancer	Overexpressed	[80,110]
USP39	Glioma	Overexpressed	[84]
OTUD3	Colorectal and lung cancer	Expression depends on tissue type	[89,91,111]
OTUB1	Colorectal cancer	Overexpressed	[90]
OTUD4	Breast, liver, and lung cancers	Overexpressed	[92]

Fbw7: F-Box/WD Repeat-Containing Protein 7, Mdm2: Mouse double minute 2 homolog, TRIM7: Tripartite motif-containing 7, SPOP: Speckle-type BTB–POZ protein, Cdc20: cell division cycle protein 20, APC/C (Cdh1): Anaphase Promoting Complex or Cyclosome (Cdc20 Homolog 1), HUWE1: HECT, UBA, and WWE Domain Containing E3 Ubiquitin Protein Ligase 1, E6AP: Ubiquitin Protein Ligase E3A, USP: Ubiquitin-Specific Protease.

4. Ubiquitination-Mediated Regulation in Neurodegeneration and Neurodegenerative Diseases

The neuronal accumulation of insoluble proteins is known the major cause of neurodegenerative diseases, including Alzheimer's disease (AD) [112], Parkinson disease (PD) [113], dementia, progressive supranuclear palsy (PSP), and frontotemporal dementia with parkinsonism linked to chromosome 17 (FTDP-17) [114]. Under normal physiological conditions, UPS, autophagy, or lysosomal degradation systems are responsible for the removal of these accumulated aggregates and misfolded proteins. Today, it is known that some enzymatic mutations in UPS are one of the main reasons for abnormal protein aggregation in neurodegenerative diseases [115].

AD is observed in around 10% of the population over 65 years of age. In the sporadic form of AD, the accumulation of specific neurotoxic proteins, hyperphosphorylated tau, and β-amyloid (Aβ) is the typical characteristic of this disease. E3-ligase Parkin has been shown to have a cytoprotective effect in AD. The study demonstrated that wild-type Parkin expression decreased the intracellular Aβ42 level through UPS and reversed impaired proteasome function [116]. Proteomic analyses showed a wide range of proteins ubiquitinated by Parkin, yet these proteins had no obvious functional association with any cellular mechanism; however, proteins including endocytic trafficking components were over-represented. Further studies are necessary for identifying whether Parkin substrates are functionally related to any known pathway or not [117]. In addition to Parkin, another decreased E3 ligase in AD patients' brains is "ER related HRD1". It promotes neural cell survival by mediating the ubiquitination of tau protein. The loss of HRD1 expression results in amyloid precursor protein (APP) accumulation and β-amyloid generation [118]. In addition, it is considered that β-amyloid-dependent oxidative stress leads to the aggresome formation of HRD1, which affects its solubility [119]. The C-terminus of Hsp70-interacting protein (CHIP) is another E3 ligase, and its binding protein Hsp70 is highly expressed in the AD patients' brain to overcome the phosphorylated tau accumulation by inducing ubiquitination [120,121]. Moreover, the inhibition of CHIP causes an increase in hyperphosphorylated and caspase-3 cleaved tau species [122]. In a recent study, Hsp70 administration is found to be sufficient against late-stage Alzheimer-type neuropathology in a mouse model [123]. Another study showed that the upregulation of CHIP by sulforaphane treatment decreased the accumulation of neurotoxic proteins in a mice model of AD [124]. Therefore, the CHIP/Hsp70 axis seems to be a promising approach for the removal of toxic proteins in the brain.

In addition to the E3 ligases, DUBs also have critical roles in AD. Ubiquitin C-terminal hydrolase L1 (UCH-L1) was found in tau tangles; the downregulation and considerable oxidative modification of this protein was observed in AD patients. Moreover, the overexpression of UCH-L1 was shown to decrease Aβ plaques and promoted the memory deficiency in an AD mice model [125,126]. Furthermore, OTUB1 (OTU deubiquitinase ubiquitin aldehyde-binding 1) was found in Aβ plaques of AD patients [127], and interestingly, USP14 was able to inhibit tau function [128].

In fact, ubiquitination or deubiquitination are not the only defective mechanisms in AD pathology; the impairment of proteasomal function is another main reason for the elevated amount of ubiquitinated protein in the cytoplasm. It is similar to toxic tau accumulation or in other words "tauopathy", and it is related to decreased proteasomal activity and an increased amount of ubiquitinated proteins in the cell. Within this context, a study has indicated that the activation of cAMP (cyclic Adenosine Monophosphate)–protein kinase A (PKA) by rolipram (phosphodiesterase type 4 (PDE4) inhibitor that increases cAMP levels) leads to the attenuation of proteasome dysfunction by a phosphorylating proteasome subunit [50].

The pathology of the second most common neurodegenerative disease PD is characterized by the accumulation of Lewy bodies, α-synuclein, and related multimers in dopaminergic neurons [129]. Parkin/PINK1 is one of the best-studied proteins in this context. Depolarization in the mitochondrial membrane potential activates PINK1 (PTEN-induced putative kinase 1) and stimulates Parkin E3-ligase activity by phosphorylating its Serine 65 residue. Loss-of-function mutations of Parkin/PINK1 were shown to be directly linked to the early onset of PDs [130,131]. Recently, iPSC (induced Pluripotent Stem Cells)-derived midbrain dopamine neurons from PD patients, harboring PINK1 and Parkin mutations, demonstrate an aberrant cytosolic accumulation of α-synuclein and mitochondrial dysfunctions [132]. Additionally, the Parkin/PINK1 pathway is associated with an increased expression of tumor necrosis factor receptor-associated factor 6 (TRAF6) in a PD patient brain [133]. The TRAF6-mediated Lys63 ubiquitination of PINK1 at Lys433 is necessary for mitochondrial regulation and proper PINK1-Parkin-TRAF6 complex formation [134]. In Parkin deficiency, PINK1 triggers mitochondrial antigen presentation in immune cells, which creates inflammatory conditions [135]. Moreover, an idea about the link between the brain and intestinal system is supported by a very recent study of the same research group. Intestinal infection in Pink1 mutant mice shows Parkinson's-like disease symptoms. Thus, the Parkin/PINK1 axis contributes to an autoimmune response that is directly related to PD etiology [136]. Similar to Parkin/PINK1, recent findings indicate the role of ubiquitination in α-synuclein pathologies. In other words, E3 ligases, namely E6-associated protein (E6AP) and Nedd4, are involved in the α-synuclein degradation process. E6AP has been found to colocalize with α-synuclein, and the overexpression of E6-AP increases α-synuclein degradation in a proteasome-dependent manner [137]. The role of Nedd4 was shown by its overexpression, which causes the hyperubiquitination of α-synuclein in contrast to other ligases. Moreover, the SCF$^{FBXO7/PARK15}$ complex has a considerable ubiquitinase activity, specifically for FBXO7. Glycogen synthase kinase 3β (GSK3β) and Translocase Of Outer Mitochondrial Membrane 20 (TOMM20) have been found as substrates of this complex during PD progression. Furthermore, Teixeria et. al. reported that FBOX7 ubiquitinates and alters GSK3β by Lys63 linkage.

Recent discoveries show the implication of DUBs in PD pathology. For instance, the elevated level of USP13 was found in PD patients [138]. Additionally, UCH-L1 and OTUB1 were shown in Lewy bodies, suggesting that UCH-L1 has the ability to stabilize α-synuclein levels in a context-dependent manner under normal or pathological conditions, while OTUB1 causes amyloid aggregation apart from its DUB activity [127,139]. On the other hand, the activation of UPS via promoting Protein Kinase A phosphorylation eventually stimulated the degradation of α-synuclein [140]. Similarly, the regulation of UPS activity by several chemicals attenuated α-synuclein oligomers [141], suggesting that the recovery of proteasomal deficiency is also critical for synucleinopathies in PD.

The most common polyQ disorders, Huntington's disease (HD), is characterized by the expansion of CAG repeats in the *huntingtin (htt)* gene as well as the formation of inclusion bodies in striatal neurons [142]. In an in vivo HD model, the genetic inhibition of E3-ligase CHIP showed enhanced neuropathology, indicating that CHIP reduces neurotoxicity by suppressing the accumulation of polyQ [143]. Several other E3 ligases, including UBE3A [144], WWP1 [145], Parkin [146], Hrd1 [147], and HACE1 [148] were also reported to target the Huntingtin protein.

Recently, one of the DUBs, YOD1, was suggested to decrease proteotoxicity in HD. Neurogenic proteins stimulated YOD1 expression, which caused a reduction in the ubiquitination of abnormal proteins [149]. In addition to ubiquitin enzymes, proteasomal activity is regulated by a member of DUBs, namely USP14. USP14 suppression enhanced proteasome activity and induced the degradation of abnormal proteins such as tau and Ataxin-3 [150]. Similarly, the promotion of proteasomal degradation by USP14 protects neuronal cells from mutant huntingtin-induced cell degeneration [151]. Huntingtin-associated protein (HAP40) is another protein affecting the proteasomal activity in HD. The overexpression of this protein stimulates mutant Htt aggregation due to the impaired UPS [152]. An in vivo study demonstrated that intraneuronal Huntington filaments but not inclusion bodies inhibit proteasome activity [153]. On the other hand, proteasomes are dynamically recruited into inclusion bodies, while these proteasomes remain catalytically active for substrates' degradation. These results seem to be controversial when compared to previous findings indicating defective proteasomes in HD. Thus, further studies are required for the clear definition of proteasome activity in HD [154].

5. Ubiquitination in Immune-Related Diseases

Ubiquitination is one of the most essential post-translational modifications, playing an important role in both innate and adaptive immunity. Notably, E3 ligases have a major role in leukocytes activation, differentiation, and development [155]. Besides, DUBs, particularly USPs, are known as an important modulator for T cell function [156]. In accordance, recent studies demonstrated that defective UPS, including aberrant E3 ligases, DUBs, or proteasomes tend to the impairment of immune regulation and thereby cause the development of multiple inflammatory or autoimmune diseases [157].

Casitas B-lineage lymphoma (Cbl) proteins, consisting of Cbl-b, Cbl-3, and c-Cbl, are a member of the RING finger-containing E3-ligase superfamily [158]. Due to their RING finger domain, they serve as an E3 ligase to various substrates, forming stable interaction with E2-conjugating enzymes and subsequently promoting ubiquitin transfer from the E2 enzyme to their substrate. In previous studies, Cbls was revealed as a major modulator in immune regulation during the early development of hematopoietic precursor cells into the effector immune cells [159]. *Cb*$^{-/-}$ transgenic mice models show that the amount of CD4+ SP thymocytes are significantly elevated by ubiquitination of the CD3 ζ chain of the T-cell receptor (TCR) complex [160]. In B cell development, *Cbl*-deficient B cells show an improvement of BCR (B-cell antigen receptor) signaling and promote several downstream processes, such as Ca^{+2} influx, Syk, and CD79A. Therefore, Cbl is known as a negative regulator of B and T cell development via ubiquitin-dependent degradation [161]. In addition, Cbls are known to regulate the activation of macrophages regarding immune disorders, particularly in cancer and obesity. For instance, *Cbl*$^{-/-}$ mice show enhanced tumor growth in colorectal cancer, and the recovery of Cbl expression increases the tumor cell phagocytosis of macrophages through the ubiquitination of surface proteins [162]. In obesity, Cbls act as negative regulators for macrophage activation by suppressing migration signals to adipose tissue. Therefore, insulin resistance and obesity are induced in *Cbl*$^{-/-}$ mice. Stimulated macrophage accumulation in adipose tissue results in the secretion of inflammatory cytokines in this model [163,164]. Although there are several studies about Cbls in macrophage regulation, molecular mechanisms underlying inflammatory diseases are still poorly understood.

Cbl-b works with other E3 ligases, which are known as Itch, to prevent the over-reactivity of T-cell-dependent peripheral tolerance. Itch is a monomeric protein, namely HECT (homologous to the E6AP carboxyl terminus), presenting intrinsic catalytic activity. Its main role in immunity is regulating immune cell development and function by mediating protein ubiquitination [155]. Recent studies show that Itch and Cbl-b collaboration leads to proteolysis, independent of ubiquitination of the TCRζ chain. Thus, the signal transduction of the TCR complex is suppressed by preventing ZAP70 and TCRζ chain binding [165]. The cooperative functions of different E3 ligases support the regulation of immune cell homeostasis by ubiquitination [166].

Von Hippel-Lindau (VHL) is an E3-ligase complex, consisting of elongin B, elongin C, cullin 2, and ring box protein 1 (RBP1). These ligases are mainly regulated by Hypoxia-inducible factor-1α (HIF-1α), which functions as a critical transcription factor in immune regulation under hypoxic conditions. Under normal conditions, the VHL ligase attaches ubiquitin tags on HIF-1α and prevents target gene activation [167]. In fact, HIF-1α is responsible for the immunosuppressive function of regulatory T cells (Treg) [168]. In VHL deficiency, HIF-1α promotes the elevation of interferon-γ (IFN-γ) production in Treg cells, which causes the emergence of the Th1-like inflammatory phenotype [169], indicating the critical role of the VHL ligase–HIF-1α axis in T cell immunity and differentiation [156]. Low oxygen tension also affects innate immunity via HIF-1α-dependent signaling. An impaired maturation of alveolar macrophages and suppression of neutrophil apoptosis were demonstrated in hypoxia or VHL-deficient conditions [170]. These findings support that the VHL/HIF-1α pathway is responsible not only for T and B cell differentiation but also myeloid cell function during an innate immune response [171]. Moreover, hypoxia and VHL ligases are commonly associated with renal carcinoma. It has been shown that VHL expression is negatively correlated with tumor malignancy through promoting immune responses [172] and in patients, mutated VHL ligase proteins were found to show more natural killer cell toxicity [173]. Taken together, VHL E3 ligases appears to be a remarkable protein for immune regulation and immune-related disease prevention.

Since ubiquitination is a reversible process, deubiquitinases play a crucial role in maintaining an effective immune response, particularly for constituting adaptive immunity [156]. USPs are the largest family of DUBs including at least 50 members; however, the most important USPs in immune regulation are USP4, USP8, USP9X, USP12, and USP19 [75,174], which are mainly involved in T cell homeostasis as positive or negative regulators. Specifically, USP4 is responsible for promoting Th17 cell differentiation and function. The most established transcription factor RORγt (RAR-related orphan receptor gamma) stabilization for Th17 cell differentiation is mediated by highly expressing USP4. Hence, USP4 was suggested as a potential target for inhibiting Th17-mediated autoimmune disorders, such as rheumatic heart disease [175]. USP8 is another important member of DUBs, which are critical for T cell maturation. The knockdown of T cell-specific USP8 showed that the maturation and proliferation of thymocytes were impaired by a Foxo1–IL-7Rα-dependent mechanism. USP8 mutant mice developed lethal colitis because of impaired T cell homeostasis [176]. Moreover, somatic mutations in the human USP8 gene were suggested to suppress the immune system, which causes the development of Cushing's Disease. These mutations are associated with the 14-3-3 binding motif (RSYSS) of USP8, which is important for its phosphorylation; thus, mutations in RSYSS motif lead to impairing its DUB activity. In patients with Cushing's disease, it was observed that adrenocorticotropic hormone (ACTH) is produced at a higher level because of USP8 mutations [177]. Similar to USP8, USP9X acts as a positive mediator for TCR signaling and T cell tolerance. The loss of USP9X leads to the decrease in T cell proliferation and T cell differentiation into helper T cells, hence affecting cytokine production. USP9X deficiency in T cells attenuates TCR signaling and promotes the activation of the NF-κB pathway because of modulating upstream signaling proteins of the NF-κB [178]. USP12 deubiquitinates some of the TCR adaptor proteins, such as LAT (Linker for Activation of T cells) and Trat1, stabilizing the TCR:CD3 complex on the cell surface and consequently preventing the lysosomal degradation of LAT and Trat1 [179]. On the other side, USP19 has a different role in immune regulation compared to other USPs discussed above, since it is related to an innate immune response with its Toll-Like Receptor (TLR)-dependent activity. In a very recent study, Usp19$^{-/-}$ mice treated with poly(I:C) or LPS showed the elevation of pro-inflammatory cytokines and type 1 interferons secretion, indicating that USP19 negatively regulates TLR3/4-mediated signaling by impairing the accession of the essential adaptor protein to TLR3/4 [180].

Cylindromatosis (CYLD) was first discovered as a tumor-suppressor protein [181]. However, it has been shown that CYLD belongs to the deubiquitinating enzyme family, and it removes the Lys63-linked polyubiquitin chain from signaling proteins located upstream of NF-κB, including TNF receptor-associated factor 2 (TRAF2), TRAF6, and NEMO [182]. It is well known that CYLD is a critical factor for T cell development, activation, and function. In fact, the essential role of CYLD in T cell development was identified with threshold activation for the positive selection of T cells via NEMO-dependent NF-κB signaling. In an in vivo study, *Cyld*-deficient mice developed a chronic inflammatory digestive disease, colitis [183]. The underlying mechanism is that CYLD targets TGF-β-activated kinase 1 (TAK1) and inhibits its ubiquitination and autoactivation. This mechanism also activates downstream kinases of TAK1 such as c-Jun N-terminal kinase (JNK) and IκB kinase β (IKKβ) due to the autoactivation of TAK1 [184,185]. In line with this information, the function of CYLD in T cell regulation is considered as a negative feedback mechanism by inhibiting activation of the TAK1 axis of TCR signaling. In transgenic mouse models of CYLD, changes in immune cell function, abnormal hepatocytes, hair, and dental defects were observed [185–187]. Furthermore, CYLD polymorphisms were specifically associated with inflammatory bowel disease (IBD) that comprised 2320 patients with Crohn's disease [188].

UPS has been shown as the main modulator for the antigen presentation in immune regulation. More specifically, oligopeptides are hydrolyzed by the proteasome after binding PA28, an activator for hydrolysis, to the ends of 20S proteasome. The process is controlled by a proteolytic cascade including aminopeptidases that are responsible for peptide production for the presentation of MHC-I (Major Histocompatibility Complex-I) [189]. Notably, the MARCH (Membrane-associated RING-CH-type finger) E3 ligase is involved in the reduction of MHC-I surface expression by mediating the polyubiquitination of MHC-I. Therefore, MARCH-mediated MHC presentation is crucial for UPS function in immune regulation [190]. Interestingly, autoantibodies, reacting as self-antigens, were detected in 20S proteasome subunits in several autoimmune diseases, such as systemic lupus erythematosus, primary Sjögren's syndrome, and myositis. In vitro studies showed that autoantibodies can inhibit proteasome activation by targeting the PA28 binding. On the other hand, proteasome levels in circulation are suggested as an important biomarker for disease progression and potential drug targets, because of their increased levels observed in autoimmune myositis, systemic lupus erythematosus, primary Sjögren's syndrome, and rheumatoid arthritis [191,192].

6. Concluding Remarks

Dysregulation of the ubiquitin–proteasome system including the positive or negative regulation of E3 ligases, DUBs, or proteasomes seriously affects cellular homeostasis and causes the development of serious pathologic conditions, such as tumor suppression or promotion in cancer, protein accumulation in neurodegenerative diseases, and forming an ineffective immune response in the body (Figure 3). Thus, recently a great deal of work has been devoted to the development of novel drugs, targeting proteins that either interfere or inhibit ubiquitination and proteasome activity in disease-dependent manner. Ongoing research studies of a wide range of molecules targeting different ligases show promising data in several pathogenic conditions; nevertheless, the translation of these data into clinical application is still a major challenge. For instance, some compounds are shown to inhibit E3 ligases generally in in vitro conditions, but the effects of such compounds remain unclear in in vivo models [193]. Proteasome-associated DUBs have been suggested as remarkable drug targets due to their lower side effects compared to proteasome-targeted drugs. Currently, DUBs-targeted inhibitors are mainly small molecules, and their development is still in the preclinical stage for inflammatory disease and cancer treatment [194]. The major restriction of DUBs inhibitors is their selectivity, as they may have complex intracellular interactions with several signaling pathways. Proteasome activity is another target in UPS-dependent therapy that became a popular idea to recover prominently in neurodegenerative diseases. Proteasome inhibitors, including bortezomib, carfilzomib, and ixazomib are well-tolerated by patients and therefore approved by the FDA for clinical applications. On the other hand, marizomib and oprozomib are currently under clinical trials [195]. In upcoming years, novel molecules targeting E3 ligases, DUBs, or proteasomes are expected to be validated for therapeutic approaches; thus, a better understanding of the molecular signaling pathways involved in ubiquitination and proteasomal degradation will allow the discovery of novel targeting molecules in cancer, neurodegenerative disease, and immune-related pathological conditions.

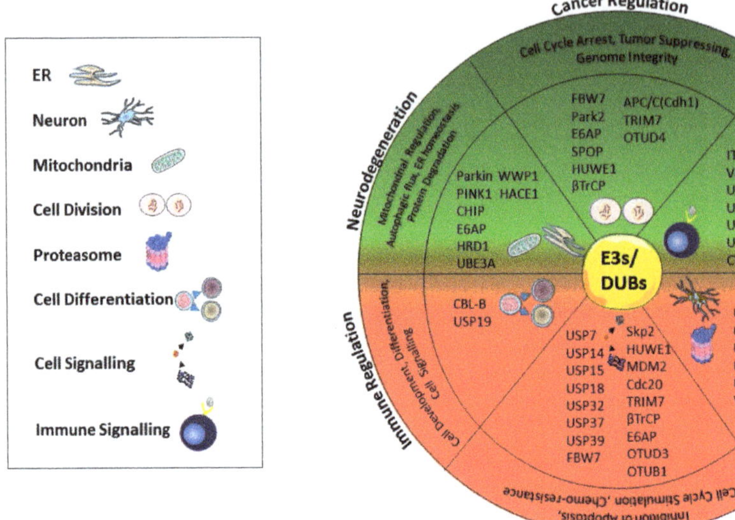

Figure 3. Summary of DUBs and E3 enzymes that play important roles in cancer progression, immune regulation, and neurodegenerative diseases. According to diseases, the enzymes positively or negatively regulate the mechanisms in mitophagy, cell differentiation, cell development, signaling, protein accumulation, neural malfunction, synaptogenesis, inhibition of apoptosis, cell cycle stimulation, mitochondrial regulation, autophagic flux, ER homeostasis, and protein degradation. E3 ligases and DUBs in the upside or downside of the semicircle indicate negative (red area) or positive regulators (green area) of these diseases.

Author Contributions: G.C., H.K., E.O.; original draft preparation, O.K., review, and editing. All authors have read and agreed to the published version of the manuscript.

Funding: This research received no external funding.

Conflicts of Interest: The authors declare no conflict of interest.

References

1. Goldstein, G.; Scheid, M.; Hammerling, U.; Schlesinger, D.H.; Niall, H.D.; Boyse, E.A. Isolation of a polypeptide that has lymphocyte-differentiating properties and is probably represented universally in living cells. *Proc. Natl. Acad. Sci. USA* **1975**, *72*, 11–15. [CrossRef] [PubMed]

2. Kocaturk, N.M.; Gozuacik, D. Crosstalk Between Mammalian Autophagy and the Ubiquitin-Proteasome System. *Front. Cell Dev. Biol.* **2018**, *6*. [CrossRef] [PubMed]

3. Gilberto, S.; Peter, M. Dynamic ubiquitin signaling in cell cycle regulation. *J. Cell Biol.* **2017**, *216*, 2259–2271. [CrossRef] [PubMed]

4. Vucic, D.; Dixit, V.M.; Wertz, I.E. Ubiquitylation in apoptosis: A post-translational modification at the edge of life and death. *Nat. Rev. Mol. Cell Biol.* **2011**, *12*, 439–452. [CrossRef] [PubMed]

5. Williams, R.L.; Urbe, S. The emerging shape of the ESCRT machinery. *Nat. Rev. Mol. Cell Biol.* **2007**, *8*, 355–368. [CrossRef]

6. Cao, J.; Yan, Q. Histone ubiquitination and deubiquitination in transcription, DNA damage response, and cancer. *Front. Oncol.* **2012**, *2*, 26. [CrossRef]

7. Scheuring, D.; Künzl, F.; Viotti, C.; Yan, M.S.; Jiang, L.; Schellmann, S.; Robinson, D.G.; Pimpl, P. Ubiquitin initiates sorting of Golgi and plasma membrane proteins into the vacuolar degradation pathway. *BMC Plant Biol.* **2012**, *12*, 164. [CrossRef]

8. Ronai, Z.e.A. Monoubiquitination in proteasomal degradation. *Proc. Natl. Acad. Sci. USA* **2016**, *113*, 8894–8896. [CrossRef]

9. Swatek, K.N.; Komander, D. Ubiquitin modifications. *Cell Res.* **2016**, *26*, 399–422. [CrossRef]

10. Chen, Z.J.; Sun, L.J. Nonproteolytic functions of ubiquitin in cell signaling. *Mol. Cell* **2009**, *33*, 275–286. [CrossRef]

11. Zhao, B.; Bhuripanyo, K.; Schneider, J.; Zhang, K.; Schindelin, H.; Boone, D.; Yin, J. Specificity of the E1-E2-E3 enzymatic cascade for ubiquitin C-terminal sequences identified by phage display. *ACS Chem. Biol.* **2012**, *7*, 2027–2035. [CrossRef] [PubMed]

12. Zheng, N.; Shabek, N. Ubiquitin Ligases: Structure, Function, and Regulation. *Annu. Rev. Biochem.* **2017**, *86*, 129–157. [CrossRef] [PubMed]

13. George, A.J.; Hoffiz, Y.C.; Charles, A.J.; Zhu, Y.; Mabb, A.M. A Comprehensive Atlas of E3 Ubiquitin Ligase Mutations in Neurological Disorders. *Front. Genet.* **2018**, *9*. [CrossRef]

14. Stewart, M.D.; Ritterhoff, T.; Klevit, R.E.; Brzovic, P.S. E2 enzymes: More than just middle men. *Cell Res.* **2016**, *26*, 423–440. [CrossRef]

15. Lecker, S.H.; Goldberg, A.L.; Mitch, W.E. Protein Degradation by the Ubiquitin–Proteasome Pathway in Normal and Disease States. *J. Am. Soc. Nephrol.* **2006**, *17*, 1807–1819. [CrossRef]

16. Benaroudj, N.; Zwickl, P.; Seemuller, E.; Baumeister, W.; Goldberg, A.L. ATP hydrolysis by the proteasome regulatory complex PAN serves multiple functions in protein degradation. *Mol. Cell* **2003**, *11*, 69–78. [CrossRef]

17. Tanaka, K. The proteasome: Overview of structure and functions. *Proc. Jpn. Acad. Ser. B Phys. Biol. Sci.* **2009**, *85*, 12–36. [CrossRef]

18. Abdul Rehman, S.A.; Kristariyanto, Y.A.; Choi, S.Y.; Nkosi, P.J.; Weidlich, S.; Labib, K.; Hofmann, K.; Kulathu, Y. MINDY-1 Is a Member of an Evolutionarily Conserved and Structurally Distinct New Family of Deubiquitinating Enzymes. *Mol. Cell* **2016**, *63*, 146–155. [CrossRef] [PubMed]

19. He, M.; Zhou, Z.; Shah, A.A.; Zou, H.; Tao, J.; Chen, Q.; Wan, Y. The emerging role of deubiquitinating enzymes in genomic integrity, diseases, and therapeutics. *Cell Biosci.* **2016**, *6*, 62. [CrossRef]

20. Zou, Q.; Jin, J.; Hu, H.; Li, H.S.; Romano, S.; Xiao, Y.; Nakaya, M.; Zhou, X.; Cheng, X.; Yang, P.; et al. USP15 stabilizes MDM2 to mediate cancer-cell survival and inhibit antitumor T cell responses. *Nat. Immunol.* **2014**, *15*, 562–570. [CrossRef]

21. Li, L.; Martinez, S.S.; Hu, W.; Liu, Z.; Tjian, R. A specific E3 ligase/deubiquitinase pair modulates TBP protein levels during muscle differentiation. *Elife* **2015**, *4*, e08536. [CrossRef] [PubMed]

22. Grou, C.P.; Pinto, M.P.; Mendes, A.V.; Domingues, P.; Azevedo, J.E. The de novo synthesis of ubiquitin: Identification of deubiquitinases acting on ubiquitin precursors. *Sci. Rep.* **2015**, *5*, 12836. [CrossRef] [PubMed]

23. Yang, Y.; Kitagaki, J.; Dai, R.M.; Tsai, Y.C.; Lorick, K.L.; Ludwig, R.L.; Pierre, S.A.; Jensen, J.P.; Davydov, I.V.; Oberoi, P.; et al. Inhibitors of ubiquitin-activating enzyme (E1), a new class of potential cancer therapeutics. *Cancer Res.* **2007**, *67*, 9472–9481. [CrossRef] [PubMed]

24. Liu, J.; Shaik, S.; Dai, X.; Wu, Q.; Zhou, X.; Wang, Z.; Wei, W. Targeting the ubiquitin pathway for cancer treatment. *Biochim. Biophys. Acta* **2015**, *1855*, 50–60. [CrossRef] [PubMed]

25. Hosseini, S.M.; Okoye, I.; Chaleshtari, M.G.; Hazhirkarzar, B.; Mohamadnejad, J.; Azizi, G.; Hojjat-Farsangi, M.; Mohammadi, H.; Shotorbani, S.S.; Jadidi-Niaragh, F. E2 ubiquitin-conjugating enzymes in cancer: Implications for immunotherapeutic interventions. *Clin. Chim. Acta* **2019**, *498*, 126–134. [CrossRef] [PubMed]

26. Ceccarelli, D.F.; Tang, X.; Pelletier, B.; Orlicky, S.; Xie, W.; Plantevin, V.; Neculai, D.; Chou, Y.C.; Ogunjimi, A.; Al-Hakim, A.; et al. An allosteric inhibitor of the human Cdc34 ubiquitin-conjugating enzyme. *Cell* **2011**, *145*, 1075–1087. [CrossRef]

27. Scheper, J.; Guerra-Rebollo, M.; Sanclimens, G.; Moure, A.; Masip, I.; González-Ruiz, D.; Rubio, N.; Crosas, B.; Meca-Cortés, O.; Loukili, N.; et al. Protein-protein interaction antagonists as novel inhibitors of non-canonical polyubiquitylation. *PLoS ONE* **2010**, *5*, e11403. [CrossRef]

28. Ushiyama, S.; Umaoka, H.; Kato, H.; Suwa, Y.; Morioka, H.; Rotinsulu, H.; Losung, F.; Mangindaan, R.E.; de Voogd, N.J.; Yokosawa, H.; et al. Manadosterols A and B, sulfonated sterol dimers inhibiting the Ubc13-Uev1A interaction, isolated from the marine sponge Lissodendryx fibrosa. *J. Nat. Prod.* **2012**, *75*, 1495–1499. [CrossRef]

29. Hodge, C.D.; Edwards, R.A.; Markin, C.J.; McDonald, D.; Pulvino, M.; Huen, M.S.; Zhao, J.; Spyracopoulos, L.; Hendzel, M.J.; Glover, J.N. Covalent Inhibition of Ubc13 Affects Ubiquitin Signaling and Reveals Active Site Elements Important for Targeting. *ACS Chem. Biol.* **2015**, *10*, 1718–1728. [CrossRef]

30. Huang, D.T.; Miller, D.W.; Mathew, R.; Cassell, R.; Holton, J.M.; Roussel, M.F.; Schulman, B.A. A unique E1-E2 interaction required for optimal conjugation of the ubiquitin-like protein NEDD8. *Nat. Struct. Mol. Biol.* **2004**, *11*, 927–935. [CrossRef]

31. Kumar, A.; Ito, A.; Hirohama, M.; Yoshida, M.; Zhang, K.Y.J. Identification of Sumoylation Inhibitors Targeting a Predicted Pocket in Ubc9. *J. Chem. Inf. Model.* **2014**, *54*, 2784–2793. [CrossRef] [PubMed]

32. Wu, K.; Chong, R.A.; Yu, Q.; Bai, J.; Spratt, D.E.; Ching, K.; Lee, C.; Miao, H.; Tappin, I.; Hurwitz, J.; et al. Suramin inhibits cullin-RING E3 ubiquitin ligases. *Proc. Natl. Acad. Sci. USA* **2016**, *113*, E2011–E2018. [CrossRef] [PubMed]

33. Zhou, B.; Li, H.; Xing, C.; Ye, H.; Feng, J.; Wu, J.; Lu, Z.; Fang, J.; Gao, S. Honokiol induces proteasomal degradation of AML1-ETO oncoprotein via increasing ubiquitin conjugase UbcH8 expression in leukemia. *Biochem. Pharm.* **2017**, *128*, 12–25. [CrossRef] [PubMed]

34. Kothayer, H.; Spencer, S.M.; Tripathi, K.; Westwell, A.D.; Palle, K. Synthesis and in vitro anticancer evaluation of some 4,6-diamino-1,3,5-triazine-2-carbohydrazides as Rad6 ubiquitin conjugating enzyme inhibitors. *Bioorg. Med. Chem. Lett.* **2016**, *26*, 2030–2034. [CrossRef]

35. Yada, M.; Hatakeyama, S.; Kamura, T.; Nishiyama, M.; Tsunematsu, R.; Imaki, H.; Ishida, N.; Okumura, F.; Nakayama, K.; Nakayama, K.I. Phosphorylation-dependent degradation of c-Myc is mediated by the F-box protein Fbw7. *EMBO J.* **2004**, *23*, 2116–2125. [CrossRef]

36. Yumimoto, K.; Nakayama, K.I. Recent insight into the role of FBXW7 as a tumor suppressor. *Semin. Cancer Biol.* **2020**. [CrossRef]

37. Xu, W.; Taranets, L.; Popov, N. Regulating Fbw7 on the road to cancer. *Semin. Cancer Biol.* **2016**, *36*, 62–70. [CrossRef]

38. Bahram, F.; von der Lehr, N.; Cetinkaya, C.; Larsson, L.G. c-Myc hot spot mutations in lymphomas result in inefficient ubiquitination and decreased proteasome-mediated turnover. *Blood* **2000**, *95*, 2104–2110. [CrossRef]

39. Yeh, C.-H.; Bellon, M.; Nicot, C. FBXW7: A critical tumor suppressor of human cancers. *Mol. Cancer* **2018**, *17*, 115. [CrossRef]

40. Reavie, L.; Buckley, S.M.; Loizou, E.; Takeishi, S.; Aranda-Orgilles, B.; Ndiaye-Lobry, D.; Abdel-Wahab, O.; Ibrahim, S.; Nakayama, K.I.; Aifantis, I. Regulation of c-Myc ubiquitination controls chronic myelogenous leukemia initiation and progression. *Cancer Cell* **2013**, *23*, 362–375. [CrossRef]

41. Jin, X.; Yang, C.; Fan, P.; Xiao, J.; Zhang, W.; Zhan, S.; Liu, T.; Wang, D.; Wu, H. CDK5/FBW7-dependent ubiquitination and degradation of EZH2 inhibits pancreatic cancer cell migration and invasion. *J. Biol. Chem.* **2017**, *292*, 6269–6280. [CrossRef] [PubMed]

42. Li, S.; Oh, Y.T.; Yue, P.; Khuri, F.R.; Sun, S.Y. Inhibition of mTOR complex 2 induces GSK3/FBXW7-dependent degradation of sterol regulatory element-binding protein 1 (SREBP1) and suppresses lipogenesis in cancer cells. *Oncogene* **2016**, *35*, 642–650. [CrossRef] [PubMed]

43. Yeh, C.-H.; Bellon, M.; Pancewicz-Wojtkiewicz, J.; Nicot, C. Oncogenic mutations in the FBXW7 gene of adult T-cell leukemia patients. *Proc. Natl. Acad. Sci. USA* **2016**, *113*, 6731–6736. [CrossRef] [PubMed]

44. Imura, S.; Tovuu, L.O.; Utsunomiya, T.; Morine, Y.; Ikemoto, T.; Arakawa, Y.; Kanamoto, M.; Iwahashi, S.; Saito, Y.; Takasu, C.; et al. Role of Fbxw7 expression in hepatocellular carcinoma and adjacent non-tumor liver tissue. *J. Gastroenterol. Hepatol.* **2014**, *29*, 1822–1829. [CrossRef] [PubMed]

45. Iwatsuki, M.; Mimori, K.; Ishii, H.; Yokobori, T.; Takatsuno, Y.; Sato, T.; Toh, H.; Onoyama, I.; Nakayama, K.I.; Baba, H.; et al. Loss of FBXW7, a cell cycle regulating gene, in colorectal cancer: Clinical significance. *Int. J. Cancer* **2010**, *126*, 1828–1837. [CrossRef] [PubMed]

46. Yokobori, T.; Mimori, K.; Iwatsuki, M.; Ishii, H.; Tanaka, F.; Sato, T.; Toh, H.; Sudo, T.; Iwaya, T.; Tanaka, Y.; et al. Copy number loss of FBXW7 is related to gene expression and poor prognosis in esophageal squamous cell carcinoma. *Int. J. Oncol.* **2012**, *41*, 253–259. [CrossRef]

47. Wertz, I.E.; Kusam, S.; Lam, C.; Okamoto, T.; Sandoval, W.; Anderson, D.J.; Helgason, E.; Ernst, J.A.; Eby, M.; Liu, J.; et al. Sensitivity to antitubulin chemotherapeutics is regulated by MCL1 and FBW7. *Nature* **2011**, *471*, 110–114. [CrossRef]

48. Devine, T.; Dai, M.S. Targeting the ubiquitin-mediated proteasome degradation of p53 for cancer therapy. *Curr. Pharm. Des.* **2013**, *19*, 3248–3262. [CrossRef]

49. Wade, M.; Li, Y.C.; Wahl, G.M. MDM2, MDMX and p53 in oncogenesis and cancer therapy. *Nat. Rev. Cancer* **2013**, *13*, 83–96. [CrossRef]

50. Wang, Z.; Kang, W.; You, Y.; Pang, J.; Ren, H.; Suo, Z.; Liu, H.; Zheng, Y. USP7: Novel Drug Target in Cancer Therapy. *Front. Pharmacol.* **2019**, *10*. [CrossRef]

51. Shaikh, M.F.; Morano, W.F.; Lee, J.; Gleeson, E.; Babcock, B.D.; Michl, J.; Sarafraz-Yazdi, E.; Pincus, M.R.; Bowne, W.B. Emerging Role of MDM2 as Target for Anti-Cancer Therapy: A Review. *Ann. Clin. Lab. Sci.* **2016**, *46*, 627–634.

52. Alfarsi, L.H.; Ansari, R.E.; Craze, M.L.; Toss, M.S.; Masisi, B.; Ellis, I.O.; Rakha, E.A.; Green, A.R. CDC20 expression in oestrogen receptor positive breast cancer predicts poor prognosis and lack of response to endocrine therapy. *Breast Cancer Res. Treat.* **2019**, *178*, 535–544. [CrossRef]

53. Cheng, S.; Castillo, V.; Sliva, D. CDC20 associated with cancer metastasis and novel mushroom-derived CDC20 inhibitors with antimetastatic activity. *Int. J. Oncol.* **2019**, *54*, 2250–2256. [CrossRef] [PubMed]

54. Han, T.; Jiang, S.; Zheng, H.; Yin, Q.; Xie, M.; Little, M.R.; Yin, X.; Chen, M.; Song, S.J.; Beg, A.A.; et al. Interplay between c-Src and the APC/C co-activator Cdh1 regulates mammary tumorigenesis. *Nat. Commun.* **2019**, *10*, 3716. [CrossRef] [PubMed]

55. Schrock, M.S.; Stromberg, B.R.; Scarberry, L.; Summers, M.K. APC/C ubiquitin ligase: Functions and mechanisms in tumorigenesis. *Semin. Cancer Biol.* **2020**. [CrossRef] [PubMed]

56. Zhong, J.; Shaik, S.; Wan, L.; Tron, A.E.; Wang, Z.; Sun, L.; Inuzuka, H.; Wei, W. SCF β-TRCP targets MTSS1 for ubiquitination-mediated destruction to regulate cancer cell proliferation and migration. *Oncotarget* **2013**, *4*, 2339–2353. [CrossRef]

57. Ma, J.; Lu, Y.; Zhang, S.; Li, Y.; Huang, J.; Yin, Z.; Ren, J.; Huang, K.; Liu, L.; Yang, K.; et al. β-Trcp ubiquitin ligase and RSK2 kinase-mediated degradation of FOXN2 promotes tumorigenesis and radioresistance in lung cancer. *Cell Death Differ.* **2018**, *25*, 1473–1485. [CrossRef]

58. Shaik, S.; Nucera, C.; Inuzuka, H.; Gao, D.; Garnaas, M.; Frechette, G.; Harris, L.; Wan, L.; Fukushima, H.; Husain, A.; et al. SCF(β-TRCP) suppresses angiogenesis and thyroid cancer cell migration by promoting ubiquitination and destruction of VEGF receptor 2. *J. Exp. Med.* **2012**, *209*, 1289–1307. [CrossRef]

59. Li, C.; Du, L.; Ren, Y.; Liu, X.; Jiao, Q.; Cui, D.; Wen, M.; Wang, C.; Wei, G.; Wang, Y.; et al. SKP2 promotes breast cancer tumorigenesis and radiation tolerance through PDCD4 ubiquitination. *J. Exp. Clin. Cancer Res.* **2019**, *38*, 76. [CrossRef]

60. Ma, J.; Chang, K.; Peng, J.; Shi, Q.; Gan, H.; Gao, K.; Feng, K.; Xu, F.; Zhang, H.; Dai, B.; et al. SPOP promotes ATF2 ubiquitination and degradation to suppress prostate cancer progression. *J. Exp. Clin. Cancer Res.* **2018**, *37*, 145. [CrossRef]

61. Fong, K.-w.; Zhao, J.C.; Song, B.; Zheng, B.; Yu, J. TRIM28 protects TRIM24 from SPOP-mediated degradation and promotes prostate cancer progression. *Nat. Commun.* **2018**, *9*, 5007. [CrossRef] [PubMed]

62. Paul, P.J.; Raghu, D.; Chan, A.L.; Gulati, T.; Lambeth, L.; Takano, E.; Herold, M.J.; Hagekyriakou, J.; Vessella, R.L.; Fedele, C.; et al. Restoration of tumor suppression in prostate cancer by targeting the E3 ligase E6AP. *Oncogene* **2016**, *35*, 6235–6245. [CrossRef] [PubMed]

63. Raghu, D.; Paul, P.J.; Gulati, T.; Deb, S.; Khoo, C.; Russo, A.; Gallo, E.; Blandino, G.; Chan, A.-L.; Takano, E.; et al. E6AP promotes prostate cancer by reducing p27 expression. *Oncotarget* **2017**, *8*, 42939–42948. [CrossRef] [PubMed]

64. Gamell, C.; Bandilovska, I.; Gulati, T.; Kogan, A.; Lim, S.C.; Kovacevic, Z.; Takano, E.A.; Timpone, C.; Agupitan, A.D.; Litchfield, C.; et al. E6AP Promotes a Metastatic Phenotype in Prostate Cancer. *iScience* **2019**, *22*, 1–15. [CrossRef] [PubMed]

65. Venuto, S.; Merla, G. E3 Ubiquitin Ligase TRIM Proteins, Cell Cycle and Mitosis. *Cells* **2019**, *8*, 510. [CrossRef]

66. Zhu, L.; Qin, C.; Li, T.; Ma, X.; Qiu, Y.; Lin, Y.; Ma, D.; Qin, Z.; Sun, C.; Shen, X.; et al. The E3 ubiquitin ligase TRIM7 suppressed hepatocellular carcinoma progression by directly targeting Src protein. *Cell Death Differ.* **2019**. [CrossRef]

67. Jin, J.; Lu, Z.; Wang, X.; Liu, Y.; Han, T.; Wang, Y.; Wang, T.; Gan, M.; Xie, C.; Wang, J.; et al. E3 ubiquitin ligase TRIM7 negatively regulates NF-kappa B signaling pathway by degrading p65 in lung cancer. *Cell. Signal.* **2020**, *69*. [CrossRef]

68. Chakraborty, A.; Diefenbacher, M.E.; Mylona, A.; Kassel, O.; Behrens, A. The E3 ubiquitin ligase Trim7 mediates c-Jun/AP-1 activation by Ras signalling. *Nat. Commun.* **2015**, *6*, 6782. [CrossRef]

69. Yang, D.; Cheng, D.; Tu, Q.; Yang, H.; Sun, B.; Yan, L.; Dai, H.; Luo, J.; Mao, B.; Cao, Y.; et al. HUWE1 controls the development of non-small cell lung cancer through down-regulation of p53. *Theranostics* **2018**, *8*, 3517–3529. [CrossRef]

70. Gong, X.; Du, D.; Deng, Y.; Zhou, Y.; Sun, L.; Yuan, S. The structure and regulation of the E3 ubiquitin ligase HUWE1 and its biological functions in cancer. *Investig. New Drugs* **2020**, *38*, 515–524. [CrossRef]

71. Fujiwara, M.; Marusawa, H.; Wang, H.Q.; Iwai, A.; Ikeuchi, K.; Imai, Y.; Kataoka, A.; Nukina, N.; Takahashi, R.; Chiba, T. Parkin as a tumor suppressor gene for hepatocellular carcinoma. *Oncogene* **2008**, *27*, 6002–6011. [CrossRef] [PubMed]

72. Lin, D.C.; Xu, L.; Chen, Y.; Yan, H.; Hazawa, M.; Doan, N.; Said, J.W.; Ding, L.W.; Liu, L.Z.; Yang, H.; et al. Genomic and Functional Analysis of the E3 Ligase PARK2 in Glioma. *Cancer Res.* **2015**, *75*, 1815–1827. [CrossRef] [PubMed]

73. Bhattacharya, S.; Ghosh, M.K. Cell Death and Deubiquitinases: Perspectives in Cancer. *Biomed. Res. Int.* **2014**, *2014*, 435197. [CrossRef] [PubMed]

74. Chen, J.; Chen, X.; Xu, D.; Yang, L.; Yang, Z.; Yang, Q.; Yan, D.; Zhang, P.; Feng, D.; Liu, J. Autophagy Induced by Proteasomal DUB Inhibitor NiPT Restricts NiPT-Mediated Cancer Cell Death. *Front. Oncol.* **2020**, *10*, 348. [CrossRef] [PubMed]

75. Fraile, J.M.; Quesada, V.; Rodriguez, D.; Freije, J.M.; Lopez-Otin, C. Deubiquitinases in cancer: New functions and therapeutic options. *Oncogene* **2012**, *31*, 2373–2388. [CrossRef]

76. Akhavantabasi, S.; Akman, H.B.; Sapmaz, A.; Keller, J.; Petty, E.M.; Erson, A.E. USP32 is an active, membrane-bound ubiquitin protease overexpressed in breast cancers. *Mamm. Genome* **2010**, *21*, 388–397. [CrossRef]

77. Wei, R.; Liu, X.; Yu, W.; Yang, T.; Cai, W.; Liu, J.; Huang, X.; Xu, G.-t.; Zhao, S.; Yang, J.; et al. Deubiquitinases in cancer. *Oncotarget* **2015**, *6*, 12872–12889. [CrossRef]

78. Lee, J.E.; Park, C.M.; Kim, J.H. USP7 deubiquitinates and stabilizes EZH2 in prostate cancer cells. *Genet. Mol. Biol.* **2020**, *43*, e20190338. [CrossRef]

79. Tan, Y.; Zhou, G.; Wang, X.; Chen, W.; Gao, H. USP18 promotes breast cancer growth by upregulating EGFR and activating the AKT/Skp2 pathway. *Int. J. Oncol.* **2018**, *53*, 371–383. [CrossRef]

80. Pan, J.; Deng, Q.; Jiang, C.; Wang, X.; Niu, T.; Li, H.; Chen, T.; Jin, J.; Pan, W.; Cai, X.; et al. USP37 directly deubiquitinates and stabilizes c-Myc in lung cancer. *Oncogene* **2015**, *34*, 3957–3967. [CrossRef]

81. Diefenbacher, M.E.; Chakraborty, A.; Blake, S.M.; Mitter, R.; Popov, N.; Eilers, M.; Behrens, A. Usp28 counteracts Fbw7 in intestinal homeostasis and cancer. *Cancer Res.* **2015**, *75*, 1181–1186. [CrossRef] [PubMed]

82. Das, S.; Ramakrishna, S.; Kim, K.S. Critical Roles of Deubiquitinating Enzymes in the Nervous System and Neurodegenerative Disorders. *Mol. Cells* **2020**. [CrossRef]

83. Yi, L.; Cui, Y.; Xu, Q.; Jiang, Y. Stabilization of LSD1 by deubiquitinating enzyme USP7 promotes glioblastoma cell tumorigenesis and metastasis through suppression of the p53 signaling pathway. *Oncol. Rep.* **2016**, *36*, 2935–2945. [CrossRef] [PubMed]

84. Ding, K.; Ji, J.; Zhang, X.; Huang, B.; Chen, A.; Zhang, D.; Li, X.; Wang, X.; Wang, J. RNA splicing factor USP39 promotes glioma progression by inducing TAZ mRNA maturation. *Oncogene* **2019**, *38*, 6414–6428. [CrossRef] [PubMed]

85. Sha, B.; Chen, X.; Wu, H.; Li, M.; Shi, J.; Wang, L.; Liu, X.; Chen, P.; Hu, T.; Li, P. Deubiquitylatinase inhibitor b-AP15 induces c-Myc-Noxa-mediated apoptosis in esophageal squamous cell carcinoma. *Apoptosis* **2019**, *24*, 826–836. [CrossRef]

86. Narayanan, S.; Cai, C.Y.; Assaraf, Y.G.; Guo, H.Q.; Cui, Q.; Wei, L.; Huang, J.J.; Ashby, C.R., Jr.; Chen, Z.S. Targeting the ubiquitin-proteasome pathway to overcome anti-cancer drug resistance. *Drug Resist. Updat* **2020**, *48*, 100663. [CrossRef]

87. Fu, Y.; Ma, G.; Liu, G.; Li, B.; Li, H.; Hao, X.; Liu, L. USP14 as a novel prognostic marker promotes cisplatin resistance via Akt/ERK signaling pathways in gastric cancer. *Cancer Med.* **2018**, *7*, 5577–5588. [CrossRef]

88. Zhao, C.; Chen, X.; Zang, D.; Lan, X.; Liao, S.; Yang, C.; Zhang, P.; Wu, J.; Li, X.; Liu, N.; et al. A novel nickel complex works as a proteasomal deubiquitinase inhibitor for cancer therapy. *Oncogene* **2016**, *35*, 5916–5927. [CrossRef]

89. Jin, Y.; Cheng, H.; Cao, J.; Shen, W. MicroRNA 32 promotes cell proliferation, migration, and suppresses apoptosis in colon cancer cells by targeting OTU domain containing 3. *J. Cell Biochem.* **2019**, *120*, 18629–18639. [CrossRef]

90. Zhou, Y.; Wu, J.; Fu, X.; Du, W.; Zhou, L.; Meng, X.; Yu, H.; Lin, J.; Ye, W.; Liu, J.; et al. OTUB1 promotes metastasis and serves as a marker of poor prognosis in colorectal cancer. *Mol. Cancer* **2014**, *13*, 258. [CrossRef]

91. Zhang, P.; Li, C.; Li, H.; Yuan, L.; Dai, H.; Peng, Z.; Deng, Z.; Chang, Z.; Cui, C.P.; Zhang, L. Ubiquitin ligase CHIP regulates OTUD3 stability and suppresses tumour metastasis in lung cancer. *Cell Death Differ.* **2020**. [CrossRef]

92. Zhao, X.; Su, X.; Cao, L.; Xie, T.; Chen, Q.; Li, J.; Xu, R.; Jiang, C. OTUD4: A Potential Prognosis Biomarker for Multiple Human Cancers. *Cancer Manag. Res.* **2020**, *12*, 1503–1512. [CrossRef] [PubMed]

93. Qureshi, A.A.; Zuvanich, E.G.; Khan, D.A.; Mushtaq, S.; Silswal, N.; Qureshi, N. Proteasome inhibitors modulate anticancer and anti-proliferative properties via NF-kB signaling, and ubiquitin-proteasome pathways in cancer cell lines of different organs. *Lipids Health Dis.* **2018**, *17*, 62. [CrossRef]

94. Wang, Z.; Wan, L.; Zhong, J.; Inuzuka, H.; Liu, P.; Sarkar, F.H.; Wei, W. Cdc20: A potential novel therapeutic target for cancer treatment. *Curr. Pharm. Des.* **2013**, *19*, 3210–3214. [CrossRef] [PubMed]

95. Garcia-Higuera, I.; Manchado, E.; Dubus, P.; Canamero, M.; Mendez, J.; Moreno, S.; Malumbres, M. Genomic stability and tumour suppression by the APC/C cofactor Cdh1. *Nat. Cell Biol.* **2008**, *10*, 802–811. [CrossRef] [PubMed]

96. Lehman, N.L.; Tibshirani, R.; Hsu, J.Y.; Natkunam, Y.; Harris, B.T.; West, R.B.; Masek, M.A.; Montgomery, K.; van de Rijn, M.; Jackson, P.K. Oncogenic regulators and substrates of the anaphase promoting complex/cyclosome are frequently overexpressed in malignant tumors. *Am. J. Pathol.* **2007**, *170*, 1793–1805. [CrossRef]

97. Belaïdouni, N.; Peuchmaur, M.; Perret, C.; Florentin, A.; Benarous, R.; Besnard-Guérin, C. Overexpression of human βTrCP1 deleted of its F box induces tumorigenesis in transgenic mice. *Oncogene* **2005**, *24*, 2271–2276. [CrossRef]

98. Frescas, D.; Pagano, M. Deregulated proteolysis by the F-box proteins SKP2 and β-TrCP: Tipping the scales of cancer. *Nat. Rev. Cancer* **2008**, *8*, 438–449. [CrossRef]

99. Chan, C.H.; Morrow, J.K.; Li, C.F.; Gao, Y.; Jin, G.; Moten, A.; Stagg, L.J.; Ladbury, J.E.; Cai, Z.; Xu, D.; et al. Pharmacological inactivation of Skp2 SCF ubiquitin ligase restricts cancer stem cell traits and cancer progression. *Cell* **2013**, *154*, 556–568. [CrossRef]

100. Wang, G.; Chan, C.H.; Gao, Y.; Lin, H.K. Novel roles of Skp2 E3 ligase in cellular senescence, cancer progression, and metastasis. *Chin. J. Cancer* **2012**, *31*, 169–177. [CrossRef]

101. Geng, C.; He, B.; Xu, L.; Barbieri, C.E.; Eedunuri, V.K.; Chew, S.A.; Zimmermann, M.; Bond, R.; Shou, J.; Li, C.; et al. Prostate cancer-associated mutations in speckle-type POZ protein (SPOP) regulate steroid receptor coactivator 3 protein turnover. *Proc. Natl. Acad. Sci. USA* **2019**, *116*, 14386–14387. [CrossRef] [PubMed]

102. Geng, C.; Kaochar, S.; Li, M.; Rajapakshe, K.; Fiskus, W.; Dong, J.; Foley, C.; Dong, B.; Zhang, L.; Kwon, O.J.; et al. SPOP regulates prostate epithelial cell proliferation and promotes ubiquitination and turnover of c-MYC oncoprotein. *Oncogene* **2017**, *36*, 4767–4777. [CrossRef] [PubMed]

103. Gamell, C.; Gulati, T.; Levav-Cohen, Y.; Young, R.J.; Do, H.; Pilling, P.; Takano, E.; Watkins, N.; Fox, S.B.; Russell, P.; et al. Reduced abundance of the E3 ubiquitin ligase E6AP contributes to decreased expression of the INK4/ARF locus in non-small cell lung cancer. *Sci. Signal.* **2017**, *10*. [CrossRef]

104. Mortensen, F.; Schneider, D.; Barbic, T.; Sladewska-Marquardt, A.; Kuhnle, S.; Marx, A.; Scheffner, M. Role of ubiquitin and the HPV E6 oncoprotein in E6AP-mediated ubiquitination. *Proc. Natl. Acad. Sci. USA* **2015**, *112*, 9872–9877. [CrossRef]

105. Senft, D.; Qi, J.; Ronai, Z.A. Ubiquitin ligases in oncogenic transformation and cancer therapy. *Nat. Rev. Cancer* **2018**, *18*, 69–88. [CrossRef] [PubMed]

106. Adhikary, S.; Marinoni, F.; Hock, A.; Hulleman, E.; Popov, N.; Beier, R.; Bernard, S.; Quarto, M.; Capra, M.; Goettig, S.; et al. The ubiquitin ligase HectH9 regulates transcriptional activation by Myc and is essential for tumor cell proliferation. *Cell* **2005**, *123*, 409–421. [CrossRef]

107. Su, C.; Wang, T.; Zhao, J.; Cheng, J.; Hou, J. Meta-analysis of gene expression alterations and clinical significance of the HECT domain-containing ubiquitin ligase HUWE1 in cancer. *Oncol. Lett.* **2019**, *18*, 2292–2303. [CrossRef]

108. Xu, L.; Lin, D.C.; Yin, D.; Koeffler, H.P. An emerging role of PARK2 in cancer. *J. Mol. Med. (Berl)* **2014**, *92*, 31–42. [CrossRef]

109. Eichhorn, P.J.; Rodón, L.; Gonzàlez-Juncà, A.; Dirac, A.; Gili, M.; Martínez-Sáez, E.; Aura, C.; Barba, I.; Peg, V.; Prat, A.; et al. USP15 stabilizes TGF-β receptor I and promotes oncogenesis through the activation of TGF-β signaling in glioblastoma. *Nat. Med.* **2012**, *18*, 429–435. [CrossRef]

110. Qin, T.; Li, B.; Feng, X.; Fan, S.; Liu, L.; Liu, D.; Mao, J.; Lu, Y.; Yang, J.; Yu, X.; et al. Abnormally elevated USP37 expression in breast cancer stem cells regulates stemness, epithelial-mesenchymal transition and cisplatin sensitivity. *J. Exp. Clin. Cancer Res.* **2018**, *37*, 287. [CrossRef]

111. Du, T.; Li, H.; Fan, Y.; Yuan, L.; Guo, X.; Zhu, Q.; Yao, Y.; Li, X.; Liu, C.; Yu, X.; et al. The deubiquitylase OTUD3 stabilizes GRP78 and promotes lung tumorigenesis. *Nat. Commun.* **2019**, *10*, 2914. [CrossRef] [PubMed]

112. Serrano-Pozo, A.; Frosch, M.P.; Masliah, E.; Hyman, B.T. Neuropathological alterations in Alzheimer disease. *Cold Spring Harb. Perspect. Med.* **2011**, *1*, a006189. [CrossRef] [PubMed]

113. Dickson, D.W. Parkinson's disease and parkinsonism: Neuropathology. *Cold Spring Harb. Perspect Med.* **2012**, *2*. [CrossRef] [PubMed]

114. Stancu, I.C.; Ferraiolo, M.; Terwel, D.; Dewachter, I. Tau Interacting Proteins: Gaining Insight into the Roles of Tau in Health and Disease. *Adv. Exp. Med. Biol.* **2019**, *1184*, 145–166. [CrossRef] [PubMed]

115. Zheng, Q.; Huang, T.; Zhang, L.; Zhou, Y.; Luo, H.; Xu, H.; Wang, X. Dysregulation of Ubiquitin-Proteasome System in Neurodegenerative Diseases. *Front. Aging Neurosci.* **2016**, *8*, 303. [CrossRef] [PubMed]

116. Rosen, K.M.; Moussa, C.E.; Lee, H.K.; Kumar, P.; Kitada, T.; Qin, G.; Fu, Q.; Querfurth, H.W. Parkin reverses intracellular beta-amyloid accumulation and its negative effects on proteasome function. *J. Neurosci. Res.* **2010**, *88*, 167–178. [CrossRef]

117. Martinez, A.; Ramirez, J.; Osinalde, N.; Arizmendi, J.M.; Mayor, U. Neuronal proteomic analysis of the ubiquitinated substrates of the disease-linked E3 ligases parkin and Ube3a. *BioMed. Res. Int.* **2018**, *2018*, 1–14. [CrossRef]

118. Kaneko, M.; Koike, H.; Saito, R.; Kitamura, Y.; Okuma, Y.; Nomura, Y. Loss of HRD1-mediated protein degradation causes amyloid precursor protein accumulation and amyloid-beta generation. *J. Neurosci.* **2010**, *30*, 3924–3932. [CrossRef]

119. Saito, R.; Kaneko, M.; Kitamura, Y.; Takata, K.; Kawada, K.; Okuma, Y.; Nomura, Y. Effects of oxidative stress on the solubility of HRD1, a ubiquitin ligase implicated in Alzheimer's disease. *PLoS ONE* **2014**, *9*, e94576. [CrossRef]

120. Sahara, N.; Murayama, M.; Mizoroki, T.; Urushitani, M.; Imai, Y.; Takahashi, R.; Murata, S.; Tanaka, K.; Takashima, A. In vivo evidence of CHIP up-regulation attenuating tau aggregation. *J. Neurochem.* **2005**, *94*, 1254–1263. [CrossRef]

121. Shimura, H.; Schwartz, D.; Gygi, S.P.; Kosik, K.S. CHIP-Hsc70 complex ubiquitinates phosphorylated tau and enhances cell survival. *J. Biol. Chem.* **2004**, *279*, 4869–4876. [CrossRef] [PubMed]

122. Dickey, C.A.; Yue, M.; Lin, W.L.; Dickson, D.W.; Dunmore, J.H.; Lee, W.C.; Zehr, C.; West, G.; Cao, S.; Clark, A.M.; et al. Deletion of the ubiquitin ligase CHIP leads to the accumulation, but not the aggregation, of both endogenous phospho- and caspase-3-cleaved tau species. *J. Neurosci.* **2006**, *26*, 6985–6996. [CrossRef] [PubMed]

123. Evgen'ev, M.; Bobkova, N.; Krasnov, G.; Garbuz, D.; Funikov, S.; Kudryavtseva, A.; Kulikov, A.; Samokhin, A.; Maltsev, A.; Nesterova, I. The Effect of Human HSP70 Administration on a Mouse Model of Alzheimer's Disease Strongly Depends on Transgenicity and Age. *J. Alzheimer's Dis.* **2019**, *67*, 1391–1404. [CrossRef]

124. Lee, S.; Choi, B.R.; Kim, J.; LaFerla, F.M.; Park, J.H.Y.; Han, J.S.; Lee, K.W.; Kim, J. Sulforaphane Upregulates the Heat Shock Protein Co-Chaperone CHIP and Clears Amyloid-β and Tau in a Mouse Model of Alzheimer's Disease. *Mol. Nutr. Food Res.* **2018**, *62*. [CrossRef] [PubMed]

125. Zhang, M.; Cai, F.; Zhang, S.; Zhang, S.; Song, W. Overexpression of ubiquitin carboxyl-terminal hydrolase L1 (UCHL1) delays Alzheimer's progression in vivo. *Sci. Rep.* **2014**, *4*, 7298. [CrossRef] [PubMed]

126. Kang, S.-J.; Kim, J.S.; Park, S.M. Ubiquitin C-terminal Hydrolase L1 Regulates Lipid Raft-dependent Endocytosis. *Exp. Neurobiol.* **2018**, *27*, 377–386. [CrossRef] [PubMed]

127. Kumari, R.; Kumar, R.; Kumar, S.; Singh, A.K.; Hanpude, P.; Jangir, D.; Maiti, T.K. Amyloid aggregates of the deubiquitinase OTUB1 are neurotoxic, suggesting that they contribute to the development of Parkinson's disease. *J. Biol. Chem.* **2020**, *295*, 3466–3484. [CrossRef]

128. Ortuno, D.; Carlisle, H.J.; Miller, S. Does inactivation of USP14 enhance degradation of proteasomal substrates that are associated with neurodegenerative diseases? *F1000Res* **2016**, *5*, 137. [CrossRef]

129. Zeng, X.S.; Geng, W.S.; Jia, J.J.; Chen, L.; Zhang, P.P. Cellular and Molecular Basis of Neurodegeneration in Parkinson Disease. *Front. Aging Neurosci.* **2018**, *10*, 109. [CrossRef]

130. Kondapalli, C.; Kazlauskaite, A.; Zhang, N.; Woodroof, H.I.; Campbell, D.G.; Gourlay, R.; Burchell, L.; Walden, H.; Macartney, T.J.; Deak, M.; et al. PINK1 is activated by mitochondrial membrane potential depolarization and stimulates Parkin E3 ligase activity by phosphorylating Serine 65. *Open Biol.* **2012**, *2*, 120080. [CrossRef]

131. Narendra, D.; Tanaka, A.; Suen, D.F.; Youle, R.J. Parkin is recruited selectively to impaired mitochondria and promotes their autophagy. *J. Cell Biol.* **2008**, *183*, 795–803. [CrossRef] [PubMed]

132. Chung, S.Y.; Kishinevsky, S.; Mazzulli, J.R.; Graziotto, J.; Mrejeru, A.; Mosharov, E.V.; Puspita, L.; Valiulahi, P.; Sulzer, D.; Milner, T.A.; et al. Parkin and PINK1 Patient iPSC-Derived Midbrain Dopamine Neurons Exhibit Mitochondrial Dysfunction and alpha-Synuclein Accumulation. *Stem Cell Rep.* **2016**, *7*, 664–677. [CrossRef] [PubMed]

133. Chung, J.Y.; Park, H.R.; Lee, S.J.; Lee, S.H.; Kim, J.S.; Jung, Y.S.; Hwang, S.H.; Ha, N.C.; Seol, W.G.; Lee, J.; et al. Elevated TRAF2/6 expression in Parkinson's disease is caused by the loss of Parkin E3 ligase activity. *Lab. Invest.* **2013**, *93*, 663–676. [CrossRef] [PubMed]

134. Murata, H.; Sakaguchi, M.; Kataoka, K.; Huh, N.H. SARM1 and TRAF6 bind to and stabilize PINK1 on depolarized mitochondria. *Mol. Biol. Cell* **2013**, *24*, 2772–2784. [CrossRef] [PubMed]

135. Matheoud, D.; Sugiura, A.; Bellemare-Pelletier, A.; Laplante, A.; Rondeau, C.; Chemali, M.; Fazel, A.; Bergeron, J.J.; Trudeau, L.E.; Burelle, Y.; et al. Parkinson's Disease-Related Proteins PINK1 and Parkin Repress Mitochondrial Antigen Presentation. *Cell* **2016**, *166*, 314–327. [CrossRef] [PubMed]

136. Matheoud, D.; Cannon, T.; Voisin, A.; Penttinen, A.M.; Ramet, L.; Fahmy, A.M.; Ducrot, C.; Laplante, A.; Bourque, M.J.; Zhu, L.; et al. Intestinal infection triggers Parkinson's disease-like symptoms in Pink1 −/− mice. *Nature* **2019**, *571*, 565–569. [CrossRef]

137. Mulherkar, S.A.; Sharma, J.; Jana, N.R. The ubiquitin ligase E6-AP promotes degradation of alpha-synuclein. *J. Neurochem.* **2009**, *110*, 1955–1964. [CrossRef]

138. Liu, X.; Hebron, M.; Shi, W.; Lonskaya, I.; Moussa, C.E.-H. Ubiquitin specific protease-13 independently regulates parkin ubiquitination and alpha-synuclein clearance in alpha-synucleinopathies. *Hum. Mol. Genet.* **2018**, *28*, 548–560. [CrossRef]

139. Bishop, P.; Rubin, P.; Thomson, A.R.; Rocca, D.; Henley, J.M. The ubiquitin C-terminal hydrolase L1 (UCH-L1) C terminus plays a key role in protein stability, but its farnesylation is not required for membrane association in primary neurons. *J. Biol. Chem.* **2014**, *289*, 36140–36149. [CrossRef]

140. Cai, C.-Z.; Zhou, H.-F.; Yuan, N.-N.; Wu, M.-Y.; Lee, S.M.-Y.; Ren, J.-Y.; Su, H.-X.; Lu, J.-J.; Chen, X.-P.; Li, M.; et al. Natural alkaloid harmine promotes degradation of alpha-synuclein via PKA-mediated ubiquitin-proteasome system activation. *Phytomedicine* **2019**, *61*, 152842. [CrossRef]

141. Myöhänen, T.T.; Norrbacka, S.; Savolainen, M.H. Prolyl oligopeptidase inhibition attenuates the toxicity of a proteasomal inhibitor, lactacystin, in the alpha-synuclein overexpressing cell culture. *Neurosci. Lett.* **2017**, *636*, 83–89. [CrossRef] [PubMed]

142. Ortega, Z.; Lucas, J.J. Ubiquitin-proteasome system involvement in Huntington's disease. *Front. Mol. Neurosci.* **2014**, *7*, 77. [CrossRef]

143. Miller, V.M.; Nelson, R.F.; Gouvion, C.M.; Williams, A.; Rodriguez-Lebron, E.; Harper, S.Q.; Davidson, B.L.; Rebagliati, M.R.; Paulson, H.L. CHIP suppresses polyglutamine aggregation and toxicity in vitro and in vivo. *J. Neurosci.* **2005**, *25*, 9152–9161. [CrossRef] [PubMed]

144. Bhat, K.P.; Yan, S.; Wang, C.E.; Li, S.; Li, X.J. Differential ubiquitination and degradation of huntingtin fragments modulated by ubiquitin-protein ligase E3A. *Proc. Natl. Acad. Sci. USA* **2014**, *111*, 5706–5711. [CrossRef] [PubMed]

145. Lin, L.; Jin, Z.; Tan, H.; Xu, Q.; Peng, T.; Li, H. Atypical ubiquitination by E3 ligase WWP1 inhibits the proteasome-mediated degradation of mutant huntingtin. *Brain Res.* **2016**, *1643*, 103–112. [CrossRef] [PubMed]

146. Rubio, I.; Rodríguez, J.; Tomas-Zapico, C.; Ruiz, C.; Casarejos, M.; Perucho, J.; Gómez, A.; Rodal, I.; Lucas, J.; Mena, M.; et al. Effects of partial suppression of parkin on huntingtin mutant R6/1 mice. *Brain Res.* **2009**, *1281*, 91–100. [CrossRef] [PubMed]

147. Yang, H.; Zhong, X.; Ballar, P.; Luo, S.; Shen, Y.; Rubinsztein, D.C.; Monteiro, M.J.; Fang, S. Ubiquitin ligase Hrd1 enhances the degradation and suppresses the toxicity of polyglutamine-expanded huntingtin. *Exp. Cell Res.* **2007**, *313*, 538–550. [CrossRef] [PubMed]

148. Rotblat, B.; Southwell, A.L.; Ehrnhoefer, D.E.; Skotte, N.H.; Metzler, M.; Franciosi, S.; Leprivier, G.; Somasekharan, S.P.; Barokas, A.; Deng, Y.; et al. HACE1 reduces oxidative stress and mutant Huntingtin toxicity by promoting the NRF2 response. *Proc. Natl. Acad. Sci. USA* **2014**, *111*, 3032–3037. [CrossRef]

149. Tanji, K.; Mori, F.; Miki, Y.; Utsumi, J.; Sasaki, H.; Kakita, A.; Takahashi, H.; Wakabayashi, K. YOD1 attenuates neurogenic proteotoxicity through its deubiquitinating activity. *Neurobiol. Dis.* **2018**, *112*, 14–23. [CrossRef]

150. Lee, B.H.; Lee, M.J.; Park, S.; Oh, D.C.; Elsasser, S.; Chen, P.C.; Gartner, C.; Dimova, N.; Hanna, J.; Gygi, S.P.; et al. Enhancement of proteasome activity by a small-molecule inhibitor of USP14. *Nature* **2010**, *467*, 179–184. [CrossRef]

151. Hyrskyluoto, A.; Bruelle, C.; Lundh, S.H.; Do, H.T.; Kivinen, J.; Rappou, E.; Reijonen, S.; Waltimo, T.; Petersén, Å.; Lindholm, D.; et al. Ubiquitin-specific protease-14 reduces cellular aggregates and protects against mutant huntingtin-induced cell degeneration: Involvement of the proteasome and ER stress-activated kinase IRE1α. *Hum. Mol. Genet.* **2014**, *23*, 5928–5939. [CrossRef] [PubMed]

152. Huang, Z.N.; Her, L.S. The Ubiquitin Receptor ADRM1 Modulates HAP40-Induced Proteasome Activity. *Mol. Neurobiol.* **2017**, *54*, 7382–7400. [CrossRef] [PubMed]

153. Diaz-Hernandez, M.; Valera, A.G.; Moran, M.A.; Gomez-Ramos, P.; Alvarez-Castelao, B.; Castano, J.G.; Hernandez, F.; Lucas, J.J. Inhibition of 26S proteasome activity by huntingtin filaments but not inclusion bodies isolated from mouse and human brain. *J. Neurochem.* **2006**, *98*, 1585–1596. [CrossRef] [PubMed]

154. Schipper-Krom, S.; Juenemann, K.; Jansen, A.H.; Wiemhoefer, A.; van den Nieuwendijk, R.; Smith, D.L.; Hink, M.A.; Bates, G.P.; Overkleeft, H.; Ovaa, H.; et al. Dynamic recruitment of active proteasomes into polyglutamine initiated inclusion bodies. *FEBS Lett.* **2014**, *588*, 151–159. [CrossRef] [PubMed]

155. Aki, D.; Li, Q.; Li, H.; Liu, Y.C.; Lee, J.H. Immune regulation by protein ubiquitination: Roles of the E3 ligases VHL and Itch. *Protein Cell* **2019**, *10*, 395–404. [CrossRef]

156. Wang, A.; Zhu, F.; Liang, R.; Li, D.; Li, B. Regulation of T cell differentiation and function by ubiquitin-specific proteases. *Cell Immunol.* **2019**, *340*, 103922. [CrossRef]

157. Wu, Y.; Kang, J.; Zhang, L.; Liang, Z.; Tang, X.; Yan, Y.; Qian, H.; Zhang, X.; Xu, W.; Mao, F. Ubiquitination regulation of inflammatory responses through NF-κB pathway. *Am. J. Transl. Res.* **2018**, *10*, 881–891.

158. Thien, C.B.; Langdon, W.Y. c-Cbl and Cbl-b ubiquitin ligases: Substrate diversity and the negative regulation of signalling responses. *Biochem. J.* **2005**, *391*, 153–166. [CrossRef]

159. Dou, H.; Buetow, L.; Hock, A.; Sibbet, G.J.; Vousden, K.H.; Huang, D.T. Structural basis for autoinhibition and phosphorylation-dependent activation of c-Cbl. *Nat. Struct. Mol. Biol.* **2012**, *19*, 184–192. [CrossRef]

160. Huang, F.; Kitaura, Y.; Jang, I.; Naramura, M.; Kole, H.H.; Liu, L.; Qin, H.; Schlissel, M.S.; Gu, H. Establishment of the major compatibility complex-dependent development of CD4+ and CD8+ T cells by the Cbl family proteins. *Immunity* **2006**, *25*, 571–581. [CrossRef]

161. Kitaura, Y.; Jang, I.K.; Wang, Y.; Han, Y.C.; Inazu, T.; Cadera, E.J.; Schlissel, M.; Hardy, R.R.; Gu, H. Control of the B cell-intrinsic tolerance programs by ubiquitin ligases Cbl and Cbl-b. *Immunity* **2007**, *26*, 567–578. [CrossRef] [PubMed]

162. Lyle, C.; Richards, S.; Yasuda, K.; Napoleon, M.A.; Walker, J.; Arinze, N.; Belghasem, M.; Vellard, I.; Yin, W.; Ravid, J.D.; et al. c-Cbl targets PD-1 in immune cells for proteasomal degradation and modulates colorectal tumor growth. *Sci. Rep.* **2019**, *9*, 20257. [CrossRef] [PubMed]

163. Abe, T.; Hirasaka, K.; Kohno, S.; Ochi, A.; Yamagishi, N.; Ohno, A.; Teshima-Kondo, S.; Nikawa, T. Ubiquitin ligase Cbl-b and obesity-induced insulin resistance. *Endocr. J.* **2014**, *61*, 529–539. [CrossRef] [PubMed]

164. Abe, T.; Hirasaka, K.; Nikawa, T. Involvement of Cbl-b-mediated macrophage inactivation in insulin resistance. *World J. Diabetes* **2017**, *8*, 97. [CrossRef]

165. Huang, H.; Jeon, M.S.; Liao, L.; Yang, C.; Elly, C.; Yates, J.R., III; Liu, Y.C. K33-linked polyubiquitination of T cell receptor-zeta regulates proteolysis-independent T cell signaling. *Immunity* **2010**, *33*, 60–70. [CrossRef]

166. O'Connor, H.F.; Lyon, N.; Leung, J.W.; Agarwal, P.; Swaim, C.D.; Miller, K.M.; Huibregtse, J.M. Ubiquitin-Activated Interaction Traps (UBAITs) identify E3 ligase binding partners. *EMBO Rep.* **2015**, *16*, 1699–1712. [CrossRef]

167. Maxwell, P.H.; Wiesener, M.S.; Chang, G.W.; Clifford, S.C.; Vaux, E.C.; Cockman, M.E.; Wykoff, C.C.; Pugh, C.W.; Maher, E.R.; Ratcliffe, P.J. The tumour suppressor protein VHL targets hypoxia-inducible factors for oxygen-dependent proteolysis. *Nature* **1999**, *399*, 271–275. [CrossRef]

168. Dang, E.V.; Barbi, J.; Yang, H.Y.; Jinasena, D.; Yu, H.; Zheng, Y.; Bordman, Z.; Fu, J.; Kim, Y.; Yen, H.R.; et al. Control of T(H)17/T(reg) balance by hypoxia-inducible factor 1. *Cell* **2011**, *146*, 772–784. [CrossRef]

169. Lee, J.H.; Elly, C.; Park, Y.; Liu, Y.C. E3 Ubiquitin Ligase VHL Regulates Hypoxia-Inducible Factor-1α to Maintain Regulatory T Cell Stability and Suppressive Capacity. *Immunity* **2015**, *42*, 1062–1074. [CrossRef]

170. Izquierdo, H.M.; Brandi, P.; Gómez, M.J.; Conde-Garrosa, R.; Priego, E.; Enamorado, M.; Martínez-Cano, S.; Sánchez, I.; Conejero, L.; Jimenez-Carretero, D.; et al. Von Hippel-Lindau Protein Is Required for Optimal Alveolar Macrophage Terminal Differentiation, Self-Renewal, and Function. *Cell Rep.* **2018**, *24*, 1738–1746. [CrossRef]

171. Palazon, A.; Goldrath, A.W.; Nizet, V.; Johnson, R.S. HIF transcription factors, inflammation, and immunity. *Immunity* **2014**, *41*, 518–528. [CrossRef] [PubMed]

172. Labrousse-Arias, D.; Martínez-Alonso, E.; Corral-Escariz, M.; Bienes-Martínez, R.; Berridy, J.; Serrano-Oviedo, L.; Conde, E.; García-Bermejo, M.L.; Giménez-Bachs, J.M.; Salinas-Sánchez, A.S.; et al. VHL promotes immune response against renal cell carcinoma via NF-κB-dependent regulation of VCAM-1. *J. Cell Biol.* **2017**, *216*, 835–847. [CrossRef] [PubMed]

173. Trotta, A.M.; Santagata, S.; Zanotta, S.; D'Alterio, C.; Napolitano, M.; Rea, G.; Camerlingo, R.; Esposito, F.; Lamantia, E.; Anniciello, A.; et al. Mutated von Hippel-Lindau-renal cell carcinoma (RCC) promotes patients specific natural killer (NK) cytotoxicity. *J. Exp. Clin. Cancer Res.* **2018**, *37*, 297. [CrossRef] [PubMed]

174. Quesada, V.; Diaz-Perales, A.; Gutierrez-Fernandez, A.; Garabaya, C.; Cal, S.; Lopez-Otin, C. Cloning and enzymatic analysis of 22 novel human ubiquitin-specific proteases. *Biochem. Biophys. Res. Commun.* **2004**, *314*, 54–62. [CrossRef] [PubMed]

175. Yang, J.; Xu, P.; Han, L.; Guo, Z.; Wang, X.; Chen, Z.; Nie, J.; Yin, S.; Piccioni, M.; Tsun, A.; et al. Cutting Edge: Ubiquitin-Specific Protease 4 Promotes Th17 Cell Function under Inflammation by Deubiquitinating and Stabilizing RORγt. *J. Immunol.* **2015**, *194*, 4094–4097. [CrossRef] [PubMed]

176. Dufner, A.; Kisser, A.; Niendorf, S.; Basters, A.; Reissig, S.; Schonle, A.; Aichem, A.; Kurz, T.; Schlosser, A.; Yablonski, D.; et al. The ubiquitin-specific protease USP8 is critical for the development and homeostasis of T cells. *Nat. Immunol.* **2015**, *16*, 950–960. [CrossRef]

177. Huang, C.; Shi, Y.; Zhao, Y. USP8 mutation in Cushing's disease. *Oncotarget* **2015**, *6*, 18240–18241. [CrossRef]

178. Park, Y.; Jin, H.S.; Liu, Y.C. Regulation of T cell function by the ubiquitin-specific protease USP9X via modulating the Carma1-Bcl10-Malt1 complex. *Proc. Natl. Acad. Sci. USA* **2013**, *110*, 9433–9438. [CrossRef]

179. Jahan, A.S.; Lestra, M.; Swee, L.K.; Fan, Y.; Lamers, M.M.; Tafesse, F.G.; Theile, C.S.; Spooner, E.; Bruzzone, R.; Ploegh, H.L.; et al. Usp12 stabilizes the T-cell receptor complex at the cell surface during signaling. *Proc. Natl. Acad. Sci. USA* **2016**, *113*, E705–E714. [CrossRef]

180. Wu, X.; Lei, C.; Xia, T.; Zhong, X.; Yang, Q.; Shu, H.B. Regulation of TRIF-mediated innate immune response by K27-linked polyubiquitination and deubiquitination. *Nat. Commun.* **2019**, *10*, 4115. [CrossRef]

181. Bignell, G.R.; Warren, W.; Seal, S.; Takahashi, M.; Rapley, E.; Barfoot, R.; Green, H.; Brown, C.; Biggs, P.J.; Lakhani, S.R.; et al. Identification of the familial cylindromatosis tumour-suppressor gene. *Nat. Genet.* **2000**, *25*, 160–165. [CrossRef] [PubMed]

182. Tsagaratou, A.; Trompouki, E.; Grammenoudi, S.; Kontoyiannis, D.L.; Mosialos, G. Thymocyte-specific truncation of the deubiquitinating domain of CYLD impairs positive selection in a NF-kappaB essential modulator-dependent manner. *J. Immunol.* **2010**, *185*, 2032–2043. [CrossRef] [PubMed]

183. Zhang, J.; Stirling, B.; Temmerman, S.T.; Ma, C.A.; Fuss, I.J.; Derry, J.M.; Jain, A. Impaired regulation of NF-kappaB and increased susceptibility to colitis-associated tumorigenesis in CYLD-deficient mice. *J. Clin. Investig.* **2006**, *116*, 3042–3049. [CrossRef] [PubMed]

184. Reiley, W.W.; Jin, W.; Lee, A.J.; Wright, A.; Wu, X.; Tewalt, E.F.; Leonard, T.O.; Norbury, C.C.; Fitzpatrick, L.; Zhang, M.; et al. Deubiquitinating enzyme CYLD negatively regulates the ubiquitin-dependent kinase Tak1 and prevents abnormal T cell responses. *J. Exp. Med.* **2007**, *204*, 1475–1485. [CrossRef]

185. Lork, M.; Verhelst, K.; Beyaert, R. CYLD, A20 and OTULIN deubiquitinases in NF-κB signaling and cell death: So similar, yet so different. *Cell Death Differ.* **2017**, *24*, 1172–1183. [CrossRef]

186. Jin, Y.J.; Wang, S.; Cho, J.; Selim, M.A.; Wright, T.; Mosialos, G.; Zhang, J.Y. Epidermal CYLD inactivation sensitizes mice to the development of sebaceous and basaloid skin tumors. *JCI Insight* **2016**, *1*. [CrossRef]

187. Nikolaou, K.; Tsagaratou, A.; Eftychi, C.; Kollias, G.; Mosialos, G.; Talianidis, I. Inactivation of the Deubiquitinase CYLD in Hepatocytes Causes Apoptosis, Inflammation, Fibrosis, and Cancer. *Cancer Cell* **2012**, *21*, 738–750. [CrossRef]
188. Cleynen, I.; Vazeille, E.; Artieda, M.; Verspaget, H.W.; Szczypiorska, M.; Bringer, M.A.; Lakatos, P.L.; Seibold, F.; Parnell, K.; Weersma, R.K.; et al. Genetic and microbial factors modulating the ubiquitin proteasome system in inflammatory bowel disease. *Gut* **2014**, *63*, 1265–1274. [CrossRef]
189. Cascio, P.; Hilton, C.; Kisselev, A.F.; Rock, K.L.; Goldberg, A.L. 26S proteasomes and immunoproteasomes produce mainly N-extended versions of an antigenic peptide. *EMBO J.* **2001**, *20*, 2357–2366. [CrossRef]
190. Lin, H.; Li, S.; Shu, H.-B. The Membrane-Associated MARCH E3 Ligase Family: Emerging Roles in Immune Regulation. *Front. Immunol.* **2019**, *10*. [CrossRef]
191. Egerer, K.; Kuckelkorn, U.; Rudolph, P.E.; Rückert, J.C.; Dörner, T.; Burmester, G.R.; Kloetzel, P.M.; Feist, E. Circulating proteasomes are markers of cell damage and immunologic activity in autoimmune diseases. *J. Rheumatol.* **2002**, *29*, 2045–2052. [PubMed]
192. Xiao, F.; Lin, X.; Tian, J.; Wang, X.; Chen, Q.; Rui, K.; Ma, J.; Wang, S.; Wang, Q.; Wang, X.; et al. Proteasome inhibition suppresses Th17 cell generation and ameliorates autoimmune development in experimental Sjögren's syndrome. *Cell Mol. Immunol.* **2017**, *14*, 924–934. [CrossRef] [PubMed]
193. Landré, V.; Rotblat, B.; Melino, S.; Bernassola, F.; Melino, G. Screening for E3-ubiquitin ligase inhibitors: Challenges and opportunities. *Oncotarget* **2014**, *5*, 7988–8013. [CrossRef] [PubMed]
194. Harrigan, J.A.; Jacq, X.; Martin, N.M.; Jackson, S.P. Deubiquitylating enzymes and drug discovery: Emerging opportunities. *Nat. Rev. Drug Discov.* **2018**, *17*, 57–78. [CrossRef] [PubMed]
195. Thibaudeau, T.A.; Smith, D.M. A Practical Review of Proteasome Pharmacology. *Pharmacol. Rev.* **2019**, *71*, 170–197. [CrossRef]

International Journal of
Molecular Sciences

Review

Targeting E3 Ubiquitin Ligases and Deubiquitinases in Ciliopathy and Cancer

Takashi Shiromizu [1], Mizuki Yuge [1], Kousuke Kasahara [2], Daishi Yamakawa [2], Takaaki Matsui [3], Yasumasa Bessho [3], Masaki Inagaki [2] and Yuhei Nishimura [1,*]

[1] Department of Integrative Pharmacology, Graduate School of Medicine, Mie University, Tsu, Mie 514-8507, Japan; tshiromizu@doc.medic.mie-u.ac.jp (T.S.); 320d030@m.mie-u.ac.jp (M.Y.)

[2] Department of Physiology, Graduate School of Medicine, Mie University, Tsu, Mie 514-5807, Japan; kkasahara@doc.medic.mie-u.ac.jp (K.K.); dyama@doc.medic.mie-u.ac.jp (D.Y.); minagaki@doc.medic.mie-u.ac.jp (M.I.)

[3] Gene Regulation Research, Division of Biological Sciences, Nara Institute of Science and Technology, Takayama, Nara 630-0192, Japan; matsui@bs.naist.jp (T.M.); ybessho@bs.naist.jp (Y.B.)

* Correspondence: yuhei@doc.medic.mie-u.ac.jp; Tel.: +81-(59)-231-5006; Fax: +81-(59)-232-1765

Received: 31 July 2020; Accepted: 17 August 2020; Published: 19 August 2020

Abstract: Cilia are antenna-like structures present in many vertebrate cells. These organelles detect extracellular cues, transduce signals into the cell, and play an essential role in ensuring correct cell proliferation, migration, and differentiation in a spatiotemporal manner. Not surprisingly, dysregulation of cilia can cause various diseases, including cancer and ciliopathies, which are complex disorders caused by mutations in genes regulating ciliary function. The structure and function of cilia are dynamically regulated through various mechanisms, among which E3 ubiquitin ligases and deubiquitinases play crucial roles. These enzymes regulate the degradation and stabilization of ciliary proteins through the ubiquitin–proteasome system. In this review, we briefly highlight the role of cilia in ciliopathy and cancer; describe the roles of E3 ubiquitin ligases and deubiquitinases in ciliogenesis, ciliopathy, and cancer; and highlight some of the E3 ubiquitin ligases and deubiquitinases that are potential therapeutic targets for these disorders.

Keywords: ubiquitin–proteasome pathway; cilia; ciliogenesis; differentiation; proliferation; ciliopathy; cancer

---

1. Introduction

Cilia are antenna-like structures that are present in a variety of vertebrate cells [1–6]. There are two broad classes of cilia: Motile and nonmotile cilia [1]. The nonmotile cilia are called primary cilia. Both motile and primary cilia contain receptors and channels that detect signals from extracellular cues, such as mechanical flow and chemical stimulation, and transduce them into the cell, where they contribute to the maintenance of proper development and homeostasis. Considering these functions, it is not surprising that dysregulation of cilia function can cause cancer and other diseases, including ciliopathies, which manifest as various disease phenotypes, such as congenital anomalies, neurodevelopmental disorders, and obesity [1,6–9].

The structure and function of cilia are dynamically and precisely regulated, enabling cells to proliferate, migrate, and differentiate in a spatiotemporally controlled manner [6,10]. The primary cilium is composed of three compartments: The basal body, the transition zone, and the axoneme [2]. The basal body is derived from the mother centriole. Both centrioles and basal body contain nine circularly arranged triplets of microtubules (A-, B-, and C-tubules). The axoneme consists of nine microtubule doublets projected from the A- and B-tubules of the basal body. A central pair of singlet microtubules is present (+2) and absent (+0) in motile and primary cilia, respectively [11–14].

Therefore, the axoneme of motile and primary cilia is described as $9 \times 2 + 2$ and $9 \times 2 + 0$, respectively. The transition zone is a short area located above the basal body characterized by Y-shaped fibers connecting the microtubule doublets to the ciliary membrane [15]. The structure of motile cilia is more complex [16]. The daughter centriole plays an important role in the formation of motile cilia [16,17].

Primary cilia are disassembled and assembled when cells enter mitosis and exit the cell cycle, respectively [18–20]. The formation of primary cilia starts with the binding of small cytoplasmic vesicles transported from the Golgi apparatus to the mother centriole and conversion from the mother centriole to the basal body. The basal body is then moved and anchored to the plasma membrane. Coiled-coil protein 110, a component of the inhibitory complex of ciliogenesis, is removed to initiate axoneme elongation [21,22]. The ciliary vesicle then fuses with the plasma membrane, and large amounts of tubulin are transported from the cytoplasm into the cilium to extend the axoneme [23]. Many signaling molecules are also transported from the cytoplasm into the cilium (anterograde) and from the cilium into the cytoplasm (retrograde) by kinesin and dynein, respectively, which are motor proteins that travel along the axoneme [24].

Various types of posttranslational modification, including phosphorylation, acetylation, and ubiquitination, are involved in the dynamic regulation of the structure and function of cilia [2,4–6,25]. Modification of proteins by attachment of ubiquitin, a highly conserved 76-amino-acid protein, is a critical step in targeting the selective degradation of proteins by proteasomes as part of the ubiquitin–proteasome system (UPS) [26]. Protein ubiquitination occurs in three steps. First, ubiquitin-activating enzymes (E1) bind to ubiquitin, which is expressed in all cell types; second, ubiquitin is transferred from E1 enzymes to ubiquitin-conjugating enzymes (E2); and finally, ubiquitin-ligating enzymes (E3) transfer the ubiquitin from E2 enzymes and ligate it to lysine residues on the target protein. To date, 2, approximately 40, and about 600 E1, E2, and E3 enzymes, respectively, have been identified in humans [26]. The selectivity of target protein ubiquitination is conferred by the combination of E2 and E3 enzymes. Protein ubiquitination is counteracted by deubiquitinase (DUB)-mediated removal of ubiquitin moieties from ubiquitinated proteins [27]. About 100 DUBs have been identified in humans. The balance between ubiquitination and deubiquitination of target proteins and their proteasomal degradation are tightly regulated processes, and dysregulation of the UPS has been detected in various disorders [28–30].

Several lines of evidence support a major role for the UPS in regulating the structure and function of cilia [4–6,31–35]. Here, we briefly review the role of cilia in phenotypes of ciliopathy and cancer. We then focus on the role of E3 ubiquitin ligases and DUBs in ciliogenesis, ciliopathy, and cancer, and suggest that these enzymes may serve as novel therapeutic targets for the development of treatments for these disorders.

2. Roles of Cilia in Ciliopathy and Cancer

Cilia play crucial roles in the development of vertebrates. In some cell types, cilia are present only transiently during a critical point in development [36], and the spatiotemporal dysregulation of cilia can therefore affect the development of many organ systems, including the central nervous, sensory, cardiovascular, digestive, metabolic, and skeletal systems [8,14,37,38]. These complex multisystem developmental disorders are collectively termed ciliopathies (Table 1).

Table 1. The roles of cilia in ciliopathy phenotypes.

Ciliopathy Phenotype	Role of Cilia in the Phenotype	References
Intellectual disability	Dysfunction of cilia in radial glial progenitors impairs the proliferation, migration, and differentiation, resulting in the disruption of cerebral cortical development and intellectual disability.	[39,40]
Retinal degeneration	Mutation of genes related to the structure and function of axoneme in photoreceptor cells impair protein (e.g., rhodopsin) transport along the axoneme, resulting in retinal degeneration	[41,42]
Craniofacial malformation	Dysfunction of cilia in cranial neural crest cells impairs the epithelial-mesenchymal transition and the formation of facial prominences, causing craniofacial malformation such as cleft lip/palate	[43,44]
Laterality disorders	Dysfunction of cilia in ventral node fails to break left-right symmetry, left or right-side morphogenesis, causing laterality disorders, such as situs inversus and heterotaxy.	[45,46]
Cystic kidney disease	Dysfunction of cilia in renal tubular cells fails to detect fluid flow, increase Ca^{2+} concentration, and suppress protein kinase A, causing renal cystogenesis through dysregulated proliferation, apoptosis, and cell polarity.	[47,48]
Obesity	Dysfunction of cilia in hypothalamic neurons and adipocyte progenitor cess fails to suppress appetite and regulate appropriate differentiation to adipocytes, respectively, causing obesity.	[49,50]
Scoliosis	Primary cilia of osteoblasts are abnormally elongated and dysfunctional in mechanotransduction, which may impair loading-induced bone adaptation and cause scoliosis	[51,52]
Respiratory distress	Mutations of genes affecting dynein arm, radial spoke, central apparatus or multiciliation impair the structure and/or function of motile cilia of epithelial cells lining most of the respiratory tract, resulting in mucus obstruction and respiratory failure.	[53,54]
Infertility	Impairment of sperm tail, which has microtubule arrangement similar to that of motile cilia, cause sperm immotility and male infertility. Dysfunction of motile and primary cilia at the reproductive tract also causes both male and female infertility.	[55,56]

2.1. Intellectual Disability

Defects in cerebral cortical development can lead to intellectual disabilities through a number of mechanisms [57]. Cortical development occurs in several steps: (1) Polarized radial glial progenitor formation, (2) radial/glial progenitor and intermediate progenitor proliferation, (3) radial/glial-guided neuronal migration, and (4) post-migratory neuronal differentiation, such as outgrowth and fasciculation of axons and dendrites [39]. The primary cilia of these progenitor cells play important roles in cerebral cortical development [40]. In Joubert syndrome, Bardet–Biedl syndrome (BBS), and oro-facial-digital syndrome, pathogenic mutations in genes regulating primary cilia function in progenitors disrupt cerebral cortical development [39], which can lead to intellectual disability [38].

2.2. Retinal Degeneration

The neuronal cell bodies of the retina are precisely organized in three major laminae: The ganglion cell, inner nuclear, and outer nuclear layers [58]. These layers are connected by neuronal projections of the cells located in each layer. Photoreceptor cells located in the outer nuclear layer capture photons and transmit signals to the brain through the inner nuclear and ganglion cell layers. Photoreceptor cells are composed of an outer segment (OS), inner segment (IS), and transition zone (also known as the connecting cilium), which connects the OS and IS [41,42]. The axoneme arises from the basal body in the IS and extends to the OS through the transition zone. Photon sensing is mainly performed by opsin proteins in the OS. Because the OS lacks protein synthesis machinery, opsin and other proteins involved in photon sensing are transported from the IS to the transition zone to the OS along the axoneme. Genes causative for ciliopathies, including Joubert syndrome, BBS, oro-facial-digital syndrome, Usher syndrome, and Meckel syndrome, are frequently involved in maintaining the structure and function of axonemes in photoreceptor cells [59]. Therefore, causative gene mutations in these ciliopathies often impair protein transport along the axoneme, resulting in retinal degeneration, and, potentially, vision loss, as is the case in retinitis pigmentosa and Leber congenital amaurosis [41,42,59].

2.3. Craniofacial Malformation

Fusion of distinct prominences, including the frontonasal, paired maxillary, and mandibular prominences, is crucial for proper craniofacial development [43]. Cranial neural crest cells (CNCCs) originate at the neural tube, undergo epithelial–mesenchymal transition, and migrate toward and proliferate in facial prominences [44]. The primary cilia of CNCCs play crucial roles in these steps through transduction of the Hedgehog and wingless-type MMTV integration site family (WNT) signaling pathways [43,44,60]. Gene mutations associated with ciliopathies affecting the function of primary cilia in CNCCs can lead to craniofacial malformations, such as cleft or lip palate, hyper/hypotelorism, micrognathia, and craniosynostosis [43,44,61,62].

2.4. Laterality Disorders

Although the human body is externally symmetrical, the visceral organs are arranged asymmetrically in a stereotyped manner [45,46,63]. Motile cilia of pit cells and nonmotile cilia of crown cells in the ventral node of the mammalian embryo play crucial roles in regulating left–right asymmetry [64,65]. When the motile cilia of pit cells generate leftward flow, the nonmotile cilia of crown cells located at the left side of the pit cells sense the flow and secrete Nodal–Gdf1 heterodimers [66,67]. In turn, the heterodimers bind to receptors in lateral plate mesoderm-derived cells and increase the expression of Nodal, Lefty2, and Pitx2, leading to left-side morphogenesis [68–71]. Accordingly, the impairment of cilia in the ventral node can cause laterality disorders, such as situs inversus and heterotaxy [45,46,72–75].

2.5. Cystic Kidney Disease

Cystic kidney disease is one of the main renal ciliopathies [36,76]. Renal tubular cells detect fluid flow through cilia. In these cells, fluid flow increases Ca^{2+} uptake through calcium channels, such as polycystin 2 [77,78], and the consequent increase in the intracellular Ca^{2+} concentration inhibits adenylate cyclase 6 and suppresses cyclic adenosine monophosphate (cAMP) signaling. The dysfunction of cilia in renal tubular cells prevents the increase in the Ca^{2+} concentration and suppression of cAMP signaling in response to fluid flow, resulting in activation of protein kinase A [47,48]. In turn, protein kinase A activation increases fluid secretion through chloride channels and deregulates multiple cellular pathways, including proliferation, apoptosis, and the polarity of renal tubular cells, leading to renal cystogenesis [79].

2.6. Obesity

Dysregulation of primary cilia in the central nervous system and peripheral tissues is associated with obesity, which often accompanies ciliopathies, such as BBS and Alström syndrome [49,50,80,81]. Obesity results from an excessive calorie intake relative to energy expenditure. In response to food intake, leptin is secreted from adipocytes and binds to receptors located in the primary cilia of anorexigenic and orexigenic neurons in the hypothalamus, resulting in increased and decreased expression of the anorexigenic peptide pro-opiomelanocortin and the orexigenic peptide Agouti-related peptide, respectively [82–84]. Genes associated with BBS and Alström syndrome regulate primary cilia in these hypothalamic neurons, and mutation of these genes can lead to obesity by failing to suppress the appetite through primary cilia-mediated leptin signaling [50].

Another contributing factor in obesity is elevated adipogenesis, resulting in an increased abundance of adipocytes [85]. Primary cilia are present in differentiating preadipocytes and play critical roles in adipogenesis [86]. Impairment of primary cilia in preadipocytes' knockdown of BBS proteins (BBS10 and BBS12) stimulates adipogenesis by activation of the glycogen synthase kinase 3 pathway and nuclear accumulation of peroxisome proliferator-activated receptor γ [86]. Knockdown of BBS12 in human mesenchymal stem cells also impairs ciliogenesis and enhances adipogenesis [87]. In contrast, knockdown of intraflagellar transport 88 in preadipocytes or mesenchymal stem cells inhibits adipogenesis by impairing the localization of insulin-like growth factor-1 receptors in primary cilia [88,89]. The precise mechanisms by which primary cilia regulate adipogenesis remain to be fully elucidated.

2.7. Scoliosis

Scoliosis is a skeletal dysfunction characterized by abnormal spine curvature. Scoliosis is associated with an impaired structure and function of cilia [51]. The primary cilium of osteocyte acts as a hub in a mechanotransduction pathway for loading-induced bone adaptation [52,90]. In general, short primary cilia of osteocytes are perpendicularly oriented to the long axis of bone [91]. In contrast, primary cilia of osteoblasts from idiopathic scoliosis (IS) patients are significantly longer than those of control samples [51]. The induction of osteogenic factors, including bone morphogenic protein 2 (BMP2) and cyclooxygenase 2 (COX2), are impaired in osteoblasts from the IS patients and osteoblasts with elongated primary cilia by lithium chloride treatment [51]. These findings suggest that elongated primary cilia in osteocytes may be dysfunctional in mechanotransduction and warrant further investigation to elucidate the molecular mechanisms of scoliosis.

2.8. Respiratory Distress

Respiratory distress, which is characterized by congestion, coughing, tachypnea, and hypoxia, is a cardinal feature of primary ciliary dyskinesia (PCD) [53]. Motile cilia of the epithelial cell lining of most of the upper and lower respiratory tracts are dysfunctional in PCD [54,92,93]. The dysfunction of motile cilia causes the impairment of mucociliary clearance and mucus obstruction, resulting in bronchiectasis and respiratory failure [53]. Most PCD follow an autosomal recessive inheritance. The mutations identified as being causative of PCD explain roughly 70% of the affected individuals [93]. These mutations impair the structure and/or function of motile cilia by affecting the dynein arm, radial spoke, central apparatus, or multiciliation [93]. However, the clinical phenotype of PCD is highly variable [53]. The relationship between the genotype and clinical phenotype remains to be fully elucidated.

2.9. Infertility

The sperm tail has microtubule arrangement ($9 \times 2 + 2$), which is similar to that of motile cilia [55]. Therefore, infertility is frequently observed in males with PCD [54]. There are some differences, however, between the sperm tail and motile cilia, including cell type-specific axonemal proteins and accessory structures specific to the sperm tail, such as the mitochondrial sheath, fibrous sheath, and outer dense fibers [55]. Mutations in genes causative of PCD are not always associated with male infertility and vice versa [55]. The impairment of the sperm tail affects sperm motility. The impairment of motile and primary cilia also affects the function of the reproductive tract in both males and females [54,56]. Structural and functional studies of cilia associated with infertility constitute an important area in reproductive research [55,56].

2.10. Roles of Primary Cilia in Cancer

Primary cilia in cultured mouse 3T3 fibroblasts and human retinal pigment epithelial (RPE1) cells can be disassembled and assembled by serum stimulation and deprivation, respectively [18,19,94]. Aurora A kinase (AURKA), one of the most important mitotic kinases for cell-cycle control [95], plays important roles in deciliation by serum stimulation [96,97]. AURKA is activated by serum stimulation through Ca^{2+}/calmodulin signaling, the non-canonical WNT pathway, and phosphatidylinositol signaling [97–100]. Serum stimulation also activates AURKA through the pathway involving epidermal growth factor receptor (EGFR), ubiquitin-specific peptidase 8 (USP8), and trichoplein (TCHP) (described in the next section) [33,34,101]. Activated AURKA phosphorylates itself and target proteins during G1 phase, which stimulates the disassembly of primary cilia [97]. Several proteins associated with AURKA and ciliogenesis have been identified, including histone deacetylase 6 [94] and nudE neurodevelopment protein 1 (NDE1) [102]. In response to serum stimulation, NDE1 localizes at the basal body and suppresses ciliogenesis by tethering dynein light chain 1 [103]. Under serum deprivation conditions, cyclin-dependent kinase 5 is activated and phosphorylates NDE1. Phosphorylated NDE1 is then recognized and ubiquitylated by the E3 ligase complex SCF^{FBXW7}, resulting in ciliogenesis [104,105]. Importantly, forced ciliation in cells growing under serum stimulation conditions can cause cell-cycle arrest [33,34,101,102,106]. These findings suggest that the primary cilium can act as a negative regulator of the cell cycle and may be a tumor suppressor organelle [3–6,9,107–109]. In fact, the suppression of primary cilia function is associated with tumorigenesis, cell proliferation, and metastasis in many cancers, including glioblastoma [110], esophageal cancer [111], colon cancer [112], cholangiocarcinoma [113,114], pancreatic ductal adenocarcinoma [115], clear cell renal carcinoma [116], prostate cancer [117], ovarian cancer [118,119], melanoma [120], and chondrosarcoma [121] (Table 2). However, primary cilia can promote tumor progression under certain conditions. In medulloblastoma and basal cell skin carcinoma caused by gain-of-function mutation of SMO, primary cilia convert the GLI transcription factors GLI2 and GLI3 to their activated forms, inducing their translocation to the nucleus, increased transcription of Hedgehog target genes, and promotion of cell proliferation [122,123]. In contrast, primary cilia of medulloblastoma and basal cell skin carcinoma caused by gain-of-function mutation of GLI2 increases the activity of GLI3 as a transcriptional repressor, resulting in suppression of the proliferation of these cancer cells [122,123]. Further work will thus be necessary to fully understand the context-dependent roles of primary cilia in cell proliferation.

Table 2. The roles of primary cilia in cancer.

Cancer Cell	The Role of PC in the Cancer	References
Glioblastoma	Inhibition of HDAC6 restores the loss of PC and suppressed the proliferation	[110]
Esophageal squamous cell carcinoma	KD of PRDX1 restores the loss of PC and suppressed the proliferation	[111]
Colon cancer	Knockout of TTLL3 causes the loss of PC and promotes tumorigenesis in colon	[112]
Cholangiocarcinoma	The number of PC is frequently reduced. Inhibition of HDAC6 restores the loss of PC and suppressed the proliferation	[113,114]
Pancreatic ductal adenocarcinoma	Inhibition of HDAC2 in Panc1 induces ciliogenesis and suppressed the proliferation	[115]
Clear cell renal carcinoma	PC is lost by inactivation of VHL tumor suppressor	[116]
Prostate cancer	KD of TACC3 restores the loss of PC and suppressed the proliferation	[117]
Epithelial ovarian cancer	The number of PC is reduced, which is associated with centrosomal localization of AURKA. KD of AURKA restores the loss of PC and suppressed the oncogenic hedgehog signaling	[118,119]
Melanoma	Deconstruction of PC is sufficient to drive metastatic formation	[120]
Chondrosarcoma	Inhibition of HDAC6 restores the loss of PC and suppressed the proliferation	[121]
Medulloblastoma, basal cell carcinoma	-	-
with GOF mutation of SMO	PC increase transcriptional activator and stimulate proliferation	[122,123]
with GOF mutation of GLI2	PC increase transcriptional suppressor and inhibit proliferation	[122,123]

PC—primary cilia; KD—knockdown; GOF—gain-of-function.

3. Roles of E3 Ubiquitin Ligases and DUBs in Ciliogenesis, Ciliopathy, and Cancer

The structure and function of cilia are dynamically regulated by many ciliary proteins through posttranslational modification [2,4–6,25]. Ubiquitination and deubiquitination of ciliary proteins by E3 ubiquitin ligase and DUBs, respectively, are crucial for the dynamic regulation of cilia [4–6,31–34]. In this section, we briefly describe the role of E3 ubiquitin ligases and DUBs in ciliogenesis, ciliopathy, and cancer (Table 3).

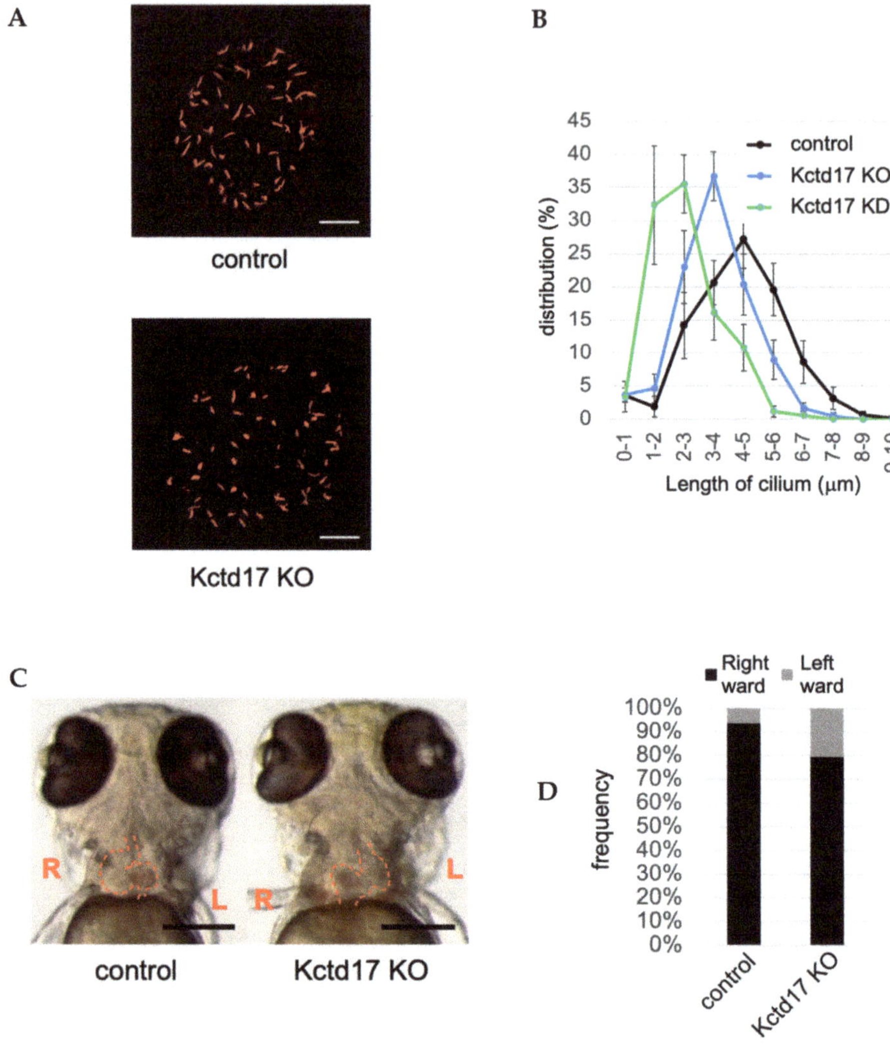

Figure 1. Suppression of Kctd17 in zebrafish impairs ciliogenesis in Kupffer's vesicle and induces situs inversus (**A**) Staining of acetylated alpha-tubulin in the cilia of Kupffer's vesicle at 12 h post-fertilization (hpf) of wild-type (control) or Kctd17 knockout (KO) zebrafish. (**B**) Distribution of the cilia length in Kupffer's vesicle at 12 hpf in control or Kctd17 KO or knockdown (KD) zebrafish. (**C**) Representative in vivo images of control and Kctd17 KO zebrafish at 3 days post-fertilization (dpf). Control and Kctd17 KO zebrafish show rightward and leftward looping of the heart, respectively. Please see Video S1. (**D**) At 3 dpf, rightward and leftward looping of the heart is observed in about 95% of the control zebrafish and about 20% of the Kctd17 KO zebrafish, respectively. Scale bar: 20 μm (**A**), 200 μm (**C**).

Table 3. The roles of E3 ubiquitin ligases and deubiquitinases in ciliogenesis, ciliopathy, and cancer.

E3 Ligase or DUB	The Role in Ciliogenesis (Substrate)	The Role in Ciliopathy and/or Cancer	References
CRL3-KCTD17	inhibited by KD of KCTD17 (TCHP)	KO of KCTD17 causes situs inversus in zebrafish	[33], this paper (Figure 1)
USP8	stimulated by KD of USP8 (TCHP)	KO of USP8 causes cystic kidney in zebrafish	[34]
		USP8 is highly expressed and oncogenic in melanoma	[124]
		Inhibition of USP8 suppresses the proliferation of glioblastoma stem cells	[125]
MARCHF7	inhibited by OE of MARCHF7 (IQCB1)	MARCHF7 promotes proliferation and invasion of cervical cancer cells	[126,127]
TRIM32	inhibited by OE of TRIM32 (IQCB1)	TRIM32 is a causative gene of BBS (BBS11)	[126,127]
		TRIM32 is oncogenic in head and neck squamous cell carcinoma and skin cancer	[128,129]
USP9X	inhibited by KD of USP9X (IQCB1)	LOF mutations in USP9X cause phenotypes related to ciliopathy	[126,130]
		USP9X is a major tumor suppressor gene in pancreatic ductal adenocarcinoma	[131]
CYLD	inhibited by KD/KO of CYLD (CEP70, MIB1)	KO/KD of CYLD causes ciliopathy-related phenotype in mouse and zebrafish	[132–134]
		LOF mutation in CYLD cause skin cancer (familial cylindromatosis)	[135]
MIB1	inhibited by OE of MIB1 (PCM1, KIAA0586)	MIB1 is oncogenic in upper urinary-tract urothelial carcinomas	[133,136,137]
CRL2-VHL	stimulated by OE of VHL (HIF1A)	KO of VHL causes cystic kidney	[138,139]
		VHL is tumor-suppressive in renal cancers	[140]

DUBs—deubiquitinases; KD—knockdown; KO—knockout; OE—overexpression; LOF—loss-of-function.

3.1. CRL3^{KCTD17}, USP8, and TCHP

TCHP, a centriolar protein originally identified as a keratin-binding protein, activates AURKA and suppresses ciliogenesis [101,141,142]. TCHP is ubiquitinated by the E3 ligase CRL3^{KCTD17}, a complex of the scaffold protein Cullin 3, RING box protein 1 (RBX1), and potassium channel tetramerization domain-containing 17 (KCTD17) [33]. Knockdown of KCTD17 in RPE1 cells suppresses ciliogenesis by stabilizing TCHP, leading to the activation of AURKA [33]. NDE1-like 1 (NDEL1), a modulator of dynein activity localized at the subdistal appendage of the mother centriole [143,144], indirectly inhibits ubiquitination of TCHP by CRL3^{KCTD17} [106]. In contrast, TCHP is deubiquitinated by USP8 after EGFR-mediated phosphorylation of USP8 at tyrosine residues 717 and 810 [34]. Knockdown of USP8 in RPE1 cells induces ciliogenesis and cell-cycle arrest even in the presence of serum [34]. These findings suggest that forced ciliogenesis by inhibition of USP8 may be a potential therapeutic strategy for cancers with a high expression of USP8 and loss of cilia. In fact, USP8 is highly expressed and plays an oncogenic role in melanoma [124], and inhibition of USP8 suppresses the proliferation of glioblastoma stem cells [125]. The precise effect of USP8 on ciliogenesis in these tumor cells remains to be elucidated. In zebrafish, knockout of Usp8 increases ciliogenesis in renal tubules and causes renal cysts [34], whereas knockout of Kctd17 impairs ciliogenesis in Kupffer's vesicle and causes situs inversus (Figure 1). Because AURKA is also associated with both cancer and ciliopathy [95,97], these findings suggest that the involvement of KCTD17 and USP8 in cancer and ciliopathy might be mediated by effects on ciliogenesis via a TCHP–AURKA pathway.

3.2. MARCHF7, TRIM32, USP9X, and IQCB1

The IQ motif containing B1 (IQCB1), also known as Nephrocystin-5, increases ciliogenesis by binding to centrosomal protein 290 (CEP290) [145]. IQCB1 is ubiquitinated by membrane-associated ring-CH-type finger 7 (MARCHF7) and the tripartite motif containing 32 (TRIM32, also known as BBS11) [126]. Overexpression of MARCHF7 or TRIM32 inhibits ciliogenesis [126]. MARCHF7 promotes proliferation and invasion of cervical cancer cells [127], and TRIM32 is also oncogenic in head and neck squamous cell carcinoma and skin cancer [128,129]. Conversely, IQCB1 is deubiquitinated and stabilized by USP9X [126]. Knockdown of USP9X inhibits ciliogenesis [126]. USP9X is a major tumor suppressor gene in pancreatic ductal adenocarcinoma [131]. These findings suggest that MARCHF7, TRIM32, and USP9X may be involved in cancer via the modulation of ciliogenesis. Mutation of TRIM32, USP9X, and their substrate IQCB1 causes various phenotypes related to ciliopathy [130,146,147].

3.3. CYLD and MIB1

Cylindromatosis (CYLD) is a member of the USP family of proteins and is expressed in centriolar satellites [148]. CYLD stimulates ciliogenesis by stabilizing centrosomal protein 70 (CEP70) and pericentriolar material 1 (PCM1) [132,133]. Stabilization of PCM1 by CYLD is mediated by deubiquitination of mindbomb E3 ubiquitin protein ligase 1 (MIB1), which is activated by ubiquitination on lysine 63 [133]. Lysine 63-ubiquitinated MIB1 then ubiquitinates PCM1 and stimulates its degradation in proteasomes. CYLD antagonizes the degradation of PCM1 by suppressing the activity of MIB1. MIB1 also ubiquitinates KIAA0586, a centrosomal protein also known as Talpid3, thereby stimulating its degradation and inhibiting ciliogenesis [136]. These findings suggest that CYLD and MIB1 positively and negatively, respectively, regulate ciliogenesis. Loss-of-function mutations in CYLD are associated with familial cylindromatosis, a condition involving multiple skin tumors [135]. MIB1 is oncogenic in upper urinary tract urothelial carcinomas [137]. Knockout or knockdown of CYLD in mice and zebrafish show phenotypes related to ciliopathy [134,149]. PCM1 and KIAA0586, both substrates of MIB1, are also associated with ciliopathies [149,150].

3.4. CRL2^{VHL}

The tumor suppressor protein von Hippel–Lindau (VHL) is a component of an E3 ubiquitin ligase complex that also contains the scaffold protein Cullin 2 and RBX1 [151]. Mutations in VHL related to formation of the E3 ubiquitin ligase complex lead to von Hippel–Lindau syndrome, which can exhibit both ciliopathy and cancer phenotypes [140,152]. CRL2^{VHL} ubiquitinates the α subunit of the transcription factor hypoxia-inducible factor 1 (HIF1α), leading to its proteasomal degradation [138,153]. VHL and HIF1α positively and negatively, respectively, regulate ciliogenesis and cancer [107,139,140,154].

4. Future Directions

4.1. Identification of E3 Ubiquitin Ligases and DUBs Related to Cilia Assembly and Disassembly

As summarized above, the UPS has been implicated as a key system for the regulation of cilia assembly and disassembly [32,33,155]. Many E3 ubiquitin ligases and DUBs other than those listed in Table 3 have been identified as regulators of cilia assembly and disassembly. These include the E3 ubiquitin ligases cyclin F [156], FBW7 [104], NEDD4L [157], MYCBP2 [157], and UBR5 [158]; and the DUBs USP14 [159] and USP33 [160]; however, other enzymes undoubtedly remain to be identified. We showed that NDEL1, a modulator of dynein activity [143,144], inhibits ubiquitination of TCHP by CRL3^{KCTD17} and suppresses ciliogenesis in RPE1 cells incubated in the presence of serum [106].

Furthermore, in the absence of serum, NDEL1 is degraded by the UPS, resulting in the disappearance of TCHP from the mother centriole and induction of ciliogenesis [106]. The proteins involved in UPS-mediated NDEL1 degradation remain unknown.

One efficient approach to identifying E3 ubiquitin ligases of a substrate of interest is two-stepped global E3 screening, in which a wheat germ cell-free expression system is used to produce more than 1000 E3 ubiquitin ligases in the first step, and the enzymes are then screened using specific small interfering RNAs in the second step [33]. Genome-wide RNAi screening and proteomic profiling approaches may identify novel UPS proteins regulating cilia assembly and disassembly [157,161,162]. It will also be important to elucidate the subcellular compartment (basal body, transition zone, and/or axoneme) in which these E3 ubiquitin ligases and DUBs are active [32]. Compartment-specific proteomic profiling could be an efficient strategy to address these questions [163,164].

4.2. Identification of E3 Ubiquitin Ligase and DUB Substrates

Although many E3 ubiquitin ligases and DUBs involved in the assembly and disassembly of cilia have been identified, the precise substrates of many of them remain unknown. For example, we performed RNAi screening of RPE1 cells and identified six DUBs, USP8, USP38, USP43, USP52, USP54, and UCHL3, as suppressors of ciliogenesis [34]. We also identified TCHP as a substrate of USP8 in the regulation of ciliogenesis [34], but the substrates for the other five DUBs are unknown. A variety of experimental approaches have been developed to identify DUB substrates [165]. Stable overexpression or knockdown of DUBs followed by quantitative proteomic analysis to detect proteins differentially expressed in control and manipulated cells identified Sec28p and NFX1-123 as substrates of Ubp3p and USP9X, respectively [166,167]. Affinity purification proteomics using tagged DUBs has identified CEP192 as a substrate of CYLD [168]. Similarly, affinity purification proteomics using antibodies that recognize the diglycine residues, a remnant present on the ε-amine of lysine following trypsin digestion of ubiquitinated proteins, was successful in identifying S100A6 and hnRNP K as SseL substrates [169]. The diglycine remnant affinity purification method can also be used to identify E3 ubiquitin ligase substrates [170,171].

Once ligated to its substrate, ubiquitin itself can be modified by ubiquitination of one or more of its seven lysine residues (K6, K11, K27, K29, K33, K48, and K63) or the N-terminal methionine [172]. Polyubiquitination can be homotypic (same linkage) or heterotypic (different linkage) [173] and it plays various roles in cell signaling regulation depending on the linkage type [173]. For example, homotypic K48 polyubiquitination is related to classical proteasomal degradation; homotypic K63 polyubiquitination regulates protein–protein interactions; homotypic K6 and K27 polyubiquitinations are involved in the DNA damage response; and homotypic K29 and K33 polyubiquitinations are linked to innate immunity. In centriolar satellites, which are electron-dense and spherical cytoplasmic granules around centrosomes, modification of the E3 ubiquitin ligase MIB1 by homotypic K63 polyubiquitination induces homotypic K48 polyubiquitination of its substrate PCM1, resulting in PCM1 proteasomal degradation [133]. PCM1 plays an indispensable role in the clustering of centriolar satellites around the centrosome to orchestrate ciliogenesis [174]. CYLD located in centriolar satellites deubiquitinates the K63 polyubiquitin chain of MIB1, thereby antagonizing MIB1-mediated degradation of PCM1 and suppressing ciliogenesis [133]. Somatostatin receptor 3 (SSTR3) and G protein-coupled receptor 161 (GPR161) are important G protein-coupled receptors (GPCRs) that regulate somatostatin and hedgehog signaling, respectively, in primary cilia. K63 polyubiquitination of these GPCRs enables them to be recognized by ciliary exit machinery [175]. β-arrestin is known to mediate the K63 polyubiquitination of SSTR3 and GPR161, but the identity of the E3 ubiquitin ligase(s) involved is unclear. One important task for the future is the development of novel tools that will enable the identification of all E3 ubiquitin ligase and DUB substrates, as well as their ubiquitination patterns, related to the assembly and disassembly of cilia [172,173,176].

4.3. Identification of Drugs Targeting E3 Ubiquitin Ligases and DUBs

The screening of compounds for effects on protein ubiquitination and deubiquitination has led to the identification of a number of E3 ubiquitin ligase- and DUB-targeting drugs [29,165,177]. Because the substrates of these enzymes have a wide variety of functions, such drugs can interfere with various signaling pathways and impair physiological functions. Targeting the interaction of E3 ubiquitin ligases, DUBs, and their substrates involved in cilia assembly and disassembly may be a fruitful approach to developing selective drugs for the treatment of cancer and ciliopathies. Several technological advances have accelerated the development of drugs targeting protein–protein interactions [178,179]. The F-box protein S-phase kinase-associated protein 2 (SKP2), a component of E3 ubiquitin ligase SCF^{SKP2}, ubiquitinates several proteins important for cell proliferation and survival, including p27^{KIP1}, p21^{CIP1}, and AKT serine/threonine kinases [180]. Some compounds have been identified that bind to a pocket in SKP2 that acts as the binding site for cyclin-dependent kinases regulatory subunit (CKS1), an accessory protein that can bind to p27^{KIP1} phosphorylated by cyclin-dependent kinase 2/cyclin E [181–183]. These chemicals inhibit the interaction between SKP2 and CKS1, resulting in selective inhibition of SKP2-mediated ubiquitination and degradation of p27^{KIP1}. Screening for compounds that disrupt the subcellular translocation of E3 ubiquitin ligases and DUBs is another potential approach to developing selective drugs. USP4 and USP15 function in both the cytosol and the nucleus. In the cytosol, they deubiquitinate proteins involved in many signaling pathways, including those important to inflammation and oxidative stress; in the nucleus, they deubiquitinate proteins regulating splicing [184]. Nuclear translocation of these DUBs is inhibited by phosphorylation of two threonine residues located in the binding sites for spliceosome-associated factor 3 (SART3), a binding partner that facilitates the nuclear translocation of USP4 and USP15 [185]. Drugs interfering with the interaction between USP4 and USP15 and SART3 may selectively inhibit the DUB functions in the nucleus. Future work should include elucidation of the structure–activity relationships for E3 ubiquitin ligase- and/or DUB-targeting compounds and the identification of druggable sites in non-catalytic regions of the enzymes. Collectively, these approaches may lead to the development of novel drugs that regulate the enzymes in a context-dependent manner.

In summary, we have described the dysregulation of cilia in ciliopathies and cancers, and how that dysregulation results from changes in ciliary protein stability regulated by the UPS. As noted, some of the E3 ubiquitin ligases and DUBs involved in the maintenance of ciliary protein stability may be therapeutic targets for the associated disorders. Indeed, small molecules targeting these E3 ubiquitin ligases and DUBs, including USP8, USP9X, CYLD, and VHL, have been successfully developed [29,177,186]. However, the role of E3 ubiquitin ligases and DUBs in disease can be context dependent [28,148,187,188]. Thus, it will be important to develop small molecule modulators of the interactions between E3 ubiquitin ligases or DUBs and their binding proteins in a context-specific manner.

Supplementary Materials: Supplementary materials can be found at http://www.mdpi.com/1422-0067/21/17/5962/s1. Video S1: Representative movies of control and Kctd17 KO zebrafish at 3 dpf.

Funding: This work was supported in part by the Japan Society for the Promotion of Science KAKENHI (18K06890 to TS, 20H03448 to K.K., 20K07356 to D.Y., 19K07318 to Y.N.), Uehara Memorial Foundation (K.K.), and Takeda Science Foundation (K.K., Y.N., and M.I.).

Acknowledgments: We sincerely apologize to all researchers whose important work could not be cited because of space considerations. We thank Anne M. O'Rourke for editing a draft of this manuscript.

Conflicts of Interest: Research in the author's laboratories was conducted in the absence of any commercial or financial relationships that could be construed as a potential conflict of interest.

References

1. Anvarian, Z.; Mykytyn, K.; Mukhopadhyay, S.; Pedersen, L.B.; Christensen, S.T. Cellular signalling by primary cilia in development, organ function and disease. *Nat. Reviews. Nephrol.* **2019**, *15*, 199–219. [CrossRef] [PubMed]

2. Malicki, J.J.; Johnson, C.A. The Cilium: Cellular Antenna and Central Processing Unit. *Trends Cell Biol.* **2017**, *27*, 126–140. [CrossRef] [PubMed]

3. Goto, H.; Inoko, A.; Inagaki, M. Cell cycle progression by the repression of primary cilia formation in proliferating cells. *Cell. Mol. Life Sci.* **2013**, *70*, 3893–3905. [CrossRef] [PubMed]

4. Izawa, I.; Goto, H.; Kasahara, K.; Inagaki, M. Current topics of functional links between primary cilia and cell cycle. *Cilia* **2015**, *4*, 12. [CrossRef] [PubMed]

5. Goto, H.; Inaba, H.; Inagaki, M. Mechanisms of ciliogenesis suppression in dividing cells. *Cell. Mol. Life Sci.* **2017**, *74*, 881–890. [CrossRef] [PubMed]

6. Nishimura, Y.; Kasahara, K.; Shiromizu, T.; Watanabe, M.; Inagaki, M. Primary cilia as signaling hubs in health and disease. *Adv. Sci.* **2019**, *6*, 1801138. [CrossRef]

7. Valente, E.M.; Rosti, R.O.; Gibbs, E.; Gleeson, J.G. Primary cilia in neurodevelopmental disorders. *Nat. Rev. Neurol.* **2014**, *10*, 27–36. [CrossRef]

8. Reiter, J.F.; Leroux, M.R. Genes and molecular pathways underpinning ciliopathies. *Nat. Rev. Mol. Cell Biol.* **2017**, *18*, 533–547. [CrossRef]

9. Liu, H.; Kiseleva, A.A.; Golemis, E.A. Ciliary signalling in cancer. *Nat. Rev. Cancer* **2018**, *18*, 511–524. [CrossRef]

10. Wang, L.; Dynlacht, B.D. The regulation of cilium assembly and disassembly in development and disease. *Development* **2018**, *145*. [CrossRef]

11. Silverman, M.A.; Leroux, M.R. Intraflagellar transport and the generation of dynamic, structurally and functionally diverse cilia. *Trends Cell Biol.* **2009**, *19*, 306–316. [CrossRef] [PubMed]

12. Loreng, T.D.; Smith, E.F. The Central Apparatus of Cilia and Eukaryotic Flagella. *Cold Spring Harb. Perspect Biol.* **2017**, *9*. [CrossRef] [PubMed]

13. Ishikawa, T. Axoneme Structure from Motile Cilia. *Cold Spring Harb. Perspect. Biol.* **2017**, *9*. [CrossRef] [PubMed]

14. Mitchison, H.M.; Valente, E.M. Motile and non-motile cilia in human pathology: From function to phenotypes. *J. Pathol.* **2017**, *241*, 294–309. [CrossRef] [PubMed]

15. Reiter, J.F.; Blacque, O.E.; Leroux, M.R. The base of the cilium: Roles for transition fibres and the transition zone in ciliary formation, maintenance and compartmentalization. *EMBO Rep.* **2012**, *13*, 608–618. [CrossRef]

16. Alieva, I.; Staub, C.; Uzbekova, S.; Uzbekov, R. A question of flagella origin for spermatids—Mother or daughter centriole? In *Flagella and Cilia Types, Strucure and Functions*; Uzbekov, R.E., Ed.; Nova Science Publishers, Inc.: New York, NY, USA, 2018; pp. 109–126.

17. Al Jord, A.; Lemaître, A.I.; Delgehyr, N.; Faucourt, M.; Spassky, N.; Meunier, A. Centriole amplification by mother and daughter centrioles differs in multiciliated cells. *Nature* **2014**, *516*, 104–107. [CrossRef]

18. Tucker, R.W.; Pardee, A.B.; Fujiwara, K. Centriole ciliation is related to quiescence and DNA synthesis in 3T3 cells. *Cell* **1979**, *17*, 527–535. [CrossRef]

19. Tucker, R.W.; Scher, C.D.; Stiles, C.D. Centriole deciliation associated with the early response of 3T3 cells to growth factors but not to SV40. *Cell* **1979**, *18*, 1065–1072. [CrossRef]

20. Rieder, C.L.; Jensen, C.G.; Jensen, L.C.W. The resorption of primary cilia during mitosis in a vertebrate (PtK1) cell line. *J. Ultrastruct. Res.* **1979**, *68*, 173–185. [CrossRef]

21. Tsang, W.Y.; Dynlacht, B.D. CP110 and its network of partners coordinately regulate cilia assembly. *Cilia* **2013**, *2*, 9. [CrossRef]

22. Yadav, S.P.; Sharma, N.K.; Liu, C.; Dong, L.; Li, T.; Swaroop, A. Centrosomal protein CP110 controls maturation of the mother centriole during cilia biogenesis. *Developement* **2016**, *143*, 1491–1501.

23. Craft, J.M.; Harris, J.A.; Hyman, S.; Kner, P.; Lechtreck, K.F. Tubulin transport by IFT is upregulated during ciliary growth by a cilium-autonomous mechanism. *J. Cell Biol.* **2015**, *208*, 223–237. [CrossRef] [PubMed]

24. Malicki, J.; Avidor-Reiss, T. From the cytoplasm into the cilium: Bon voyage. *Organogenesis* **2014**, *10*, 138–157. [CrossRef] [PubMed]

25. Mirvis, M.; Stearns, T.; James Nelson, W. Cilium structure, assembly, and disassembly regulated by the cytoskeleton. *Biochem. J.* **2018**, *475*, 2329–2353. [CrossRef]
26. Clague, M.J.; Heride, C.; Urbé, S. The demographics of the ubiquitin system. *Trends Cell Biol.* **2015**, *25*, 417–426. [CrossRef]
27. Leznicki, P.; Kulathu, Y. Mechanisms of regulation and diversification of deubiquitylating enzyme function. *J. Cell Sci.* **2017**, *130*, 1997–2006. [CrossRef]
28. Popovic, D.; Vucic, D.; Dikic, I. Ubiquitination in disease pathogenesis and treatment. *Nat. Med.* **2014**, *20*, 1242–1253. [CrossRef]
29. Harrigan, J.A.; Jacq, X.; Martin, N.M.; Jackson, S.P. Deubiquitylating enzymes and drug discovery: Emerging opportunities. *Nat. Rev. Drug Discov.* **2018**, *17*, 57–78. [CrossRef]
30. Senft, D.; Qi, J.; Ronai, Z.A. Ubiquitin ligases in oncogenic transformation and cancer therapy. *Nat. Rev. Cancer* **2018**, *18*, 69–88. [CrossRef]
31. Shearer, R.F.; Saunders, D.N. Regulation of primary cilia formation by the ubiquitin-proteasome system. *Biochem. Soc. Trans.* **2016**, *44*, 1265–1271. [CrossRef]
32. Hossain, D.; Tsang, W.Y. The role of ubiquitination in the regulation of primary cilia assembly and disassembly. *Semin. Cell Dev. Biol.* **2019**, *93*, 145–152. [CrossRef] [PubMed]
33. Kasahara, K.; Kawakami, Y.; Kiyono, T.; Yonemura, S.; Kawamura, Y.; Era, S.; Matsuzaki, F.; Goshima, N.; Inagaki, M. Ubiquitin-proteasome system controls ciliogenesis at the initial step of axoneme extension. *Nat. Commun.* **2014**, *5*, 5081. [CrossRef] [PubMed]
34. Kasahara, K.; Aoki, H.; Kiyono, T.; Wang, S.; Kagiwada, H.; Yuge, M.; Tanaka, T.; Nishimura, Y.; Mizoguchi, A.; Goshima, N.; et al. EGF receptor kinase suppresses ciliogenesis through activation of USP8 deubiquitinase. *Nat. Commun.* **2018**, *9*, 758. [CrossRef] [PubMed]
35. Toulis, V.; Marfany, G. By the Tips of Your Cilia: Ciliogenesis in the Retina and the Ubiquitin-Proteasome System. *Adv. Exp. Med. Biol* **2020**, *1233*, 303–310. [PubMed]
36. Pazour, G.J.; Quarmby, L.; Smith, A.O.; Desai, P.B.; Schmidts, M. Cilia in cystic kidney and other diseases. *Cell Signal.* **2020**, *69*, 109519. [CrossRef] [PubMed]
37. Hildebrandt, F.; Benzing, T.; Katsanis, N. Ciliopathies. *N. Engl. J. Med.* **2011**, *364*, 1533–1543. [CrossRef] [PubMed]
38. Braun, D.A.; Hildebrandt, F. Ciliopathies. *Cold Spring Harb. Perspect. Biol.* **2017**, *9*. [CrossRef] [PubMed]
39. Guo, J.; Higginbotham, H.; Li, J.; Nichols, J.; Hirt, J.; Ghukasyan, V.; Anton, E.S. Developmental disruptions underlying brain abnormalities in ciliopathies. *Nat. Commun.* **2015**, *6*, 7857. [CrossRef]
40. Youn, Y.H.; Han, Y.G. Primary Cilia in Brain Development and Diseases. *Am. J. Pathol.* **2018**, *188*, 11–22. [CrossRef]
41. Bujakowska, K.M.; Liu, Q.; Pierce, E.A. Photoreceptor Cilia and Retinal Ciliopathies. *Cold Spring Harb. Perspect. Biol.* **2017**, *9*. [CrossRef]
42. Bachmann-Gagescu, R.; Neuhauss, S.C. The photoreceptor cilium and its diseases. *Curr. Opin. Genet. Dev.* **2019**, *56*, 22–33. [CrossRef] [PubMed]
43. Schock, E.N.; Brugmann, S.A. Discovery, Diagnosis, and Etiology of Craniofacial Ciliopathies. *Cold Spring Harb. Perspect. Biol.* **2017**, *9*. [CrossRef]
44. Cortés, C.R.; Metzis, V.; Wicking, C. Unmasking the ciliopathies: Craniofacial defects and the primary cilium. *Wiley Interdiscip. Rev. Dev. Biol.* **2015**, *4*, 637–653. [CrossRef] [PubMed]
45. Dasgupta, A.; Amack, J.D. Cilia in vertebrate left-right patterning. *Philos. Trans. R. Soc. Lond. B. Biol. Sci.* **2016**, *1710*, 20150410. [CrossRef] [PubMed]
46. Grimes, D.T. Making and breaking symmetry in development, growth and disease. *Development* **2019**, *146*. [CrossRef]
47. Ye, H.; Wang, X.; Constans, M.M.; Sussman, C.R.; Chebib, F.T.; Irazabal, M.V.; Young, W.F., Jr.; Harris, P.C.; Kirschner, L.S.; Torres, V.E. The regulatory 1α subunit of protein kinase A modulates renal cystogenesis. *Am. J. Physiol. Ren. Physiol.* **2017**, *313*, 677–686. [CrossRef]
48. Cornec-Le Gall, E.; Alam, A.; Perrone, R.D. Autosomal dominant polycystic kidney disease. *Lancet* **2019**, *393*, 919–935. [CrossRef]
49. Mariman, E.C.; Vink, R.G.; Roumans, N.J.; Bouwman, F.G.; Stumpel, C.T.; Aller, E.E.; van Baak, M.A.; Wang, P. The cilium: A cellular antenna with an influence on obesity risk. *Br. J. Nutr.* **2016**, *116*, 576–592. [CrossRef]

50. Engle, S.E.; Bansal, R.; Antonellis, P.J.; Berbari, N.F. Cilia signaling and obesity. *Semin. Cell Dev. Biol.* **2020**. Available online: https://www.sciencedirect.com/science/article/abs/pii/S1084952119301831 (accessed on 19 August 2020).

51. Oliazadeh, N.; Gorman, K.F.; Eveleigh, R.; Bourque, G.; Moreau, A. Identification of Elongated Primary Cilia with Impaired Mechanotransduction in Idiopathic Scoliosis Patients. *Sci. Rep.* **2017**, *7*, 44260. [CrossRef]

52. Qin, L.; Liu, W.; Cao, H.; Xiao, G. Molecular mechanosensors in osteocytes. *Bone Res.* **2020**, *8*, 23. [CrossRef]

53. Sagel, S.D.; Davis, S.D.; Campisi, P.; Dell, S.D. Update of respiratory tract disease in children with primary ciliary dyskinesia. *Proc. Am. Thorac. Soc.* **2011**, *8*, 438–443. [CrossRef] [PubMed]

54. Mirra, V.; Werner, C.; Santamaria, F. Primary Ciliary Dyskinesia: An Update on Clinical Aspects, Genetics, Diagnosis, and Future Treatment Strategies. *Front. Pediatr.* **2017**, *5*, 135. [CrossRef] [PubMed]

55. Sironen, A.; Shoemark, A.; Patel, M.; Loebinger, M.R.; Mitchison, H.M. Sperm defects in primary ciliary dyskinesia and related causes of male infertility. *Cell. Mol. Life Sci.* **2020**, *77*, 2029–2048. [CrossRef] [PubMed]

56. Girardet, L.; Augière, C.; Asselin, M.-P.; Belleannée, C. Primary cilia: Biosensors of the male reproductive tract. *Andrology* **2019**, *7*, 588–602. [CrossRef] [PubMed]

57. Juric-Sekhar, G.; Hevner, R.F. Malformations of Cerebral Cortex Development: Molecules and Mechanisms. *Annu. Rev. Pathol.* **2019**, *14*, 293–318. [CrossRef] [PubMed]

58. Vecino, E.; Rodriguez, F.D.; Ruzafa, N.; Pereiro, X.; Sharma, S.C. Glia-neuron interactions in the mammalian retina. *Prog. Retin. Eye Res.* **2016**, *51*, 1–40. [CrossRef]

59. May-Simera, H.; Nagel-Wolfrum, K.; Wolfrum, U. Cilia—The sensory antennae in the eye. *Prog. Retin. Eye Res.* **2017**, *60*, 144–180. [CrossRef]

60. Jeong, J.; Mao, J.; Tenzen, T.; Kottmann, A.H.; McMahon, A.P. Hedgehog signaling in the neural crest cells regulates the patterning and growth of facial primordia. *Genes Dev.* **2004**, *18*, 937–951. [CrossRef]

61. Tobin, J.L.; Di Franco, M.; Eichers, E.; May-Simera, H.; Garcia, M.; Yan, J.; Quinlan, R.; Justice, M.J.; Hennekam, R.C.; Briscoe, J.; et al. Inhibition of neural crest migration underlies craniofacial dysmorphology and Hirschsprung's disease in Bardet-Biedl syndrome. *Proc. Natl. Acad. Sci. USA* **2008**, *105*, 6714–6719. [CrossRef]

62. Brugmann, S.A.; Allen, N.C.; James, A.W.; Mekonnen, Z.; Madan, E.; Helms, J.A. A primary cilia-dependent etiology for midline facial disorders. *Hum. Mol. Genet.* **2010**, *19*, 1577–1592. [CrossRef]

63. Hirokawa, N.; Tanaka, Y.; Okada, Y. Left-right determination: Involvement of molecular motor KIF3, cilia, and nodal flow. *Cold Spring Harb. Perspect. Biol.* **2009**, *1*, a000802. [CrossRef] [PubMed]

64. Kawasumi, A.; Nakamura, T.; Iwai, N.; Yashiro, K.; Saijoh, Y.; Belo, J.A.; Shiratori, H.; Hamada, H. Left-right asymmetry in the level of active Nodal protein produced in the node is translated into left-right asymmetry in the lateral plate of mouse embryos. *Dev. Biol.* **2011**, *353*, 321–330. [CrossRef] [PubMed]

65. Shiratori, H.; Hamada, H. TGFbeta signaling in establishing left-right asymmetry. *Semin. Cell Dev. Biol.* **2014**, *32*, 80–84. [CrossRef] [PubMed]

66. Marques, S.; Borges, A.C.; Silva, A.C.; Freitas, S.; Cordenonsi, M.; Belo, J.A. The activity of the Nodal antagonist Cerl-2 in the mouse node is required for correct L/R body axis. *Genes Dev.* **2004**, *18*, 2342–2347. [CrossRef]

67. Nakamura, T.; Saito, D.; Kawasumi, A.; Shinohara, K.; Asai, Y.; Takaoka, K.; Dong, F.; Takamatsu, A.; Belo, J.A.; Mochizuki, A.; et al. Fluid flow and interlinked feedback loops establish left-right asymmetric decay of Cerl2 mRNA. *Nat. Commun.* **2012**, *3*, 1322. [CrossRef]

68. Logan, M.; Pagan-Westphal, S.M.; Smith, D.M.; Paganessi, L.; Tabin, C.J. The transcription factor Pitx2 mediates situs-specific morphogenesis in response to left-right asymmetric signals. *Cell* **1998**, *94*, 307–317. [CrossRef]

69. Saijoh, Y.; Adachi, H.; Sakuma, R.; Yeo, C.Y.; Yashiro, K.; Watanabe, M.; Hashiguchi, H.; Mochida, K.; Ohishi, S.; Kawabata, M.; et al. Left-right asymmetric expression of lefty2 and nodal is induced by a signaling pathway that includes the transcription factor FAST2. *Mol. Cell* **2000**, *5*, 35–47. [CrossRef]

70. Shiratori, H.; Sakuma, R.; Watanabe, M.; Hashiguchi, H.; Mochida, K.; Sakai, Y.; Nishino, J.; Saijoh, Y.; Whitman, M.; Hamada, H. Two-step regulation of left-right asymmetric expression of Pitx2: Initiation by nodal signaling and maintenance by Nkx2. *Mol. Cell* **2001**, *7*, 137–149. [CrossRef]

71. Botilde, Y.; Yoshiba, S.; Shinohara, K.; Hasegawa, T.; Nishimura, H.; Shiratori, H.; Hamada, H. Cluap1 localizes preferentially to the base and tip of cilia and is required for ciliogenesis in the mouse embryo. *Dev. Biol.* **2013**, *381*, 203–212. [CrossRef]

72. Ware, S.M.; Aygun, M.G.; Hildebrandt, F. Spectrum of clinical diseases caused by disorders of primary cilia. *Proc. Am. Thorac. Soc.* **2011**, *8*, 444–450. [CrossRef]

73. Oud, M.M.; Lamers, I.J.; Arts, H.H. Ciliopathies: Genetics in Pediatric Medicine. *J. Pediatr. Genet.* **2017**, *6*, 18–29. [CrossRef] [PubMed]

74. Matsui, T.; Bessho, Y. Left-right asymmetry in zebrafish. *Cell. Mol. Life Sci.* **2012**, *69*, 3069–3077. [CrossRef] [PubMed]

75. Matsui, T.; Ishikawa, H.; Bessho, Y. Cell collectivity regulation within migrating cell cluster during Kupffer's vesicle formation in zebrafish. *Front. Cell Dev. Biol.* **2015**, *3*, 27. [CrossRef] [PubMed]

76. Devlin, L.A.; Sayer, J.A. Renal ciliopathies. *Curr. Opin. Genet. Dev.* **2019**, *56*, 49–60. [CrossRef]

77. Mangolini, A.; de Stephanis, L.; Aguiari, G. Role of calcium in polycystic kidney disease: From signaling to pathology. *World J. Nephrol* **2016**, *5*, 76–83. [CrossRef]

78. Avasthi, P.; Maser, R.L.; Tran, P.V. Primary Cilia in Cystic Kidney Disease. *Results Probl. Cell Differ.* **2017**, *60*, 281–321.

79. Malekshahabi, T.; Khoshdel Rad, N.; Serra, A.L.; Moghadasali, R. Autosomal dominant polycystic kidney disease: Disrupted pathways and potential therapeutic interventions. *J. Cell. Physiol.* **2019**, *234*, 12451–12470. [CrossRef]

80. Oh, E.C.; Vasanth, S.; Katsanis, N. Metabolic regulation and energy homeostasis through the primary Cilium. *Cell Metab.* **2015**, *21*, 21–31. [CrossRef]

81. Vaisse, C.; Reiter, J.F.; Berbari, N.F. Cilia and Obesity. *Cold Spring Harb. Perspect. Biol.* **2017**, *9*, a028217. [CrossRef]

82. Ernst, M.B.; Wunderlich, C.M.; Hess, S.; Paehler, M.; Mesaros, A.; Koralov, S.B.; Kleinridders, A.; Husch, A.; Munzberg, H.; Hampel, B.; et al. Enhanced Stat3 activation in POMC neurons provokes negative feedback inhibition of leptin and insulin signaling in obesity. *J. Neurosci.* **2009**, *29*, 11582–11593. [CrossRef]

83. Mesaros, A.; Koralov, S.B.; Rother, E.; Wunderlich, F.T.; Ernst, M.B.; Barsh, G.S.; Rajewsky, K.; Bruning, J.C. Activation of Stat3 signaling in AgRP neurons promotes locomotor activity. *Cell Metab.* **2008**, *7*, 236–248. [CrossRef] [PubMed]

84. Han, Y.M.; Kang, G.M.; Byun, K.; Ko, H.W.; Kim, J.; Shin, M.S.; Kim, H.K.; Gil, S.Y.; Yu, J.H.; Lee, B.; et al. Leptin-promoted cilia assembly is critical for normal energy balance. *J. Clin. Investig.* **2014**, *124*, 2193–2197. [CrossRef]

85. Sebo, Z.L.; Rodeheffer, M.S. Assembling the adipose organ: Adipocyte lineage segregation and adipogenesis in vivo. *Development* **2019**, *146*. [CrossRef] [PubMed]

86. Marion, V.; Stoetzel, C.; Schlicht, D.; Messaddeq, N.; Koch, M.; Flori, E.; Danse, J.M.; Mandel, J.L.; Dollfus, H. Transient ciliogenesis involving Bardet-Biedl syndrome proteins is a fundamental characteristic of adipogenic differentiation. *Proc. Natl. Acad. Sci. USA* **2009**, *106*, 1820–1825. [CrossRef]

87. Marion, V.; Mockel, A.; De Melo, C.; Obringer, C.; Claussmann, A.; Simon, A.; Messaddeq, N.; Durand, M.; Dupuis, L.; Loeffler, J.P.; et al. BBS-induced ciliary defect enhances adipogenesis, causing paradoxical higher-insulin sensitivity, glucose usage, and decreased inflammatory response. *Cell Metab.* **2012**, *16*, 363–377. [CrossRef] [PubMed]

88. Zhu, D.; Shi, S.; Wang, H.; Liao, K. Growth arrest induces primary-cilium formation and sensitizes IGF-1-receptor signaling during differentiation induction of 3T3-L1 preadipocytes. *J. Cell Sci.* **2009**, *122*, 2760–2768. [CrossRef]

89. Dalbay, M.T.; Thorpe, S.D.; Connelly, J.T.; Chapple, J.P.; Knight, M.M. Adipogenic Differentiation of hMSCs is Mediated by Recruitment of IGF-1r Onto the Primary Cilium Associated With Cilia Elongation. *Stem Cells* **2015**, *33*, 1952–1961. [CrossRef]

90. Spasic, M.; Jacobs, C.R. Primary cilia: Cell and molecular mechanosensors directing whole tissue function. *Semin. Cell Dev. Biol.* **2017**, *71*, 42–52. [CrossRef]

91. Uzbekov, R.E.; Maurel, D.B.; Aveline, P.C.; Pallu, S.; Benhamou, C.L.; Rochefort, G.Y. Centrosome fine ultrastructure of the osteocyte mechanosensitive primary cilium. *Microsc. Microanal.* **2012**, *18*, 1430–1441. [CrossRef] [PubMed]

92. Takeuchi, K.; Kitano, M.; Ishinaga, H.; Kobayashi, M.; Ogawa, S.; Nakatani, K.; Masuda, S.; Nagao, M.; Fujisawa, T. Recent advances in primary ciliary dyskinesia. *Auris Nasus Larynx* **2016**, *43*, 229–236. [CrossRef] [PubMed]

93. Horani, A.; Ferkol, T.W.; Dutcher, S.K.; Brody, S.L. Genetics and biology of primary ciliary dyskinesia. *Paediatr. Respir. Rev.* **2016**, *18*, 18–24. [CrossRef] [PubMed]

94. Pugacheva, E.N.; Jablonski, S.A.; Hartman, T.R.; Henske, E.P.; Golemis, E.A. HEF1-dependent Aurora A activation induces disassembly of the primary cilium. *Cell* **2007**, *129*, 1351–1363. [CrossRef] [PubMed]

95. Otto, T.; Sicinski, P. Cell cycle proteins as promising targets in cancer therapy. *Nat. Rev. Cancer* **2017**, *17*, 93–115. [CrossRef] [PubMed]

96. Liang, Y.; Meng, D.; Zhu, B.; Pan, J. Mechanism of ciliary disassembly. *Cell. Mol. Life Sci.* **2016**, *73*, 1787–1802. [CrossRef] [PubMed]

97. Korobeynikov, V.; Deneka, A.Y.; Golemis, E.A. Mechanisms for nonmitotic activation of Aurora-A at cilia. *Biochem. Soc. Trans.* **2017**, *45*, 37–49. [CrossRef] [PubMed]

98. Plotnikova, O.V.; Nikonova, A.S.; Loskutov, Y.V.; Kozyulina, P.Y.; Pugacheva, E.N.; Golemis, E.A. Calmodulin activation of Aurora-A kinase (AURKA) is required during ciliary disassembly and in mitosis. *Mol. Biol. Cell* **2012**, *23*, 2658–2670. [CrossRef]

99. Lee, K.H.; Johmura, Y.; Yu, L.R.; Park, J.E.; Gao, Y.; Bang, J.K.; Zhou, M.; Veenstra, T.D.; Yeon Kim, B.; Lee, K.S. Identification of a novel Wnt5a-CK1varepsilon-Dvl2-Plk1-mediated primary cilia disassembly pathway. *EMBO J.* **2012**, *31*, 3104–3117. [CrossRef]

100. Plotnikova, O.V.; Seo, S.; Cottle, D.L.; Conduit, S.; Hakim, S.; Dyson, J.M.; Mitchell, C.A.; Smyth, I.M. INPP5E interacts with AURKA, linking phosphoinositide signaling to primary cilium stability. *J. Cell Sci.* **2015**, *128*, 364–372. [CrossRef]

101. Inoko, A.; Matsuyama, M.; Goto, H.; Ohmuro-Matsuyama, Y.; Hayashi, Y.; Enomoto, M.; Ibi, M.; Urano, T.; Yonemura, S.; Kiyono, T.; et al. Trichoplein and Aurora A block aberrant primary cilia assembly in proliferating cells. *J. Cell Biol.* **2012**, *197*, 391–405. [CrossRef]

102. Gabriel, E.; Wason, A.; Ramani, A.; Gooi, L.M.; Keller, P.; Pozniakovsky, A.; Poser, I.; Noack, F.; Telugu, N.S.; Calegari, F.; et al. CPAP promotes timely cilium disassembly to maintain neural progenitor pool. *EMBO J.* **2016**, *35*, 803–819. [CrossRef]

103. Pazour, G.J.; Wilkerson, C.G.; Witman, G.B. A dynein light chain is essential for the retrograde particle movement of intraflagellar transport (IFT). *J. Cell Biol.* **1998**, *141*, 979–992. [CrossRef] [PubMed]

104. Maskey, D.; Marlin, M.C.; Kim, S.; Kim, S.; Ong, E.C.; Li, G.; Tsiokas, L. Cell cycle-dependent ubiquitylation and destruction of NDE1 by CDK5-FBW7 regulates ciliary length. *EMBO J.* **2015**, *34*, 2424–2440. [CrossRef] [PubMed]

105. Nikonova, A.S.; Golemis, E.A. The tumor suppressor FBW7 controls ciliary length. *EMBO J.* **2015**, *34*, 2388–2390. [CrossRef] [PubMed]

106. Inaba, H.; Goto, H.; Kasahara, K.; Kumamoto, K.; Yonemura, S.; Inoko, A.; Yamano, S.; Wanibuchi, H.; He, D.; Goshima, N.; et al. Ndel1 suppresses ciliogenesis in proliferating cells by regulating the trichoplein-Aurora A pathway. *J. Cell Biol.* **2016**, *212*, 409–423. [CrossRef]

107. Fabbri, L.; Bost, F.; Mazure, N.M. Primary Cilium in Cancer Hallmarks. *Int. J. Mol. Sci.* **2019**, *20*, 1336. [CrossRef]

108. Higgins, M.; Obaidi, I.; McMorrow, T. Primary cilia and their role in cancer. *Oncol. Lett.* **2019**, *17*, 3041–3047. [CrossRef]

109. Peixoto, E.; Richard, S.; Pant, K.; Biswas, A.; Gradilone, S.A. The primary cilium: Its role as a tumor suppressor organelle. *Biochem. Pharm.* **2020**, *175*, 113906. [CrossRef] [PubMed]

110. Urdiciain, A.; Erausquin, E.; Meléndez, B.; Rey, J.A.; Idoate, M.A.; Castresana, J.S. Tubastatin A, an inhibitor of HDAC6, enhances temozolomide-induced apoptosis and reverses the malignant phenotype of glioblastoma cells. *Int. J. Oncol.* **2019**, *54*, 1797–1808. [CrossRef]

111. Chen, Q.; Li, J.; Yang, X.; Ma, J.; Gong, F.; Liu, Y. Prdx1 promotes the loss of primary cilia in esophageal squamous cell carcinoma. *BMC Cancer* **2020**, *20*, 372. [CrossRef]

112. Rocha, C.; Papon, L.; Cacheux, W.; Marques Sousa, P.; Lascano, V.; Tort, O.; Giordano, T.; Vacher, S.; Lemmers, B.; Mariani, P.; et al. Tubulin glycylases are required for primary cilia, control of cell proliferation and tumor development in colon. *EMBO J.* **2014**, *33*, 2247–2260. [CrossRef]

113. Gradilone, S.A.; Radtke, B.N.; Bogert, P.S.; Huang, B.Q.; Gajdos, G.B.; LaRusso, N.F. HDAC6 inhibition restores ciliary expression and decreases tumor growth. *Cancer Res.* **2013**, *73*, 2259–2270. [CrossRef] [PubMed]

114. Mansini, A.P.; Peixoto, E.; Thelen, K.M.; Gaspari, C.; Jin, S.; Gradilone, S.A. The cholangiocyte primary cilium in health and disease. *Biochim. Biophys. Acta* **2018**, *1864*, 1245–1253. [CrossRef] [PubMed]

115. Kobayashi, T.; Nakazono, K.; Tokuda, M.; Mashima, Y.; Dynlacht, B.D.; Itoh, H. HDAC2 promotes loss of primary cilia in pancreatic ductal adenocarcinoma. *EMBO Rep.* **2017**, *18*, 334–343. [CrossRef] [PubMed]

116. Esteban, M.A.; Harten, S.K.; Tran, M.G.; Maxwell, P.H. Formation of primary cilia in the renal epithelium is regulated by the von Hippel-Lindau tumor suppressor protein. *J. Am. Soc. Nephrol.* **2006**, *17*, 1801–1806. [CrossRef]

117. Qie, Y.; Wang, L.; Du, E.; Chen, S.; Lu, C.; Ding, N.; Yang, K.; Xu, Y. TACC3 promotes prostate cancer cell proliferation and restrains primary cilium formation. *Exp. Cell Res.* **2020**, *390*, 111952. [CrossRef]

118. Bhattacharya, R.; Kwon, J.; Ali, B.; Wang, E.; Patra, S.; Shridhar, V.; Mukherjee, P. Role of hedgehog signaling in ovarian cancer. *Clin. Cancer Res.* **2008**, *14*, 7659–7666. [CrossRef]

119. Egeberg, D.L.; Lethan, M.; Manguso, R.; Schneider, L.; Awan, A.; Jorgensen, T.S.; Byskov, A.G.; Pedersen, L.B.; Christensen, S.T. Primary cilia and aberrant cell signaling in epithelial ovarian cancer. *Cilia* **2012**, *1*, 15. [CrossRef]

120. Zingg, D.; Debbache, J.; Peña-Hernández, R.; Antunes, A.T.; Schaefer, S.M.; Cheng, P.F.; Zimmerli, D.; Haeusel, J.; Calçada, R.R.; Tuncer, E.; et al. EZH2-Mediated Primary Cilium Deconstruction Drives Metastatic Melanoma Formation. *Cancer Cell* **2018**, *34*, 69–84. [CrossRef]

121. Xiang, W.; Guo, F.; Cheng, W.; Zhang, J.; Huang, J.; Wang, R.; Ma, Z.; Xu, K. HDAC6 inhibition suppresses chondrosarcoma by restoring the expression of primary cilia. *Oncol. Rep.* **2017**, *38*, 229–236. [CrossRef]

122. Wong, S.Y.; Seol, A.D.; So, P.L.; Ermilov, A.N.; Bichakjian, C.K.; Epstein, E.H., Jr.; Dlugosz, A.A.; Reiter, J.F. Primary cilia can both mediate and suppress Hedgehog pathway-dependent tumorigenesis. *Nat. Med.* **2009**, *15*, 1055–1061. [CrossRef]

123. Han, Y.G.; Kim, H.J.; Dlugosz, A.A.; Ellison, D.W.; Gilbertson, R.J.; Alvarez-Buylla, A. Dual and opposing roles of primary cilia in medulloblastoma development. *Nat. Med.* **2009**, *15*, 1062–1065. [CrossRef] [PubMed]

124. Jeong, M.; Lee, E.W.; Seong, D.; Seo, J.; Kim, J.H.; Grootjans, S.; Kim, S.Y.; Vandenabeele, P.; Song, J. USP8 suppresses death receptor-mediated apoptosis by enhancing FLIP(L) stability. *Oncogene* **2017**, *36*, 458–470. [CrossRef] [PubMed]

125. MacLeod, G.; Bozek, D.A.; Rajakulendran, N.; Monteiro, V.; Ahmadi, M.; Steinhart, Z.; Kushida, M.M.; Yu, H.; Coutinho, F.J.; Cavalli, F.M.G.; et al. Genome-Wide CRISPR-Cas9 Screens Expose Genetic Vulnerabilities and Mechanisms of Temozolomide Sensitivity in Glioblastoma Stem Cells. *Cell Rep.* **2019**, *27*, 971–986. [CrossRef] [PubMed]

126. Das, A.; Qian, J.; Tsang, W.Y. USP9X counteracts differential ubiquitination of NPHP5 by MARCH7 and BBS11 to regulate ciliogenesis. *PLoS Genet.* **2017**, *13*, e1006791. [CrossRef] [PubMed]

127. Hu, J.; Meng, Y.; Zeng, J.; Zeng, B.; Jiang, X. Ubiquitin E3 Ligase MARCH7 promotes proliferation and invasion of cervical cancer cells through VAV2-RAC1-CDC42 pathway. *Oncol. Lett.* **2018**, *16*, 2312–2318. [CrossRef] [PubMed]

128. Kano, S.; Miyajima, N.; Fukuda, S.; Hatakeyama, S. Tripartite motif protein 32 facilitates cell growth and migration via degradation of Abl-interactor 2. *Cancer Res.* **2008**, *68*, 5572–5580. [CrossRef]

129. Horn, E.J.; Albor, A.; Liu, Y.; El-Hizawi, S.; Vanderbeek, G.E.; Babcock, M.; Bowden, G.T.; Hennings, H.; Lozano, G.; Weinberg, W.C.; et al. RING protein Trim32 associated with skin carcinogenesis has anti-apoptotic and E3-ubiquitin ligase properties. *Carcinogenesis* **2004**, *25*, 157–167. [CrossRef]

130. Reijnders, M.R.; Zachariadis, V.; Latour, B.; Jolly, L.; Mancini, G.M.; Pfundt, R.; Wu, K.M.; van Ravenswaaij-Arts, C.M.; Veenstra-Knol, H.E.; Anderlid, B.M.; et al. De Novo Loss-of-Function Mutations in USP9X Cause a Female-Specific Recognizable Syndrome with Developmental Delay and Congenital Malformations. *Am. J. Hum. Genet.* **2016**, *98*, 373–381. [CrossRef]

131. Pérez-Mancera, P.A.; Rust, A.G.; van der Weyden, L.; Kristiansen, G.; Li, A.; Sarver, A.L.; Silverstein, K.A.T.; Grützmann, R.; Aust, D.; Rümmele, P.; et al. The deubiquitinase USP9X suppresses pancreatic ductal adenocarcinoma. *Nature* **2012**, *486*, 266–270. [CrossRef]

132. Yang, Y.; Ran, J.; Liu, M.; Li, D.; Li, Y.; Shi, X.; Meng, D.; Pan, J.; Ou, G.; Aneja, R.; et al. CYLD mediates ciliogenesis in multiple organs by deubiquitinating Cep70 and inactivating HDAC6. *Cell Res.* **2014**, *24*, 1342–1353. [CrossRef]

133. Douanne, T.; André-Grégoire, G.; Thys, A.; Trillet, K.; Gavard, J.; Bidère, N. CYLD Regulates Centriolar Satellites Proteostasis by Counteracting the E3 Ligase MIB1. *Cell Rep.* **2019**, *27*, 1657–1665. [CrossRef]

134. Tse, W.K.F. Importance of deubiquitinases in zebrafish craniofacial development. *Biochem. Biophys. Res. Commun.* **2017**, *487*, 813–819. [CrossRef] [PubMed]

135. Bignell, G.R.; Warren, W.; Seal, S.; Takahashi, M.; Rapley, E.; Barfoot, R.; Green, H.; Brown, C.; Biggs, P.J.; Lakhani, S.R.; et al. Identification of the familial cylindromatosis tumour-suppressor gene. *Nat. Genet.* **2000**, *25*, 160–165. [CrossRef]

136. Wang, L.; Lee, K.; Malonis, R.; Sanchez, I.; Dynlacht, B.D. Tethering of an E3 ligase by PCM1 regulates the abundance of centrosomal KIAA0586/Talpid3 and promotes ciliogenesis. *Elife* **2016**, *5*, e12950. [CrossRef]

137. Lei, Y.; Li, Z.; Qi, L.; Tong, S.; Li, B.; He, W.; Chen, M. The Prognostic Role of Ki-67/MIB-1 in Upper Urinary-Tract Urothelial Carcinomas: A Systematic Review and Meta-Analysis. *J. Endourol.* **2015**, *29*, 1302–1308. [CrossRef]

138. Ang, S.O.; Chen, H.; Hirota, K.; Gordeuk, V.R.; Jelinek, J.; Guan, Y.; Liu, E.; Sergueeva, A.I.; Miasnikova, G.Y.; Mole, D.; et al. Disruption of oxygen homeostasis underlies congenital Chuvash polycythemia. *Nat. Genet.* **2002**, *32*, 614–621. [CrossRef] [PubMed]

139. Schermer, B.; Ghenoiu, C.; Bartram, M.; Müller, R.U.; Kotsis, F.; Höhne, M.; Kühn, W.; Rapka, M.; Nitschke, R.; Zentgraf, H.; et al. The von Hippel-Lindau tumor suppressor protein controls ciliogenesis by orienting microtubule growth. *J. Cell Biol.* **2006**, *175*, 547–554. [CrossRef] [PubMed]

140. Kaelin, W.G., Jr. The VHL Tumor Suppressor Gene: Insights into Oxygen Sensing and Cancer. *Trans. Am. Clin. Clim. Assoc.* **2017**, *128*, 298–307.

141. Nishizawa, M.; Izawa, I.; Inoko, A.; Hayashi, Y.; Nagata, K.; Yokoyama, T.; Usukura, J.; Inagaki, M. Identification of trichoplein, a novel keratin filament-binding protein. *J. Cell Sci.* **2005**, *118*, 1081–1090. [CrossRef] [PubMed]

142. Ibi, M.; Zou, P.; Inoko, A.; Shiromizu, T.; Matsuyama, M.; Hayashi, Y.; Enomoto, M.; Mori, D.; Hirotsune, S.; Kiyono, T.; et al. Trichoplein controls microtubule anchoring at the centrosome by binding to Odf2 and ninein. *J. Cell Sci.* **2011**, *124*, 857–864. [CrossRef]

143. Sasaki, S.; Shionoya, A.; Ishida, M.; Gambello, M.J.; Yingling, J.; Wynshaw-Boris, A.; Hirotsune, S. A LIS1/NUDEL/cytoplasmic dynein heavy chain complex in the developing and adult nervous system. *Neuron* **2000**, *28*, 681–696. [CrossRef]

144. Niethammer, M.; Smith, D.S.; Ayala, R.; Peng, J.; Ko, J.; Lee, M.S.; Morabito, M.; Tsai, L.H. NUDEL is a novel Cdk5 substrate that associates with LIS1 and cytoplasmic dynein. *Neuron* **2000**, *28*, 697–711. [CrossRef]

145. Barbelanne, M.; Song, J.; Ahmadzai, M.; Tsang, W.Y. Pathogenic NPHP5 mutations impair protein interaction with Cep290, a prerequisite for ciliogenesis. *Hum. Mol. Genet.* **2013**, *22*, 2482–2494. [CrossRef]

146. Chiang, A.P.; Beck, J.S.; Yen, H.J.; Tayeh, M.K.; Scheetz, T.E.; Swiderski, R.E.; Nishimura, D.Y.; Braun, T.A.; Kim, K.Y.; Huang, J.; et al. Homozygosity mapping with SNP arrays identifies TRIM32, an E3 ubiquitin ligase, as a Bardet-Biedl syndrome gene (BBS11). *Proc. Natl. Acad. Sci. USA* **2006**, *103*, 6287–6292. [CrossRef] [PubMed]

147. Otto, E.A.; Loeys, B.; Khanna, H.; Hellemans, J.; Sudbrak, R.; Fan, S.; Muerb, U.; O'Toole, J.F.; Helou, J.; Attanasio, M.; et al. Nephrocystin-5, a ciliary IQ domain protein, is mutated in Senior-Loken syndrome and interacts with RPGR and calmodulin. *Nat. Genet.* **2005**, *37*, 282–288. [CrossRef] [PubMed]

148. Yang, Y.; Zhou, J. CYLD—A deubiquitylase that acts to fine-tune microtubule properties and functions. *J. Cell Sci.* **2016**, *129*, 2289–2295. [CrossRef]

149. Farooqi, I.S. Genetic and hereditary aspects of childhood obesity. *Best Pract. Res. Clin. Endocrinol. Metab.* **2005**, *19*, 359–374. [CrossRef]

150. Stephen, L.A.; Tawamie, H.; Davis, G.M.; Tebbe, L.; Nürnberg, P.; Nürnberg, G.; Thiele, H.; Thoenes, M.; Boltshauser, E.; Uebe, S.; et al. TALPID3 controls centrosome and cell polarity and the human ortholog KIAA0586 is mutated in Joubert syndrome (JBTS23). *Elife* **2015**, *4*, e08077. [CrossRef]

151. Kamura, T.; Koepp, D.M.; Conrad, M.N.; Skowyra, D.; Moreland, R.J.; Iliopoulos, O.; Lane, W.S.; Kaelin, W.G., Jr.; Elledge, S.J.; Conaway, R.C.; et al. Rbx1, a component of the VHL tumor suppressor complex and SCF ubiquitin ligase. *Science* **1999**, *284*, 657–661. [CrossRef]

152. Kuehn, E.W.; Walz, G.; Benzing, T. Von hippel-lindau: A tumor suppressor links microtubules to ciliogenesis and cancer development. *Cancer Res.* **2007**, *67*, 4537–4540. [CrossRef]

153. Ohh, M.; Park, C.W.; Ivan, M.; Hoffman, M.A.; Kim, T.-Y.; Huang, L.E.; Pavletich, N.; Chau, V.; Kaelin, W.G. Ubiquitination of hypoxia-inducible factor requires direct binding to the β-domain of the von Hippel–Lindau protein. *Nat. Cell Biol.* **2000**, *2*, 423–427. [CrossRef]

154. Hayashi, Y.; Yokota, A.; Harada, H.; Huang, G. Hypoxia/pseudohypoxia-mediated activation of hypoxia-inducible factor-1α in cancer. *Cancer Sci.* **2019**, *110*, 1510–1517. [CrossRef] [PubMed]

155. Huang, K.; Diener, D.R.; Rosenbaum, J.L. The ubiquitin conjugation system is involved in the disassembly of cilia and flagella. *J. Cell Biol.* **2009**, *186*, 601–613. [CrossRef] [PubMed]

156. D'Angiolella, V.; Donato, V.; Vijayakumar, S.; Saraf, A.; Florens, L.; Washburn, M.P.; Dynlacht, B.; Pagano, M. SCF(Cyclin F) controls centrosome homeostasis and mitotic fidelity through CP110 degradation. *Nature* **2010**, *466*, 138–142. [CrossRef] [PubMed]

157. Mick, D.U.; Rodrigues, R.B.; Leib, R.D.; Adams, C.M.; Chien, A.S.; Gygi, S.P.; Nachury, M.V. Proteomics of Primary Cilia by Proximity Labeling. *Dev. Cell* **2015**, *35*, 497–512. [CrossRef] [PubMed]

158. Shearer, R.F.; Frikstad, K.M.; McKenna, J.; McCloy, R.A.; Deng, N.; Burgess, A.; Stokke, T.; Patzke, S.; Saunders, D.N. The E3 ubiquitin ligase UBR5 regulates centriolar satellite stability and primary cilia. *Mol. Biol. Cell* **2018**, *29*, 1542–1554. [CrossRef] [PubMed]

159. Massa, F.; Tammaro, R.; Prado, M.A.; Cesana, M.; Lee, B.H.; Finley, D.; Franco, B.; Morleo, M. The deubiquitinating enzyme Usp14 controls ciliogenesis and Hedgehog signaling. *Hum. Mol. Genet.* **2019**, *28*, 764–777. [CrossRef]

160. Li, J.; D'Angiolella, V.; Seeley, E.S.; Kim, S.; Kobayashi, T.; Fu, W.; Campos, E.I.; Pagano, M.; Dynlacht, B.D. USP33 regulates centrosome biogenesis via deubiquitination of the centriolar protein CP110. *Nature* **2013**, *495*, 255–259. [CrossRef]

161. Wheway, G.; Schmidts, M.; Mans, D.A.; Szymanska, K.; Nguyen, T.T.; Racher, H.; Phelps, I.G.; Toedt, G.; Kennedy, J.; Wunderlich, K.A.; et al. An siRNA-based functional genomics screen for the identification of regulators of ciliogenesis and ciliopathy genes. *Nat. Cell Biol.* **2015**, *17*, 1074–1087. [CrossRef]

162. Kim, J.H.; Ki, S.M.; Joung, J.G.; Scott, E.; Heynen-Genel, S.; Aza-Blanc, P.; Kwon, C.H.; Kim, J.; Gleeson, J.G.; Lee, J.E. Genome-wide screen identifies novel machineries required for both ciliogenesis and cell cycle arrest upon serum starvation. *Biochim. Biophys. Acta* **2016**, *1863*, 1307–1318. [CrossRef]

163. Kohli, P.; Höhne, M.; Jüngst, C.; Bertsch, S.; Ebert, L.K.; Schauss, A.C.; Benzing, T.; Rinschen, M.M.; Schermer, B. The ciliary membrane-associated proteome reveals actin-binding proteins as key components of cilia. *EMBO Rep.* **2017**, *18*, 1521–1535. [CrossRef] [PubMed]

164. Lundberg, E.; Borner, G.H.H. Spatial proteomics: A powerful discovery tool for cell biology. *Nat. Rev. Mol. Cell Biol.* **2019**, *20*, 285–302. [CrossRef] [PubMed]

165. Kliza, K.; Husnjak, K. Resolving the Complexity of Ubiquitin Networks. *Front. Mol. Biosci.* **2020**, *7*, 21. [CrossRef] [PubMed]

166. Poulsen, J.W.; Madsen, C.T.; Young, C.; Kelstrup, C.D.; Grell, H.C.; Henriksen, P.; Juhl-Jensen, L.; Nielsen, M.L. Comprehensive profiling of proteome changes upon sequential deletion of deubiquitylating enzymes. *J. Proteom.* **2012**, *75*, 3886–3897. [CrossRef] [PubMed]

167. Chen, X.; Lu, D.; Gao, J.; Zhu, H.; Zhou, Y.; Gao, D.; Zhou, H. Identification of a USP9X Substrate NFX1-123 by SILAC-Based Quantitative Proteomics. *J. Proteome Res.* **2019**, *18*, 2654–2665. [CrossRef] [PubMed]

168. Gomez-Ferreria, M.A.; Bashkurov, M.; Mullin, M.; Gingras, A.-C.; Pelletier, L. CEP192 interacts physically and functionally with the K63-deubiquitinase CYLD to promote mitotic spindle assembly. *Cell Cycle* **2012**, *11*, 3555–3558. [CrossRef]

169. Nakayasu, E.S.; Sydor, M.A.; Brown, R.N.; Sontag, R.L.; Sobreira, T.J.P.; Slysz, G.W.; Humphrys, D.R.; Skarina, T.; Onoprienko, O.; Di Leo, R.; et al. Identification of Salmonella Typhimurium Deubiquitinase SseL Substrates by Immunoaffinity Enrichment and Quantitative Proteomic Analysis. *J. Proteome Res.* **2015**, *14*, 4029–4038. [CrossRef]

170. Iconomou, M.; Saunders, D.N. Systematic approaches to identify E3 ligase substrates. *Biochem. J.* **2016**, *473*, 4083–4101. [CrossRef]

171. O'Connor, H.F.; Huibregtse, J.M. Enzyme-substrate relationships in the ubiquitin system: Approaches for identifying substrates of ubiquitin ligases. *Cell. Mol. Life Sci.* **2017**, *74*, 3363–3375. [CrossRef]

172. Mattern, M.; Sutherland, J.; Kadimisetty, K.; Barrio, R.; Rodriguez, M.S. Using Ubiquitin Binders to Decipher the Ubiquitin Code. *Trends Biochem. Sci.* **2019**, *44*, 599–615. [CrossRef]

173. Huang, Q.; Zhang, X. Emerging Roles and Research Tools of Atypical Ubiquitination. *Proteomics* **2020**, *20*, 1900100. [CrossRef] [PubMed]

174. Dammermann, A.; Merdes, A. Assembly of centrosomal proteins and microtubule organization depends on PCM-1. *J. Cell Biol.* **2002**, *159*, 255–266. [CrossRef] [PubMed]

175. Shinde, S.R.; Nager, A.R.; Nachury, M.V. Lysine63-linked ubiquitin chains earmark GPCRs for BBSome-mediated removal from cilia. *bioRxiv* **2020**. Available online: https://www.biorxiv.org/content/10.1101/2020.03.04.977090v1 (accessed on 17 August 2020).

176. Bhagwat, S.R.; Hajela, K.; Kumar, A. Proteolysis to Identify Protease Substrates: Cleave to Decipher. *Proteomics* **2018**, *18*, e1800011. [CrossRef] [PubMed]

177. Wu, H.Q.; Baker, D.; Ovaa, H. Small molecules that target the ubiquitin system. *Biochem Soc. Trans.* **2020**, *48*, 479–497. [CrossRef] [PubMed]

178. Ran, X.; Gestwicki, J.E. Inhibitors of protein–protein interactions (PPIs): An analysis of scaffold choices and buried surface area. *Curr. Opin. Chem. Biol.* **2018**, *44*, 75–86. [CrossRef]

179. Andrei, S.A.; Sijbesma, E.; Hann, M.; Davis, J.; O'Mahony, G.; Perry, M.W.D.; Karawajczyk, A.; Eickhoff, J.; Brunsveld, L.; Doveston, R.G.; et al. Stabilization of protein-protein interactions in drug discovery. *Expert Opin. Drug Discov.* **2017**, *12*, 925–940. [CrossRef]

180. Huang, X.; Dixit, V.M. Drugging the undruggables: Exploring the ubiquitin system for drug development. *Cell Res.* **2016**, *26*, 484–498. [CrossRef]

181. Wu, L.; Grigoryan, A.V.; Li, Y.; Hao, B.; Pagano, M.; Cardozo, T.J. Specific small molecule inhibitors of Skp2-mediated p27 degradation. *Chem. Biol.* **2012**, *19*, 1515–1524. [CrossRef]

182. Ganoth, D.; Bornstein, G.; Ko, T.K.; Larsen, B.; Tyers, M.; Pagano, M.; Hershko, A. The cell-cycle regulatory protein Cks1 is required for SCF(Skp2)-mediated ubiquitinylation of p27. *Nat. Cell Biol.* **2001**, *3*, 321–324. [CrossRef]

183. Spruck, C.; Strohmaier, H.; Watson, M.; Smith, A.P.; Ryan, A.; Krek, T.W.; Reed, S.I. A CDK-independent function of mammalian Cks1: Targeting of SCF(Skp2) to the CDK inhibitor p27Kip1. *Mol. Cell* **2001**, *7*, 639–650. [CrossRef]

184. Das, T.; Shin, S.C.; Song, E.J.; Kim, E.E. Regulation of Deubiquitinating Enzymes by Post-Translational Modifications. *Int. J. Mol. Sci.* **2020**, *21*, 4028. [CrossRef] [PubMed]

185. Das, T.; Kim, E.E.; Song, E.J. Phosphorylation of USP15 and USP4 Regulates Localization and Spliceosomal Deubiquitination. *J. Mol. Biol* **2019**, *431*, 3900–3912. [CrossRef]

186. Yamanaka, S.; Sato, Y.; Oikawa, D.; Goto, E.; Fukai, S.; Tokunaga, F.; Takahashi, H.; Sawasaki, T. Subquinocin, a small molecule inhibitor of CYLD and USP-family deubiquitinating enzymes, promotes NF-κB signaling. *Biochem. Biophys. Res. Commun.* **2020**, *524*, 1–7. [CrossRef] [PubMed]

187. Dufner, A.; Knobeloch, K.P. Ubiquitin-specific protease 8 (USP8/UBPy): A prototypic multidomain deubiquitinating enzyme with pleiotropic functions. *Biochem. Soc. Trans.* **2019**, *47*, 1867–1879. [CrossRef]

188. Murtaza, M.; Jolly, L.A.; Gecz, J.; Wood, S.A. La FAM fatale: USP9X in development and disease. *Cell. Mol. Life Sci.* **2015**, *72*, 2075–2089. [CrossRef] [PubMed]

International Journal of
Molecular Sciences

Review

Role of the Ubiquitin Proteasome System in the Regulation of Blood Pressure: A Review

Osamu Yamazaki, Daigoro Hirohama, Kenichi Ishizawa and Shigeru Shibata *

Division of Nephrology, Department of Internal Medicine, Teikyo University School of Medicine,
Tokyo 173-8605, Japan; osamu195-tky@umin.ac.jp (O.Y.); hirohama-tky@umin.ac.jp (D.H.);
ken.ishizawa@gmail.com (K.I.)
* Correspondence: shigeru.shibata@med.teikyo-u.ac.jp; Tel.: +81-3-3964-1211

Received: 10 July 2020; Accepted: 24 July 2020; Published: 28 July 2020

Abstract: The kidney and the vasculature play crucial roles in regulating blood pressure. The ubiquitin proteasome system (UPS), a multienzyme process mediating covalent conjugation of the 76-amino acid polypeptide ubiquitin to a substrate protein followed by proteasomal degradation, is involved in multiple cellular processes by regulating protein turnover in various tissues. Increasing evidence demonstrates the roles of UPS in blood pressure regulation. In the kidney, filtered sodium is reabsorbed through diverse sodium transporters and channels along renal tubules, and studies conducted till date have provided insights into the complex molecular network through which ubiquitin ligases modulate sodium transport in different segments. Components of these pathways include ubiquitin ligase neuronal precursor cell-expressed developmentally downregulated 4-2, Cullin-3, and Kelch-like 3. Moreover, accumulating data indicate the roles of UPS in blood vessels, where it modulates nitric oxide bioavailability and vasoconstriction. Cullin-3 not only regulates renal salt reabsorption but also controls vascular tone using different adaptor proteins that target distinct substrates in vascular smooth muscle cells. In endothelial cells, UPS can also contribute to blood pressure regulation by modulating endothelial nitric oxide synthase. In this review, we summarize current knowledge regarding the role of UPS in blood pressure regulation, focusing on renal sodium reabsorption and vascular function.

Keywords: blood pressure; renal salt reabsorption; vascular function; ubiquitin proteasome system

---

1. Introduction

Hypertension is not only one of the most frequent diseases in the world, but it is also a key risk factor for cardiovascular disease and renal dysfunction. The kidney plays a pivotal role in the regulation of body fluid levels and blood pressure (BP), and an impaired kidney function comprises a major mechanism that alters the salt sensitivity of BP [1]. Because renal salt handling is critical for maintaining an independent life for terrestrial mammals, these animals have developed highly differentiated diverse tubule cells that are involved in the transport of sodium and other ions. The major renal sodium transporters and channels include Na^+/H^+ exchanger isoform 3 (NHE3) in the proximal tubule (PT), Na^+-K^+-$2Cl^-$ cotransporter (NKCC2) in the thick ascending limb (TAL), Na^+-Cl^- cotransporter (NCC) in the distal convoluted tubule (DCT), and epithelial sodium channel (ENaC) and Cl^-/HCO_3^- exchanger pendrin in the connecting tubule (CNT) and the collecting duct (CD). The significance of several of these transporters and their regulators in the renal nephron has been confirmed by the monogenic hypertensive or hypotensive disorders [2–4], as well as by the clinical efficacy of the pharmacological agents that block these sodium transport mechanisms.

In addition to the role of the kidney, it is well known that the dysregulation of vascular function significantly contributes to BP elevation [5,6]. The arterial wall consists of intimal endothelial cells, vascular smooth muscle cells, and adventitia. Vascular endothelial cells (VECs) play vital roles in regulating diverse biological functions by secreting various vasoactive factors, including nitric oxide (NO). NO, a strong vasodilator that tightly modulates vascular function, is primarily produced by endothelial NO synthase (eNOS) in endothelial cells [7]. Studies have demonstrated that both genetic and pharmacological ablation of eNOS elicits significant BP elevations [8–10]. Vascular smooth muscle cells (VSMCs) also play important roles in controlling the tonus of blood vessels, thereby regulating BP levels [11].

Ubiquitylation is a stepwise process involving three classes of enzymes. Ubiquitin-activating enzymes (E1s) activate the ubiquitin molecule combined with ATP hydrolysis [12]. Ubiquitin is then transferred to ubiquitin-conjugating enzymes (E2s) with an active cysteine [13,14]. Following this, ubiquitin is transferred to substrates via the ubiquitin protein ligases (E3s). Humans have only one E1, ~40 E2s, and 500–1000 of E3s [15–17]. Two types of E3s exist, termed the homologous to the E6-AP C terminus and the really interesting new gene (RING). E3s provide substrate specificity to the ubiquitin system and recognize multiple substrates through different protein–protein interactions, thus regulating multiple cellular processes, including DNA damage repair, cell cycle progression, development, and signal transduction. Given that the ubiquitin proteasome system (UPS) enables adaptation to physiological challenges by controlling the protein abundance of target substrates, the involvement of UPS in BP regulation has attracted extensive research attention. In this article, we review the role of UPS in BP homeostasis, especially focusing on sodium transporters of the kidney and vascular functions (Figure 1).

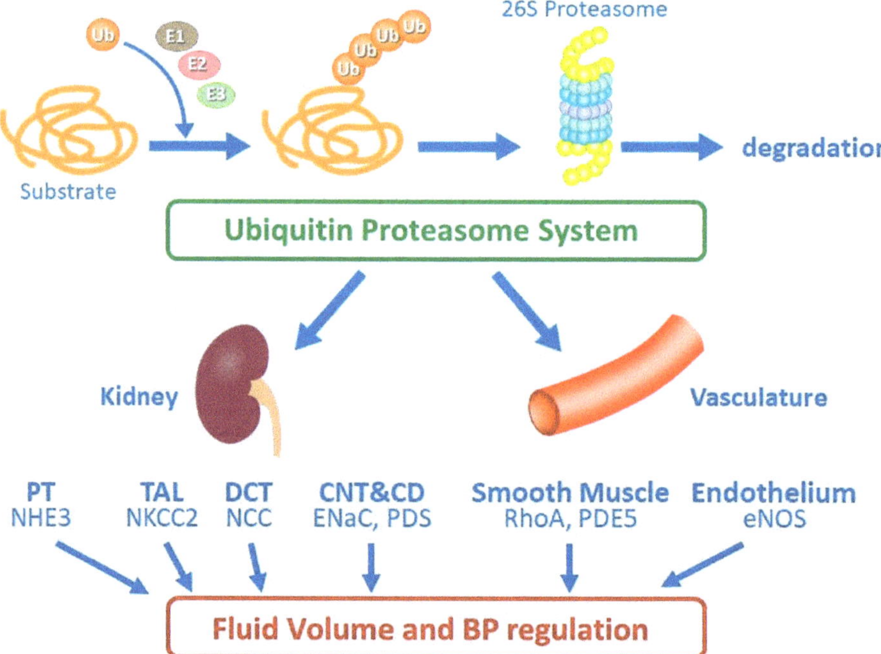

Figure 1. The role of ubiquitin proteasome system in fluid volume and blood pressure regulation. PT, proximal tubule; TAL, thick ascending limb; DCT, distal convoluted tubule; CNT, connecting tubule; CD, collecting duct.

2. Role of UPS in the Regulation of Tubular Function in the Kidney

2.1. Proximal Tubule

Among the salt transport mechanisms in the PT, NHE3 has a major role in sodium reabsorption in this segment [18,19]. Human NHE3 contains a PY motif that binds to ubiquitin ligase neuronal precursor cell-expressed developmentally downregulated 4-2 (NEDD4-2), and this interaction can modulate cell surface expression and internalization of NHE3 [20], although it is unclear whether NHE3 is directly ubiquitylated by Nedd4-2. It is interesting to note that this interaction appears to be exclusive to humans and several primates, which is because the PY motif in NHE3 was not identified in other mammals in the alignment analysis. Hatanaka et al. reported that angiotensin II signaling alters NHE3 levels, thereby regulating salt sensitivity [21]. Using subtotal nephrectomized mice, they showed that NHE3 abundance was lower in subtotal nephrectomized mice receiving azilsartan, an angiotensin II receptor 1 (AT1R) blocker, than in those receiving vehicle and that lactacystin, a proteasome inhibitor, blocked the azilsartan-induced decrease in NHE3 expression. These data indicate that NHE3 levels are regulated by UPS that are modulated by AT1R signaling. It currently remains unknown whether the interaction between NHE3 and NEDD4-2 is regulated by angiotensin II.

2.2. Thick Ascending Limb

NKCC2, a target of loop diuretics such as furosemide, regulates sodium reabsorption in the TAL [22]. Regarding the UPS-mediated modulation of NKCC2, Wu et al. reported a role of UPS in the regulation of NKCC2 abundance in a high-salt condition [23]. They used a cytochrome P450 4F2 transgenic mouse model, which exhibited an increased production of 20-hydroxyeicosatetraenoic acid (20-HETE), a regulator of vascular tone and renal sodium reabsorption, by blocking Ca^{2+}-activated K^+ channels [23]. Compared with wild-type mice, the transgenic mice displayed a profound decrease in renal NKCC2 abundance in response to a high-salt diet. This effect was not accompanied by the changes in NKCC2 mRNA expression but increased the abundance of ubiquitylated NKCC2. NKCC2 interacted with NEDD4-2, suggesting a role of this ubiquitin enzyme in the regulation of NKCC2 abundance. Another study reported that dibutyryl cyclic GMP (db-cGMP), a cell-permeable cGMP analog, decreased NKCC2 levels by increasing NKCC2 ubiquitylation and proteasomal degradation in rats [24]. In that study, db-cGMP induced a significant reduction in surface NKCC2 levels in suspensions of rat medullary TALs, which was inhibited in the presence of the proteasome inhibitor MG132. Furthermore, that study reported that NKCC2 levels were constitutively ubiquitylated and that the process was promoted by db-cGMP [24]. Pathways that modulate NKCC2 ubiquitylation at the downstream of cGMP signaling remains to be determined. Given that phosphorylation can regulate the interaction between substrates and ubiquitin ligases, roles of cGMP-dependent kinase might be worth exploring in future studies.

2.3. Distal Convoluted Tubule

2.3.1. KLHL3-Mediated WNK4 Ubiquitylation and NCC

NCC, a target of thiazide diuretics, modulates sodium reabsorption in the DCT, and accumulating evidence has demonstrated its importance in controlling BP [25]. Familial hyperkalemic hypertension, also known as pseudohypoaldosteronism type II (PHAII) or Gordon syndrome, is characterized by salt-sensitive hypertension, hyperkalemia, and metabolic acidosis [26–28]. The phenotypes in these patients can be reversed by thiazide treatment, thus suggesting the involvement of NCC in the pathogenesis of PHAII. Till date, mutations in four genes are known to cause PHAII, which include serine-threonine kinase with-no-lysine (WNK) 1 and WNK4 and Cullin 3 (CUL3) and Kelch-like 3 (KLHL3), the components of the cullin-RING ubiquitin ligase (CRL) complex [3,28,29]. WNKs are substrates for the KLHL3-CUL3 ligase complex. WNKs phosphorylate the downstream kinases STE20/SPS1-related proline-alanine-rich protein kinase (SPAK) and oxidative stress-responsive 1 [30], which in turn increase the levels of phosphorylated NCC, an active form of NCC. We and others have identified by mass spectrometry and co-immunoprecipitation that KLHL3 normally binds to WNK1 and WNK4 [4,31–33]. KLHL3-WNK4 binding induces ubiquitylation in at least 15 specific sites, leading to reduced WNK4 levels [4]. KLHL3 is phosphorylated at serine 433 (KLHL3^{S433}) in the Kelch domain, which is regulated by angiotensin II–protein kinase C signaling [34]. Of interest, this site is recurrently mutated in independent PHAII families, and phosphorylation or single amino acid substitution of this site impairs the binding of KLHL3 with WNKs, resulting in its accumulation and activation of downstream signaling. It has also been reported that Akt and protein kinase A (PKA), key downstream substrates of insulin and vasopressin signaling, respectively, increase phosphorylated KLHL3^{S433} [35]. In addition, calcium-sensing receptor signaling can modulate KLHL3-WNK4-SPAK pathway by phosphorylating KLHL3 and WNK4 [36,37]. Conversely, phosphatase calcineurin is capable of dephosphorylating KLHL3 phosphorylation at KLHL3^{S433} [38]. These mechanisms probably play important roles in several pathological conditions such as low-K$^+$-induced BP elevation and hypertension associated with obese diabetes mellitus [39,40]. CRLs are activated by neddylation of cullin. It has been demonstrated that CUL3 is also neddylated, and that its neddylation status is regulated by multisubunit deneddylase COP9 signalosome [41,42].

2.3.2. NEDD4-2-Mediated Ubiquitylation and NCC

Accumulating data also indicate that the ubiquitin ligase NEDD4-2 regulates NCC. Arroyo et al. demonstrated that in cultured cells, NEDD4-2 interacts with NCC, resulting in its ubiquitylation and reduced cell surface expression [43]. They also observed that serum/glucocorticoid-regulated kinase 1 (Sgk1) prevented the NEDD4-2-mediated deactivation of NCC in a kinase-dependent manner, indicating that Sgk1 is also involved in the NEDD4-2-mediated NCC regulation [43]. The role of Sgk1 in regulating NEDD4-2 and NCC has been demonstrated in vivo in Sgk1 knockout mice [44]. In another study, tetracycline-inducible, nephron-specific NEDD4-2 knockout mice exhibited increased NCC protein levels and salt-sensitive hypertension [45]. The mRNA expression of NCC remained unchanged, suggesting that NEDD4-2 regulates NCC abundance at the post-transcriptional level. Roy et al. reported that NEDD4-2 regulates NCC function through WNK1 [46]. They identified two alternatively spliced exons within a proline-rich region of WNK1 that contain PY motifs. NEDD4-2 binds to the PY motifs of WNK1, ubiquitylating WNK1 and targeting it for proteasomal degradation [46]. Dysregulation of NEDD4-2 has been implicated in the pathophysiology of salt-sensitive hypertension in a model of chronic kidney disease, which resulted in NCC activation through WNK1/SPAK [47]. In a recent study, Wu et al. reported that NEDD4-2 modulated NCC levels through a mechanism involving basolateral K$^+$ channel Kir4.1 (KCNJ10) [48]. The authors observed that kidney-specific deletion of NEDD4-2 hyperpolarized the DCT membrane, accompanied by the increase in NCC abundance. These changes were abolished in kidney-specific NEDD4-2/KCNJ10 double-knockout mice, leading to the suppression of NCC and blunted thiazide-induced natriuresis [48]. These data demonstrate a role of Kir4.1 in the NEDD4-2-mediated regulation of NCC.

2.4. Connecting Tubule and Collecting Duct

2.4.1. NEDD4-2-Mediated Ubiquitylation and ENaC

ENaC, consisting of three subunits, α, β, and γ, is a primary regulator of sodium reabsorption in the CNT and CD [49,50]. It has been reported that gain-of-function mutations of SCNN1B and SCNN1G cause Liddle's syndrome, which is characterized by salt-sensitive hypertension, hypokalemia, metabolic alkalosis, and low aldosterone levels [51–54]. This phenotype is induced by the disruption or elimination of PY motifs in the β- and γ-subunits of ENaC. Provided that NEDD4-2 ubiquitylates ENaC and regulates its membrane expression and activity [55–58], these mutations cause both increased channel expression and intrinsic activity with a consequent increase of sodium reabsorption. When the renin-angiotensin-aldosterone system (RAAS) is inactivated, NEDD4-2 continuously ubiquitylates

ENaC and downregulates ENaC abundance. When RAAS is activated, aldosterone-induced Sgk1 phosphorylates NEDD4-2, resulting in the recruitment of 14-3-3 adaptor proteins. These proteins inhibit the association between NEDD4-2 and ENaC, thereby leading to the elevation of ENaC levels [59,60]. Consistently, several studies have demonstrated that mice lacking functional NEDD4-2 exhibit high levels of ENaC and salt-sensitive hypertension [61,62]. In humans, several reports indicated that common variants in *NEDD4L* (encoding NEDD4-2) are associated with BP disorder [63–66].

2.4.2. NEDD4-2 and Pendrin

Although there is limited information available regarding the role of UPS in the intercalated cells (ICs) of CNT and CD, a recent study has demonstrated a role of NEDD4-2 in regulating electrolyte transport mechanisms in these cells [67]. Nanami et al. examined the phenotype of IC-specific NEDD4-2 knockout mice and found that these mice displayed increased pendrin abundance and Cl^-/HCO_3^- transport in the ICs, accompanied by the elevation of BP [67]. Furthermore, pendrin gene ablation was found to eliminate the BP increase observed in global NEDD4-2 knockout mice. These data indicate that the ubiquitin ligase NEDD4-2 in ICs is also involved in electrolyte transport and regulation of BP.

3. Role of UPS in the Regulation of Vascular Function

3.1. Proteasome Inhibitors and Cardiovascular Disorders

It is well known that the vasculature is an important determinant of BP. UPS ubiquitously regulates tissue function and can regulate BP through its effect on blood vessels. Proteasome inhibitors have been clinically used as therapeutic agents for multiple myeloma. Carfilzomib, the first irreversible proteasome inhibitor, was found to bind selectively to its target, the chymotrypsin-like activity of the 20S proteasome [68]. It exhibited a higher efficacy in the treatment of patients with relapsed and/or refractory multiple myeloma when applied in combination with dexamethasone with or without lenalidomide [69,70]. Since its approval during the year 2010, there have been increasing reports of carfilzomib-associated cardiovascular adverse events, including hypertension. A systematic review and meta-analysis showed that hypertension (12.2%) was most common among carfilzomib-associated cardiovascular adverse events [71], supporting the involvement of UPS in BP control.

3.2. Vascular Endothelial Cells

With respect to the mechanisms of carfilzomib-associated hypertension, vascular endothelial dysfunction may play a vital role [71–73]. It is known that carfilzomib elicits renal toxic effects as well as microangiopathy, which is believed to be mediated by endothelial dysfunction [74–76]. The key feature of vascular endothelial dysfunction is the decreased NO bioavailability, which is caused due to low NO production and/or increased consumption. Provided that endothelial eNOS is responsible for most of the vascular NO produced [77], its dysfunction results in the impairment of endothelium-dependent vasodilatation [78]. Tetrahydrobiopterin (BH4) is known as an essential cofactor for eNOS-mediated NO synthesis [79]. GTP cyclohydrolase (GTPCH), the rate-limiting enzyme involved in BH4 synthesis, has been reported to be regulated by UPS, and that cigarette smoke extract diminished GTPCH abundance that was inhibited by the proteasomal inhibitor MG132 [80]. This BH4 depletion in turn induced eNOS uncoupling with the loss of NO generation and increased superoxide production, resulting in VEC dysfunction [80]. There are also data indicating that UPS-mediated degradation of GTPCH is associated with oxidative stress in angiotensin II-induced hypertension [81] and diabetes mellitus [82]. It was observed that angiotensin II induced the proteasomal degradation of GTPCH via tyrosine nitration of an important regulatory subunit of 26S proteasome, which was triggered by NADPH oxidase activation and generation of free radicals [81]. In another study, streptozotocin-induced diabetic mice displayed reduced eNOS activity, which was restored by the administration of a proteasome inhibitor through the inhibition of the proteasome-dependent GTPCH reduction [82]. These results imply that the UPS-mediated degradation of GTPCH underlies VCE dysfunction through eNOS regulation. In fact, there have been several reports demonstrating that proteasome inhibitors can improve the function of VECs [83–85]. The role of UPS in endothelial function may vary depending on the disease state and stage, and further studies are required to investigate the role of UPS in VECs.

3.3. Vascular Smooth Muscle Cells

The UPS in VSMCs can also regulate BP. Peroxisome proliferator-activated receptor gamma (PPARγ) is a nuclear regulator superfamily of transcription factors, which is an important regulator of lipid and glucose metabolism. PPARγ is expressed in numerous tissues, including VSMCs. Importantly, studies have shown that mutations (P467L or V290M) in the ligand-binding domain of PPARγ cause not only insulin resistance but also early-onset hypertension [86,87], indicating its role in BP regulation. Moreover, dominant negative mice model of PPARγ (S-P467L) in VSMCs developed arterial stiffness and vascular dysfunction, accompanied by hypertension [88,89]. These results indicate that PPARγ in VSMCs may play an essential role in regulating BP.

Recent studies have suggested that the effect of PPARγ in VSMCs is mediated by its downstream effector molecule, Rho-related BTB domain-containing protein 1 (RhoBTB1) [90]. RhoBTB1, a new subfamily of Rho GTPases [91], is expressed in various tissues [92]. Several genome-wide association studies have demonstrated that RhoBTB1 loci are associated with BP [93,94]. RhoBTB1 interacts with the N-terminal of CUL3 through its first BTB domain [95]. Recently, Mukohda et al. demonstrated that RhoBTB1 protects against hypertension and arterial stiffness by restoring the activity of phosphodiesterase 5 (PDE5) [89]. They generated tamoxifen-inducible and VSMC-specific RhoBTB1 transgenic mice (S-RhoBTB1) and found that Rho-BTB1 expression was reduced in S-P467L mice, whereas S-P467L/S-RhoBTB1 mice exhibited the restoration of RhoBTB1 expression and improvement of vasocontraction in VSMCs, which was accompanied by the reduced PDE5 activity, leading to the attenuation of hypertension. In addition, tadalafil, a PDE5 inhibitor, reduced BP in the S-P467L/S-RhoBTB1 mice. It is interesting to note that RhoBTB1 promoted PDE5 ubiquitylation in the presence of CUL3, which was blunted upon treatment with an inhibitor (MLN4924) of neddylation, a modification that is required for CUL3 activation. The authors concluded that RhoBTB1 is involved in the PPARγ-mediated regulation of BP by regulating PDE5 activity through CUL3-dependent ubiquitylation [89]. Accumulating data indicate that phosphodiesterase 3 (PDE3), another member of the phosphodiesterase family, also critically regulates BP. Recent studies have demonstrated that six missense mutations of PDE3A in six unrelated families with Mendelian hypertension exhibit severe salt-independent but age-dependent hypertension [96]. In vitro analyses of mesenchymal stem cell-derived VSMCs demonstrated that the mutations increased the PKA-mediated PDE3A phosphorylation and resulted in gain of function, with increased cAMP-hydrolytic activity [96]. Whether PDE3 is regulated through UPS needs to be determined.

CUL3 also mediates the ubiquitylation and degradation of RhoA by interacting with a BTB domain-containing adaptor, BACURD [97], which regulates vascular contraction. It has been demonstrated that hypertension-causing mutations in CUL3 impair RhoA ubiquitylation [98] and that selective expression of mutant CUL3 in VSMCs results in augmented RhoA signaling and vascular dysfunction, leading to elevation of BP [99,100].

4. Conclusions

In this review article, we have summarized the current evidence regarding the role of UPS in BP regulation, especially focusing on sodium reabsorption in the kidney and vascular functions (Figure 1). In the kidney, sodium reabsorption regulated by NEDD4-2 has been well characterized in principal cells and has been extensively analyzed in other nephron segments. Studies have also demonstrated the emerging roles of other mechanisms including CUL3 and KLHL3. In addition, accumulating evidence reveals the involvement of vascular functions in UPS-mediated BP regulation. Given that

UPS is present ubiquitously and elicits multiple functions, future investigation is necessary for the complete elucidation of the precise role of UPS in modulating BP.

Author Contributions: Writing–Original Draft Preparation, O.Y. and D.H.; Writing–Review & Editing, K.I. and S.S. All authors have read and agreed to the published version of the manuscript.

Funding: This work was supported in part by JSPS KAKENHI grants 19H03678 (S.S.).

Conflicts of Interest: The authors declare no conflict of interest.

Abbreviations

20-HETE	20-Hydroxyeicosatetraenoic acid
AT1R	Angiotensin II type 1 receptor
BH4	Tetrahydrobiopterin
BP	Blood pressure
CD	Collecting duct
CNT	Connecting tubule
CUL3	Cullin 3
DCT	Distal convoluted tubule
E1	Ubiquitin-activating enzyme
E2	Ubiquitin-conjugating enzyme
E3	Ubiquitin protein ligase
ENaC	Epithelial sodium channel
eNOS	Endothelial NO synthase
GTPCH	GTP cyclohydrolase
IC	Intercalated cell
KLHL3	Kelch-like 3
NCC	Na^+-Cl^- cotransporter
NEDD4-2	Neuronal precursor cell-expressed developmentally downregulated 4-2
NHE3	Na^+/H^+ exchanger isoform 3
NKCC2	Na^+-K^+-$2Cl^-$ cotransporter
NO	Nitric oxide
PDE	Phosphodiesterase
PHAII	Pseudohypoaldosteronism type II
PKA	Protein kinase A
PPARγ	Peroxisome proliferator-activated receptor gamma
PT	Proximal tubule
RhoBTB1	Rho-related BTB domain-containing protein 1
RING	Really interesting new gene
Sgk1	Serum/glucocorticoid-regulated kinase 1
SPAK	STE20/SPS1-related proline-alanine-rich protein kinase
TAL	Thick ascending limb
UPS	Ubiquitin proteasome system
VEC	Vascular endothelial cell
VSMC	Vascular smooth muscle cell
WNK	With-no-lysine

References

1. Guyton, A.C. The surprising kidney-fluid mechanism for pressure control–its infinite gain! *Hypertension* **1990**, *16*, 725–730. [CrossRef] [PubMed]

2. Lifton, R.P.; Gharavi, A.G.; Geller, D.S. Molecular mechanisms of human hypertension. *Cell* **2001**, *104*, 545–556. [CrossRef]

3. Boyden, L.M.; Choi, M.; Choate, K.A.; Nelson-Williams, C.J.; Farhi, A.; Toka, H.R.; Tikhonova, I.R.; Bjornson, R.; Mane, S.M.; Colussi, G.; et al. Mutations in kelch-like 3 and cullin 3 cause hypertension and electrolyte abnormalities. *Nature* **2012**, *482*, 98–102. [CrossRef]

4.   Shibata, S.; Zhang, J.; Puthumana, J.; Stone, K.L.; Lifton, R.P. Kelch-like 3 and Cullin 3 regulate electrolyte homeostasis via ubiquitination and degradation of WNK4. *Proc. Natl. Acad. Sci. USA* **2013**, *110*, 7838–7843. [CrossRef] [PubMed]

5.   Linder, L.; Kiowski, W.; Bühler, F.R.; Lüscher, T.F. Indirect evidence for release of endothelium-derived relaxing factor in human forearm circulation in vivo. Blunted response in essential hypertension. *Circulation* **1990**, *81*, 1762–1767. [PubMed]

6.   Panza, J.A.; Quyyumi, A.A.; Brush, J.E., Jr.; Epstein, S.E. Abnormal endothelium-dependent vascular relaxation in patients with essential hypertension. *N. Engl. J. Med.* **1990**, *323*, 22–27. [CrossRef]

7.   Palmer, R.M.; Ashton, D.S.; Moncada, S. Vascular endothelial cells synthesize nitric oxide from L-arginine. *Nature* **1988**, *333*, 664–666. [CrossRef]

8.   Huang, P.L.; Huang, Z.; Mashimo, H.; Bloch, K.D.; Moskowitz, M.A.; Bevan, J.A.; Fishman, M.C. Hypertension in mice lacking the gene for endothelial nitric oxide synthase. *Nature* **1995**, *377*, 239–242. [CrossRef]

9.   Leonard, A.M.; Chafe, L.L.; Montani, J.P.; Van Vliet, B.N. Increased salt-sensitivity in endothelial nitric oxide synthase-knockout mice. *Am. J. Hypertens* **2006**, *19*, 1264–1269. [CrossRef]

10.   Lahera, V.; Salom, M.G.; Miranda-Guardiola, F.; Moncada, S.; Romero, J.C. Effects of NG-nitro-L-arginine methyl ester on renal function and blood pressure. *Am. J. Physiol.* **1991**, *261*, F1033–F1037. [CrossRef]

11.   Owens, G.K.; Kumar, M.S.; Wamhoff, B.R. Molecular regulation of vascular smooth muscle cell differentiation in development and disease. *Physiol. Rev.* **2004**, *84*, 767–801. [CrossRef]

12.   Haas, A.L.; Warms, J.V.; Rose, I.A. Ubiquitin adenylate: Structure and role in ubiquitin activation. *Biochemistry* **1983**, *22*, 4388–4394. [PubMed]

13.   Jentsch, S. The ubiquitin-conjugation system. *Annu. Rev. Genet.* **1992**, *26*, 179–207. [CrossRef] [PubMed]

14.   Hershko, A.; Ciechanover, A. The ubiquitin system. *Annu. Rev. Biochem.* **1998**, *67*, 425–479. [PubMed]

15.   Clague, M.J.; Heride, C.; Urbé, S. The demographics of the ubiquitin system. *Trends Cell Biol.* **2015**, *25*, 417–426. [CrossRef] [PubMed]

16.   Van Wijk, S.J.; Timmers, H.T. The family of ubiquitin-conjugating enzymes (E2s): Deciding between life and death of proteins. *FASEB J.* **2010**, *24*, 981–993. [CrossRef] [PubMed]

17.   Nakayama, K.I.; Nakayama, K. Ubiquitin ligases: Cell-cycle control and cancer. *Nat. Rev. Cancer* **2006**, *6*, 369–381. [CrossRef] [PubMed]

18.   Kocinsky, H.S.; Girardi, A.C.; Biemesderfer, D.; Nguyen, T.; Mentone, S.; Orlowski, J.; Aronson, P.S. Use of phospho-specific antibodies to determine the phosphorylation of endogenous Na+/H+ exchanger NHE3 at PKA consensus sites. *Am. J. Physiol. Renal Physiol.* **2005**, *289*, F249–F258.

19.   Thomson, R.B.; Wang, T.; Thomson, B.R.; Tarrats, L.; Girardi, A.; Mentone, S.; Soleimani, M.; Kocher, O.; Aronson, P.S. Role of PDZK1 in membrane expression of renal brush border ion exchangers. *Proc. Natl. Acad. Sci. USA* **2005**, *102*, 13331–13336.

20.   No, Y.R.; He, P.; Yoo, B.K.; Yun, C.C. Unique regulation of human Na+/H+ exchanger 3 (NHE3) by Nedd4-2 ligase that differs from non-primate NHE3s. *J. Biol. Chem.* **2014**, *289*, 18360–18372. [CrossRef]

21.   Hatanaka, M.; Kaimori, J.Y.; Yamamoto, S.; Matsui, I.; Hamano, T.; Takabatake, Y.; Ecelbarger, C.M.; Takahara, S.; Isaka, Y.; Rakugi, H. Azilsartan Improves Salt Sensitivity by Modulating the Proximal Tubular Na+-H+ Exchanger-3 in Mice. *PLoS ONE* **2016**, *11*, e0147786. [CrossRef]

22.   Russell, J.M. Sodium-potassium-chloride cotransport. *Physiol. Rev.* **2000**, *80*, 211–276. [PubMed]

23.   Wu, J.; Liu, X.; Lai, G.; Yang, X.; Wang, L.; Zhao, Y. Synergistical effect of 20-HETE and high salt on NKCC2 protein and blood pressure via ubiquitin-proteasome pathway. *Hum. Genet.* **2013**, *132*, 179–187. [PubMed]

24.   Ares, G.R. cGMP induces degradation of NKCC2 in the thick ascending limb via the ubiquitin-proteasomal system. *Am. J. Physiol. Renal Physiol.* **2019**, *316*, F838–F846. [PubMed]

25.   Gamba, G. The thiazide-sensitive Na+-Cl- cotransporter: Molecular biology, functional properties, and regulation by WNKs. *Am. J. Physiol. Renal Physiol.* **2009**, *297*, F838–F848.

26.   Mayan, H.; Munter, G.; Shaharabany, M.; Mouallem, M.; Pauzner, R.; Holtzman, E.J.; Farfel, Z. Hypercalciuria in familial hyperkalemia and hypertension accompanies hyperkalemia and precedes hypertension: Description of a large family with the Q565E WNK4 mutation. *J. Clin. Endocrinol. Metab.* **2004**, *89*, 4025–4030.

27.   Mayan, H.; Vered, I.; Mouallem, M.; Tzadok-Witkon, M.; Pauzner, R.; Farfel, Z. Pseudohypoaldosteronism type II: Marked sensitivity to thiazides, hypercalciuria, normomagnesemia, and low bone mineral density. *J. Clin. Endocrinol. Metab.* **2002**, *87*, 3248–3254.

28. Wilson, F.H.; Disse-Nicodeme, S.; Choate, K.A.; Ishikawa, K.; Nelson-Williams, C.; Desitter, I.; Gunel, M.; Milford, D.V.; Lipkin, G.W.; Achard, J.M.; et al. Human hypertension caused by mutations in WNK kinases. *Science* **2001**, *293*, 1107–1112.

29. Louis-Dit-Picard, H.; Barc, J.; Trujillano, D.; Miserey-Lenkei, S.; Bouatia-Naji, N.; Pylypenko, O.; Beaurain, G.; Bonnefond, A.; Sand, O.; Simian, C.; et al. KLHL3 mutations cause familial hyperkalemic hypertension by impairing ion transport in the distal nephron. *Nat. Genet.* **2012**, *44*, 456–460, S451–S453.

30. Vitari, A.C.; Deak, M.; Morrice, N.A.; Alessi, D.R. The WNK1 and WNK4 protein kinases that are mutated in Gordon's hypertension syndrome phosphorylate and activate SPAK and OSR1 protein kinases. *Biochem. J.* **2005**, *391*, 17–24.

31. Ohta, A.; Schumacher, F.R.; Mehellou, Y.; Johnson, C.; Knebel, A.; Macartney, T.J.; Wood, N.T.; Alessi, D.R.; Kurz, T. The CUL3-KLHL3 E3 ligase complex mutated in Gordon's hypertension syndrome interacts with and ubiquitylates WNK isoforms: Disease-causing mutations in KLHL3 and WNK4 disrupt interaction. *Biochem. J.* **2013**, *451*, 111–122. [PubMed]

32. Wakabayashi, M.; Mori, T.; Isobe, K.; Sohara, E.; Susa, K.; Araki, Y.; Chiga, M.; Kikuchi, E.; Nomura, N.; Mori, Y.; et al. Impaired KLHL3-mediated ubiquitination of WNK4 causes human hypertension. *Cell Rep.* **2013**, *3*, 858–868. [PubMed]

33. Wu, G.; Peng, J.B. Disease-causing mutations in KLHL3 impair its effect on WNK4 degradation. *FEBS Lett.* **2013**, *587*, 1717–1722.

34. Shibata, S.; Arroyo, J.P.; Castaneda-Bueno, M.; Puthumana, J.; Zhang, J.; Uchida, S.; Stone, K.L.; Lam, T.T.; Lifton, R.P. Angiotensin II signaling via protein kinase C phosphorylates Kelch-like 3, preventing WNK4 degradation. *Proc. Natl. Acad. Sci. USA* **2014**, *111*, 15556–15561.

35. Yoshizaki, Y.; Mori, Y.; Tsuzaki, Y.; Mori, T.; Nomura, N.; Wakabayashi, M.; Takahashi, D.; Zeniya, M.; Kikuchi, E.; Araki, Y.; et al. Impaired degradation of WNK by Akt and PKA phosphorylation of KLHL3. *Biochem. Biophys. Res. Commun.* **2015**, *467*, 229–234.

36. Castañeda-Bueno, M.; Arroyo, J.P.; Zhang, J.; Puthumana, J.; Yarborough, O., 3rd; Shibata, S.; Rojas-Vega, L.; Gamba, G.; Rinehart, J.; Lifton, R.P. Phosphorylation by PKC and PKA regulate the kinase activity and downstream signaling of WNK4. *Proc. Natl. Acad. Sci. USA* **2017**, *114*, E879–E886.

37. Bazúa-Valenti, S.; Rojas-Vega, L.; Castañeda-Bueno, M.; Barrera-Chimal, J.; Bautista, R.; Cervantes-Pérez, L.G.; Vázquez, N.; Plata, C.; Murillo-de-Ozores, A.R.; González-Mariscal, L.; et al. The Calcium-Sensing Receptor Increases Activity of the Renal NCC through the WNK4-SPAK Pathway. *J. Am. Soc. Nephrol.* **2018**, *29*, 1838–1848. [PubMed]

38. Ishizawa, K.; Wang, Q.; Li, J.; Yamazaki, O.; Tamura, Y.; Fujigaki, Y.; Uchida, S.; Lifton, R.P.; Shibata, S. Calcineurin dephosphorylates Kelch-like 3, reversing phosphorylation by angiotensin II and regulating renal electrolyte handling. *Proc. Natl. Acad. Sci. USA* **2019**, *116*, 3155–3160.

39. Ishizawa, K.; Xu, N.; Loffing, J.; Lifton, R.P.; Fujita, T.; Uchida, S.; Shibata, S. Potassium depletion stimulates Na-Cl cotransporter via phosphorylation and inactivation of the ubiquitin ligase Kelch-like 3. *Biochem. Biophys. Res. Commun.* **2016**, *480*, 745–751. [CrossRef]

40. Ishizawa, K.; Wang, Q.; Li, J.; Xu, N.; Nemoto, Y.; Morimoto, C.; Fujii, W.; Tamura, Y.; Fujigaki, Y.; Tsukamoto, K.; et al. Inhibition of Sodium Glucose Cotransporter 2 Attenuates the Dysregulation of Kelch-Like 3 and NaCl Cotransporter in Obese Diabetic Mice. *J. Am. Soc. Nephrol.* **2019**, *30*, 782–794. [CrossRef]

41. McCormick, J.A.; Yang, C.L.; Zhang, C.; Davidge, B.; Blankenstein, K.I.; Terker, A.S.; Yarbrough, B.; Meermeier, N.P.; Park, H.J.; McCully, B.; et al. Hyperkalemic hypertension-associated cullin 3 promotes WNK signaling by degrading KLHL3. *J. Clin. Investig.* **2014**, *124*, 4723–4736. [CrossRef] [PubMed]

42. Cornelius, R.J.; Si, J.; Cuevas, C.A.; Nelson, J.W.; Gratreak, B.D.K.; Pardi, R.; Yang, C.L.; Ellison, D.H. Renal COP9 Signalosome Deficiency Alters CUL3-KLHL3-WNK Signaling Pathway. *J. Am. Soc. Nephrol.* **2018**, *29*, 2627–2640. [CrossRef] [PubMed]

43. Arroyo, J.P.; Lagnaz, D.; Ronzaud, C.; Vazquez, N.; Ko, B.S.; Moddes, L.; Ruffieux-Daidie, D.; Hausel, P.; Koesters, R.; Yang, B.; et al. Nedd4-2 modulates renal Na+-Cl- cotransporter via the aldosterone-SGK1-Nedd4-2 pathway. *J. Am. Soc. Nephrol.* **2011**, *22*, 1707–1719. [CrossRef] [PubMed]

44. Faresse, N.; Lagnaz, D.; Debonneville, A.; Ismailji, A.; Maillard, M.; Fejes-Toth, G.; Náray-Fejes-Tóth, A.; Staub, O. Inducible kidney-specific Sgk1 knockout mice show a salt-losing phenotype. *Am. J. Physiol. Renal Physiol.* **2012**, *302*, F977–F985. [CrossRef] [PubMed]

45. Ronzaud, C.; Loffing-Cueni, D.; Hausel, P.; Debonneville, A.; Malsure, S.R.; Fowler-Jaeger, N.; Boase, N.A.; Perrier, R.; Maillard, M.; Yang, B.; et al. Renal tubular NEDD4-2 deficiency causes NCC-mediated salt-dependent hypertension. *J. Clin. Investig.* **2013**, *123*, 657–665. [CrossRef] [PubMed]

46. Roy, A.; Al-Qusairi, L.; Donnelly, B.F.; Ronzaud, C.; Marciszyn, A.L.; Gong, F.; Chang, Y.P.; Butterworth, M.B.; Pastor-Soler, N.M.; Hallows, K.R.; et al. Alternatively spliced proline-rich cassettes link WNK1 to aldosterone action. *J. Clin. Investig.* **2015**, *125*, 3433–3448. [CrossRef]

47. Furusho, T.; Sohara, E.; Mandai, S.; Kikuchi, H.; Takahashi, N.; Fujimaru, T.; Hashimoto, H.; Arai, Y.; Ando, F.; Zeniya, M.; et al. Renal TNFα activates the WNK phosphorylation cascade and contributes to salt-sensitive hypertension in chronic kidney disease. *Kidney Int.* **2020**, *97*, 713–727. [CrossRef]

48. Wu, P.; Su, X.T.; Gao, Z.X.; Zhang, D.D.; Duan, X.P.; Xiao, Y.; Staub, O.; Wang, W.H.; Lin, D.H. Renal Tubule Nedd4-2 Deficiency Stimulates Kir4.1/Kir5.1 and Thiazide-Sensitive NaCl Cotransporter in Distal Convoluted Tubule. *J. Am. Soc. Nephrol.* **2020**, *31*, 1226–1242. [CrossRef]

49. Rossier, B.C. Epithelial sodium channel (ENaC) and the control of blood pressure. *Curr. Opin. Pharmacol.* **2014**, *15*, 33–46. [CrossRef]

50. Pearce, D.; Soundararajan, R.; Trimpert, C.; Kashlan, O.B.; Deen, P.M.; Kohan, D.E. Collecting duct principal cell transport processes and their regulation. *Clin. J. Am. Soc. Nephrol.* **2015**, *10*, 135–146. [CrossRef]

51. Schild, L.; Canessa, C.M.; Shimkets, R.A.; Gautschi, I.; Lifton, R.P.; Rossier, B.C. A mutation in the epithelial sodium channel causing Liddle disease increases channel activity in the Xenopus laevis oocyte expression system. *Proc. Natl. Acad. Sci. USA* **1995**, *92*, 5699–5703. [CrossRef]

52. Hansson, J.H.; Nelson-Williams, C.; Suzuki, H.; Schild, L.; Shimkets, R.; Lu, Y.; Canessa, C.; Iwasaki, T.; Rossier, B.; Lifton, R.P. Hypertension caused by a truncated epithelial sodium channel gamma subunit: Genetic heterogeneity of Liddle syndrome. *Nat. Genet.* **1995**, *11*, 76–82. [CrossRef] [PubMed]

53. Tamura, H.; Schild, L.; Enomoto, N.; Matsui, N.; Marumo, F.; Rossier, B.C. Liddle disease caused by a missense mutation of beta subunit of the epithelial sodium channel gene. *J. Clin. Investig.* **1996**, *97*, 1780–1784. [CrossRef] [PubMed]

54. Staub, O.; Dho, S.; Henry, P.; Correa, J.; Ishikawa, T.; McGlade, J.; Rotin, D. WW domains of Nedd4 bind to the proline-rich PY motifs in the epithelial Na+ channel deleted in Liddle's syndrome. *EMBO J.* **1996**, *15*, 2371–2380. [CrossRef]

55. Staub, O.; Gautschi, I.; Ishikawa, T.; Breitschopf, K.; Ciechanover, A.; Schild, L.; Rotin, D. Regulation of stability and function of the epithelial Na+ channel (ENaC) by ubiquitination. *EMBO J.* **1997**, *16*, 6325–6336. [CrossRef] [PubMed]

56. Lu, C.; Pribanic, S.; Debonneville, A.; Jiang, C.; Rotin, D. The PY motif of ENaC, mutated in Liddle syndrome, regulates channel internalization, sorting and mobilization from subapical pool. *Traffic* **2007**, *8*, 1246–1264. [CrossRef]

57. Wiemuth, D.; Ke, Y.; Rohlfs, M.; McDonald, F.J. Epithelial sodium channel (ENaC) is multi-ubiquitinated at the cell surface. *Biochem. J.* **2007**, *405*, 147–155. [CrossRef]

58. Zhou, R.; Patel, S.V.; Snyder, P.M. Nedd4-2 catalyzes ubiquitination and degradation of cell surface ENaC. *J. Biol. Chem.* **2007**, *282*, 20207–20212. [CrossRef]

59. Lang, F.; Pearce, D. Regulation of the epithelial Na+ channel by the mTORC2/SGK1 pathway. *Nephrol. Dial. Transplant.* **2016**, *31*, 200–205. [CrossRef]

60. Bhalla, V.; Daidie, D.; Li, H.; Pao, A.C.; LaGrange, L.P.; Wang, J.; Vandewalle, A.; Stockand, J.D.; Staub, O.; Pearce, D. Serum- and glucocorticoid-regulated kinase 1 regulates ubiquitin ligase neural precursor cell-expressed, developmentally down-regulated protein 4-2 by inducing interaction with 14-3-3. *Mol. Endocrinol.* **2005**, *19*, 3073–3084. [CrossRef]

61. Shi, P.P.; Cao, X.R.; Sweezer, E.M.; Kinney, T.S.; Williams, N.R.; Husted, R.F.; Nair, R.; Weiss, R.M.; Williamson, R.A.; Sigmund, C.D.; et al. Salt-sensitive hypertension and cardiac hypertrophy in mice deficient in the ubiquitin ligase Nedd4-2. *Am. J. Physiol. Renal Physiol.* **2008**, *295*, F462–F470. [CrossRef] [PubMed]

62. Minegishi, S.; Ishigami, T.; Kino, T.; Chen, L.; Nakashima-Sasaki, R.; Araki, N.; Yatsu, K.; Fujita, M.; Umemura, S. An isoform of Nedd4-2 is critically involved in the renal adaptation to high salt intake in mice. *Sci. Rep.* **2016**, *6*, 27137. [CrossRef] [PubMed]

63. Dunn, D.M.; Ishigami, T.; Pankow, J.; von Niederhausern, A.; Alder, J.; Hunt, S.C.; Leppert, M.F.; Lalouel, J.M.; Weiss, R.B. Common variant of human NEDD4L activates a cryptic splice site to form a frameshifted transcript. *J. Hum. Genet.* **2002**, *47*, 665–676. [CrossRef] [PubMed]

64. Fouladkou, F.; Alikhani-Koopaei, R.; Vogt, B.; Flores, S.Y.; Malbert-Colas, L.; Lecomte, M.C.; Loffing, J.; Frey, F.J.; Frey, B.M.; Staub, O. A naturally occurring human Nedd4-2 variant displays impaired ENaC regulation in Xenopus laevis oocytes. *Am. J. Physiol. Renal Physiol.* **2004**, *287*, F550–F561. [CrossRef] [PubMed]

65. Araki, N.; Umemura, M.; Miyagi, Y.; Yabana, M.; Miki, Y.; Tamura, K.; Uchino, K.; Aoki, R.; Goshima, Y.; Umemura, S.; et al. Expression, transcription, and possible antagonistic interaction of the human Nedd4L gene variant: Implications for essential hypertension. *Hypertension* **2008**, *51*, 773–777. [CrossRef]

66. Luo, F.; Wang, Y.; Wang, X.; Sun, K.; Zhou, X.; Hui, R. A functional variant of NEDD4L is associated with hypertension, antihypertensive response, and orthostatic hypotension. *Hypertension* **2009**, *54*, 796–801. [CrossRef]

67. Nanami, M.; Pham, T.D.; Kim, Y.H.; Yang, B.; Sutliff, R.L.; Staub, O.; Klein, J.D.; Lopez-Cayuqueo, K.I.; Chambrey, R.; Park, A.Y.; et al. The Role of Intercalated Cell Nedd4-2 in BP Regulation, Ion Transport, and Transporter Expression. *J. Am. Soc. Nephrol.* **2018**, *29*, 1706–1719. [CrossRef]

68. Kuhn, D.J.; Chen, Q.; Voorhees, P.M.; Strader, J.S.; Shenk, K.D.; Sun, C.M.; Demo, S.D.; Bennett, M.K.; van Leeuwen, F.W.; Chanan-Khan, A.A.; et al. Potent activity of carfilzomib, a novel, irreversible inhibitor of the ubiquitin-proteasome pathway, against preclinical models of multiple myeloma. *Blood* **2007**, *110*, 3281–3290. [CrossRef]

69. Stewart, A.K.; Rajkumar, S.V.; Dimopoulos, M.A.; Masszi, T.; Špička, I.; Oriol, A.; Hájek, R.; Rosiñol, L.; Siegel, D.S.; Mihaylov, G.G.; et al. Carfilzomib, lenalidomide, and dexamethasone for relapsed multiple myeloma. *N. Engl. J. Med.* **2015**, *372*, 142–152. [CrossRef]

70. Dimopoulos, M.A.; Moreau, P.; Palumbo, A.; Joshua, D.; Pour, L.; Hájek, R.; Facon, T.; Ludwig, H.; Oriol, A.; Goldschmidt, H.; et al. Carfilzomib and dexamethasone versus bortezomib and dexamethasone for patients with relapsed or refractory multiple myeloma (ENDEAVOR): A randomised, phase 3, open-label, multicentre study. *Lancet Oncol.* **2016**, *17*, 27–38. [CrossRef]

71. Waxman, A.J.; Clasen, S.; Hwang, W.T.; Garfall, A.; Vogl, D.T.; Carver, J.; O'Quinn, R.; Cohen, A.D.; Stadtmauer, E.A.; Ky, B.; et al. Carfilzomib-Associated Cardiovascular Adverse Events: A Systematic Review and Meta-analysis. *JAMA Oncol.* **2018**, *4*, e174519. [CrossRef] [PubMed]

72. Gavazzoni, M.; Vizzardi, E.; Gorga, E.; Bonadei, I.; Rossi, L.; Belotti, A.; Rossi, G.; Ribolla, R.; Metra, M.; Raddino, R. Mechanism of cardiovascular toxicity by proteasome inhibitors: New paradigm derived from clinical and pre-clinical evidence. *Eur. J. Pharmacol.* **2018**, *828*, 80–88. [CrossRef] [PubMed]

73. Chen-Scarabelli, C.; Corsetti, G.; Pasini, E.; Dioguardi, F.S.; Sahni, G.; Narula, J.; Gavazzoni, M.; Patel, H.; Saravolatz, L.; Knight, R.; et al. Spasmogenic Effects of the Proteasome Inhibitor Carfilzomib on Coronary Resistance, Vascular Tone and Reactivity. *EBioMedicine* **2017**, *21*, 206–212. [CrossRef] [PubMed]

74. Rosenthal, A.; Luthi, J.; Belohlavek, M.; Kortüm, K.M.; Mookadam, F.; Mayo, A.; Fonseca, R.; Bergsagel, P.L.; Reeder, C.B.; Mikhael, J.R.; et al. Carfilzomib and the cardiorenal system in myeloma: An endothelial effect? *Blood Cancer J.* **2016**, *6*, e384. [CrossRef]

75. Yui, J.C.; Van Keer, J.; Weiss, B.M.; Waxman, A.J.; Palmer, M.B.; D'Agati, V.D.; Kastritis, E.; Dimopoulos, M.A.; Vij, R.; Bansal, D.; et al. Proteasome inhibitor associated thrombotic microangiopathy. *Am. J. Hematol.* **2016**, *91*, E348–E352. [CrossRef]

76. Hobeika, L.; Self, S.E.; Velez, J.C. Renal thrombotic microangiopathy and podocytopathy associated with the use of carfilzomib in a patient with multiple myeloma. *BMC Nephrol.* **2014**, *15*, 156. [CrossRef]

77. Förstermann, U.; Münzel, T. Endothelial nitric oxide synthase in vascular disease: From marvel to menace. *Circulation* **2006**, *113*, 1708–1714. [CrossRef]

78. Vanhoutte, P.M.; Zhao, Y.; Xu, A.; Leung, S.W. Thirty Years of Saying NO: Sources, Fate, Actions, and Misfortunes of the Endothelium-Derived Vasodilator Mediator. *Circ Res.* **2016**, *119*, 375–396. [CrossRef]

79. Alp, N.J.; Channon, K.M. Regulation of endothelial nitric oxide synthase by tetrahydrobiopterin in vascular disease. *Arterioscler Thromb. Vasc. Biol.* **2004**, *24*, 413–420. [CrossRef]

80. Abdelghany, T.M.; Ismail, R.S.; Mansoor, F.A.; Zweier, J.R.; Lowe, F.; Zweier, J.L. Cigarette smoke constituents cause endothelial nitric oxide synthase dysfunction and uncoupling due to depletion of tetrahydrobiopterin with degradation of GTP cyclohydrolase. *Nitric. Oxide* **2018**, *76*, 113–121. [CrossRef]

81. Xu, J.; Wang, S.; Wu, Y.; Song, P.; Zou, M.H. Tyrosine nitration of PA700 activates the 26S proteasome to induce endothelial dysfunction in mice with angiotensin II-induced hypertension. *Hypertension* **2009**, *54*, 625–632. [CrossRef] [PubMed]

82. Xu, J.; Wu, Y.; Song, P.; Zhang, M.; Wang, S.; Zou, M.H. Proteasome-dependent degradation of guanosine 5′-triphosphate cyclohydrolase I causes tetrahydrobiopterin deficiency in diabetes mellitus. *Circulation* **2007**, *116*, 944–953. [CrossRef] [PubMed]

83. Lorenz, M.; Wilck, N.; Meiners, S.; Ludwig, A.; Baumann, G.; Stangl, K.; Stangl, V. Proteasome inhibition prevents experimentally-induced endothelial dysfunction. *Life Sci.* **2009**, *84*, 929–934. [CrossRef] [PubMed]

84. De Martin, R.; Hoeth, M.; Hofer-Warbinek, R.; Schmid, J.A. The transcription factor NF-kappa B and the regulation of vascular cell function. *Arterioscler Thromb. Vasc. Biol.* **2000**, *20*, E83–E88. [PubMed]

85. Takaoka, M.; Ohkita, M.; Matsumura, Y. Pathophysiological role of proteasome-dependent proteolytic pathway in endothelin-1-related cardiovascular diseases. *Curr. Vasc Pharmacol.* **2003**, *1*, 19–26. [CrossRef]

86. Barroso, I.; Gurnell, M.; Crowley, V.E.; Agostini, M.; Schwabe, J.W.; Soos, M.A.; Maslen, G.L.; Williams, T.D.; Lewis, H.; Schafer, A.J.; et al. Dominant negative mutations in human PPARgamma associated with severe insulin resistance, diabetes mellitus and hypertension. *Nature* **1999**, *402*, 880–883. [CrossRef]

87. Savage, D.B.; Tan, G.D.; Acerini, C.L.; Jebb, S.A.; Agostini, M.; Gurnell, M.; Williams, R.L.; Umpleby, A.M.; Thomas, E.L.; Bell, J.D.; et al. Human metabolic syndrome resulting from dominant-negative mutations in the nuclear receptor peroxisome proliferator-activated receptor-gamma. *Diabetes* **2003**, *52*, 910–917. [CrossRef]

88. Rahim, N.G.; Harismendy, O.; Topol, E.J.; Frazer, K.A. Genetic determinants of phenotypic diversity in humans. *Genome Biol.* **2008**, *9*, 215. [CrossRef]

89. Mukohda, M.; Fang, S.; Wu, J.; Agbor, L.N.; Nair, A.R.; Ibeawuchi, S.C.; Hu, C.; Liu, X.; Lu, K.T.; Guo, D.F.; et al. RhoBTB1 protects against hypertension and arterial stiffness by restraining phosphodiesterase 5 activity. *J. Clin. Investig.* **2019**, *129*, 2318–2332. [CrossRef]

90. Pelham, C.J.; Ketsawatsomkron, P.; Groh, S.; Grobe, J.L.; de Lange, W.J.; Ibeawuchi, S.R.; Keen, H.L.; Weatherford, E.T.; Faraci, F.M.; Sigmund, C.D. Cullin-3 regulates vascular smooth muscle function and arterial blood pressure via PPARγ and RhoA/Rho-kinase. *Cell Metab.* **2012**, *16*, 462–472. [CrossRef]

91. Rivero, F.; Dislich, H.; Glockner, G.; Noegel, A.A. The Dictyostelium discoideum family of Rho-related proteins. *Nucleic Acids Res.* **2001**, *29*, 1068–1079. [CrossRef] [PubMed]

92. Berthold, J.; Schenkova, K.; Ramos, S.; Miura, Y.; Furukawa, M.; Aspenstrom, P.; Rivero, F. Characterization of RhoBTB-depedent Cul3 ubiquitin ligase complexes—evidence for an autoregulatory mechanism. *Exp. Cell Res.* **2008**, *314*, 3453–3465. [CrossRef] [PubMed]

93. Newton-Cheh, C.; Johnson, T.; Gateva, V.; Tobin, M.D.; Bochud, M.; Coin, L.; Najjar, S.S.; Zhao, J.H.; Heath, S.C.; Eyheramendy, S.; et al. Genome-wide association study identifies eight loci associated with blood pressure. *Nat. Genet.* **2009**, *41*, 666–676. [CrossRef] [PubMed]

94. Evangelou, E.; Warren, H.R.; Mosen-Ansorena, D.; Mifsud, B.; Pazoki, R.; Gao, H.; Ntritsos, G.; Dimou, N.; Cabrera, C.P.; Karaman, I.; et al. Genetic analysis of over 1 million people identifies 535 new loci associated with blood pressure traits. *Nat. Genet.* **2018**, *50*, 1412–1425. [CrossRef] [PubMed]

95. Ramos, S.; Khademi, F.; Somesh, B.P.; Rivero, F. Genomic organization and expression profile of the small GTPases of the RhoBTB family in human and mouse. *Gene* **2002**, *298*, 147–157. [CrossRef]

96. Maass, P.G.; Aydin, A.; Luft, F.C.; Schächterle, C.; Weise, A.; Stricker, S.; Lindschau, C.; Vaegler, M.; Qadri, F.; Toka, H.R.; et al. PDE3A mutations cause autosomal dominant hypertension with brachydactyly. *Nat. Genet.* **2015**, *47*, 647–653. [CrossRef]

97. Chen, Y.; Yang, Z.; Meng, M.; Zhao, Y.; Dong, N.; Yan, H.; Liu, L.; Ding, M.; Peng, H.B.; Shao, F. Cullin mediates degradation of RhoA through evolutionarily conserved BTB adaptors to control actin cytoskeleton structure and cell movement. *Mol. Cell* **2009**, *35*, 841–855. [CrossRef]

98. Ibeawuchi, S.R.; Agbor, L.N.; Quelle, F.W.; Sigmund, C.D. Hypertension-causing Mutations in Cullin3 Protein Impair RhoA Protein Ubiquitination and Augment the Association with Substrate Adaptors. *J. Biol. Chem.* **2015**, *290*, 19208–19217. [CrossRef]

99. Agbor, L.N.; Ibeawuchi, S.C.; Hu, C.; Wu, J.; Davis, D.R.; Keen, H.L.; Quelle, F.W.; Sigmund, C.D. Cullin-3 mutation causes arterial stiffness and hypertension through a vascular smooth muscle mechanism. *JCI Insight.* **2016**, *1*, e91015. [CrossRef]
100. Abdel Khalek, W.; Rafael, C.; Loisel-Ferreira, I.; Kouranti, I.; Clauser, E.; Hadchouel, J.; Jeunemaitre, X. Severe Arterial Hypertension from Cullin 3 Mutations Is Caused by Both Renal and Vascular Effects. *J. Am. Soc. Nephrol.* **2019**, *30*, 811–823. [CrossRef]

International Journal of
Molecular Sciences

Review

The Regulatory Role of T Cell Responses in Cardiac Remodeling Following Myocardial Infarction

Tabito Kino [1], Mohsin Khan [2] and Sadia Mohsin [1],*

[1]   Cardiovascular Research Center, Lewis Katz School of Medicine, Temple University, Philadelphia, PA 19140, USA; tun60782@temple.edu
[2]   Center for Metabolic Disease Research, Lewis Katz School of Medicine, Temple University, Philadelphia, PA 19140, USA; mohsin.khan@temple.edu
*   Correspondence: sadia.mohsin@temple.edu; Tel.: +1-215-707-3152; Fax: +1-215-707-5737

Received: 6 June 2020; Accepted: 13 July 2020; Published: 16 July 2020

Abstract: Ischemic injury to the heart causes cardiomyocyte and supportive tissue death that result in adverse remodeling and formation of scar tissue at the site of injury. The dying cardiac tissue secretes a variety of cytokines and chemokines that trigger an inflammatory response and elicit the recruitment and activation of cardiac immune cells to the injury site. Cell-based therapies for cardiac repair have enhanced cardiac function in the injured myocardium, but the mechanisms remain debatable. In this review, we will focus on the interactions between the adoptively transferred stem cells and the post-ischemic environment, including the active components of the immune/inflammatory response that can mediate cardiac outcome after ischemic injury. In particular, we highlight how the adaptive immune cell response can mediate tissue repair following cardiac injury. Several cell-based studies have reported an increase in pro-reparative T cell subsets after stem cell transplantation. Paracrine factors secreted by stem cells polarize T cell subsets partially by exogenous ubiquitination, which can induce differentiation of T cell subset to promote tissue repair after myocardial infarction (MI). However, the mechanism behind the polarization of different subset after stem cell transplantation remains poorly understood. In this review, we will summarize the current status of immune cells within the heart post-MI with an emphasis on T cell mediated reparative response after ischemic injury.

Keywords: regulatory T cells; ubiquitin; mesenchymal stem cell; cortical bone derived stem cell; myocardial infarction

---

1. Introduction

Acute MI is the most severe manifestation of coronary artery disease, which causes more than 2.4 million deaths in the USA, more than 4 million deaths in Europe and Northern Asia [1]. During cardiac ischemic events, the heart undergoes deleterious changes that result in cardiac remodeling of the left ventricular (LV) resulting in both structural and functional alternations. The ischemia in the heart triggers an inflammatory response that leads to the formation of a collagen-rich-scar, which is replaced from necrotic tissue to prevent cardiac rupture. Therefore, it is reasonable to conclude that the healing process is tightly coupled with the inflammatory microenvironment of the infarcted heart [2,3].

The cells of the immune system and their secreted factors play crucial roles in the initiation, progression, and resolution of inflammation following MI. Immune cell subsets contribute to both damage and repair of cardiac tissue specifically in regard to scar formation and LV remodeling [4]. Various types of inflammatory cells are recruited to the damaged area in a temporal fashion, where they remove necrotic tissue and promote scar formation [5]. The participation of T cells in myocardial inflammation and repair has been observed in experimental rodent models. In particular, regulatory T cells (Tregs) mainly mediate organ-specific regenerative programs [6–8]. T cell reactivity can benefit myocardial healing by promoting reparative fibrosis in a postmitotic organ [9]. However, sustained T cell responses in the heart can lead to adverse remodeling and contribute to the progression of ischemic heart failure (HF) at later chronic stages [10]. Temporal and spatial regulation from these biphasic immune cell populations is essential to maintain reparative processes [11]. Importantly, focusing on T cells, including Tregs, can be a clue to reveal the reparative mechanism. Moreover, they can be a target of therapy for patients with ischemic heart disease (IHD).

Pharmacotherapy was traditionally promoted in patients with IHD. After surviving from acute coronary syndrome (ACS), optimal medical therapy (OMT) is a golden standard to prevent cardiovascular death [12]. However, OMT cannot promote a regenerative effect in the ischemic area. To date, target therapies are improving and include specific antibodies and the exogenous ubiquitin helping in reducing the scar area in rodent models after cardiac injury [13,14]. In addition, stem cell-based therapies had developed with improvement in cardiac function, however, the overall beneficial effects are relatively modest with fundamental mechanisms of stem cell-mediated repair being largely unknown. This review aims to summarize evidence regarding the role of T cell responses in myocardial remodeling following MI, including how stem cell therapies can be used to mediate the ubiquitination state of T cells.

2. Immune Cell Response Post-Ischemic Injury

After MI, the rapid and uncontrolled cellular death and release of intracellular contents into the intercellular space are initiated via necrosis. Necrosis of the ischemic area triggers an inflammatory response in the heart with the infiltration of cells including neutrophils, macrophages, monocytes, T cells, and B cells to clear dead cells and cellular debris [15]. In the first stage, the injured myocardium releases damage associated molecular patterns (DAMPs), which bind toll-like receptors (TLRs), and initiate the production of cytokines/chemokines to induce the activation and recruitment of neutrophils and $Ly6C^{high}$ monocytes. Some $Ly6C^{high}$ monocytes differentiate into M1 macrophages, which contain a pro-inflammatory secretome enriched in interleukin (IL)-1β, tumor necrosis factor (TNF)-α, and IL-6 [11]. In the second stage, $Ly6C^{low}$ monocytes and M2-like macrophages with high expression of IL-10, transforming growth factor (TGF)-β and vascular endothelial growth factor (VEGF) come in with the anti-inflammatory function much needed after injury [16]. After this phase, fibroblasts are activated and move into the infarcted area. Cytokines/chemokines and growth factors play a critical role in the differentiation of fibroblasts into myofibroblasts [17]. Myofibroblasts, a major producer of extracellular matrix (ECM) proteins, produce matrix metalloproteinases (MMPs) and tissue inhibitors of metalloproteinase (TIMPs). Therefore, a highly regulated balance between MMPs and TIMPs is essential in the maintenance of ECM homeostasis [18]. Figure 1 shows the mechanisms of the innate and adaptive immune response after cardiac damage and tissue repair.

Monocytes and macrophages have a reparative effect in ischemic heart tissue after MI. As their activation mainly occurs in acute phase, other components also regulate repair mechanism from acute to chronic phase. The adaptive immune response, especially T cells, participate in those cascades including myofibroblasts transition. The adverse cardiac remodeling may occur when there is an inappropriate regulation of inflammation. However, the mechanism of adverse cardiac remodeling is unclear and is linked to chronic inflammation. There is a dire need of accumulated studies in chronic phase to illustrate cardiac remodeling following MI including immune response to better understand the wound healing response with potential clinical application.

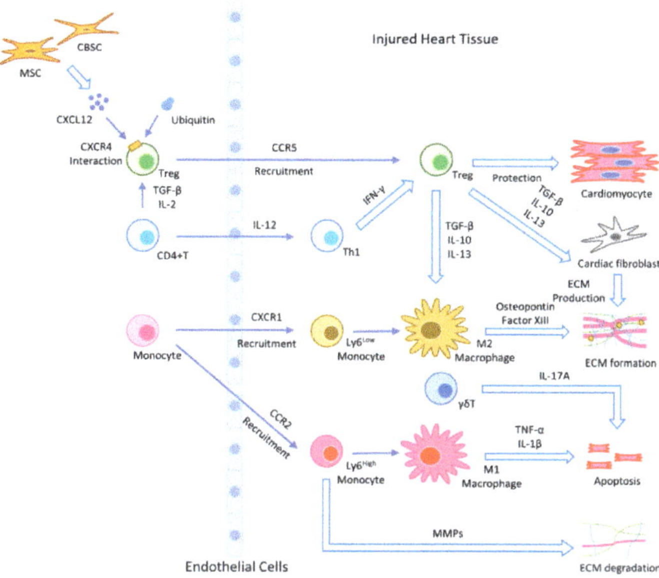

Figure 1. Illustration of the wound healing process after stem cell therapy: activated monocytes/macrophages produce various chemokines and cytokines that initiate inflammation and cell migration. Chemokines produced at the site of infarction induce CCR2-dependent migration of proinflammatory Ly6C^{high} monocytes, which secrete pro-inflammatory cytokines and produce a MMP to degrade ECM. Ly6C^{high} pro-inflammatory monocytes differentiate into classically activated M1 macrophages that express IL-1β and TNF-α. The migration of anti-inflammatory Ly6C^{low} monocytes is mediated via CXCR1. Ly6C^{low} reparative monocytes can differentiate into alternatively activated M2 macrophages. MI induces the activation and proliferation of CD4+ T cells in the heart-draining lymph nodes, which express high levels of IFN-γ at the site of infarction. Additionally, γδT cells participate in apoptosis. Tregs produce IL-10, IL-13, and TGF-β and play an important role in the resolution of inflammation and cardiac repair following MI. MSCs and CBSCs produce chemokine. CXCL12 is a ligand of CXCR4, and exogenous ubiquitin interacts with CXCR4. They participate in cardioprotective function with Tregs.

3. The Role of the Adaptive Immune Response Focused on T Cells after Cardiac Damage

CD4+ T helper (Th) cells provide proper immune cell homeostasis and host defense but CD4+ T cells have also been shown to promote autoimmune and inflammatory diseases. The original description of Th cells included Th1 and Th2 cells; however, additional Th cell subsets were discovered later, including Th17 and Tregs, which are characterized by specific cytokine profiles [19]. In the healthy myocardium, there are about 10^3–10^4 CD4+ T cells per heart assessed by a flow cytometry analysis [20]. Based on histological observation, CD4+ T cells accumulate in the infarct zone within 2 min after reperfusion [21]. In a permanent coronary occlusion injury model, infiltrating CD4+ T cells influx gradually and peak at day 7. Th1 cells and Tregs are the predominant subsets of CD4+ T cells, whereas Th2 cells and Th17 cells are minor populations. CD8+ T cells, γδT cells, and B cells also peaked on day 7 post-MI [22]. In clinical studies, patients with ACS have significantly greater activation of Th1 and Th17, and a prominent decrease in Treg numbers [23,24].

During scar maturation, the inflammatory response enters the chronic phase, which is characterized by low-grade, persistent inflammation. Few studies have reported the chronic inflammatory response after MI. One pilot study describes a global expansion of CD4+ T cells, especially Th2 and Th17 subsets, and a small number of CD8+ T cells at 8 weeks post-MI. This study also showed that depleting CD4+ T cells prevented adverse ventricular remodeling [10]. Another study demonstrated that tissue-specific T-cell responses with predominantly Th1 cells and cytotoxic CD8+ T cells phenotypes are present in ischemic failing human hearts, which contribute to the progression of HF from T cell receptor (TCR) sequencing [25]. According to these studies, CD4+ and CD8+ T cells participate in chronic stage after cardiac injury, but further studies need to be performed to reveal the regulation of adaptive immune responses.

IL-17A producing Th17 effector cells have an important role after MI. Th17 cells can be primed through activation of conventional type 2 dendritic cells (DCs) after MI [26]. IL-17A is primarily produced by γδT cells in the infarcted heart [27] and are involved in late remodeling stages after MI by promoting a sustained infiltration of neutrophils and macrophages and upregulation of pro-inflammatory cytokines leading to cardiomyocyte death and fibrosis [28] via the MMP/TIMP signaling pathway [29]. In a study using IL-17A deficient mice, there was no difference in infarct size compared with WT mice at day 1. However, improved survival and attenuated LV dilation were observed over 28 days post MI [28]. Additionally, several reports showed that IL-17 can directly activate MMP-1 in human cardiac fibrosis via p38 mitogen activated protein kinase (MAPK) and extracellular regulated kinase (ERK) dependent activator protein 1, nuclear factor (NF)-κB activation [30]. Additionally, increased expression of downstream target genes including IL-6, TNF, chemokine ligand (CCL) 20 and C-X-C motif chemokine ligand (CXCL) 1 were also upregulated [31]. In clinical studies, both circulating Th17 cells and IL-17 levels are increased in patients with ACS and stable angina compared to healthy controls [32,33]. Most studies conclude that a high IL-17A circulating level is associated with poor prognosis. One study, however, found that low serum levels of IL-17 are associated with a higher risk of major cardiovascular events in patients with acute MI [34]. Conclusively, the role of Th17/IL-17A in the context of MI has not yet been adequately addressed.

CD8+ T-cells also have been involved in both beneficial and detrimental cardiac remodeling. CD8+ T cells, which have angiotensin II receptors, infiltrate the peri-infarct myocardium 7 days after MI [35]. CD8+ T cells were characterized by upregulated IL-10 and downregulated IL-2 and INF-γ production, which have been shown to reduce ischemic heart injury. On the other hand, another study reported CD8+ T cells activation eliciting determinantal effects on cardiomyocytes in vitro [36]. The cytotoxic activity against healthy cardiomyocytes was myocyte-specific, which suggested major histocompatibility complex (MHC) class I and an antigen-specific cytotoxic response. CD8+ T cells may be detrimental to the cardiac remodeling process by inducing direct cytotoxic effects on healthy cardiomyocytes and amplifying neutrophil and macrophage-mediated inflammation, thus resulting in increased LV dilation and decreased cardiac function. However, CD8+ T cell actions would be indirectly beneficial by decreasing MMP-mediated collagen turnover, facilitating scar formation and decreasing incidence of cardiac rupture [37]. Moreover, CD8+ T cells can reduce cardiac fibrosis and improve cardiac function after injury in mice by the ablation of fibroblast activation protein [38].

Adaptive immune response focused on T cells play an important role to regulate inflammation from the acute to chronic phase. However, the detailed inflammation mechanism remains to be investigated as each T cells subsets are dynamic. Many more studies are needed to correlate the role of different T cell subset dynamics in regard to species, age, genetic modifications, injury models, and postoperative days. Conducting cell-specific transcriptome or proteome analyses of the temporal dynamics of cardiac immune cell accumulations following MI may contribute to identify the mechanism of inflammation after cardiac tissue damage including cytokines/chemokines. Moreover, we must corroborate rodent models in a clinical setting to benefit patients with cardiac injury. Clinically relevant studies are always challenging due to inherent variability in between the sample and tissue availability at multiple time points after injury. Most of the clinically relevant data is on human cells that are usually acquired from blood samples, so they are not necessarily a reflective of ischemic heart tissue conditions. However, currently, we can use human tissue cells under specific environments including ischemic cardiac injury from the cell bank, which can help us solve a complicated puzzle of the wound healing process after MI.

4. The Role of the Adaptive Immune System in Cardiac Tissue Repair

After the inflammatory phase, dying neutrophils are cleared by local macrophages that switch their phenotypic polarization to support healing [39]. Tregs can be a potential solution for tissue repair as they can terminate the pro-inflammatory phase and initiate the anti-inflammatory or pro-reparative phase at the site of tissue injury. Natural Tregs are generated in the thymus during fetal development and the first few years of life, while induced Treg cells can be developed later in the periphery from naive CD4+ T cells. Tregs express a specific transcription factor called FoxP3, for the forkhead/winged helix transcription factor, crucial for their development and functions [40]. Few resident Tregs are present in the healthy myocardium, but they can rapidly infiltrate after acute ischemia and have been shown to peak by day 7 after MI [41,42]. Tregs can suppress effector activities of differentiated CD4+ T cells, CD8+ T cells, Th17 cells, and the function of natural killer and B cells [43]. They also influence wound healing processes by modulating monocyte/macrophage differentiation. Some studies have suggested that TGF-β plays a critical role in the induction of FoxP3 expressions and is a main regulator of Tregs, in vivo and in vitro [44].

The ablation of Tregs with CD25 antibody enhances the numbers of both inflammatory myeloid cells and lymphocytes associated with M1 macrophage polarization and delay the healing response [41]. Tregs also promote the differentiation of recruited Ly6C^{high} monocytes toward anti-inflammatory M2 macrophages in the myocardium by secreting IL-10, IL-13, and TGF-β, which can directly stimulate fibroblasts in the myocardium as shown in Figure 1 [45]. Tregs also induced the expression of mediators in macrophages such as osteopontin and transglutaminase factor XIII; these factors are well known to contribute to myocardial healing [36]. Tregs attenuate myocardial ischemia/reperfusion (I/R) injury through a CD39 dependent mechanism involving Akt and ERK pathway [46], and modulated matrix-preserving cardiac fibroblast phenotype, reducing expression of α-smooth muscle actin and MMP-3, and attenuating contraction of fibroblast-populated collagen pads in the later phase [47]. Based on these findings, Tregs play an essential role in cardiac tissue repair.

Tregs have a potential to be a main target to improve cardiac function in ischemic heart tissue, however, reparative ability of Tregs decrease during aging [48]. It is important to investigate factors that can increase the life of a reparative Tregs. Treg capacity in neonatal rodent models is considered much higher than adults, but not fully explored. Therefore, focusing on rejuvenating factors identified from neonatal models may become a clue to develop our understanding of the cardiac repair mechanism in elderly patients.

5. The Role of the Adaptive Immune System Focused on T Cells Associated with Ubiquitin

Ubiquitin is important for regulating protein turn over via the ubiquitin–proteasome system (UPS). The UPS regulates fundamental cell functions including mitosis, DNA replication and repair, cell differentiation, transcriptional regulation, and receptor internalization, which all play a role in heart biology [49]. UPS is not only important in protein degradation, but it is also involved in multiple inflammatory processes such as NF-κB activation [50]. In the heart, several ubiquitin ligases (E3 enzymes) are involved in cardiac physiology, such as atrogin 1, muscle RING finger family (MuRF) 1, and murine double minute (MDM) 2 [49]. Consequently, UPS regulates cardiac signal transduction pathways and transcription factors such as calcineurin, which are associated with regulation of pathological hypertrophic growth exhibited post-MI.

Regarding T cells, ubiquitination plays a crucial role in the regulation of TCR-proximal signaling. It also regulates the initial activation and subsequent differentiation of T cells [51], TCR-stimulated endocytosis, and degradation of the linker for activation of T cells (LAT); although the underlying mechanism is poorly defined [52,53].

There are few studies associated with ubiquitin and T cells directly. Recently, however, some studies implied that exogenous ubiquitin participated in cardioprotective function with Tregs via C-X-C motif chemokine receptor (CXCR) 4. Exogenous ubiquitin plays a protective role in attenuating the cardiac inflammatory response and decreasing infarct size after I/R injury. Moreover, exogenous ubiquitin increased the expression of MMP-2 and MMP-9, which can increase ECM degradation and can contribute to a reduction in infarct size [54]. Extracellular ubiquitin interacts with CXCR4 and affects the proliferation of cardiac fibrosis via the ERK 1/2 pathway [55,56]. Another study revealed that a CXCR4 antagonist promotes tissue repair after MI by enhancing Treg mobilization and immune-regulatory function in mice [57]. In a clinical study, combined computed tomography and a positron emission tomography can detect CXCR4 as a surrogate for T cells in humans after MI [9].

There is no direct evidence of exogenous ubiquitin of Tregs via CXCR4. Moreover, exogenous ubiquitin was not illustrated in the T cell ubiquitination even if it is a component of the UPS. However, exogenous ubiquitin has a therapeutic potential and a clue to identify the mechanism between the UPS and the Treg phenotype/recruitments. Future accumulated studies about CXCR4 may contribute to adapt evidence of rodent models to humans.

6. Stem Cell Derived Therapies for Cardiac Injury

Current therapeutic and interventional options for the treatment of acute MI focus on the prevention or reverting adverse cardiac remodeling [58,59]. Current therapeutic options can help with functional output, however, it cannot regenerate the lost myocytes due to an ischemic episode. It is now well established that cardiomyocytes in the mature heart cannot proliferate much due to their limited self-regenerative capacity [60]. Exogenous stem cells therapies gain prominence in the last decade due to their ability to repair damaged organs. However, the mechanism by which the organ function improves is debatable and involves minimal regeneration of myocyte warranting investigation of other reparative processes responsible for improvement in function. Other biological processes that can be involved in the wound healing process are immune cells that are one of the key players in a wound healing process after injury. Recently, significant advances have been made in MI treatment using mesenchymal stem cells (MSCs) due to their angiogenic, antiapoptotic, anti-inflammatory, and cardioprotective effects [61–64]. MSCs have been tested for their ability to differentiate into several lineages in vitro and also tested in animal models to mediate tissue repair [61]. They have been reported to exert profound immunoregulatory effects on DCs, Tregs, and monocytes or macrophages via paracrine effects [65–69].

The potential for MSCs to induce and increase proliferation of Tregs has been shown via a wide range of credible direct and indirect mechanisms. A direct contact between human MSCs and purified CD4+ T cells has been shown to be essential for the induction of Treg as elimination of contact by a semipermeable membrane greatly compromises FoxP3 expression [70]. Soluble factors such as TGF-β1 are a potent paracrine factor that induce FoxP3 expression [71]. A significant increase in the number of FoxP3+ Treg was observed when human CD4+ T cells were cocultured with dental pulp MSCs, which was in turn reduced when TGF-β1 production was blocked [72]. Recently, MSCs have been shown to promote the immune regulatory effects through the release of extracellular vesicles (EVs) and has been suggested as a mechanism of the Treg proliferation. As exosomes payload is rich in mRNA, miRNA, and protein cargo, and have the potential to regulate immune cell gene transcription, intracellular signaling, and effector function [73]. MSC-EV also conditioned human DCs and has shown to have increased secretion of IL-10 and TGF-β leading to greater Tregs induction in pancreatic islet antigen-specific stimulation assays of T cells with type 1 diabetes [74].

Similarly, a novel stem cell population isolated from the cortical bone has demonstrated improved cardiomyocytes survival, cardiac function, and attenuation of remodeling compared with other stem cells because their secretome is enriched in pro-reparative factors [75]. Additionally, CBSCs also show unique expression of immunomodulatory proteins and cytokine production, which affect cardiomyocytes and immune cell populations. Intramyocardial injections of CBSCs resulted in CBSC engraftment and a decrease in the frequency of apoptosis overall 7 days after MI [76]. Concurrently, T cell and macrophage recruitment to the tissue was increased in CBSC treated animals. CBSCs produce cytokines and growth factors that are known to promote T cell and macrophage growth, chemotaxis, differentiation, and survival such as: TIMP-1, CXCL12, and macrophage colony stimulating factor

(M-CSF). CXCL12, as a ligand of CXCR4 [77], strongly attracts lymphocytes [78] and has also been shown to protect cardiomyocytes from apoptosis [79]. M-CSF is known to induce the differentiation of bone marrow cells to macrophages, induce an immunosuppressive phenotype, and induce the production of CCL2 [80]. CCL2 is a potent chemotactic signal molecule for monocytes and lymphocytes, including T cells [81], and plays a crucial role in healing infarcts [82].

Clinical applications have demonstrated the ability of bone marrow-derived cells [83–85], cardiac-derived cells [86,87], and MSCs [88–90] to offer moderate functional benefits when transplanted after cardiac injury. Although, they show sustained modest beneficial effects and bring new knowledge for the cardiac repair mechanism, the basic biology and interactions of stem cells with the host cells and its immune environment are still not fully understood and need to be studied in details. We must identify the key regulators of wound healing processes that are delivered or activated after cell therapy. Moreover, we should focus on the time of delivery of stem cells as the delivery of stem cell factors in a timely manner can be a key for modulating the immune response that effects the wound healing process.

7. Conclusions

Accumulative evidence has shown that the adaptive immune response is involved in post-ischemic cardiac remodeling after MI. From the acute to chronic phase, T cells have played an important role in the mediation of tissue repair following cardiac injury. Tregs can terminate the pro-inflammatory phase and initiate the anti-inflammatory or regenerative phase, promoting the differentiation of Ly6C^{high} monocytes toward M2 macrophages in the myocardium by secreting pre reparative cytokines including IL-10, IL-13, and TGF-β. They also stimulate fibroblasts directly in the myocardium. Paracrine signaling to and from immune cells and stem cells can be key in understanding the wound healing process after cardiac injury. Stem cells can induce Tregs via direct and indirect mechanisms. However, the mechanism behind polarization of different T cell subsets after stem cells transplantation remains poorly understood. If we focus on the UPS and its effect on cell therapy, it can give us some clues on how to modulate a pro-reparative phenotype especially associated with Tregs. This may potentially unravel some mechanisms that can augment cardiac healing after ischemic injury. The development of new therapeutic strategies targeting the adaptive immune system in IHD and via stem cells and its interplay with the UPS can contribute to be a more effective treatment in patients with heart disease.

Author Contributions: T.K. performed collection of references and wrote the current manuscript. M.K. and S.M. supervised review scope and aided in the preparation of the current manuscript. All authors have read and agreed to the published version of the manuscript.

Funding: The work was supported by National Institute of Health grant HL137850 and American Heart association grant 15SDG25550038 to S.M. and NIH HL135177, AHA Scientific Development Grant 15SDG22680018 and H1801 WW Smith Charitable Trust to M.K.

Acknowledgments: We thank all Mohsin Lab members for their input.

Conflicts of Interest: The authors declare no conflict of interest.

Abbreviations

ACS	Acute coronary syndrome
APC	Antigen presenting cell
CCL	Chemokine ligand
CXCL	C-X-C motif chemokine ligand
CXCR	C-X-C motif chemokine receptor
DAMP	Damage associated molecular pattern
DC	Dendritic cell
ECM	Extracellular matrix
ERK	Extracellular regulated kinase
EV	Extracellular vesicle
HF	Heart failure
I/R	Ischemia / reperfusion

IFN	Interferon
IHD	Ischemic heart disease
IL	Interleukin
LAT	Linker for activation of T cells
LV	Left ventricular
MAPK	Mitogen activated protein kinase
M-CSF	Macrophage colony stimulating factor
MDM	Murine double minute
MHC	Major histocompatibility complex
MSC	Mesenchymal stem cell
MI	Myocardial infarction
MMP	Matrix metalloproteinase
MuRF	Muscle RING finger family
NF	Nuclear factor
OMT	Optical medical therapy
TCR	T cell receptor
TGF	Transforming growth factor
TLR	Toll like receptors
Th	T helper
TIMP	Tissue inhibitors of metalloproteinase
TNF	Tumor necrosis factor
Treg	Regulatory T cell
VEGF	Vascular endothelial growth factor

References

1. Nichols, M.; Townsend, N.; Scarborough, P.; Rayner, M. Cardiovascular disease in Europe 2014: Epidemiological update. *Eur. Heart J.* **2014**, *35*, 2950–2959. [CrossRef]
2. Westman, P.C.; Lipinski, M.J.; Luger, D.; Waksman, R.; Bonow, R.O.; Wu, E.; Epstein, S.E. Inflammation as a Driver of Adverse Left Ventricular Remodeling After Acute Myocardial Infarction. *J. Am. Coll. Cardiol.* **2016**, *67*, 2050–2060. [CrossRef]
3. Prabhu, S.D.; Frangogiannis, N.G. The Biological Basis for Cardiac Repair After Myocardial Infarction: From Inflammation to Fibrosis. *Circ. Res.* **2016**, *119*, 91–112. [CrossRef]
4. Epelman, S.; Liu, P.P.; Mann, D.L. Role of innate and adaptive immune mechanisms in cardiac injury and repair. *Nat. Rev. Immunol.* **2015**, *15*, 117–129. [CrossRef]
5. Harty, M.; Neff, A.W.; King, M.W.; Mescher, A.L. Regeneration or scarring: An immunologic perspective. *Dev. Dyn.* **2003**, *226*, 268–279. [CrossRef]
6. Maisel, A.; Cesario, D.; Baird, S.; Rehman, J.; Haghighi, P.; Carter, S. Experimental autoimmune myocarditis produced by adoptive transfer of splenocytes after myocardial infarction. *Circ. Res.* **1998**, *82*, 458–463. [CrossRef]
7. Hofmann, U.; Beyersdorf, N.; Weirather, J.; Podolskaya, A.; Bauersachs, J.; Ertl, G.; Kerkau, T.; Frantz, S. Activation of CD4+ T lymphocytes improves wound healing and survival after experimental myocardial infarction in mice. *Circulation* **2012**, *125*, 1652–1663. [CrossRef]
8. Ramos, G.C.; Dalbo, S.; Leite, D.P.; Goldfeder, E.; Carvalho, C.R.; Vaz, N.M.; Assreuy, J. The autoimmune nature of post-infarct myocardial healing: Oral tolerance to cardiac antigens as a novel strategy to improve cardiac healing. *Autoimmunity* **2012**, *45*, 233–244. [CrossRef]
9. Rieckmann, M.; Delgobo, M.; Gaal, C.; Büchner, L.; Steinau, P.; Reshef, D.; Gil-Cruz, C.; Horst, E.N.T.; Kircher, M.; Reiter, T.; et al. Myocardial infarction triggers cardioprotective antigen-specific T helper cell responses. *J. Clin. Investig.* **2019**, *130*, 4922–4936. [CrossRef]
10. Bansal, S.S.; Ismahil, M.A.; Goel, M.; Patel, B.; Hamid, T.; Rokosh, G.; Prabhu, S.D. Activated T Lymphocytes are Essential Drivers of Pathological Remodeling in Ischemic Heart Failure. *Circ. Heart Fail.* **2017**, *10*, e003688. [CrossRef]
11. Wagner, M.J.; Khan, M.; Mohsin, S. Healing the Broken Heart; The Immunomodulatory Effects of Stem Cell Therapy. *Front. Immunol.* **2020**, *11*, 639. [CrossRef]

12. Boden, W.E.; O'Rourke, R.A.; Teo, K.K.; Hartigan, P.M.; Maron, D.J.; Kostuk, W.J.; Knudtson, M.; Dada, M.; Casperson, P.; Harris, C.L.; et al. Optimal medical therapy with or without PCI for stable coronary disease. *N. Engl. J. Med.* **2007**, *356*, 1503–1516. [CrossRef]

13. Wernly, B.; Paar, V.; Aigner, A.; Pilz, P.M.; Podesser, B.K.; Förster, M.; Jung, C.; Hofbauer, J.P.; Tockner, B.; Wimmer, M.; et al. Anti-CD3 Antibody Treatment Reduces Scar Formation in a Rat Model of Myocardial Infarction. *Cells* **2020**, *9*, 295. [CrossRef]

14. Daniels, C.R.; Foster, C.R.; Yakoob, S.; Dalal, S.; Joyner, W.L.; Singh, M.; Singh, K. Exogenous ubiquitin modulates chronic β-adrenergic receptor-stimulated myocardial remodeling: Role in Akt activity and matrix metalloproteinase expression. *Am. J. Physiol. Heart Circ. Physiol.* **2012**, *303*, H1459–H1468. [CrossRef]

15. Frangogiannis, N.G.; Smith, C.W.; Entman, M.L. The inflammatory response in myocardial infarction. *Cardiovasc. Res.* **2002**, *53*, 31–47. [CrossRef]

16. Saparov, A.; Ogay, V.; Nurgozhin, T.; Chen, W.C.W.; Mansurov, N.; Issabekova, A.; Zhakupova, J. Role of the immune system in cardiac tissue damage and repair following myocardial infarction. *Inflamm. Res.* **2017**, *66*, 739–751. [CrossRef]

17. Bujak, M.; Frangogiannis, N.G. The role of TGF-beta signaling in myocardial infarction and cardiac remodeling. *Cardiovasc. Res.* **2007**, *74*, 184–195. [CrossRef]

18. Fan, D.; Takawale, A.; Lee, J.; Kassiri, Z. Cardiac fibroblasts, fibrosis and extracellular matrix remodeling in heart disease. *Fibrogenes. Tissue Repair* **2012**, *5*, 15. [CrossRef]

19. Raphael, I.; Nalawade, S.; Eagar, T.N.; Forsthuber, T.G. T cell subsets and their signature cytokines in autoimmune and inflammatory diseases. *Cytokine* **2015**, *74*, 5–17. [CrossRef]

20. Ramos, G.C.; van den Berg, A.; Nunes-Silva, V.; Weirather, J.; Peters, L.; Burkard, M.; Friedrich, M.; Pinnecker, J.; Abesser, M.; Heinze, K.G.; et al. Myocardial aging as a T-cell-mediated phenomenon. *Proc. Natl. Acad. Sci. USA* **2017**, *114*, E2420–E2429. [CrossRef]

21. Yang, Z.; Day, Y.J.; Toufektsian, M.C.; Xu, Y.; Ramos, S.I.; Marshall, M.A.; French, B.A.; Linden, J. Myocardial infarct-sparing effect of adenosine A2A receptor activation is due to its action on CD4+ T lymphocytes. *Circulation* **2006**, *114*, 2056–2064. [CrossRef]

22. Yan, X.; Anzai, A.; Katsumata, Y.; Matsuhashi, T.; Ito, K.; Endo, J.; Yamamoto, T.; Takeshima, A.; Shinmura, K.; Shen, W.; et al. Temporal dynamics of cardiac immune cell accumulation following acute myocardial infarction. *J. Mol. Cell. Cardiol.* **2013**, *62*, 24–35. [CrossRef] [PubMed]

23. Methe, H.; Brunner, S.; Wiegand, D.; Nabauer, M.; Koglin, J.; Edelman, E.R. Enhanced T-helper-1 lymphocyte activation patterns in acute coronary syndromes. *J. Am. Coll. Cardiol.* **2005**, *45*, 1939–1945. [CrossRef]

24. Cheng, X.; Yu, X.; Ding, Y.J.; Fu, Q.Q.; Xie, J.J.; Tang, T.T.; Yao, R.; Chen, Y.; Liao, Y.H. The Th17/Treg imbalance in patients with acute coronary syndrome. *Clin. Immunol.* **2008**, *127*, 89–97. [CrossRef]

25. Tang, T.T.; Zhu, Y.C.; Dong, N.G.; Zhang, S.; Cai, J.; Zhang, L.X.; Han, Y.; Xia, N.; Nie, S.F.; Zhang, M.; et al. Pathologic T-cell response in ischaemic failing hearts elucidated by T-cell receptor sequencing and phenotypic characterization. *Eur. Heart J.* **2019**, *40*, 3924–3933. [CrossRef] [PubMed]

26. Van der Borght, K.; Scott, C.L.; Nindl, V.; Bouche, A.; Martens, L.; Sichien, D.; Van Moorleghem, J.; Vanheerswynghels, M.; De Prijck, S.; Saeys, Y.; et al. Myocardial Infarction Primes Autoreactive T Cells through Activation of Dendritic Cells. *Cell Rep.* **2017**, *18*, 3005–3017. [CrossRef] [PubMed]

27. Liao, Y.H.; Xia, N.; Zhou, S.F.; Tang, T.T.; Yan, X.X.; Lv, B.J.; Nie, S.F.; Wang, J.; Iwakura, Y.; Xiao, H.; et al. Interleukin-17A contributes to myocardial ischemia/reperfusion injury by regulating cardiomyocyte apoptosis and neutrophil infiltration. *J. Am. Coll. Cardiol.* **2012**, *59*, 420–429. [CrossRef]

28. Yan, X.; Shichita, T.; Katsumata, Y.; Matsuhashi, T.; Ito, H.; Ito, K.; Anzai, A.; Endo, J.; Tamura, Y.; Kimura, K.; et al. Deleterious effect of the IL-23/IL-17A axis and gammadeltaT cells on left ventricular remodeling after myocardial infarction. *J. Am. Heart Assoc.* **2012**, *1*, e004408. [CrossRef]

29. Feng, W.; Li, W.; Liu, W.; Wang, F.; Li, Y.; Yan, W. IL-17 induces myocardial fibrosis and enhances RANKL/OPG and MMP/TIMP signaling in isoproterenol-induced heart failure. *Exp. Mol. Pathol.* **2009**, *87*, 212–218. [CrossRef]

30. Cortez, D.M.; Feldman, M.D.; Mummidi, S.; Valente, A.J.; Steffensen, B.; Vincenti, M.; Barnes, J.L.; Chandrasekar, B. IL-17 stimulates MMP-1 expression in primary human cardiac fibroblasts via p38 MAPK- and ERK1/2-dependent C/EBP-beta, NF-kappaB, and AP-1 activation. *Am. J. Physiol. Heart Circ. Physiol.* **2007**, *293*, H3356–H3365. [CrossRef]

31. Chang, S.L.; Hsiao, Y.W.; Tsai, Y.N.; Lin, S.F.; Liu, S.H.; Lin, Y.J.; Lo, L.W.; Chung, F.P.; Chao, T.F.; Hu, Y.F.; et al. Interleukin-17 enhances cardiac ventricular remodeling via activating MAPK pathway in ischemic heart failure. *J. Mol. Cell. Cardiol.* **2018**, *122*, 69–79. [CrossRef]

32. Zhu, F.; Wang, Q.; Guo, C.; Wang, X.; Cao, X.; Shi, Y.; Gao, F.; Ma, C.; Zhang, L. IL-17 induces apoptosis of vascular endothelial cells: A potential mechanism for human acute coronary syndrome. *Clin. Immunol.* **2011**, *141*, 152–160. [CrossRef]

33. Hashmi, S.; Zeng, Q.T. Role of interleukin-17 and interleukin-17-induced cytokines interleukin-6 and interleukin-8 in unstable coronary artery disease. *Coron. Artery Dis.* **2006**, *17*, 699–706. [CrossRef]

34. Simon, T.; Taleb, S.; Danchin, N.; Laurans, L.; Rousseau, B.; Cattan, S.; Montely, J.M.; Dubourg, O.; Tedgui, A.; Kotti, S.; et al. Circulating levels of interleukin-17 and cardiovascular outcomes in patients with acute myocardial infarction. *Eur. Heart J.* **2013**, *34*, 570–577. [CrossRef]

35. Curato, C.; Slavic, S.; Dong, J.; Skorska, A.; Altarche-Xifro, W.; Miteva, K.; Kaschina, E.; Thiel, A.; Imboden, H.; Wang, J.; et al. Identification of noncytotoxic and IL-10-producing CD8+AT2R+ T cell population in response to ischemic heart injury. *J. Immunol.* **2010**, *185*, 6286–6293. [CrossRef]

36. Varda-Bloom, N.; Leor, J.; Ohad, D.G.; Hasin, Y.; Amar, M.; Fixler, R.; Battler, A.; Eldar, M.; Hasin, D. Cytotoxic T lymphocytes are activated following myocardial infarction and can recognize and kill healthy myocytes in vitro. *J. Mol. Cell. Cardiol.* **2000**, *32*, 2141–2149. [CrossRef] [PubMed]

37. Ilatovskaya, D.V.; Pitts, C.; Clayton, J.; Domondon, M.; Troncoso, M.; Pippin, S.; DeLeon-Pennell, K.Y. CD8(+) T-cells negatively regulate inflammation post-myocardial infarction. *Am. J. Physiol. Heart. Circ. Physiol.* **2019**, *317*, H581–H596. [CrossRef] [PubMed]

38. Aghajanian, H.; Kimura, T.; Rurik, J.G.; Hancock, A.S.; Leibowitz, M.S.; Li, L.; Scholler, J.; Monslow, J.; Lo, A.; Han, W.; et al. Targeting cardiac fibrosis with engineered T cells. *Nature* **2019**, *573*, 430–433. [CrossRef] [PubMed]

39. Harel-Adar, T.; Ben Mordechai, T.; Amsalem, Y.; Feinberg, M.S.; Leor, J.; Cohen, S. Modulation of cardiac macrophages by phosphatidylserine-presenting liposomes improves infarct repair. *Proc. Natl. Acad. Sci. USA* **2011**, *108*, 1827–1832. [CrossRef]

40. Santos-Zas, I.; Lemarie, J.; Tedgui, A.; Ait-Oufella, H. Adaptive Immune Responses Contribute to Post-ischemic Cardiac Remodeling. *Front. Cardiovasc. Med.* **2018**, *5*, 198. [CrossRef]

41. Weirather, J.; Hofmann, U.D.; Beyersdorf, N.; Ramos, G.C.; Vogel, B.; Frey, A.; Ertl, G.; Kerkau, T.; Frantz, S. Foxp3+ CD4+ T cells improve healing after myocardial infarction by modulating monocyte/macrophage differentiation. *Circ. Res.* **2014**, *115*, 55–67. [CrossRef] [PubMed]

42. Zacchigna, S.; Martinelli, V.; Moimas, S.; Colliva, A.; Anzini, M.; Nordio, A.; Costa, A.; Rehman, M.; Vodret, S.; Pierro, C.; et al. Paracrine effect of regulatory T cells promotes cardiomyocyte proliferation during pregnancy and after myocardial infarction. *Nat. Commun.* **2018**, *9*, 2432. [CrossRef] [PubMed]

43. Sakaguchi, S.; Yamaguchi, T.; Nomura, T.; Ono, M. Regulatory T cells and immune tolerance. *Cell* **2008**, *133*, 775–787. [CrossRef] [PubMed]

44. Shevach, E.M. Mechanisms of foxp3+ T regulatory cell-mediated suppression. *Immunity* **2009**, *30*, 636–645. [CrossRef]

45. Hofmann, U.; Frantz, S. Role of lymphocytes in myocardial injury, healing, and remodeling after myocardial infarction. *Circ. Res.* **2015**, *116*, 354–367. [CrossRef] [PubMed]

46. Xia, N.; Jiao, J.; Tang, T.T.; Lv, B.J.; Lu, Y.Z.; Wang, K.J.; Zhu, Z.F.; Mao, X.B.; Nie, S.F.; Wang, Q.; et al. Activated regulatory T-cells attenuate myocardial ischaemia/reperfusion injury through a CD39-dependent mechanism. *Clin. Sci.* **2015**, *128*, 679–693. [CrossRef]

47. Saxena, A.; Dobaczewski, M.; Rai, V.; Haque, Z.; Chen, W.; Li, N.; Frangogiannis, N.G. Regulatory T cells are recruited in the infarcted mouse myocardium and may modulate fibroblast phenotype and function. *Am. J. Physiol. Heart Circ. Physiol.* **2014**, *307*, H1233–H1242. [CrossRef]

48. Jagger, A.; Shimojima, Y.; Goronzy, J.J.; Weyand, C.M. Regulatory T cells and the immune aging process: A mini-review. *Gerontology* **2014**, *60*, 130–137. [CrossRef]

49. Pagan, J.; Seto, T.; Pagano, M.; Cittadini, A. Role of the ubiquitin proteasome system in the heart. *Circ. Res.* **2013**, *112*, 1046–1058. [CrossRef]

50. Iwai, K. Diverse roles of the ubiquitin system in NF-kappaB activation. *Biochim. Biophys. Acta* **2014**, *1843*, 129–136. [CrossRef]

51. Hu, H.; Sun, S.C. Ubiquitin signaling in immune responses. *Cell Res.* **2016**, *26*, 457–483. [CrossRef] [PubMed]

52. Balagopalan, L.; Barr, V.A.; Sommers, C.L.; Barda-Saad, M.; Goyal, A.; Isakowitz, M.S.; Samelson, L.E. c-Cbl-mediated regulation of LAT-nucleated signaling complexes. *Mol. Cell. Biol.* **2007**, *27*, 8622–8636. [CrossRef] [PubMed]

53. Balagopalan, L.; Ashwell, B.A.; Bernot, K.M.; Akpan, I.O.; Quasba, N.; Barr, V.A.; Samelson, L.E. Enhanced T-cell signaling in cells bearing linker for activation of T-cell (LAT) molecules resistant to ubiquitylation. *Proc. Natl. Acad. Sci. USA* **2011**, *108*, 2885–2890. [CrossRef] [PubMed]

54. Scofield, S.L.C.; Dalal, S.; Lim, K.A.; Thrasher, P.R.; Daniels, C.R.; Peterson, J.M.; Singh, M.; Singh, K. Exogenous ubiquitin reduces inflammatory response and preserves myocardial function 3 days post-ischemia-reperfusion injury. *Am. J. Physiol. Heart Circ. Physiol.* **2019**, *316*, H617–H628. [CrossRef]

55. Saini, V.; Marchese, A.; Majetschak, M. CXC chemokine receptor 4 is a cell surface receptor for extracellular ubiquitin. *J. Biol. Chem.* **2010**, *285*, 15566–15576. [CrossRef]

56. Scofield, S.L.C.; Daniels, C.R.; Dalal, S.; Millard, J.A.; Singh, M.; Singh, K. Extracellular ubiquitin modulates cardiac fibroblast phenotype and function via its interaction with CXCR4. *Life Sci.* **2018**, *211*, 8–16. [CrossRef]

57. Wang, Y.; Dembowsky, K.; Chevalier, E.; Stuve, P.; Korf-Klingebiel, M.; Lochner, M.; Napp, L.C.; Frank, H.; Brinkmann, E.; Kanwischer, A.; et al. C-X-C Motif Chemokine Receptor 4 Blockade Promotes Tissue Repair After Myocardial Infarction by Enhancing Regulatory T Cell Mobilization and Immune-Regulatory Function. *Circulation* **2019**, *139*, 1798–1812. [CrossRef]

58. Ibanez, B.; James, S.; Agewall, S.; Antunes, M.J.; Bucciarelli-Ducci, C.; Bueno, H.; Caforio, A.L.P.; Crea, F.; Goudevenos, J.A.; Halvorsen, S.; et al. 2017 ESC Guidelines for the management of acute myocardial infarction in patients presenting with ST-segment elevation: The Task Force for the management of acute myocardial infarction in patients presenting with ST-segment elevation of the European Society of Cardiology (ESC). *Eur. Heart J.* **2018**, *39*, 119–177. [CrossRef]

59. Yancy, C.W.; Jessup, M.; Bozkurt, B.; Butler, J.; Casey, D.E., Jr.; Colvin, M.M.; Drazner, M.H.; Filippatos, G.S.; Fonarow, G.C.; Givertz, M.M.; et al. 2017 ACC/AHA/HFSA Focused Update of the 2013 ACCF/AHA Guideline for the Management of Heart Failure: A Report of the American College of Cardiology/American Heart Association Task Force on Clinical Practice Guidelines and the Heart Failure Society of America. *Circulation* **2017**, *136*, e137–e161. [CrossRef]

60. Soonpaa, M.H.; Field, L.J. Survey of studies examining mammalian cardiomyocyte DNA synthesis. *Circ. Res.* **1998**, *83*, 15–26. [CrossRef]

61. Pittenger, M.F.; Martin, B.J. Mesenchymal stem cells and their potential as cardiac therapeutics. *Circ. Res.* **2004**, *95*, 9–20. [CrossRef] [PubMed]

62. Leistner, D.M.; Fischer-Rasokat, U.; Honold, J.; Seeger, F.H.; Schachinger, V.; Lehmann, R.; Martin, H.; Burck, I.; Urbich, C.; Dimmeler, S.; et al. Transplantation of progenitor cells and regeneration enhancement in acute myocardial infarction (TOPCARE-AMI): Final 5-year results suggest long-term safety and efficacy. *Clin. Res. Cardiol.* **2011**, *100*, 925–934. [CrossRef]

63. Caplan, A.I.; Dennis, J.E. Mesenchymal stem cells as trophic mediators. *J. Cell. Biochem.* **2006**, *98*, 1076–1084. [CrossRef] [PubMed]

64. Chen, S.L.; Fang, W.W.; Ye, F.; Liu, Y.H.; Qian, J.; Shan, S.J.; Zhang, J.J.; Chunhua, R.Z.; Liao, L.M.; Lin, S.; et al. Effect on left ventricular function of intracoronary transplantation of autologous bone marrow mesenchymal stem cell in patients with acute myocardial infarction. *Am. J. Cardiol.* **2004**, *94*, 92–95. [CrossRef] [PubMed]

65. English, K.; Mahon, B.P. Allogeneic mesenchymal stem cells: Agents of immune modulation. *J. Cell. Biochem.* **2011**, *112*, 1963–1968. [CrossRef]

66. Akiyama, K.; Chen, C.; Wang, D.; Xu, X.; Qu, C.; Yamaza, T.; Cai, T.; Chen, W.; Sun, L.; Shi, S. Mesenchymal-stem-cell-induced immunoregulation involves FAS-ligand-/FAS-mediated T cell apoptosis. *Cell Stem Cell* **2012**, *10*, 544–555. [CrossRef]

67. Kishore, R.; Khan, M. More Than Tiny Sacks: Stem Cell Exosomes as Cell-Free Modality for Cardiac Repair. *Circ. Res.* **2016**, *118*, 330–343. [CrossRef] [PubMed]

68. Zhao, J.; Li, X.; Hu, J.; Chen, F.; Qiao, S.; Sun, X.; Gao, L.; Xie, J.; Xu, B. Mesenchymal stromal cell-derived exosomes attenuate myocardial ischaemia-reperfusion injury through miR-182-regulated macrophage polarization. *Cardiovasc. Res.* **2019**, *115*, 1205–1216. [CrossRef] [PubMed]

69. Sun, X.; Shan, A.; Wei, Z.; Xu, B. Intravenous mesenchymal stem cell-derived exosomes ameliorate myocardial inflammation in the dilated cardiomyopathy. *Biochem. Biophys. Res. Commun.* **2018**, *503*, 2611–2618. [CrossRef] [PubMed]

70. English, K.; Ryan, J.M.; Tobin, L.; Murphy, M.J.; Barry, F.P.; Mahon, B.P. Cell contact, prostaglandin E(2) and transforming growth factor beta 1 play non-redundant roles in human mesenchymal stem cell induction of CD4+CD25(High) forkhead box P3+ regulatory T cells. *Clin. Exp. Immunol.* **2009**, *156*, 149–160. [CrossRef]

71. Hong, J.W.; Lim, J.H.; Chung, C.J.; Kang, T.J.; Kim, T.Y.; Kim, Y.S.; Roh, T.S.; Lew, D.H. Immune Tolerance of Human Dental Pulp-Derived Mesenchymal Stem Cells Mediated by CD4(+)CD25(+)FoxP3(+) Regulatory T-Cells and Induced by TGF-beta1 and IL-10. *Yonsei Med. J.* **2017**, *58*, 1031–1039. [CrossRef] [PubMed]

72. Zhao, Z.G.; Xu, W.; Sun, L.; You, Y.; Li, F.; Li, Q.B.; Zou, P. Immunomodulatory function of regulatory dendritic cells induced by mesenchymal stem cells. *Immunol. Investig.* **2012**, *41*, 183–198. [CrossRef] [PubMed]

73. Rani, S.; Ryan, A.E.; Griffin, M.D.; Ritter, T. Mesenchymal Stem Cell-derived Extracellular Vesicles: Toward Cell-free Therapeutic Applications. *Mol. Ther.* **2015**, *23*, 812–823. [CrossRef] [PubMed]

74. Favaro, E.; Carpanetto, A.; Caorsi, C.; Giovarelli, M.; Angelini, C.; Cavallo-Perin, P.; Tetta, C.; Camussi, G.; Zanone, M.M. Human mesenchymal stem cells and derived extracellular vesicles induce regulatory dendritic cells in type 1 diabetic patients. *Diabetologia* **2016**, *59*, 325–333. [CrossRef] [PubMed]

75. Duran, J.M.; Makarewich, C.A.; Sharp, T.E.; Starosta, T.; Zhu, F.; Hoffman, N.E.; Chiba, Y.; Madesh, M.; Berretta, R.M.; Kubo, H.; et al. Bone-derived stem cells repair the heart after myocardial infarction through transdifferentiation and paracrine signaling mechanisms. *Circ. Res.* **2013**, *113*, 539–552. [CrossRef] [PubMed]

76. Hobby, A.R.H.; Sharp, T.E., III; Berretta, R.M.; Borghetti, G.; Feldsott, E.; Mohsin, S.; Houser, S.R. Cortical bone-derived stem cell therapy reduces apoptosis after myocardial infarction. *Am. J. Physiol. Heart Circ. Physiol.* **2019**, *317*, H820–H829. [CrossRef] [PubMed]

77. Karpova, D.; Bonig, H. Concise Review: CXCR4/CXCL12 Signaling in Immature Hematopoiesis–Lessons From Pharmacological and Genetic Models. *Stem Cells* **2015**, *33*, 2391–2399. [CrossRef]

78. Bleul, C.C.; Fuhlbrigge, R.C.; Casasnovas, J.M.; Aiuti, A.; Springer, T.A. A highly efficacious lymphocyte chemoattractant, stromal cell-derived factor 1 (SDF-1). *J. Exp. Med.* **1996**, *184*, 1101–1109. [CrossRef]

79. Zhang, M.; Mal, N.; Kiedrowski, M.; Chacko, M.; Askari, A.T.; Popovic, Z.B.; Koc, O.N.; Penn, M.S. SDF-1 expression by mesenchymal stem cells results in trophic support of cardiac myocytes after myocardial infarction. *FASEB J.* **2007**, *21*, 3197–3207. [CrossRef]

80. Ushach, I.; Zlotnik, A. Biological role of granulocyte macrophage colony-stimulating factor (GM-CSF) and macrophage colony-stimulating factor (M-CSF) on cells of the myeloid lineage. *J. Leukoc. Biol.* **2016**, *100*, 481–489. [CrossRef]

81. Carr, M.W.; Roth, S.J.; Luther, E.; Rose, S.S.; Springer, T.A. Monocyte chemoattractant protein 1 acts as a T-lymphocyte chemoattractant. *Proc. Natl. Acad. Sci. USA* **1994**, *91*, 3652–3656. [CrossRef] [PubMed]

82. Dewald, O.; Zymek, P.; Winkelmann, K.; Koerting, A.; Ren, G.; Abou-Khamis, T.; Michael, L.H.; Rollins, B.J.; Entman, M.L.; Frangogiannis, N.G. CCL2/Monocyte Chemoattractant Protein-1 regulates inflammatory responses critical to healing myocardial infarcts. *Circ. Res.* **2005**, *96*, 881–889. [CrossRef] [PubMed]

83. Yousef, M.; Schannwell, C.M.; Kostering, M.; Zeus, T.; Brehm, M.; Strauer, B.E. The BALANCE Study: Clinical benefit and long-term outcome after intracoronary autologous bone marrow cell transplantation in patients with acute myocardial infarction. *J. Am. Coll. Cardiol.* **2009**, *53*, 2262–2269. [CrossRef] [PubMed]

84. Hare, J.M.; Fishman, J.E.; Gerstenblith, G.; DiFede Velazquez, D.L.; Zambrano, J.P.; Suncion, V.Y.; Tracy, M.; Ghersin, E.; Johnston, P.V.; Brinker, J.A.; et al. Comparison of allogeneic vs autologous bone marrow-derived mesenchymal stem cells delivered by transendocardial injection in patients with ischemic cardiomyopathy: The POSEIDON randomized trial. *JAMA* **2012**, *308*, 2369–2379. [CrossRef] [PubMed]

85. Patila, T.; Lehtinen, M.; Vento, A.; Schildt, J.; Sinisalo, J.; Laine, M.; Hammainen, P.; Nihtinen, A.; Alitalo, R.; Nikkinen, P.; et al. Autologous bone marrow mononuclear cell transplantation in ischemic heart failure: A prospective, controlled, randomized, double-blind study of cell transplantation combined with coronary bypass. *J. Heart Lung Transplant.* **2014**, *33*, 567–574. [CrossRef]

86. Chugh, A.R.; Beache, G.M.; Loughran, J.H.; Mewton, N.; Elmore, J.B.; Kajstura, J.; Pappas, P.; Tatooles, A.; Stoddard, M.F.; Lima, J.A.; et al. Administration of cardiac stem cells in patients with ischemic cardiomyopathy: The SCIPIO trial: Surgical aspects and interim analysis of myocardial function and viability by magnetic resonance. *Circulation* **2012**, *126*, S54–S64. [CrossRef]

87. Malliaras, K.; Makkar, R.R.; Smith, R.R.; Cheng, K.; Wu, E.; Bonow, R.O.; Marban, L.; Mendizabal, A.; Cingolani, E.; Johnston, P.V.; et al. Intracoronary cardiosphere-derived cells after myocardial infarction: Evidence of therapeutic regeneration in the final 1-year results of the CADUCEUS trial (CArdiosphere-Derived aUtologous stem CElls to reverse ventricUlar dySfunction). *J. Am. Coll. Cardiol.* **2014**, *63*, 110–122. [CrossRef]

88. Karantalis, V.; DiFede, D.L.; Gerstenblith, G.; Pham, S.; Symes, J.; Zambrano, J.P.; Fishman, J.; Pattany, P.; McNiece, I.; Conte, J.; et al. Autologous mesenchymal stem cells produce concordant improvements in regional function, tissue perfusion, and fibrotic burden when administered to patients undergoing coronary artery bypass grafting: The Prospective Randomized Study of Mesenchymal Stem Cell Therapy in Patients Undergoing Cardiac Surgery (PROMETHEUS) trial. *Circ. Res.* **2014**, *114*, 1302–1310. [CrossRef] [PubMed]
89. Ascheim, D.D.; Gelijns, A.C.; Goldstein, D.; Moye, L.A.; Smedira, N.; Lee, S.; Klodell, C.T.; Szady, A.; Parides, M.K.; Jeffries, N.O.; et al. Mesenchymal precursor cells as adjunctive therapy in recipients of contemporary left ventricular assist devices. *Circulation* **2014**, *129*, 2287–2296. [CrossRef] [PubMed]
90. Bartolucci, J.; Verdugo, F.J.; Gonzalez, P.L.; Larrea, R.E.; Abarzua, E.; Goset, C.; Rojo, P.; Palma, I.; Lamich, R.; Pedreros, P.A.; et al. Safety and Efficacy of the Intravenous Infusion of Umbilical Cord Mesenchymal Stem Cells in Patients With Heart Failure: A Phase 1/2 Randomized Controlled Trial (RIMECARD Trial [Randomized Clinical Trial of Intravenous Infusion Umbilical Cord Mesenchymal Stem Cells on Cardiopathy]). *Circ. Res.* **2017**, *121*, 1192–1204. [CrossRef]

International Journal of
Molecular Sciences

Review

Protein Degradation and the Pathologic Basis of Phenylketonuria and Hereditary Tyrosinemia

Neha Sarodaya [1], Bharathi Suresh [1], Kye-Seong Kim [1,2,*] and Suresh Ramakrishna [1,2,*]

[1] Graduate School of Biomedical Science and Engineering, Hanyang University, Seoul 04763, Korea; neyha19@gmail.com (N.S.); bharathi.suri@gmail.com (B.S.)
[2] College of Medicine, Hanyang University, Seoul 04763, Korea
* Correspondence: ks66kim@hanyang.ac.kr (K.-S.K.); suri28@hanyang.ac.kr or suresh.ramakris@gmail.com (S.R.)

Received: 27 May 2020; Accepted: 13 July 2020; Published: 15 July 2020

Abstract: A delicate intracellular balance among protein synthesis, folding, and degradation is essential to maintaining protein homeostasis or proteostasis, and it is challenged by genetic and environmental factors. Molecular chaperones and the ubiquitin proteasome system (UPS) play a vital role in proteostasis for normal cellular function. As part of protein quality control, molecular chaperones recognize misfolded proteins and assist in their refolding. Proteins that are beyond repair or refolding undergo degradation, which is largely mediated by the UPS. The importance of protein quality control is becoming ever clearer, but it can also be a disease-causing mechanism. Diseases such as phenylketonuria (PKU) and hereditary tyrosinemia-I (HT1) are caused due to mutations in *PAH* and *FAH* gene, resulting in reduced protein stability, misfolding, accelerated degradation, and deficiency in functional proteins. Misfolded or partially unfolded proteins do not necessarily lose their functional activity completely. Thus, partially functional proteins can be rescued from degradation by molecular chaperones and deubiquitinating enzymes (DUBs). Deubiquitination is an important mechanism of the UPS that can reverse the degradation of a substrate protein by covalently removing its attached ubiquitin molecule. In this review, we discuss the importance of molecular chaperones and DUBs in reducing the severity of PKU and HT1 by stabilizing and rescuing mutant proteins.

Keywords: deubiquitination; inhibitors; protein quality control; proteolysis; protein stabilization

---

1. Introduction: Overview of Phenylketonuria and Hereditary Tyrosinemia

Phenylalanine hydroxylase (PAH) and fumarylacetoacetate hydroxylase (FAH) are two highly regulated liver enzymes that catalyze the rate-limiting step in phenylalanine and tyrosine metabolism [1,2]. Mammalian PAH (phenylalanine 4-monooxygenase, E.C. 1.14.16.1) catalyzes the stereospecific hydroxylation of L-phenylalanine into L-tyrosine using tetrahydrobiopterin (BH4), non-heme iron, and dioxygen as co-substrates in the cytosol of the liver and kidney [3]. PAH facilitates oxidation of excess L-phenylalanine into carbon dioxide and water, and is the major enzyme degrading 75% of L-phenylalanine from the diet [2]. PAH assembles as a homotetrameric protein, each subunit composed of N-terminal regulatory domain for allosteric activation by Phe, a central catalytic domain, and C-terminal helix responsible for tetramer formation [4,5].

Likewise, FAH is the last enzyme in the tyrosine catabolism pathway, and it catalyzes the hydrolysis of fumarylacetoacetate into fumarate and acetoacetate as the final step in phenylalanine and tyrosine degradation. FAH is a cytosolic dimer that consists of two α–β domains; 300 residues of the C-terminal domain form the active site that binds to Ca^{2+} and participates in intermolecular interactions at the dimer interface; 120 residues of the N-terminal domain play the regulatory role [6,7]. The FAH dimer is solely considered to be catalytically active [7]. The human *FAH* gene occupies chromosome 15q23–q25, spans 30–35 kb, and contains 14 exons [8], whereas the *PAH* gene is located on chromosome 12q at position 23.2, spans 90 kb, and contains 13 exons [9].

In 1932, Grace Medes discovered 4-hydroxyphenylpyruvate in the urine of a 49-year-old man and described it as "tyrosinosis" [10]. In the 1960s, the condition was referred to as hereditary tyrosinemia type-I (HT1), and it was later understood to result from FAH deficiency [11–14]. Deficiency of this enzyme leads to the accumulation of upstream metabolites such as fumarylacetoacetate (FAA) and maleylacetoacetate, which are subsequently converted into succinylacetone. FAA and succinylacetone are both genotoxic and carcinogenic [15]. Similarly, in 1934, Dr. Asbjørn Følling recognized elevated levels of phenylketonuric acid in the urine of two mentally retarded siblings and named the condition "phenylpyruvic oligophrenia" or phenylketonuria (PKU) [3]. Elevated levels of blood phenylalanine and its metabolites, such as keto acid and phenylpyruvate, along with reduced blood tyrosine levels, are the characteristics of PKU and its milder variant hyperphenylalaninemia (HPA). PKU is classified as classical PKU (plasma Phe levels > 1200 μM), mild or atypical or variant PKU (600–1200 μM), and non-PKU mild HPA (120–600 μM) [16,17]. PKU is associated with mental retardation, epilepsy, brain damage, and neurological and behavioral problems due to the accumulation of phenylalanine byproducts. Tyrosine is the precursor for multiple molecules; therefore, tyrosine deficiency leads to deficiency of catecholamine neurotransmitters, melanin, and L-thyroxine [3,18].

HT1 pathogenicity is largely unknown; however, missense mutations in the *FAH* gene may influence catalytic activity, protein stability, and/or protein homeostasis and monomer-dimer equilibrium [7]. Despite being studied extensively since years, the pathophysiology of PKU is not fully elucidated. Mutation-driven PAH protein instability, misfolding, and aggregation are the hallmark associated with the disease resulting in subsequent protein turnover [19–21]. The regulation of L-Phe by PAH is a complex mechanism associated with transition between oligomeric state, changes in conformation, phosphorylation and substrate activation, and cofactor inhibition [4,5]. The newly discovered crystal structure supports the notion that PAH exists in two native states: resting state-PAH (RS-PAH) and activated-PAH (A-PAH). The RS-PAH and A-PAH was determined by X-ray crystallography and small-angle X-ray scattering respectively [4]. The RS-PAH has low affinity for Phe and helps maintain the basal level of Phe essentially available for cellular functions. Also, BH4 is complexed with RS-PAH, thus acting as a negative regulator for L-Phe activation [4,22]. BH4 serves as a pharmacological chaperone stabilizing PAH and increasing the steady state level of enzyme [20]. As the concentration of Phe increases, excess Phe acts as an activator and binds to A-PAH allosterically, shifting the equilibrium from RS-PAH to A-PAH. Binding of Phe induces large conformational change and dimerization of regulatory domain of the enzyme, thus exposing the active site for the conversion of Phe to Tyr [4,5,22]. BH4 and Phe binding drives the newly synthesized, partially folded, PAH into equilibrium of native structure [22]. PKU disease-associated alleles affect several different operations (like allosteric activation by Phe, stabilization by BH4) that join forces for efficient degradation of excess Phe. Therefore, it is important to maintain the PAH structure equilibrium which is hampered due to disease-associated mutations.

The crystal structure of the PAH tetramer providing information about PAH allostery and BH4 associated stability was recently discovered [22,23]. The allosterically activated form of PAH is majorly responsible for the conversion of phenylalanine to tyrosine; however, stability calculations are not possible for this form as its high resolution structure is not yet available. Nonetheless, certain experimental reports suggested increased aggregation, high instability, and accelerated degradation of the PAH mutant expressed in Enu$^{1/1}$ and Enu$^{1/2}$ heteroallelic mouse model, primary hepatocytes and COS-7 cells [24–26]. The mutant PAH proteins (e.g., p.V106A) expressed in Enu$^{1/1}$ mouse model, are also known to be highly ubiquitinated in vitro and in vivo, targeting it for proteasome-mediated degradation and selective autophagy [24].

To combat the pathogenic accumulation of defective proteins, the cells are equipped with the protein quality control (PQC) system, mainly including molecular chaperones and the ubiquitin proteasomal system (UPS). The supplementation with cofactor BH4, also acting as a pharmacological chaperone, stabilizes the PAH tetramer structure, providing a rationale for the BH4-responsive *PAH*-variants [20]. When *PAH* variants are co-expressed with GroEL/ES bacterial chaperone in *Escherichia coli*, decrease in dimer portions, increase in tetramer formation, and increase in residual activity were observed. These results suggest that co-expression with GroEL/ES bacterial chaperone might affect the PAH folding in *Escherichia coli* [19]. These results indicate that molecular chaperones have the potential to prevent protein misfolding and help to stabilize a range of mutant proteins. The proteins that cannot be stabilized by chaperones undergo degradation to avoid its interaction with other native and non-native proteins [27]. UPS is the major cellular degradation pathway, responsible for degrading more than 80% of intracellular proteins [28]. The proteins have to be tagged with ubiquitin moiety, in order to be degraded by the UPS. The PAH and FAH protein is reported to be ubiquitinated [7,24,28–30] and the variants are prone to aggregation and/or degradation [22]. However, the *PAH* and *FAH* variants exert some amount of residual activity depending upon the severity of mutation. Therefore, certain *PAH* and *FAH* mutants with folding defects are still functional, but they nonetheless suffer rapid degradation [21,31–33]. The degradative system therefore needs a way to differentiate between lethal defects and negligible defects. In this review, we discuss different strategies for stabilizing and increasing the concentrations of those functional mutant proteins, that display instability and folding defects and which are conjugated with ubiquitin molecule for degradation. We propose recruiting members of the ubiquitin proteasomal system (UPS) and protein quality control (PQC) chaperones into therapeutic endeavors to rescue functional misfolded proteins from accelerated degradation.

2. Etiology

More than 1000 mutations result in PKU [16], and more than 100 mutations result in HT1, and most of them are missense mutations [7]. In the *PAH* gene, 60.5% of mutations are missense mutations, and in the *FAH* gene, 45% of mutations are missense mutations [8,9]. The severity of a mutation depends on its effect on the resulting enzyme's conformation and function. In other words, the genotypic effect on the clinical phenotype is variable [3,33]. Most of the mutations causing PKU and HT1 result in PAH and FAH instability, leading to misfolding and loss of function [20,22,33–35].

Because both enzymes are biallelic, it is possible to have many disease-causing mutations, resulting in a compound heterozygote [36,37]. Mutations in the *PAH* and *FAH* genes are known to decrease catalytic activity and reduce the kinetic stability of the enzymes, inducing accelerated degradation [7,38].

3. Epidemiology

PKU and HT1 are autosomal recessive traits that affect 1 in 2500 to 100,000 births [9,39] and 1 in 100,000 births [40], respectively. PKU has high prevalence in the United Kingdom, Turkey, and Ireland and is rare in Thailand, whereas HT1 is present worldwide except for Central America and Oceania. The most common mutation in PKU is p.Arg408Trp, which is frequently found in Russia and East European countries such as Hungary, the Czech Republic, Slovakia, and Croatia [41,42] and Baltic countries such as Estonia, Lithuania, and Latvia [42]. Mutations such as p.Arg241Cys, p.Arg243Gln, and Ex6-96A>G are frequent in the Chinese population [43], and p.Pro281Leu is common in Iranians [44]. Other common variants include p.Arg261Gln, p.Tyr414Cys, and p.Ser349Pro [42]. More than 40 mutations have been found in the *FAH* gene, and some of the most common are D233V in Turks, W262X in Finns [45], p.Gly64His in Asians, p.1 Met>Val in Saudi Arabians [8], and c.974C>T in Europeans and Caucasians [46].

4. Protein Quality Control

Eukaryotic cells possess a robust complement of proteins to monitor and maintain a healthy proteome for their survival. Proteome integrity is maintained under the scrutiny of the PQC system [47]. The maturation from a nascent polypeptide to a functional protein is crucial to its function and involves a multistep process, including proper post-translational modification. The folding process for some proteins starts during their synthesis itself, which is called co-translational folding, whereas other proteins fold in the cytoplasm or the endoplasmic reticulum (ER) and mitochondria after synthesis [48]. The fundamentals of protein folding are also governed by the cellular environment and its over-crowding [49]. The hydrophobic patches in a polypeptide are buried in the native state. Exposure leads to the formation of intermediates that can interact inappropriately with other molecules. Thus, several studies suggest that protein folding is initiated by composing a folding nucleus within a primary structure around which the remaining polypeptide folds. The most important requirement for a correct folding pattern is the interaction between the hydrophobic and polar residues during nucleation, which encourages the structure to be packed correctly [50].

The protein-folding mechanism is much more complex for larger proteins than for smaller ones. Evidence from various studies indicates that, during protein folding, some proteins attain the native structure, whereas others cannot, for reasons such as a non-native interaction that leads to intermediates or a transiently folded protein state. Therefore, large proteins are assembled from diverse segments or domains that are folded simultaneously and independently, ensuring the proper folding of each segment, so that they can correctly interact with one another to form a highly stable and compact, native, three-dimensional protein structure. In other words, for large protein complexes such as proteasomes and ribosomes, the folding pathway involves a two state mechanism [51,52].

The ability of a protein to fold correctly de novo, though thermodynamically favorable, is often hampered by transcriptional or translational errors, destabilizing mutations, or stress conditions such as heat, oxygen radicals, aging, or environmental threats, giving rise to misfolded proteins and off-pathway aggregates [53–55]. The misfolded proteins can exhibit either loss of function, characterized by protein dysfunction and a propensity for degradation, or gain of function, characterized by protein aggregates that cause the misfolding of other proteins through inappropriate interactions [50,56]. Cells are rescued from the dangers of misfolded proteins by the PQC system, which keeps proteins under the constant surveillance of molecular chaperones and induces the rapid degradation of misfolded proteins through the UPS or autophagy-driven lysosomal proteolysis [57,58] (Figure 1). PQC relies on three parallel strategies whereby misfolded proteins are refolded, degraded, or delivered to a quality-control compartment capable of sequestering them, such as the juxta nuclear quality control, insoluble protein deposit, aggresome, or ER-associated degradation (ERAD) vesicles [48].

4.1. Molecular Chaperones for Folding/Refolding

Molecular chaperones are present in all cell organelles and can be categorized as "folding helpers" and "holding types." A folding helper assists in polypeptide folding during translation and partially unfolds misfolded intermediates [59]. The holding type, on the other hand, guards an unfolded or misfolded protein from aggregation and degradation, and presents it to a folding helper [60]. They transiently interact with mutant proteins to protect them from interacting with normal proteins and buy time to refold them into their native conformation [27]. Most mutated proteins show a prolonged interaction with the chaperones compared with their corresponding correctly folded protein, because the chaperones need time to fold them properly [61]. However, the severity of the mutation determines whether the chaperones refold and rescue the proteins or direct them for degradation. Therefore, the ectopic expression of chaperones increases the chance that chaperones will interact with misfolded proteins and restore their active form, as observed in previous studies [62,63]. In other words, chaperones along with other members of PQC, play a decisive role in maintaining a pool of functional mutant proteins.

Molecular chaperones also help certain newly synthesized proteins to fold efficiently and in a biologically relevant time frame. Chaperones such as HSP70/40 attach to a partially folded polypeptide during synthesis by a ribosome and stabilize it to its correct native conformation [64] (Figure 1). Most proteins fold to a native form in the cytosol and do not need any further assistance from chaperones, but certain proteins always need assistance from chaperones, such as HSP90 and the TCP-1 ring complex (TRiC) [65]. Likewise, chaperones such as HSP70 and HSP90 and some chaperonins (such as HSP60s) assist in the refolding of misfolded proteins through the ATP and co-factor binding and release cycle. If a misfolded protein cannot be completely refolded, chaperones help target it for degradation to remove the toxic conformation. Co-chaperones such as the Bcl-2-associates-athanogene domain direct the interaction between a molecular chaperone and the protein degradation system [66,67].

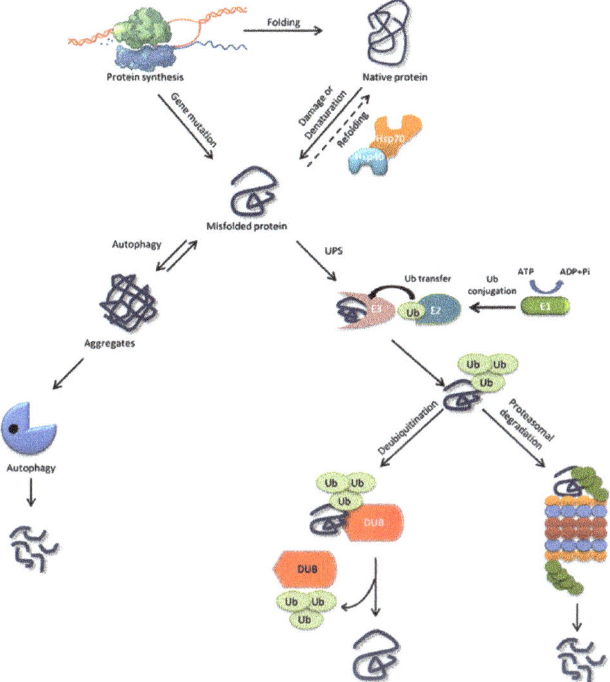

Figure 1. Protein folding, misfolding, and degradation. Protein folding starts during the ribosomal translation process and attain the native conformation to execute cellular processes. The native folded proteins are often misfolded due to mutations and other environmental factors. Molecular chaperones catalyze the folding/refolding events, disaggregation of the protein aggregates, and targeting the protein for degradation. Aggregates are typically degraded by autophagy, whereas the ubiquitin proteasome system (UPS) degrades the destabilized/misfolded proteins by covalent attachment of a ubiquitin molecule assisted by E1-E2-E3 enzymes. The ubiquitinated proteins are recognized by the 26S proteasome and are degraded. However, the ubiquitin moiety is cleaved off by Deubiquitinating enzymes (DUBs) and the protein can be rescued from the degradation cycle.

Several classes of molecular chaperones are distinguished by molecular weight and mode of action, such as Hsp70, Hsp90, Hsp60, and Hsp40, and bind to the hydrophobic region of a non-native protein [64,68].

4.1.1. Hsp70

The 70 kDa heat shock protein (Hsp70) and its homologs heat shock cognate 70 (Hsc70) in the cytosol and BiP/GRP78 in the ER are some of the most abundant chaperones engaged in a plethora of folding and refolding processes. Hsp70 consists of two domains: an N-terminal nucleotide-binding domain (NBD) of ~40 kDa and a C-terminal substrate-binding domain (SBD) of ~30 kDa, connected by a hydrophobic linker. The Hsp70 cycle can use co-chaperones such as Hsp40 to recognize and transfer a substrate protein to Hsp70, J proteins to stimulate ATP hydrolysis in the SBD, and nuclear exchange factor proteins to change ATP to ADP in the NBD. ADP-Hsp70 holds a substrate protein in an unfolded state, until it has no exposed hydrophobic patches and spontaneous folding is achieved [60,69].

4.1.2. Hsp90

Hsp90 is a homodimer, and each of its monomers consists of a highly conserved amino-terminal domain (NTD) of ~25 kDa, a middle domain of ~40 kDa, and a C-terminal dimerization domain of ~12 kDa [70,71]. Hsp90 is also an ATP-dependent chaperone present in almost all the compartments of a eukaryotic cell. Hsp90 regulates the stability and maturation of >300 proteins that are key players in many biological processes, such as immune response, telomere maintenance, cancer development, steroid signaling, and vesicular transport [72]. To recognize its enormous number of client proteins, Hsp90 interacts with more than 20 co-chaperones. Hop, also known as p60, and STI1 mediate the transfer of client proteins from Hsp70 to Hsp90, and Cdc37 binds to kinase clients that inhibit Hsp90's ATPase activity. Co-chaperone Aha1 enhances Hsp90's ATPase activity by binding between HSP90's middle domain catalytic loop and the NTD nucleotide-binding site, facilitating the transition to a more stable closed transformation [72]. Unlike other chaperones, Hsp90 binds to partially folded intermediate conformations rather than to fully denatured proteins [73,74].

4.1.3. Hsp60

Hsp60 or chaperonins are 800–900 kDa double ring cylindrical complexes that originate in the mitochondria, but migrate to the cytosol under cellular stress. Cytosolic Hsp60 is also called TRiC or the TCP1 complex and usually has an 8-membered ring, whereas mitochondrial Hsp60 has a 7-membered ring modulated by a lid structure made of co-chaperone Hsp10. Client proteins enter the central cavity where the apical domain of Hsp60 exposes its hydrophobic residues for substrate binding. Following substrate binding, Hsp60 subunits undergo excessive conformational changes, ATP is hydrolyzed, and the folded protein is released through the Hsp10 lid [75,76].

4.1.4. Hsp40

Hsp40, also known as J protein, is mostly studied in its function as a co-chaperone with Hsp70. The Hsp40 family is divided into 3 types depending on the location of the J chain. In types I and II, the J-chain is located at the N-terminal, whereas in type-III, the J chain can be located anywhere in the protein sequence. The type-I protein additionally possesses two zinc finger motifs. The J chain, which is 70-amino-acid-residue long, stimulates the ATPase activity of Hsp70. Hsp40 is known to recognize and bind misfolded proteins and guide them to Hsp70 for folding. Hsp40 has been shown to play a crucial role in neurodegenerative diseases and cancers; however, its mechanism for recognizing non-proteins and modulating Hsp70 activity is poorly understood [77].

4.2. Ubiquitin Proteasome System

Misfolded or destabilized proteins undergo intracellular proteolysis through two main pathways: the UPS and the autophagy-lysosome pathway. More than 80% of native and misfolded intracellular proteins undergo UPS-mediated degradation [28]. The UPS functions in both the nucleus and the cytosol to recycle and degrade soluble proteins. On the other hand, autophagy functions only in the cytoplasm and generally eliminates large, insoluble aggregates and degenerated organelles that escaped from the UPS [68]. A detailed explanation of the autophagy-lysosome pathway has been provided elsewhere [78]. In this review, we focus mainly on UPS-mediated protein degradation and its association with chaperones during the regulation of PAH and FAH proteins.

Targeting a substrate protein for UPS degradation requires the covalent attachment of Ub molecule/s, which is called ubiquitination. Ubiquitination is achieved when three enzyme families work consecutively: ubiquitin-activating enzyme (E1) activates ubiquitin by ATP hydrolysis and forms a thioester link between its own cysteine residue and the C-terminal carboxyl group of Ub. Second, ubiquitin-conjugating enzyme (E2) receives Ub from E1 by a trans-thiolation reaction in which Ub binds to the cysteine residue of E2. Third, the ubiquitin-protein ligase (E3) and E2 together position the target protein substrate and attach the ubiquitin moiety to the ε-amino group in a lysine residue on the target protein. The ubiquitin-conjugated proteins are then recognized and degraded by the 26S proteasome unless the ubiquitin chains are removed by a crucial set of enzymes called deubiquitinating enzymes (DUBs) [79,80] (Figure 1).

The role of the UPS has mainly been characterized in neurodegenerative diseases [81]. In gain-of-function disorders such as Alzheimer's, Parkinson's, Huntington's, and Creutzfeldt-Jakob diseases, the misfolded proteins accumulate to form aggregates and fail to undergo proteasomal degradation because of dysfunction in the UPS. In Alzheimer's disease, for example, the amyloid precursor protein is cleaved into amyloid β (Aβ) peptides to form intraneuronal neurofibrillary tangles. The Aβ peptides inhibit UPS-mediated degradation by binding to the catalytic core of the 26S proteasome, thereby inhibiting its chymotrypsin-like activity [82]. Interestingly, the molecular chaperones Hsp70 and Hsp40 reduce the accumulation of Aβ aggregates [83]. Therefore, the UPS and molecular chaperones balance each other's functions.

5. Deubiquitinating Enzymes Regulate Molecular Chaperones

Approximately 100 putative DUBs have been identified in humans. These large ubiquitin-cleaving proteases are classified into seven families: ubiquitin-specific proteases, ubiquitin C-terminal hydrolases, ovarian tumor proteases, Machado-Joseph disease domain proteases, Jab1/Mpn/Mov34 metalloenzymes, monocyte chemotactic protein-induced proteases, and zinc finger with UFM1-85 specific peptidase domain proteins, but most of their functions and substrates have not yet been characterized [84]. Some of the important functions of DUBs in the ubiquitin pathway include generating free ubiquitin monomers by processing inactive ubiquitin precursors, acting as an E3 ligase antagonist by cleaving the ubiquitin molecule from the substrate proteins, and maintaining a ubiquitin pool by recycling cleaved ubiquitin molecules. DUBs are known to be involved in physiological processes and thus are predicted to be involved in cancers [85], neurodegeneration [84], and infectious diseases [86]. Interestingly DUBs also regulate the members of another major degradation pathway in PQC called autophagy. Misfolded proteins are recognized by molecular chaperones in the HSP family, which coordinates with the UPS for protein refolding and the removal of misfolded proteins [87].

Ubiquitination and deubiquitination both play crucial roles in the dynamic regulation of different stages of the autophagic process. To induce proper autophagy, post-translational modification of its initiators is essential. It is a well-organized game of "on" and "off" between the E3 ligase and DUBs in controlling autophagy signals [87]. Numerous E3-chaperone complexes work in parallel to target misfolded proteins. For example, the E3 ligase carboxy-terminus of Hsc70 interacting protein (CHIP) tightly regulates the function of Hsp70/Hsp90 to orchestrate cellular protein folding and degradation. Ubiquitination of substrate proteins is antagonized by DUBs, allowing misfolded proteins to escape from degradation [88]. A growing body of evidence suggests crosstalk between the DUBs and the HSPs as well. For instance, proteasome-bound USP14 protein was found to interact with molecular chaperone Hsc70 to modulate autophagy in neuroblastoma cells. Striatal neuronal cells expressing mutant huntingtin protein had a defect in autophagosome maturation that was influenced by Hsc70 and proteasome free-USP14, indicating a link between the proteasome-independent function of USP14 and Hsc70 in mediating crosstalk among autophagy, ER stress signaling, and the proteasome [89]. Similarly, the DUB USP19 has two major isoforms. One isoform contains a transmembrane domain at its C-terminus and is associated with ERAD for an unfolded protein response; the other isoform contains an EEVD extension at the C-terminus that interacts with CHIP. The N-terminus of both isoforms interacts with the Hsp90 chaperone. The regulatory function of USP19 was recently confirmed in a study demonstrating that USP19 interacts directly with chaperone Hsp90 and upregulates the aggregation of poly-Q containing the proteins Ataxin-3 and Huntingtin, which causes spinocerebellar ataxia type-3 and Huntington's disease, respectively [90]. Direct evidence indicates that chaperone Hsp90 enhances USP19 DUB activity by promoting its substrate recognition [91]. Hsp90 recruits misfolded proteins for refolding, and should the protein fail to refold, the co-chaperone CHIP ubiquitinates the misfolded protein for degradation with the help of Hsp90, or the misfolded protein is deubiquitinated by USP19, allowing it to avoid degradation and promoting aggregation [90]. This process is perfectly synchronized as a defense mechanism against proteins whose aggregation is cytotoxic to the cells. However, to enhance the rescue of functional mutant proteins, understanding the regulatory mechanism of DUBs and molecular chaperones is beneficial.

6. Rapid Degradation of Misfolded PAH and FAH Proteins

More than 1000 variants in the human *PAH* gene are recorded in the locus-specific database *PAH*vdb (http://www.biopku.org/home/pah.asp), and certain missense mutations in the regulatory and catalytic domains cause protein instability and folding defects of the PAH protein, resulting in its rapid degradation and loss of function [31,32,92]. Thus, PKU was generally considered to be the paradigm of misfolded metabolic diseases [93]. The destabilized mutants of PAH are precisely degraded by the cellular PQC system. Mutation-dependent destabilization and accelerated proteolytic degradation are the main pathogenic mechanisms in PKU [94]. PAH is reported to be a substrate for Ub-conjugating enzyme and is likely degraded by the UPS. Døskeland et al. demonstrated that PAH isolated from rat liver is conjugated with mono- and multi-/poly-ubiquitination at its catalytic domain [29]. More recently, in an ENU$^{1/2}$ heteroallelic mouse model of HPA, mutant PAH was highly ubiquitinated, which corresponded with an increased rate of degradation [24]. The mutant proteins were degraded more rapidly than the wild type enzyme [62]. The wild type is reported to have a half-life of 2 days in rat liver and 7–8 h in hepatoma cells; in contrast, mutants are degraded rapidly, due to the destabilization of their protein structure [29]. Molecular chaperones such as DNAJC12/HSP70 play a role in processing mutant PAH for UPS-mediated degradation or ubiquitin-mediated autophagy [28,30].

Likewise, more than 100 mutations of the *FAH* gene cause HT1. Like PAH, most of the mutations produce FAH destabilization, causing the enzyme to be rushed to the aggregation pathway. When cells expressing the FAH protein were subjected to the proteasomal inhibitor MG132, FAH protein levels were restored. Therefore, the FAH protein undergoes proteasomal degradation [7]. FAH is also conjugated with Ub at multiple lysine residues according to the PhosphoSitePlus (www.phosphosite.org) database. However, no evidence indicates the type of ubiquitin linkage and whether it targets FAH for degradation or tags it for further cellular processes. The reduced activity and deficiency of FAH found in HT1 could result from the rapid degradation of destabilized mutant proteins [95], similar to PAH in PKU.

7. Residual Catalytic Activity of PAH and FAH Can Be Rescued by Deubiquitination or Molecular Chaperones

Certain cases of PKU result from genetic mutations that impede the normal folding of the wild type PAH protein, leading to reduced or no enzyme activity. Genotype-based prediction of metabolic phenotypes, including patients with homozygosity and those with functional hemizygosity, has been studied for several years [16]. Two alleles, both with severe mutations in the *PAH* gene, produce an enzyme with little or no enzyme activity, whereas the presence of two mild mutations or one severe and one mild mutation produces high residual enzyme activity, producing HPA or mild PKU (>30% activity compared with wild type PAH) [93]. Certain combinations of mutations in the genotype and their predicted residual enzyme activity have already been reported [96,97]. Some mutations characterized by high residual activity were found to be responsive to natural co-factor BH4 [98]. BH4 responsiveness has a multifactorial basis, including intragenic polymorphisms and non-genetic factors. The main molecular mechanism underlying BH4 responsiveness is its chaperone-like effect on PAH, whereby it protects PAH protein integrity and rescues it from Ub-dependent degradation [99].

It is increasingly apparent that molecular chaperones could help mutant PAH proteins that are partially functional serve their purpose and help to prevent the pathogenic mechanisms that underlie genetic diseases.

Given the importance of chaperones, mutations in the chaperones themselves can be lethal. Multiple diseases are associated with mutations in the regulating chaperones. For example, a missense mutation in the equatorial domain of HSP60 causes spastic paraplegia, and mutation in tubulin-specific chaperone E causes hypoparathyroidism, mental retardation, and facial dysmorphism [100]. Similarly, in PKU an autosomal recessive mutation in DNAJC12, a PAH co-chaperone, reduced the activity of wild type PAH, leading to HPA. DNAJC12 is involved in PAH folding and interacts with the monoubiquitinated *PAH* variant, marking it for the Ub-dependent proteasomal/autophagy degradation system. Further studies are ongoing to elucidate the role of DNAJC12 in regulating PAH and PAH mutants [25,101]. Gene therapy and the ectopic expression of wild type chaperones might help to restore the partially functional mutant proteins [102,103].

Some patients with HT1 who are treated with NTBC (2-[2-nitro-4-(trifluoromethyl) benzoyl] cyclohexane-1,3-dione) also suffer from chronic hepatopathy and the development of hepatocellular carcinoma [104]. In a murine model of HT1, chaperones such as HSPB and HSPA were found to be associated with the anti-apoptotic proteins BCL-2 and BAG in the hepatocarcinogenetic process [105]. However, the role of molecular chaperones in FAH protein stability and degradation needs to be investigated.

Another system that can be targeted to rescue defective proteins is the UPS. As discussed in the previous section, the UPS is mainly driven by E1-E2-E3 enzymes that tag substrate proteins with ubiquitin molecules to mark them for degradation via the 26S proteasome, and DUBs can reverse that process. The role of DUBs in disease regulation has been imagined ever since their discovery, because they are involved in almost all cellular processes [106]. In PQC, the ubiquitin-mediated proteolytic pathway is a dynamic system responsible for regulating the fate of many proteins. In loss-of-function diseases, saving functional misfolded proteins from degradation can be a better alternative than dealing with a deficiency of proteins caused by rapid degradation. DUBs can rescue proteins from degradation by cleaving their degradative signals. Thus, DUBs act as proofreaders for mis-tagged substrate proteins and prevent them from degradation. In that way, DUBs could be used to curb protein misfolding diseases. It is unsurprising that direct evidence on this point is sparse. Most studies dealing with diseases related to protein folding problems aim to clear the misfolded proteins from the cells rapidly, and thus they target DUBs or the proteasome via specific inhibitors to prevent the pathogenesis of defective protein accumulation [107,108]. However, in diseases such as PKU and HT1, artificial manipulation of those systems could prove advantageous and pave the way for new therapeutic approaches. Nonetheless, the regulation of the proteostasis is not possible for those missense mutations which are present at the active site of the enzyme and other mutations causing truncation and splice variant. Therefore, controlling the proteostasis might be favorable only to the missense mutations that are located outside the active site.

PAH and FAH enzyme proteins are ubiquitinated and degraded by the Ub-dependent system, and therefore PAH and FAH mutants with high residual enzyme activity could be deubiquitinated by DUBs, which might suffice to create an adequate supply of functional protein. However, the mutations in certain genotypes can show dramatically different disease severities. Thus, in targeting DUBs as therapeutics for diseases with misfolded protein, it is important to understand the genotype-phenotype correlation and the allelic combination of mutations present in the genotype. A great deal of work remains to be done to improve understanding of how DUBs, molecular chaperones, and their combination can help to regulate enzyme deficiencies.

8. Current Treatments and Currently Ongoing Research

8.1. Dietary Treatment

PKU is the most common inborn error of metabolism; in it, the accumulation of L-Phe causes clinical features such as mental retardation, eczema, microcephaly, and behavioral problems. Blood Phe levels, which depend on the enzymatic deficiency, dictate the severity of the clinical phenotype [37]. It has been more than 65 years since the successful establishment of dietary restrictions to treat PKU, but that diet has to be maintained for life [109]. The PKU diet is low in Phe and is supplemented with a special medicinal formula to supply vitamins, amino acids (except Phe), and minerals. However, dietary restrictions are cumbersome economically and socially and can lead to nutritional deficiencies. Moreover, adhering to dietary restrictions is difficult for older patients, resulting in increased blood Phe levels. Large neutral amino acids (LNAAs) can be included in the diet to reduce the absorption of Phe in the brain by competing with the transporter across the blood–brain and gastrointestinal barriers. Because inadequate evidence supports the long-term outcomes of LNAA treatment, it is used only as a short-term therapy. Nevertheless, dietary treatment is the foundation of PKU management upon which novel improved therapies are being developed [109,110].

Dietary restrictions are the most common treatment for both PKU and HT1 patients. Patients with HT1 are recommended to consume a low tyr/Phe diet and Nitisinone [111].

8.2. Nitisinone

Nitisinone (NTBC, Orfadin®, Swedish Orphan Biovitrum, Stockholm, Sweden) has been an effective treatment for HT1 since 1992. It inhibits an enzyme, 4-hydroxyphenylpyruvate dioxygenase (HPPD), upstream of *FAH*. Nitisinone is a well-tolerated drug, but its drawbacks include the development of corneal lesions in rats, transient thrombocytopenia, and leucopenia [104,112]. Recently, clustered regularly interspaced short palindromic repeats (CRISPR)—CRISPR-associated (Cas) systems mediated gene correction has been demonstrated in vivo and ex vivo. In a mouse model of HT1, 1 of 250 liver cells was corrected by the hydrodynamic injection of a donor oligonucleotide and a plasmid co-expressing gRNA and Cas9. However, NTBC supplementation with the CRISPR components showed better therapeutic results in mice than either treatment alone [113]. In ex vivo experiments, hepatocyte cells were collected from individual HT1 mouse models, corrected using CRISPR/Cas9 components, and implanted back to the organism. That method of replacing the mutated genotype has high efficiency, but it still requires cycles of NTBC treatment [114].

8.3. Enzyme Therapy

In 1980, phenylalanine ammonia lyase (PAL) was recognized as a potential treatment for PKU, because of its ability to metabolize excess Phe into less toxic products, trans-cinnamic acid, and ammonia, regardless of genotype. Enzyme replacement therapy and enzyme substitution are two other treatment strategies adopted. In enzyme replacement therapy, a functional PAH enzyme and its co-factor BH4 are delivered by orthotopic liver transplantation. Enzyme replacement is a daunting process that is useful only for the few PKU patients who need a liver transplant. Enzyme substitution by PAL is less troublesome and does not require the co-factor, because it acts directly as a substitute for deficient PAH. When injected into humans, PAL triggers a host-immune response and is degraded by proteases. To overcome that problem, PAL is conjugated with polyethylene glycol (PEG-PAL). PEG-PAL, now called pegvaliase, is in clinical trials. In 2018, pegvaliase (trade name Palynziq®) was approved by the FDA in the USA. However, severe adverse events were frequently observed in the clinical trials. The safety of pegvaliase is also a concern during pregnancy [115–117].

8.4. Gene Therapy

Gene therapy is a promising technique for the treatment of PKU and has been studied by several researchers. Recently, a recombinant adeno-associated virus (rAAV) containing the *PAH* gene was delivered to a mouse model of PKU. But the rAAV vector could not permanently correct liver PAH because the vector was not integrated into the genome of the hepatocytes, leading to loss of the vector during subsequent hepatocyte regeneration. When the vector containing *PAH*- and co-factor-synthesizing genes was injected into muscle, it was able to metabolize Phe to Tyr, but low gene transfer means that this approach needs further improvement [118,119]. Inactive Cas9 (dCas9) fused with the *FokI* endonuclease (*FokI*-dCas9) has recently been used to correct the p.Arg408Trp mutation in the *PAH* gene. The frequency of the corrected allele was 21.4%, making *FokI*-dCas9 a promising strategy to treat PKU [120]. Several laboratories have achieved a certain degree of success in correcting the PAH deficiencies, however none has progressed to human trials yet.

8.5. Tetrahydrobiopterin

The supplementation of co-factor BH4 in some PKU patients with high residual activity reduced their Phe levels. The metabolic response to BH4 improves the stability of the misfolded PAH enzyme and increases enzyme activity by reducing proteolysis. In 2007, a synthetic form of BH4, sapropterin dihydrochloride, was approved in multiple countries as an adjuvant therapy. The treatment requires that the mutant enzyme possess some amount of residual activity, which is most commonly found in patients with milder forms of PKU. Most patients with severe forms of PKU, in which the enzyme activity is null, do not respond to sapropterin treatment. Sapropterin acts as a molecular chaperone to assist with PAH folding and stability. A reduction in blood Phe of 30% or more from baseline is considered to be a sapropterin response. Chaperone therapy is a promising new approach in the treatment of PKU and HPA [110,121].

8.6. Microbe Therapy

Development of bacteria-based drug has been focused over the past decade. In PKU, the exploitation of gut microbiome to facilitate the degradation of Phe from the diet has shown promising results in clinical trials. Synlogic (https://www.synlogictx.com/), a Massachusetts-based biotech, reprogrammed *Escherichia coli* Nissle (EcN)—that has been isolated from human microbiome. The treatment, dubbed SYNB1618, aim to target two pathways. One Phe degradation pathway, where gene *sltA* encoding Phe ammonia lyase was inserted into the EcN chromosome. Gene *sltA* encodes a cytosolic protein, hence *pheP* gene that encodes a Phe transporter was also engineered in EcN. The second pathway was the insertion of *Proteus mirabilis* pma, which encodes L-amino acid deaminase having higher Phe break down capability. Currently SYNB1618 is in phase II clinical trial to evaluate its potential to lower blood Phe in patients with PKU [122,123] (https://clinicaltrials.gov/ Identifier: NCT03516487).

8.7. mRNA Therapy

Moderna (https://www.modernatx.com/), a clinical stage biotechnology company pioneering messenger RNA (mRNA) therapeutics and vaccines, have developed mRNA-3283, that encodes the human *PAH* to restore the intracellular enzyme activity in patients with PKU at an effective dose amount encapsulated within a liposome nanoparticle; mRNA-3283 is currently in preclinical development.

8.8. Small Molecule THERAPY

Agios Pharmaceuticals 9 (https://www.agios.com/) has developed a small molecule therapy on the hypothesis that majority of mutations in PAH prevent it from folding into its native tetrameric conformation. The small molecule stabilizes some of these mutant enzymes and has demonstrated significant reduction in blood Phe in severe PKU pre-clinical models.

8.9. Red Blood Cell Therapy

Recently, the lab grown red blood cells were genetically engineered to produce the enzyme PAL. Researchers at Rubius Therapeutics (https://www.rubiustx.com/) developed RTX-134, a Red Cell Therapeutic™ (RCT) product candidate, by inserting the gene encoding PAL enzyme into the blood cells to degrade toxic levels of Phe in the bloodstream. RTX-134 entered the clinical trials; however, it was recently reported that RTX-134 failed to generate any meaningful signals of efficacy.

The current and ongoing treatment and research are summarized in the Tables 1 and 2. No optimal treatment for PKU and HT1 has yet been developed, and thus novel strategies need to be explored. It is necessary to consider each step involved—folding, assembly, refolding, and degradation—in the cellular handling of mutant PAH and FAH proteins.

Table 1. Therapeutic regimes in phenylketonuria (PKU) and hereditary tyrosinemia-I (HT1).

Disorder	Treatment	Advantages	Disadvantages	Stage of Development	Reference
PKU	Dietary Treatment	• Mainstay of treatment for PKU • Successful in curtailing intellectual disability and achieving near normal IQ	• Compliance due to unpalatability • Nutritional deficiency • Expensive	Clinical application	[109,124]
	LNAA	• Reduce cerebral Phe concentrations • Effective in maintaining acceptable plasma Phe concentrations	• Compliance to restricted diet • Unsatisfactory organoleptic properties • Suitable only for adults	FDA approved	
	Enzyme Therapy (Palynziq)	• PAL is a monomer and requires no cofactors	• Suitable only for adults • May cause anaphylaxis • Injection site reactions	FDA approved	[125,126]
	Gene Therapy	• Supplement or replace defective *PAH* gene	• Immune rejection of adenovirus-transduced hepatocytes • High dose needed • Gender dependent effect	Research	[118–120]
	BH4 or sapropterin dihydrochloride (Kuvan)	• Improves stability of enzyme • Increase enzyme activity	• Effective for BH4 responsive PKU	FDA approved	[42,127,128]
HT1	Nitisinone (NTBC)	• Well tolerant drug • Posses long half life	• Development of corneal lesions in rats, transient thrombocytopenia, and leucopenia	FDA approved	[104,112,129]
	Gene Therapy	• Supplement or replace defective *FAH* gene	• Requires NTBC treatment	Research	[113,114]

Table 2. Emerging trends in PKU management.

Disorder	Treatment	Biotech/Pharmaceutical Company	Stage of Development
PKU	Microbe therapy (SYNB1618)	Synlogic (https://www.synlogictx.com/)	Phase II clinical trail
	mRNA therapy (mRNA-3283)	Moderna (https://www.modernatx.com/)	Preclinical development
	Small molecule therapy	Agios Pharmaceuticals (https://www.agios.com/), Camp4 (https://www.camp4tx.com/)	Preclinical development
	Red blood cell therapy (RTX-134)	Rubius Therapeutics (https://www.rubiustx.com/)	Discontinued in Phase 1b trial

9. Molecular and Chemical Chaperones as Current Therapeutics for PKU and HT1

Different technologies have been exploited to discover strategies for treating the most common inborn errors of metabolism, PKU and HT1. For PKU, pharmacological chaperones are currently being studied to functionally and structurally rescue misfolded proteins. Among all the compounds discovered to date, only sapropterin dihydrochloride, a synthetic form of BH4, and dietary restriction are approved and used together for tetrahydrobiopterin-responsive PKU [42]. Side effects associated with sapropterin include rhinorrhea, pharyngolaryngeal pain, diarrhea, and lower than normal Phe levels in patients younger than 6 years [127]. Enzyme replacement therapy is another strategy for treating PKU. PEG-PAL (pegvaliase) can metabolize Phe and is approved for the treatment of patients with uncontrolled blood Phe levels. However, participants in the clinical trials had adverse events, and discontinuation of the drug is recommended during pregnancy and breastfeeding [128].

For HT1, only NTBC (Nitisinone) in conjunction with a low tyrosine and phenylalanine diet has been widely used. NTBC inhibits HPPD, an enzyme upstream of FAH, but it only partially protects against liver dysfunction [105].

Molecular Chaperones

Currently the therapeutic application of chemical chaperones and pharmacological chaperones is being used to rehabilitate misfolded proteins and restore mutant protein activity. Chemical chaperones are small-molecular-weight compounds that act as artificial chaperones to stabilize the native conformation of proteins. Chemical chaperones such as glycerol and trimethylamine N-oxide correct temperature-sensitive protein folding abnormalities in cystic fibrosis transmembrane regulator (CFTR, ΔF508), p53, viral oncogene protein pp60src, and ubiquitin activating enzyme E1. However, chemical chaperones have low specificity, which produces undesired effects, and the concentrations needed to increase protein function are so high that they are toxic to cells. Therefore, chemical chaperones are not generally used in clinical practice. On the other hand, much work has been done on pharmacological chaperones, which work at low concentrations and have high specificity [130]. This strategy has already proved effective in restoring mutant PAH activity in the form of sapropterin [131]. Pharmacological chaperones to treat Fabry disease and Pompe disease show increased enzyme activity and decreased substrate accumulation and are already in clinical trials [130]. Similar candidate molecules to stabilize PAH and FAH should be identified by using high-throughput screening of drug libraries and studying thermal protein stability. Chaperones could be one of the best candidates to manipulate because of their diverse role in protein folding, assembly, and stability. Pharmacological chaperones alone or in combination with members of the UPS system (such as DUBs and E3 ligases) could be a promising therapeutic strategy for rescuing enzyme function.

10. Alternative or Synergistic Approaches for Treating Diseases Caused by Destabilizing Missense Mutations

10.1. Screening Specific DUBs for PAH and FAH Proteins

The protein misfolding and mis-assembly in PKU and HT1 cause rapid protein degradation, which it is important to preempt. The UPS is one major pathway for intracellular protein degradation. DUBs play a central role in ubiquitin signaling and protein homeostasis. The development of DUB inhibitors has been showcased as a promising strategy for treating cancers and other diseases by enhancing the degradation of DUB substrates. No one has reported on the role of DUBs in misfolding diseases, in which treatments need to stabilize and increase the concentration of mutated proteins. DUBs can recognize different chain linkages that are specific for particular functions, such as the K6, K11, and K48 chains for proteasomal degradation and K63 chains for lysosomal targeting, DNA repair, and NF-κB activation. Thus, DUBs are highly specific in their action [106].

Any misfolded protein is tagged with mono-/poly-ubiquitination chains that mark them for proteasomal degradation. Those degradation tags can be recognized and removed by DUBs, thus aborting the protein degradation cycle. Therefore, we hypothesize that DUBs could play a decisive role in rescuing functional but misfolded PAH and FAH proteins from degradation. Instead of DUB inhibitors, drugs to enhance the function of DUBs that specifically regulate PAH and FAH protein degradation need to be identified. Various DUBs might be suitable for interacting with PAH and FAH proteins to dissociate ubiquitin molecules and maintain the relatively small amount of functional protein needed in the correct subcellular destination to prevent disease phenotypes.

Ernst et al. proposed a highly active viral DUB derived from the protease domain of Epstein-Barr virus, the BPLF1 gene called EBV-DUB, which removed ubiquitin chains and stabilized misfolded proteins in 293T cells. EBV-DUB was less toxic to cells than proteasome inhibitors, because no ubiquitylated proteins accumulated. Because the ubiquitylated proteins did not accumulate, the pool of free Ub was also maintained in the cells. Proteasomal inhibitors reduce protein synthesis with even a short exposure, whereas EBV-DUB did not immediately block the translation, again giving it an advantage over proteasomal inhibitors [132]. Inhibiting proteasomal degradation can derange many other cellular processes. Of all the attempts to stabilize the PAH and FAH mutant proteins, this approach should be an important focus of research and surveillance.

10.2. Modified PROTAC Technology to Use DUBs to Stabilize Partially Functional PAH and FAH Proteins

Small-molecule induced protein degradation using proteolysis-targeting chimeras (PROTACs) is an emerging technology targeting a broad range of proteins. This technology is based on event-driven pharmacology in the sense that it degrades a target protein as soon as the drug transiently binds to the target. After binding and degrading a protein, the PROTACs can again serve their function for multiple rounds of activity. PROTACs are bifunctional molecules that bind to the protein of interest with one end and to an E3 ligase with the other. The bound E3 ligase then attracts the E2 enzyme to transfer Ub to the protein of interest, targeting it for proteasomal degradation (Figure 2). Thus, this technology uses proximity-induced ubiquitination and degradation and has minimal off-target effects at a low concentration. PROTACs have already proved successful in an acute myeloid leukemia xenograft model and a disseminated lymphoma mouse model [133,134]. Therefore, they might also be used to remove null mutant PAH and FAH proteins from cells.

The recent success of small-molecule PROTACs has opened the door to a wide range of applications. This technique to induce protein degradation might also rescue proteins from degradation with some modifications. One end of the PROTAC could be designed to bind to the protein of interest, while the other, instead of binding to the E3 ligase, could be designed to bind to specific DUBs to regulate the target protein. Binding to the PROTAC would thus bring a ubiquitin-tagged misfolded protein and its DUB into close proximity, allowing the DUB to recognize and remove the ubiquitin tag and aborting the degradation (Figure 2). DUBs are usually high-molecular-weight proteins, which might give the modified PROTAC less permeability. Therefore, small molecular ligands that can recruit a specific DUB could be used instead. Due to their high specificity and low toxicity, PROTACs can be modified into selective rescuers of misfolded proteins that would otherwise be rapidly degraded. Therefore, efforts should focus on identifying a specific DUB candidate or small molecule ligand to attract DUBs that regulate PAH and FAH proteins and designing a PROTAC suitable for freezing the degradation process.

10.3. Inhibitor of Ubiquitin

Many studies have reported the development of novel inhibitors to target druggable enzymes of the UPS. However, the most important molecule, ubiquitin, has not yet been targeted. Recently, Nguyen et al. discovered a compound, Congo red, with the ability to bind to the recognition site and disrupt the binding activity of ubiquitin, thus making it unavailable for ubiquitination. Congo red inhibits the conjugation of K48- and K63-linked polyubiquitination. Inhibiting the ubiquitin itself inhibits all ubiquitin-mediated proteasomal/autophagic degradation, and it can therefore be used to enrich the population of all functional mutant proteins in a cell [135]. The ubiquitin inhibitor,

Congo red, by itself or in combination with another strategy, could be important in supporting the survival of functional but misfolded PAH and FAH proteins.

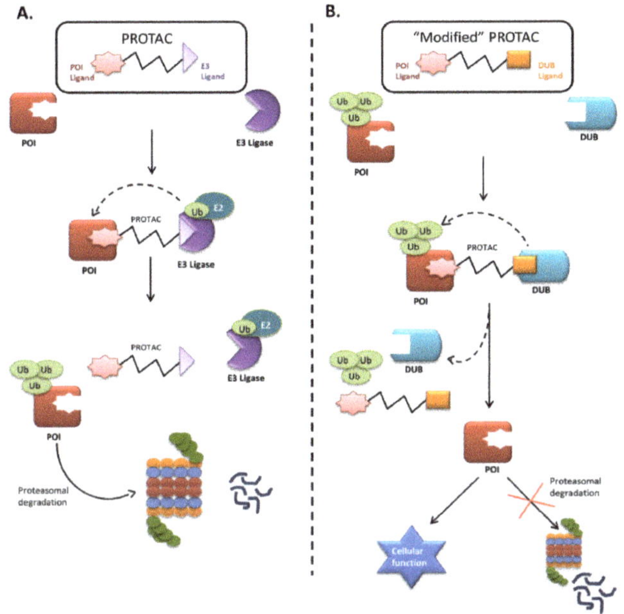

Figure 2. Mechanism of proteolysis-targeting chimera (PROTAC) and "Modified" PROTAC. (**A**) Proteolysis-targeting chimaera (PROTAC) are bifunctional molecules, whose one end binds to the protein of interest (POI) while the other recruits E3 ligase forming a ternary complex. The E3 ligase induces proximity-induced ubiquitination of POI by transferring the ubiquitin (Ub) molecules from E2 enzyme to the POI, thus facilitating its degradation; (**B**) Representation of a hypothetical figure where technology can be modified to rescue the functional misfolded proteins undergoing rapid degradation in inherited metabolic disorders like PKU and HT1 having partial functions, causing deficiency of available protein for cellular functions. Hence, a PROTAC can be designed whose one end binds to the misfolded POI and the other binds to a ligand that can recruit the regulatory DUB. The DUBs will cleave off the ubiquitin molecule, avoiding the protein degradation, and will help to maintain a pool of protein required for normal cellular function.

10.4. E1, E2, E3 Inhibitors

Most UPS-associated inhibitors block a specific upstream component, such as E1, E2, E3, or DUBs. Those inhibitors have proved to be successful anticancer drugs, some of which are currently in clinical trials. However, E3, the last enzyme in the ubiquitin cascade, which is responsible for transferring Ub to the substrate protein, is more specific than E1 and E2, and thus an excellent target. A nutlin inhibitor of MDM2-p53, an LCL161 inhibitor of XIAP, and an ALRN6924 inhibitor of MDM2-p53 are some of the E3 enzyme inhibitors currently in clinical trials [136]. Blocking the E3 ligase with specific inhibitors will block the ubiquitination and eventual degradation of the substrate protein, which is exactly what is needed to treat enzyme deficiencies caused by the rapid degradation of functional but misfolded proteins.

11. Conclusions

Although a substantial body of data has been collected on the cellular mechanisms of PAH and FAH protein stability and folding, there remains a major gap in our understanding of how critical components of the PQC system recognize folding defects. The present crystal structure of PAH tetramer increments our understanding of the equilibrium between the alternate native PAH structures (RS-PAH and A-PAH), allosteric binding, and activation by Phe, and the negative regulation and stabilization by BH4 expands our knowledge of the repertoire of disease-associated PAH. There are several excellent reviews describing the structural dynamics and function of the PAH protein. However, our review is the first attempt describing the role of ubiquitination and/or molecular chaperones in PAH protein turnover. Current efforts to identify successful therapeutic strategies to treat PKU and HT1 still depend on dietary restrictions. Enzyme replacement therapy, gene correction by CRISPR/Cas9, and mRNA therapy are some recently developed breakthrough therapies associated with substantial limitations or problems. PKU and HT1 are complex diseases, with many cellular entities interacting with the substrate proteins (PAH and FAH) and their corresponding mutants, and thus it is laborious to study each of them. However, DUBs and molecular chaperones play a significant role in the triage of misfolded ubiquitinated proteins. Molecular chaperones are already being tested as therapeutic targets, but DUBs are also an attractive target. Even though research supporting the role of DUBs in rescuing the functional mutant proteins is in its infancy, future studies should be directed toward identifying and modulating protein-specific DUBs. As PKU and HT-1 are complex trait diseases, the genotype phenotype correlation plays a significant role in the development of patient-tailored prognostic and therapeutic strategies. Molecular chaperones and DUBs, alone or in combination with other therapies already established for PKU and HT1, have a bright future in treating these proteopathies. Efficient technologies and multidisciplinary methodologies are needed to explore the strong link between molecular chaperones and DUBs and how it can be used to enrich functional mutant proteins and provide new insights into proteopathy medicine.

Author Contributions: S.R. and K.-S.K. conceived the idea. N.S. searched the literature and wrote the manuscript. B.S. reviewed and edited the manuscript. All authors have read and agreed to the published version of the manuscript.

Funding: This work was supported by a grant from the National Research Foundation of Korea (NRF) (2018M3A9H3022412).

Acknowledgments: The authors thank all the members of Suri's laboratory for their helpful insights.

Conflicts of Interest: The authors declare no conflicts of interest.

Abbreviations

PKU	Phenylketonuria
HT1	Hereditary tyrosinemia type 1
PAH	Phenylalanine hydroxylase
FAH	Fumarylacetoacetate hydroxylase
BH4	Tetrahydrobiopterin
PQC	Protein quality control
UPS	Ubiquitin proteasome system
HSP	Heat shock protein
DUBs	Deubiquitinating enzymes
ERAD	Endoplasmic reticulum-associated degradation
PROCTAC	Proteolysis-targeting chimaeras
NTBC	2-[2-nitro-4-(trifluoromethyl)benzoyl]cyclohexane-1,3-dione
A-PAH	Activated PAH
RS-PAH	Resting state PAH

References

1. Weiss, A.K.H.; Loeffler, J.R.; Liedl, K.R.; Gstach, H.; Jansen-Dürr, P. The fumarylacetoacetate hydrolase (FAH) superfamily of enzymes: Multifunctional enzymes from microbes to mitochondria. *Biochem. Soc. Trans.* **2018**, *46*, 295–309. [CrossRef]

2. Flydal, M.I.; Martinez, A. Phenylalanine hydroxylase: Function, structure, and regulation. *IUBMB Life* **2013**, *65*, 341–349. [CrossRef]

3. Williams, R.A.; Mamotte, C.D.; Burnett, J.R. Phenylketonuria: An inborn error of phenylalanine metabolism. *Clin. Biochem. Rev.* **2008**, *29*, 31–41. [PubMed]

4. Arturo, E.C.; Gupta, K.; Héroux, A.; Stith, L.; Cross, P.J.; Parker, E.J.; Loll, P.J.; Jaffe, E.K. First structure of full-length mammalian phenylalanine hydroxylase reveals the architecture of an autoinhibited tetramer. *Proc. Natl. Acad. Sci. USA* **2016**, *113*, 2394–2399. [CrossRef]

5. Arturo, E.C.; Gupta, K.; Hansen, M.R.; Borne, E.; Jaffe, E.K. Biophysical characterization of full-length human phenylalanine hydroxylase provides a deeper understanding of its quaternary structure equilibrium. *J. Biol. Chem.* **2019**, *294*, 10131–10145. [CrossRef] [PubMed]

6. Timm, D.E.; Mueller, H.A.; Bhanumoorthy, P.; Harp, J.M.; Bunick, G.J. Crystal structure and mechanism of a carbon-carbon bond hydrolase. *Structure* **1999**, *7*, 1023–1033. [CrossRef]

7. Macias, I.; Laín, A.; Bernardo-Seisdedos, G.; Gil, D.; Gonzalez, E.; Falcon-Perez, J.M.; Millet, O. Hereditary tyrosinemia type I-associated mutations in fumarylacetoacetate hydrolase reduce the enzyme stability and increase its aggregation rate. *J. Biol. Chem.* **2019**, *294*, 13051–13060. [CrossRef]

8. Angileri, F.; Bergeron, A.; Morrow, G.; Lettre, F.; Gray, G.; Hutchin, T.; Ball, S.; Tanguay, R.M. Geographical and Ethnic Distribution of Mutations of the Fumarylacetoacetate Hydrolase Gene in Hereditary Tyrosinemia Type 1. *JIMD Rep.* **2015**, *19*, 43–58. [CrossRef]

9. Scriver, C.R. The PAH gene, phenylketonuria, and a paradigm shift. *Hum. Mutat.* **2007**, *28*, 831–845. [CrossRef]

10. Medes, G. A new error of tyrosine metabolism: Tyrosinosis. The intermediary metabolism of tyrosine and phenylalanine. *Biochem. J.* **1932**, *26*, 917–940. [CrossRef]

11. Lindblad, B.; Lindstedt, S.; Steen, G. On the enzymic defects in hereditary tyrosinemia. *Proc. Natl. Acad. Sci. USA* **1977**, *74*, 4641–4645. [CrossRef]

12. Fällström, S.-P.; Lindblad, B.; Lindstedt, S.; Steen, G. Hereditary tyrosinemia-fumarylacetoacetase deficiency. *Pediatric Res.* **1979**, *13*, 78. [CrossRef]

13. Berger, R.; Smit, G.P.; Stoker-de Vries, S.A.; Duran, M.; Ketting, D.; Wadman, S.K. Deficiency of fumarylacetoacetase in a patient with hereditary tyrosinemia. *Clin. Chim. Acta* **1981**, *114*, 37–44. [CrossRef]

14. Kvittingen, E.; Jellum, E.; Stokke, O. Assay of fumarylacetoacetate fumarylhydrolase in human liver—Deficient activity in a case of hereditary tyrosinemia. *Clin. Chim. Acta* **1981**, *115*, 311–319. [CrossRef]

15. Cassiman, D.; Zeevaert, R.; Holme, E.; Kvittingen, E.A.; Jaeken, J. A novel mutation causing mild, atypical fumarylacetoacetase deficiency (Tyrosinemia type I): A case report. *Orphanet J. Rare Dis.* **2009**, *4*, 28. [CrossRef] [PubMed]

16. Garbade, S.F.; Shen, N.; Himmelreich, N.; Haas, D.; Trefz, F.K.; Hoffmann, G.F.; Burgard, P.; Blau, N. Allelic phenotype values: A model for genotype-based phenotype prediction in phenylketonuria. *Genet. Med.* **2019**, *21*, 580–590. [CrossRef]

17. Richards, D.Y.; Winn, S.R.; Dudley, S.; Nygaard, S.; Mighell, T.L.; Grompe, M.; Harding, C.O. AAV-Mediated CRISPR/Cas9 Gene Editing in Murine Phenylketonuria. *Mol. Ther. Methods Clin. Dev.* **2020**, *17*, 234–245. [CrossRef]

18. van Spronsen, F.J.; van Rijn, M.; Bekhof, J.; Koch, R.; Smit, P.G. Phenylketonuria: Tyrosine supplementation in phenylalanine-restricted diets. *Am. J. Clin. Nutr.* **2001**, *73*, 153–157. [CrossRef]

19. Pecimonova, M.; Kluckova, D.; Csicsay, F.; Reblova, K.; Krahulec, J.; Procházkova, D.; Skultety, L.; Kadasi, L.; Soltysova, A. Structural and Functional Impact of Seven Missense Variants of Phenylalanine Hydroxylase. *Genes* **2019**, *10*, 459. [CrossRef]

20. Flydal, M.I.; Alcorlo-Pagés, M.; Johannessen, F.G.; Martínez-Caballero, S.; Skjærven, L.; Fernandez-Leiro, R.; Martinez, A.; Hermoso, J.A. Structure of full-length human phenylalanine hydroxylase in complex with tetrahydrobiopterin. *Proc. Natl. Acad. Sci. USA* **2019**, *116*, 11229–11234. [CrossRef] [PubMed]

21. Wettstein, S.; Underhaug, J.; Perez, B.; Marsden, B.D.; Yue, W.W.; Martinez, A.; Blau, N. Linking genotypes database with locus-specific database and genotype-phenotype correlation in phenylketonuria. *Eur. J. Hum. Genet.* **2015**, *23*, 302–309. [CrossRef] [PubMed]

22. Jaffe, E.K. New protein structures provide an updated understanding of phenylketonuria. *Mol. Genet. Metab.* **2017**, *121*, 289–296. [CrossRef] [PubMed]

23. Patel, D.; Kopec, J.; Fitzpatrick, F.; McCorvie, T.J.; Yue, W.W. Structural basis for ligand-dependent dimerization of phenylalanine hydroxylase regulatory domain. *Sci. Rep.* **2016**, *6*, 23748. [CrossRef] [PubMed]

24. Sarkissian, C.N.; Ying, M.; Scherer, T.; Thöny, B.; Martinez, A. The mechanism of BH4 -responsive hyperphenylalaninemia—As it occurs in the ENU1/2 genetic mouse model. *Hum. Mutat.* **2012**, *33*, 1464–1473. [CrossRef] [PubMed]

25. Jung-Kc, K.; Himmelreich, N.; Prestegård, K.S.; Shi, T.S.; Scherer, T.; Ying, M.; Jorge-Finnigan, A.; Thöny, B.; Blau, N.; Martinez, A. Phenylalanine hydroxylase variants interact with the co-chaperone DNAJC12. *Hum. Mutat.* **2019**, *40*, 483–494. [CrossRef] [PubMed]

26. Eichinger, A.; Danecka, M.K.; Möglich, T.; Borsch, J.; Woidy, M.; Büttner, L.; Muntau, A.C.; Gersting, S.W. Secondary BH4 deficiency links protein homeostasis to regulation of phenylalanine metabolism. *Hum. Mol. Genet.* **2018**, *27*, 1732–1742. [CrossRef]

27. Waters, P.J. Degradation of mutant proteins, underlying "loss of function" phenotypes, plays a major role in genetic disease. *Curr. Issues Mol. Biol.* **2001**, *3*, 57–65.

28. Wang, J.; Maldonado, M.A. The ubiquitin-proteasome system and its role in inflammatory and autoimmune diseases. *Cell Mol. Immunol.* **2006**, *3*, 255–261. [PubMed]

29. Døskeland, A.P.; Flatmark, T. Conjugation of phenylalanine hydroxylase with polyubiquitin chains catalysed by rat liver enzymes. *Biochim. Biophys. Acta* **2001**, *1547*, 379–386. [CrossRef]

30. Blau, N.; Martinez, A.; Hoffmann, G.F.; Thöny, B. DNAJC12 deficiency: A new strategy in the diagnosis of hyperphenylalaninemias. *Mol. Genet. Metab.* **2018**, *123*, 1–5. [CrossRef] [PubMed]

31. Scheller, R.; Stein, A.; Nielsen, S.V.; Marin, F.I.; Gerdes, A.M.; Di Marco, M.; Papaleo, E.; Lindorff-Larsen, K.; Hartmann-Petersen, R. Toward mechanistic models for genotype-phenotype correlations in phenylketonuria using protein stability calculations. *Hum. Mutat.* **2019**, *40*, 444–457. [CrossRef] [PubMed]

32. Gersting, S.W.; Kemter, K.F.; Staudigl, M.; Messing, D.D.; Danecka, M.K.; Lagler, F.B.; Sommerhoff, C.P.; Roscher, A.A.; Muntau, A.C. Loss of function in phenylketonuria is caused by impaired molecular motions and conformational instability. *Am. J. Hum. Genet.* **2008**, *83*, 5–17. [CrossRef]

33. Bergeron, A.; D'Astous, M.; Timm, D.E.; Tanguay, R.M. Structural and functional analysis of missense mutations in fumarylacetoacetate hydrolase, the gene deficient in hereditary tyrosinemia type 1. *J. Biol. Chem.* **2001**, *276*, 15225–15231. [CrossRef] [PubMed]

34. Stone, W.L.; Basit, H.; Los, E. Phenylketonuria. In *StatPearls [Internet]*; StatPearls Publishing: St. Petersburg, FL, USA, 2019.

35. Shi, Z.; Sellers, J.; Moult, J. Protein stability and in vivo concentration of missense mutations in phenylalanine hydroxylase. *Proteins* **2012**, *80*, 61–70. [CrossRef]

36. Sandhu, I.S.; Maksim, N.J.; Amouzougan, E.A.; Gallion, B.W.; Raviele, A.L.; Ooi, A. Sustained NRF2 activation in hereditary leiomyomatosis and renal cell cancer (HLRCC) and in hereditary tyrosinemia type 1 (HT1). *Biochem. Soc. Trans.* **2015**, *43*, 650–656. [CrossRef] [PubMed]

37. Regier, D.S.; Greene, C.L. *Phenylalanine Hydroxylase Deficiency*; Adam, M.P., Ardinger, H.H., Pagon, R.A., Wallace, S.E., Bean, L.J.H., Stephens, K., Amemiya, A., Eds.; GeneReviews®; University of Washington: Seattle, WA, USA, 1993.

38. Underhaug, J.; Aubi, O.; Martinez, A. Phenylalanine hydroxylase misfolding and pharmacological chaperones. *Curr. Top. Med. Chem.* **2012**, *12*, 2534–2545. [CrossRef]

39. Muntau, A.C.; du Moulin, M.; Feillet, F. Diagnostic and therapeutic recommendations for the treatment of hyperphenylalaninemia in patients 0–4 years of age. *Orphanet J. Rare Dis.* **2018**, *13*, 173. [CrossRef]

40. Sheth, J.J.; Ankleshwaria, C.M.; Pawar, R.; Sheth, F.J. Identification of Novel Mutations in FAH Gene and Prenatal Diagnosis of Tyrosinemia in Indian Family. *Case Rep. Genet.* **2012**, *2012*, 428075. [CrossRef]

41. Loeber, J.G. Neonatal screening in Europe; the situation in 2004. *J. Inherit. Metab. Dis.* **2007**, *30*, 430–438. [CrossRef]

42. Gundorova, P.; Stepanova, A.A.; Kuznetsova, I.A.; Kutsev, S.I.; Polyakov, A.V. Genotypes of 2579 patients with phenylketonuria reveal a high rate of BH4 non-responders in Russia. *PLoS ONE* **2019**, *14*, e0211048. [CrossRef]

43. Zhang, X.; Chen, H.X.; Li, C.; Zhang, G.; Liao, S.Y.; Peng, Z.C.; Lai, X.P.; Wang, L.L. Rapid detection of PAH gene mutations in Chinese people. *BMC Med. Genet.* **2019**, *20*, 135. [CrossRef] [PubMed]

44. Biglari, A.; Saffari, F.; Rashvand, Z.; Alizadeh, S.; Najafipour, R.; Sahmani, M. Mutations of the phenylalanine hydroxylase gene in Iranian patients with phenylketonuria. *Springerplus* **2015**, *4*, 542. [CrossRef] [PubMed]

45. Dursun, A.; Ozgül, R.K.; Sivri, S.; Tokatlı, A.; Güzel, A.; Mesci, L.; Kılıç, M.; Aliefendioglu, D.; Ozçay, F.; Gündüz, M.; et al. Mutation spectrum of fumarylacetoacetase gene and clinical aspects of tyrosinemia type I disease. *JIMD Rep.* **2011**, *1*, 17–21. [CrossRef] [PubMed]

46. Yasir Zahoor, M.; Cheema, H.A.; Ijaz, S.; Fayyaz, Z. Genetic Analysis of Tyrosinemia Type 1 and Fructose-1, 6 Bisphosphatase Deficiency Affected Pakistani Cohorts. *Fetal Pediatr. Pathol.* **2019**, 1–11. [CrossRef]

47. Pilla, E.; Schneider, K.; Bertolotti, A. Coping with Protein Quality Control Failure. *Annu. Rev. Cell Dev. Biol.* **2017**, *33*, 439–465. [CrossRef]

48. Wolff, S.; Weissman, J.S.; Dillin, A. Differential scales of protein quality control. *Cell* **2014**, *157*, 52–64. [CrossRef]

49. Zhou, H.X. Influence of crowded cellular environments on protein folding, binding, and oligomerization: Biological consequences and potentials of atomistic modeling. *FEBS Lett.* **2013**, *587*, 1053–1061. [CrossRef]

50. Ashaq, I.; Shajrul, A.; Akbar, M.; Rashid, F. Protein misfolding diseases: In perspective of gain and loss of function. In *Proteostasis and Chaperone Surveillance*; Springer: Berlin, Germany, 2015; pp. 105–118.

51. Lodish, H.; Berk, A.; Zipursky, S.L.; Matsudaira, P.; Baltimore, D.; Darnell, J. Hierarchical structure of proteins. In *Molecular Cell Biology*, 4th ed.; WH Freeman: New York, NY, USA, 2000.

52. Alberts, B.; Johnson, A.; Lewis, J.; Raff, M.; Roberts, K.; Walter, P. The shape and structure of proteins. In *Molecular Biology of the Cell*, 4th ed.; Garland Science: New York, NY, USA, 2002.

53. Hartl, F.U.; Hayer-Hartl, M. Converging concepts of protein folding in vitro and in vivo. *Nat. Struct. Mol. Biol.* **2009**, *16*, 574–581. [CrossRef]

54. Chen, B.; Retzlaff, M.; Roos, T.; Frydman, J. Cellular strategies of protein quality control. *Cold Spring Harb. Perspect. Biol.* **2011**, *3*, a004374. [CrossRef] [PubMed]

55. Hartl, F.U. Protein Misfolding Diseases. *Annu. Rev. Biochem.* **2017**, *86*, 21–26. [CrossRef]

56. Sahni, N.; Yi, S.; Taipale, M.; Fuxman Bass, J.I.; Coulombe-Huntington, J.; Yang, F.; Peng, J.; Weile, J.; Karras, G.I.; Wang, Y.; et al. Widespread macromolecular interaction perturbations in human genetic disorders. *Cell* **2015**, *161*, 647–660. [CrossRef] [PubMed]

57. Fredrickson, E.K.; Gardner, R.G. Selective destruction of abnormal proteins by ubiquitin-mediated protein quality control degradation. *Semin. Cell Dev. Biol.* **2012**, *23*, 530–537. [CrossRef]

58. Tyedmers, J.; Mogk, A.; Bukau, B. Cellular strategies for controlling protein aggregation. *Nat. Rev. Mol. Cell Biol.* **2010**, *11*, 777–788. [CrossRef]

59. Duncan, E.J.; Cheetham, M.E.; Chapple, J.P.; van der Spuy, J. The role of HSP70 and its co-chaperones in protein misfolding, aggregation and disease. *Subcell. Biochem.* **2015**, *78*, 243–273. [CrossRef]

60. Finka, A.; Sharma, S.K.; Goloubinoff, P. Multi-layered molecular mechanisms of polypeptide holding, unfolding and disaggregation by HSP70/HSP110 chaperones. *Front. Mol. Biosci.* **2015**, *2*, 29. [CrossRef] [PubMed]

61. Jørgensen, M.M.; Jensen, O.N.; Holst, H.U.; Hansen, J.J.; Corydon, T.J.; Bross, P.; Bolund, L.; Gregersen, N. Grp78 is involved in retention of mutant low density lipoprotein receptor protein in the endoplasmic reticulum. *J. Biol. Chem.* **2000**, *275*, 33861–33868. [CrossRef] [PubMed]

62. Gámez, A.; Pérez, B.; Ugarte, M.; Desviat, L.R. Expression analysis of phenylketonuria mutations. Effect on folding and stability of the phenylalanine hydroxylase protein. *J. Biol. Chem.* **2000**, *275*, 29737–29742. [CrossRef]

63. Bross, P.; Andresen, B.S.; Gregersen, N. Impaired folding and subunit assembly as disease mechanism: The example of medium-chain acyl-CoA dehydrogenase deficiency. *Prog. Nucleic Acid Res. Mol. Biol.* **1998**, *58*, 301–337. [CrossRef]

64. Hartl, F.U.; Bracher, A.; Hayer-Hartl, M. Molecular chaperones in protein folding and proteostasis. *Nature* **2011**, *475*, 324–332. [CrossRef]

65. Hartl, F.U.; Hayer-Hartl, M. Molecular chaperones in the cytosol: From nascent chain to folded protein. *Science* **2002**, *295*, 1852–1858. [CrossRef]

66. Kriegenburg, F.; Jakopec, V.; Poulsen, E.G.; Nielsen, S.V.; Roguev, A.; Krogan, N.; Gordon, C.; Fleig, U.; Hartmann-Petersen, R. A chaperone-assisted degradation pathway targets kinetochore proteins to ensure genome stability. *PLoS Genet.* **2014**, *10*, e1004140. [CrossRef] [PubMed]

67. Demand, J.; Alberti, S.; Patterson, C.; Höhfeld, J. Cooperation of a ubiquitin domain protein and an E3 ubiquitin ligase during chaperone/proteasome coupling. *Curr. Biol.* **2001**, *11*, 1569–1577. [CrossRef]

68. Ciechanover, A.; Kwon, Y.T. Protein Quality Control by Molecular Chaperones in Neurodegeneration. *Front. Neurosci.* **2017**, *11*, 185. [CrossRef]

69. Radons, J. The human HSP70 family of chaperones: Where do we stand? *Cell Stress Chaperones* **2016**, *21*, 379–404. [CrossRef]

70. Li, J.; Soroka, J.; Buchner, J. The Hsp90 chaperone machinery: Conformational dynamics and regulation by co-chaperones. *Biochim. Biophys. Acta* **2012**, *1823*, 624–635. [CrossRef] [PubMed]

71. Karagöz, G.E.; Rüdiger, S.G. Hsp90 interaction with clients. *Trends Biochem. Sci.* **2015**, *40*, 117–125. [CrossRef]

72. Taipale, M.; Jarosz, D.F.; Lindquist, S. HSP90 at the hub of protein homeostasis: Emerging mechanistic insights. *Nat. Rev. Mol. Cell. Biol.* **2010**, *11*, 515–528. [CrossRef] [PubMed]

73. Röhl, A.; Rohrberg, J.; Buchner, J. The chaperone Hsp90: Changing partners for demanding clients. *Trends Biochem. Sci.* **2013**, *38*, 253–262. [CrossRef] [PubMed]

74. Pearl, L.H.; Prodromou, C. Structure and mechanism of the Hsp90 molecular chaperone machinery. *Annu. Rev. Biochem.* **2006**, *75*, 271–294. [CrossRef]

75. Horwich, A.L.; Fenton, W.A. Chaperonin-assisted protein folding: A chronologue. *Q. Rev. Biophys.* **2020**, *53*, e4. [CrossRef]

76. Dun, M.D.; Aitken, R.J.; Nixon, B. The role of molecular chaperones in spermatogenesis and the post-testicular maturation of mammalian spermatozoa. *Hum. Reprod. Update* **2012**, *18*, 420–435. [CrossRef] [PubMed]

77. Cyr, D.M.; Ramos, C.H. Specification of Hsp70 function by Type I and Type II Hsp40. *Subcell. Biochem.* **2015**, *78*, 91–102. [CrossRef]

78. Yim, W.W.; Mizushima, N. Lysosome biology in autophagy. *Cell Discov.* **2020**, *6*, 6. [CrossRef] [PubMed]

79. Sarodaya, N.; Karapurkar, J.; Kim, K.S.; Hong, S.H.; Ramakrishna, S. The Role of Deubiquitinating Enzymes in Hematopoiesis and Hematological Malignancies. *Cancers* **2020**, *12*, 1103. [CrossRef] [PubMed]

80. Haq, S.; Ramakrishna, S. Deubiquitylation of deubiquitylases. *Open Biol.* **2017**, *7*, 170016. [CrossRef] [PubMed]

81. Zheng, Q.; Huang, T.; Zhang, L.; Zhou, Y.; Luo, H.; Xu, H.; Wang, X. Dysregulation of Ubiquitin-Proteasome System in Neurodegenerative Diseases. *Front. Aging Neurosci.* **2016**, *8*, 303. [CrossRef]

82. Díaz-Villanueva, J.F.; Díaz-Molina, R.; García-González, V. Protein Folding and Mechanisms of Proteostasis. *Int. J. Mol. Sci.* **2015**, *16*, 17193–17230. [CrossRef]

83. Campanella, C.; Pace, A.; Caruso Bavisotto, C.; Marzullo, P.; Marino Gammazza, A.; Buscemi, S.; Palumbo Piccionello, A. Heat Shock Proteins in Alzheimer's Disease: Role and Targeting. *Int. J. Mol. Sci.* **2018**, *19*, 2603. [CrossRef]

84. Das, S.; Ramakrishna, S.; Kim, K.S. Critical Roles of Deubiquitinating Enzymes in the Nervous System and Neurodegenerative Disorders. *Mol. Cells* **2020**, *43*, 203–214. [CrossRef]

85. Poondla, N.; Chandrasekaran, A.P.; Kim, K.S.; Ramakrishna, S. Deubiquitinating enzymes as cancer biomarkers: New therapeutic opportunities? *BMB Rep.* **2019**, *52*, 181–189. [CrossRef]

86. Nanduri, B.; Suvarnapunya, A.E.; Venkatesan, M.; Edelmann, M.J. Deubiquitinating enzymes as promising drug targets for infectious diseases. *Curr. Pharm. Des.* **2013**, *19*, 3234–3247. [CrossRef]

87. Wang, Y.-T.; Chen, G.-C. The role of ubiquitin system in autophagy. In *Autophagy in Current Trends in Cellular Physiology and Pathology*; In Tech: Rijeka, Croatia, 2016.

88. Kevei, É.; Pokrzywa, W.; Hoppe, T. Repair or destruction-an intimate liaison between ubiquitin ligases and molecular chaperones in proteostasis. *FEBS Lett.* **2017**, *591*, 2616–2635. [CrossRef] [PubMed]

89. Srinivasan, V.; Bruelle, C.; Scifo, E.; Pham, D.D.; Soliymani, R.; Lalowski, M.; Lindholm, D. Dynamic Interaction of USP14 with the Chaperone HSC70 Mediates Crosstalk between the Proteasome, ER Signaling, and Autophagy. *iScience* **2020**, *23*, 100790. [CrossRef] [PubMed]

90. He, W.T.; Zheng, X.M.; Zhang, Y.H.; Gao, Y.G.; Song, A.X.; van der Goot, F.G.; Hu, H.Y. Cytoplasmic Ubiquitin-Specific Protease 19 (USP19) Modulates Aggregation of Polyglutamine-Expanded Ataxin-3 and Huntingtin through the HSP90 Chaperone. *PLoS ONE* **2016**, *11*, e0147515. [CrossRef]

91. Lee, J.G.; Kim, W.; Gygi, S.; Ye, Y. Characterization of the deubiquitinating activity of USP19 and its role in endoplasmic reticulum-associated degradation. *J. Biol. Chem.* **2014**, *289*, 3510–3517. [CrossRef] [PubMed]

92. Gallego, D.; Leal, F.; Gámez, A.; Castro, M.; Navarrete, R.; Sanchez-Lijarcio, O.; Vitoria, I.; Bueno-Delgado, M.; Belanger-Quintana, A.; Morais, A.; et al. Pathogenic variants of DNAJC12 and evaluation of the encoded cochaperone as a genetic modifier of hyperphenylalaninemia. *Hum. Mutat.* **2020**, *41*, 1329–1338. [CrossRef] [PubMed]

93. Blau, N.; Erlandsen, H. The metabolic and molecular bases of tetrahydrobiopterin-responsive phenylalanine hydroxylase deficiency. *Mol. Genet. Metab.* **2004**, *82*, 101–111. [CrossRef] [PubMed]

94. Pey, A.L.; Stricher, F.; Serrano, L.; Martinez, A. Predicted effects of missense mutations on native-state stability account for phenotypic outcome in phenylketonuria, a paradigm of misfolding diseases. *Am. J. Hum. Genet.* **2007**, *81*, 1006–1024. [CrossRef]

95. Yang, S.; Siepka, S.M.; Cox, K.H.; Kumar, V.; de Groot, M.; Chelliah, Y.; Chen, J.; Tu, B.; Takahashi, J.S. Tissue-specific FAH deficiency alters sleep-wake patterns and results in chronic tyrosinemia in mice. *Proc. Natl. Acad. Sci. USA* **2019**, *116*, 22229–22236. [CrossRef]

96. Shen, N.; Heintz, C.; Thiel, C.; Okun, J.G.; Hoffmann, G.F.; Blau, N. Co-expression of phenylalanine hydroxylase variants and effects of interallelic complementation on in vitro enzyme activity and genotype-phenotype correlation. *Mol. Genet. Metab.* **2016**, *117*, 328–335. [CrossRef]

97. Liang, Y.; Huang, M.Z.; Cheng, C.Y.; Chao, H.K.; Fwu, V.T.; Chiang, S.H.; Hsiao, K.J.; Niu, D.M.; Su, T.S. The mutation spectrum of the phenylalanine hydroxylase (PAH) gene and associated haplotypes reveal ethnic heterogeneity in the Taiwanese population. *J. Hum. Genet.* **2014**, *59*, 145–152. [CrossRef] [PubMed]

98. Cerreto, M.; Cavaliere, P.; Carluccio, C.; Amato, F.; Zagari, A.; Daniele, A.; Salvatore, F. Natural phenylalanine hydroxylase variants that confer a mild phenotype affect the enzyme's conformational stability and oligomerization equilibrium. *Biochim. Biophys. Acta* **2011**, *1812*, 1435–1445. [CrossRef] [PubMed]

99. Pey, A.L.; Pérez, B.; Desviat, L.R.; Martínez, M.A.; Aguado, C.; Erlandsen, H.; Gámez, A.; Stevens, R.C.; Thórólfsson, M.; Ugarte, M.; et al. Mechanisms underlying responsiveness to tetrahydrobiopterin in mild phenylketonuria mutations. *Hum. Mutat.* **2004**, *24*, 388–399. [CrossRef] [PubMed]

100. Barral, J.M.; Broadley, S.A.; Schaffar, G.; Hartl, F.U. Roles of molecular chaperones in protein misfolding diseases. *Semin. Cell Dev. Biol.* **2004**, *15*, 17–29. [CrossRef]

101. Anikster, Y.; Haack, T.B.; Vilboux, T.; Pode-Shakked, B.; Thöny, B.; Shen, N.; Guarani, V.; Meissner, T.; Mayatepek, E.; Trefz, F.K.; et al. Biallelic Mutations in DNAJC12 Cause Hyperphenylalaninemia, Dystonia, and Intellectual Disability. *Am. J. Hum. Genet.* **2017**, *100*, 257–266. [CrossRef]

102. Valenzuela, V.; Jackson, K.L.; Sardi, S.P.; Hetz, C. Gene Therapy Strategies to Restore ER Proteostasis in Disease. *Mol. Ther.* **2018**, *26*, 1404–1413. [CrossRef]

103. Ebrahimi-Fakhari, D.; Wahlster, L.; McLean, P.J. Molecular chaperones in Parkinson's disease—Present and future. *J. Parkinsons Dis.* **2011**, *1*, 299–320. [CrossRef]

104. Aktuglu-Zeybek, A.C.; Zubarioglu, T. Nitisinone: A review. *Orphan Drugs Res. Rev.* **2017**, *7*, 25–35. [CrossRef]

105. Angileri, F.; Morrow, G.; Roy, V.; Orejuela, D.; Tanguay, R.M. Heat shock response associated with hepatocarcinogenesis in a murine model of hereditary tyrosinemia type I. *Cancers* **2014**, *6*, 998–1019. [CrossRef] [PubMed]

106. Hanpude, P.; Bhattacharya, S.; Dey, A.K.; Maiti, T.K. Deubiquitinating enzymes in cellular signaling and disease regulation. *IUBMB Life* **2015**, *67*, 544–555. [CrossRef] [PubMed]

107. Lee, B.H.; Lee, M.J.; Park, S.; Oh, D.C.; Elsasser, S.; Chen, P.C.; Gartner, C.; Dimova, N.; Hanna, J.; Gygi, S.P.; et al. Enhancement of proteasome activity by a small-molecule inhibitor of USP14. *Nature* **2010**, *467*, 179–184. [CrossRef] [PubMed]

108. Seneci, P. Targeting Proteasomal Degradation of Soluble, Misfolded Proteins: Ubi Major. In *Chemical Modulators of Protein Misfolding and Neurodegenerative Disease*; Academic Press: Milan, Italy, 2015; pp. 73–94. [CrossRef]

109. Al Hafid, N.; Christodoulou, J. Phenylketonuria: A review of current and future treatments. *Transl. Pediatr.* **2015**, *4*, 304–317. [CrossRef] [PubMed]

110. Strisciuglio, P.; Concolino, D. New Strategies for the Treatment of Phenylketonuria (PKU). *Metabolites* **2014**, *4*, 1007–1017. [CrossRef]

111. Malik, S.; NiMhurchadha, S.; Jackson, C.; Eliasson, L.; Weinman, J.; Roche, S.; Walter, J. Treatment Adherence in Type 1 Hereditary Tyrosinaemia (HT1): A Mixed-Method Investigation into the Beliefs, Attitudes and Behaviour of Adolescent Patients, Their Families and Their Health-Care Team. *JIMD Rep.* **2015**, *18*, 13–22. [CrossRef] [PubMed]

112. Chinsky, J.M.; Singh, R.; Ficicioglu, C.; van Karnebeek, C.D.M.; Grompe, M.; Mitchell, G.; Waisbren, S.E.; Gucsavas-Calikoglu, M.; Wasserstein, M.P.; Coakley, K.; et al. Diagnosis and treatment of tyrosinemia type I: A US and Canadian consensus group review and recommendations. *Genet. Med.* **2017**, *19*. [CrossRef]

113. Yin, H.; Xue, W.; Chen, S.; Bogorad, R.L.; Benedetti, E.; Grompe, M.; Koteliansky, V.; Sharp, P.A.; Jacks, T.; Anderson, D.G. Genome editing with Cas9 in adult mice corrects a disease mutation and phenotype. *Nat. Biotechnol.* **2014**, *32*, 551–553. [CrossRef]

114. VanLith, C.; Guthman, R.; Nicolas, C.T.; Allen, K.; Du, Z.; Joo, D.J.; Nyberg, S.L.; Lillegard, J.B.; Hickey, R.D. Curative Ex Vivo Hepatocyte-Directed Gene Editing in a Mouse Model of Hereditary Tyrosinemia Type 1. *Hum. Gene Ther.* **2018**, *29*, 1315–1326. [CrossRef]

115. Sarkissian, C.N.; Gámez, A. Phenylalanine ammonia lyase, enzyme substitution therapy for phenylketonuria, where are we now? *Mol. Genet. Metab.* **2005**, *86* (Suppl. 1), S22–S26. [CrossRef]

116. Thomas, J.; Levy, H.; Amato, S.; Vockley, J.; Zori, R.; Dimmock, D.; Harding, C.O.; Bilder, D.A.; Weng, H.H.; Olbertz, J.; et al. Pegvaliase for the treatment of phenylketonuria: Results of a long-term phase 3 clinical trial program (PRISM). *Mol. Genet. Metab.* **2018**, *124*, 27–38. [CrossRef]

117. Sarkissian, C.N.; Gámez, A.; Wang, L.; Charbonneau, M.; Fitzpatrick, P.; Lemontt, J.F.; Zhao, B.; Vellard, M.; Bell, S.M.; Henschell, C.; et al. Preclinical evaluation of multiple species of PEGylated recombinant phenylalanine ammonia lyase for the treatment of phenylketonuria. *Proc. Natl. Acad. Sci. USA* **2008**, *105*, 20894–20899. [CrossRef]

118. Alexander, I.E.; Cunningham, S.C.; Logan, G.J.; Christodoulou, J. Potential of AAV vectors in the treatment of metabolic disease. *Gene Ther.* **2008**, *15*, 831–839. [CrossRef] [PubMed]

119. Rebuffat, A.; Harding, C.O.; Ding, Z.; Thöny, B. Comparison of adeno-associated virus pseudotype 1, 2, and 8 vectors administered by intramuscular injection in the treatment of murine phenylketonuria. *Hum. Gene Ther.* **2010**, *21*, 463–477. [CrossRef] [PubMed]

120. Pan, Y.; Shen, N.; Jung-Klawitter, S.; Betzen, C.; Hoffmann, G.F.; Hoheisel, J.D.; Blau, N. CRISPR RNA-guided FokI nucleases repair a PAH variant in a phenylketonuria model. *Sci. Rep.* **2016**, *6*, 35794. [CrossRef]

121. Lachmann, R. Sapropterin hydrochloride: Enzyme enhancement therapy for phenylketonuria. *Ther. Adv. Endocrinol. Metab.* **2011**, *2*, 127–133. [CrossRef] [PubMed]

122. Charbonneau, M.R.; Isabella, V.M.; Li, N.; Kurtz, C.B. Developing a new class of engineered live bacterial therapeutics to treat human diseases. *Nat. Commun.* **2020**, *11*, 1–11.

123. Alteri, C.J. Can the Microbiome Deliver? A Proof-of-Concept Engineered E. coli PKU Therapeutic. *Cell Host Microbe* **2019**, *25*, 473–474. [CrossRef]

124. Kochhar, J.S.; Chan, S.Y.; Ong, P.S.; Kang, L. Clinical therapeutics for phenylketonuria. *Drug Deliv. Transl. Res.* **2012**, *2*, 223–237. [CrossRef]

125. Patrawala, M.; Kuruvilla, M.; Li, H. Successful desensitization of Pegvaliase (Palynziq®) in a patient with phenylketonuria. *Mol. Genet. Metab. Rep.* **2020**, *23*, 100575. [CrossRef]

126. Markham, A. Pegvaliase: First Global Approval. *BioDrugs* **2018**, *32*, 391–395. [CrossRef]

127. Longo, N.; Arnold, G.L.; Pridjian, G.; Enns, G.M.; Ficicioglu, C.; Parker, S.; Cohen-Pfeffer, J.L. Long-term safety and efficacy of sapropterin: The PKUDOS registry experience. *Mol. Genet. Metab.* **2015**, *114*, 557–563. [CrossRef]

128. Rohr, F.; Kritzer, A.; Harding, C.O.; Viau, K.; Levy, H.L. Discontinuation of Pegvaliase therapy during maternal PKU pregnancy and postnatal breastfeeding: A case report. *Mol. Genet. Metab. Rep.* **2020**, *22*, 100555. [CrossRef] [PubMed]

129. Das, A.M. Clinical utility of nitisinone for the treatment of hereditary tyrosinemia type-1 (HT-1). *Appl. Clin. Genet.* **2017**, *10*, 43–48. [CrossRef]

130. Muntau, A.C.; Leandro, J.; Staudigl, M.; Mayer, F.; Gersting, S.W. Innovative strategies to treat protein misfolding in inborn errors of metabolism: Pharmacological chaperones and proteostasis regulators. *J. Inherit. Metab. Dis.* **2014**, *37*, 505–523. [CrossRef] [PubMed]

131. Lichter-Konecki, U.; Vockley, J. Phenylketonuria: Current Treatments and Future Developments. *Drugs* **2019**, *79*, 495–500. [CrossRef]

132. Ernst, R.; Claessen, J.H.; Mueller, B.; Sanyal, S.; Spooner, E.; van der Veen, A.G.; Kirak, O.; Schlieker, C.D.; Weihofen, W.A.; Ploegh, H.L. Enzymatic blockade of the ubiquitin-proteasome pathway. *PLoS Biol.* **2011**, *8*, e1000605. [CrossRef]

133. Wang, Y.; Jiang, X.; Feng, F.; Liu, W.; Sun, H. Degradation of proteins by PROTACs and other strategies. *Acta Pharm. Sin. B* **2020**, *10*, 207–238. [CrossRef]

134. Smith, B.E.; Wang, S.L.; Jaime-Figueroa, S.; Harbin, A.; Wang, J.; Hamman, B.D.; Crews, C.M. Differential PROTAC substrate specificity dictated by orientation of recruited E3 ligase. *Nat. Commun.* **2019**, *10*, 131. [CrossRef] [PubMed]

135. Nguyen, T.; Ho, M.; Kim, K.; Yun, S.I.; Mizar, P.; Easton, J.W.; Lee, S.S.; Kim, K.K. Suppression of the Ubiquitin Pathway by Small Molecule Binding to Ubiquitin Enhances Doxorubicin Sensitivity of the Cancer Cells. *Molecules* **2019**, *24*, 1073. [CrossRef] [PubMed]
136. Landré, V.; Rotblat, B.; Melino, S.; Bernassola, F.; Melino, G. Screening for E3-ubiquitin ligase inhibitors: Challenges and opportunities. *Oncotarget* **2014**, *5*, 7988–8013. [CrossRef]

International Journal of
Molecular Sciences

Review

Specificity in Ubiquitination Triggered by Virus Infection

Haidong Gu * and Behdokht Jan Fada

Department of Biological Sciences, Wayne State University, Detroit, MI 48202, USA; ga2861@wayne.edu
* Correspondence: haidong.gu@wayne.edu; Tel.: +1-313-577-6402

Received: 7 May 2020; Accepted: 5 June 2020; Published: 8 June 2020

Abstract: Ubiquitination is a prominent posttranslational modification, in which the ubiquitin moiety is covalently attached to a target protein to influence protein stability, interaction partner and biological function. All seven lysine residues of ubiquitin, along with the N-terminal methionine, can each serve as a substrate for further ubiquitination, which effectuates a diverse combination of mono- or poly-ubiquitinated proteins with linear or branched ubiquitin chains. The intricately composed ubiquitin codes are then recognized by a large variety of ubiquitin binding domain (UBD)-containing proteins to participate in the regulation of various pathways to modulate the cell behavior. Viruses, as obligate parasites, involve many aspects of the cell pathways to overcome host defenses and subjugate cellular machineries. In the virus-host interactions, both the virus and the host tap into the rich source of versatile ubiquitination code in order to compete, combat, and co-evolve. Here, we review the recent literature to discuss the role of ubiquitin system as the infection progresses in virus life cycle and the importance of ubiquitin specificity in the regulation of virus-host relation.

Keywords: ubiquitin code; virus infection; virus-host interaction

1. Introduction

Ubiquitination is a reversible post-translational modification (PTM) that regulates protein functions in almost every aspect of a cell's life. Conjugation of a 76-amino acid small protein, ubiquitin (Ub), to a target protein changes the stability, quantity and activity of that target. Ubiquitination was first discovered as a type of histone H2A modification about 43 years ago [1,2]. Later, it became clear that protein ubiquitination triggers selective degradation of the target and plays essential roles in many cellular processes [3]. After decades of efforts to delineate the regulation of ubiquitin enzymes and the network of ubiquitin linkage, now we know that the complex ubiquitin code is a major form of cellular communication that conveys distinct signals in cell pathways such as receptor signaling transport, DNA damage response, cell cycle progression, and stress responses (for recent reviews, see references [4–6]).

Viruses are obligate parasites that closely interact with the host to subjugate many cellular machineries to achieve viral replication. Naturally, they have evolved to manipulate and take advantage of the ubiquitin system so as to redirect the cellular pathways in their own favor. Host cells also adopt special code of ubiquitination to mount immune responses against viral infection. It is fascinating to witness the unveiling of ubiquitin regulation in cell anti-viral defenses as well as viral counteractions in recent years. In this review we will focus on the current advances of understanding the specificity of ubiquitination in the constant battles between the virus and its host. Through delineating the complex ubiquitin code on both the virus and host factors, we hope to understand the significance of ubiquitination specificity in virus-host interaction and to search for potential targets useful for prophylactic and therapeutic treatment of viral diseases.

2. Enzymes in the Ubiquitin System

Ubiquitination is carried out by a cascade of enzymatic reactions. E1 activating enzyme uses ATP to activate the C-terminal carboxyl group of ubiquitin, which then forms a thioester bond with the cysteine residue in the active site of an E2 ubiquitin conjugating enzyme. In the final step, an E3 ubiquitin ligase transfers the ubiquitin from the E2-Ub to a specific substrate. Resulted from the serial reactions, an isopeptide bond is formed between the carboxyl group of ubiquitin and the ε-amino group of a lysine residue of the substrate protein [3].

The human genome has two E1 activating enzymes, about 40 E2 conjugating enzymes and over 600 E3 ubiquitin ligases [7]. E1 serves to distinguish ubiquitin from other ubiquitin-like proteins [8]. E2s are small proteins of about 150 amino acids, whose selectivity is generally realized via their interactions with different E3s. The diverse array of E3s are categorized into three classes based on their (i) catalytic domain and (ii) the mechanism of ubiquitination reaction. The most common type of E3s is the really interesting new gene (RING) finger containing E3 ubiquitin ligases that can simultaneously bind to both the substrate and E2-Ub and directly transfer ubiquitin from the E2 to the substrate. The HECT (homologous to the E6AP carboxyl terminus) domain containing E3s and RING-between-RING (RBR) type E3s both transfer ubiquitin from E2 to form E3-Ub and then to the substrate in a two-step fashion [7,9].

Deubiquitinases (DUBs) are enzymes that remove ubiquitin from the substrate proteins. There are about 100 DUBs in humans that are classified into seven families based on their evolutionary conservation. Six of the DUB families are cysteine proteases and the last one is Zn-dependent metalloproteases [10,11].

3. Biochemical Diversity of Ubiquitin Code

The combination of E3 and DUB actions creates a sophisticated ubiquitin code [12], for which a meticulously balanced tuning is the key. This gives the cell enormous capacity to differentiate special signals and regulate the protein abundance, distribution and function in cell activities. Incidental errors in ubiquitination or deubiquitination can have dramatic physiological consequences leading to serious diseases such as cancer, inflammatory disease and neurodegeneration [13,14].

The specificity of ubiquitination is mainly controlled by: (i) substrate selection, (ii) lysine prioritization in the substrate, and (iii) lysine linkage in polyubiquitin chain. Selecting a substrate from the whole cell proteome and then prioritizing the lysine residues of the substrate are generally achieved through the different combinations of E2 and E3 in the network of specific E3 scaffold complex. The decision of one or multiple lysine residues to be used for monoubiquitination diversifies the possible outcome of this substrate. Moreover, all seven lysine residues (K6, K11, K27, K29, K33, K48 and K63) of the ubiquitin attached to the substrate can receive additional ubiquitin in multiple rounds of reactions. This leads to the formation of a variety of polyubiquitin chain on the substrate, including homotypic chain (only one particular lysine residue is used), heterotypic chain (different lysine residues are used), or branched chain that involves different numbers of lysine residues in each round of reaction. In addition, the N-terminal methionine residue (M1) of ubiquitin can also serve as the substrate to receive ubiquitin, which forms a unique head-to-tail type of linear ubiquitin chain [15–17]. To add more complexity in the ubiquitin code, non-lysine residues of the substrate protein, such as cysteine, serine and threonine can also be used in ubiquitination [18], and ubiquitin itself undergoes PTM to further increase the coding diversity of ubiquitination [4].

Regulatory signals embedded in the different architecture of ubiquitin are decoded by the UBD (ubiquitin binding domain) containing ubiquitin binding proteins. There are more than 20 families of UBD containing proteins in human proteome that recognize ubiquitin topologies and convey the signals to different cellular functions, the mechanisms of which are reviewed in reference [19].

4. Functional Specificity and Complexity of Ubiquitin Code

In recent years, a growing number of specific ubiquitin linkages have been allocated to particular cellular functions, demystifying the sophisticated ubiquitin language in the regulation of cell activities. The K48-linked ubiquitin chain is the most common signal that triggers the proteasome- dependent degradation of the substrates, conserved from yeast to humans [20–22]. However, the second most abundant form of ubiquitin chain, K63-linked chain, has little access to proteasomes [23], but is known for non-proteolytic signaling in DNA damage response, cell trafficking, autophagy and immune responses [12]. Interestingly, recent data showed that K63 ubiquitination can serve as the seed for the K48/K63 branched chain, which is then prone to target the substrate for proteasomal degradation [24], implying for more functional diversity form the complex combination of ubiquitin code.

Both K48 and K63 also form unanchored polyubiquitin chains important for intrinsic/innate immune reactions. While the K48 chains unattached to any target protein are known to activate IKKε (inhibitor of nuclear factor kappa-B kinase subunit epsilon) to promote STAT1 phosphorylation and the subsequent type I Interferon (IFN)/ISG expression [25], unanchored K63 chains are more versatile. For example, free K63 polyubiquitin chain interacts with RIG-I and activates IRF3 and NF-κB in a cell free assay, whereas in vivo it was found to activate TAK1 and upregulate NF-κB upon viral infection [26,27]. Beyond the IFN pathway, K63 chain detached from the substrate by a deubiquitinase at the proteasome can stimulate autophagy-dependent aggresome clearance, which is a critical cellular event to remove aberrant protein aggregates [28].

Mass spectrometry analysis has also captured a few other ubiquitin linkages, such as K11 and K29, whose levels are elevated in the presence of proteasome inhibitor, suggesting their involvement in proteasome degradation [21]. Recent data, however, indicated that the K11 chain may have dual roles in modulating protein stability under different circumstances. On one hand, K11 chain assembled by the cullin-containing anaphase-promoting complex (APC/C) marks the substrates for proteasomal degradation to control the cell cycle [29,30]. On the other hand, K11 chain attached to β-catenin leads to protein stabilization, not degradation [31]. In both cases, the same UBE2S is used as E2. Whether it is the difference in cell type, E3 enzyme type or substrate type that causes such opposite effects will need further investigation.

Due to the unevenness of linkage abundancy and the lack of linkage specific antibodies, much of the atypical ubiquitin signals, including K6, K27, K29 and K33, is understudied. Recent advancement of affimer technology, in which high affinity affimers binding to specific ubiquitin linkages can be screened from a library of randomized peptides, proves to be a valuable tool in deciphering ubiquitin code [32]. Along with the conventional mutagenesis assay and available antibodies, more details of special ubiquitin linkage and enzymes that control the linkage formation are being revealed every day. For example, it is now known that the low abundance K6 chain can be assembled by HUWE1, a HECT E3 ubiquitin ligase, or PARKIN, a RBR ubiquitin ligase, to be involved in DNA damage response or mitophagy, respectively [32,33]. It would be more interesting to find out whether these two events are correlated in the same cell and how they are coordinated.

The ubiquitin M1 linkage was first discovered when a mutant ubiquitin lacking all seven lysine residues was found to form polyubiquitin chain in vitro [34]. So far, the only E3 that catalyzes the head-to-tail true peptide bond of the M1 ubiquitin chain is the linear ubiquitin chain assembly complex (LUBAC), which contains two RING finger proteins, HOIL-1L and HOIP, and an adapter protein SHARPIN [34–36]. Many of the LUBAC substrates, such as NEMO, RIPK1, and TRADD, are components of TNF inflammatory signaling, indicating a critical role of M1 ubiquitin chain in innate immune responses [37,38]

5. Ubiquitination in Every Step of Viral Infection

With the critical role of ubiquitin modification in many aspects of a cell's life, viruses, as obligate parasites, are found to exploit the system in every imaginable manner. They can simply adopt an existing ubiquitin regulatory pathway to hitchhike, or they can hijack the ubiquitin system for selective advantage towards the virus. They can change the specificity of cellular E3s or DUBs to modulate a network of viral/cellular substrates, or they can encode viral ubiquitin-like modifiers or ubiquitinating/deubiquitinating enzymes to modify a whole new set of substrates. Here we will follow the footsteps of virus infection and summarize how viruses manipulate host ubiquitin system to progress in infection in their own "creative" ways.

5.1. Role of Ubiquitination in Virus Entry

Virus entry is the very first step to start an infection. At this point all a virus can rely on is its virion proteins, so to deliver the viral genome into a proper cell machinery by the few existing proteins is paramount for initiating a successful infection. Although some viruses inject the genome by direct penetration or membrane fusion, most viruses, with or without an envelope, enter a susceptible cell via endocytosis, the major cell path to uptake substances from the environment [39,40]. Membrane receptor ubiquitination often promotes receptor endocytosis, which facilitates the removal of membrane receptor to extinguish the signal transduction [41]. Viruses broadly exploit the ubiquitination regulated endocytosis for the virus entry (Figure 1).

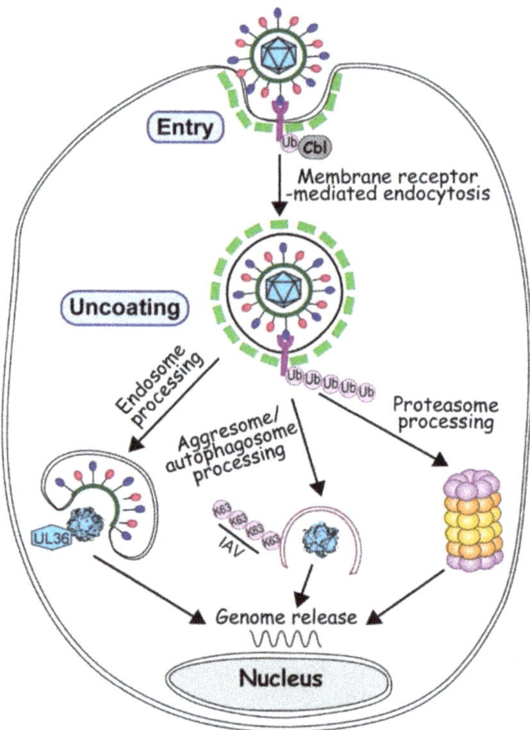

Figure 1. Involvement of ubiquitin in virus entry and uncoating. Virus entry is depicted by a generic virus (with red and blue ellipsoids attached to a green circle representing viral envelope and glycoproteins, and cyan polygon representing viral neucleocapsid) interacting with the membrane receptor colored in magenta. Ubiquitination of membrane receptor promotes virus entry via endocytosis (with green dashed lines representing the clathrin coat). Upon entry, viruses can employ endosome, autophagosome or proteasome to process the virion and release the viral genome, as described in the text.

One example of virus adopting the endocytosis pathway for entry is to use the T cell immunoglobulin and mucin (TIM) family receptors as viral receptors. TIMs are type I transmembrane proteins that function as receptors that recognize phosphatidylserine or phosphatidylethanolamine exposed on apoptotic cells and clear them by phagocytosis [42,43]. TIMs are found to facilitate the entry of a few RNA viruses in a series of recent reports [44–47]. For example, TIM-1 serves as a Dengue virus receptor to facilitate the entry via endocytosis, and ubiquitination at K338 and K346 of TIM-1 is required for an effective virus uptake [47]. Likely, these RNA viruses, with limited virion proteins, simply take advantage of the existing receptor endocytosis promoted by ubiquitination to enhance the viral entry. The specific interaction between these RNA viruses and the extra cellular IgV-like domain and mucin domain of TIM may be the determining factor for the viral selectivity. Whether TIM recognizing apoptosis-associated phospholipid has anything to do with host apoptotic effects in intrinsic defenses remains to be investigated.

Another better studied example is the involvement of Cbl in the entry process of herpesviruses. Cbl is a RING-type E3 ubiquitin ligase, which serves as an adaptor to transduce signals in the various receptor tyrosine kinase pathways. Upon signal recognition, Cbl phosphorylation triggers its association to the receptor to ubiquitinate and remove the latter from the membrane [48–50]. In both the entry process of herpes simplex virus (HSV-1) and Kaposi's sarcoma herpesvirus (KSHV), Cbl is found necessary in reducing the amount of membrane receptor via endocytosis to help the internalization of the virus [51–53]. Although it has been reported that an E2 conjugating enzyme CIN85 associated with Cbl interacts with an HSV-1 immediate early protein ICP0 (infected cell protein 0) [54], which is a viral RING-type E3 ubiquitin ligase by itself [55], it is not yet clear whether Cbl activity is controlled by ICP0 for a selective viral enhancement.

5.2. Role of Ubiquitination in Virus Uncoating

Virus uncoating is the critical step to reveal the viral genome for replication. It is frequently associated with proteolytic cleavage and proteasome/lysosome degradation for the purpose of complete or partial removal of the capsid. Ubiquitination-mediated protein degradation naturally has an irreplaceable role in the uncoating process for many viruses.

Adenovirus (AdV), with its very complex capsid but no envelope, has long been the interest of scientists studying the stepwise viral uncoating [56,57]. Recent investigations showed broad participation of the ubiquitin system in AdV disassembly and trafficking, events occurring after the virus uptake but before its docking at nuclear pore to release the genome. On one hand, the conformational change of protein VI triggers the exposure of PPxY motif, which is recognized by a HECT-type E3 ubiquitin ligase, NEDD4. Ubiquitination of pVI helps AdV to rupture endosomal membrane and escape from the autophagy degradation [58,59]. On the other hand, a RING-type E3 ligase Mib1 is newly identified to be necessary for AdV uncoating. It is not yet clear what Mib1 ubiquitinates in this case, but most likely it is related to the microtubule interaction that moves the partially disassembled AdV towards the nucleus [60]. Unlike AdV that avoids being taken up by autophagosomes on its way to uncoat the genome, Influenza A virus (IAV) actually approaches an aggresome-autophagy pathway to help release its genome [61]. HDAC6 dependent aggresome-autophagy pathway uses HDAC6 to bridge the interactions between microtubules and various ubiquitin enzymes during the aggresome and autophagosome formation. With the recruitedDUB, unanchored K63 polyubiquitin chain is generated to stimulate the degradation of misfolded proteins [62]. Unanchored ubiquitin chains are also found inside the IAV virions. Upon entry, these polyubiquitin chains interact with HDAC6-dependent aggresome degradation machinery. Via microtubule movement, RNA segments of IAV are transported towards the nucleus and released into the nuclear pore [61]. In these two cases, both AdV and IAV take the strategy of mimicry and utilize the existing cellular pathways (Figure 1). Evidently, the specificity in choosing a particular pathway to uncoat relies on the structure and composition of the virion, which at this point of infection still is the only tool the virus can use.

Unlike other virus families, herpesviruses all contain a proteinaceous layer of tegument in the virions, which gives this family of viruses more capability in the manipulation of cell ubiquitin system in the early infection. A DUB activity has been identified in the N-terminus of the largest HSV-1 tegument protein, the 3146 amino acid protein UL36, which is conserved in all herpesviruses [63].

The C-terminus of UL36 has multiple sites to interact with capsid proteins. Therefore partial deletion of UL36 impairs its incorporation into the progeny virions, making them defective in nucleus targeting and genome release in the next round of infection [64]. Due to the difficulties in biochemical and genetic manipulation of such big protein, how UL36 affects viral uncoating is still unclear. Likely it involves a complex interaction with proteasome because proteasomal inhibition also blocks the capsid transport and genome release [65].

5.3. Role of Ubiquitination in Virus Replication

Once the genome is uncoated, viral expression and replication start immediately, which produces pathogen associated molecular patterns (PAMPs) that can be recognized by the host. In an attempt to suppress the virus, the host cell instantly mobilizes and synthesizes defensive molecules to mount cellular intrinsic and innate immune responses. Many well-characterized ubiquitination regulations are associated with the IFN signaling pathways (Figure 2), including the aforementioned K48, K63 and M1 linkages.

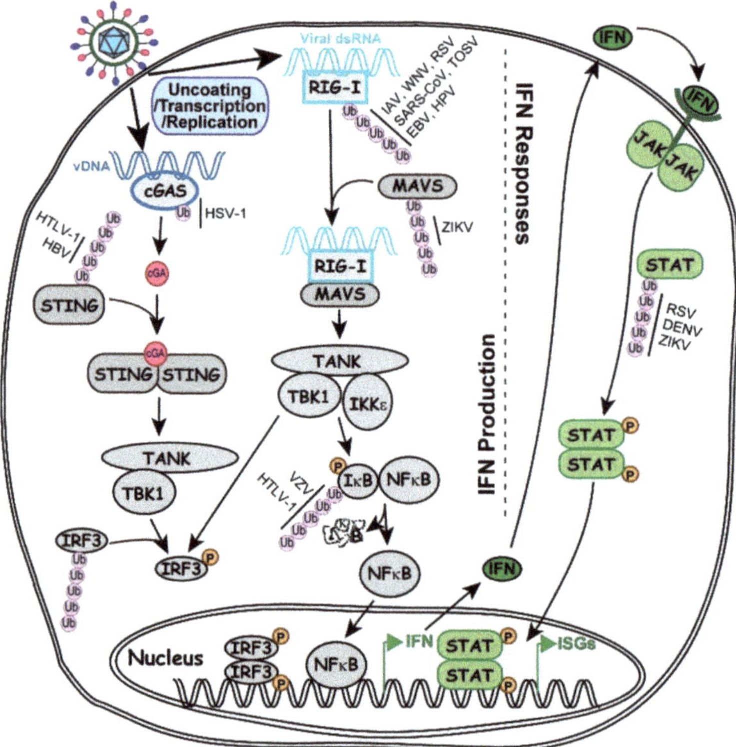

Figure 2. Involvement of ubiquitin in the key signaling pathways important for IFN production and IFN responses. Illustrated on the left side of the grey dashed line, abnormal presence of vDNA (in dark blue) or viral dsRNA (in light blue) is detected by DNA sensor cGAS (Cyclic GMP-AMP synthase) or RNA sensor RIG-I (Retinoic acid-inducible gene I), which triggers the activation of their respective signaling to induce IFN production as described in the text. On the right side of the grey dashed line, secreted IFN interacts with IFN receptor depicted in green Y-shaped line. IFN and receptor work in either autocrine or paracrine fashion to induce the activation of JAK/STAT pathway and promote ISG expression. Steps of IFN production and responses involving ubiquitination and viral regulations of the ubiquitination process are discussed in the text.

IFN was initially discovered due to its broad interference in viral infection [66,67], such as involving RNA degradation and translation shutdown to inhibit viral replication [68]. Various pathogen or damage associated molecular patterns can trigger the production of IFN, and other inflammatory cytokines, through a wide range of specific receptors and effectors and the diverse ubiquitin code [69,70]. Virus related molecular patterns generally associate with the DNA sensor- or RNA sensor-mediated IFN production, as discussed in the following subsections.

5.3.1. Ubiquitin Specificity in DNA Induced IFN Production and Viral Counteractions

Cyclic GMP-AMP synthase (cGAS) is a cytosolic protein sensing the endogenous damaged DNA or exogenous invading DNA to produce cGAMP [71], which in turn binds to stimulator of interferon genes (STING) to activate STING and transduce the signal to TANK-binding kinase 1 (TBK1) so as to phosphorylate the transcription factor IRF3. The phosphorylated IRF3 dimerizes and translocates into the nucleus to activate the transcription of type I IFN [70].

Each component in this pathway is under extensive regulation by the ubiquitin system, involving the tripartite motif (TRIM) family of RING-type E3 ubiquitin ligases [72] and some other E3s. For example, TRIM56 monoubiquitinates cGAS at K335 to enhance its binding affinity to DNA and upregulates the production of type I IFN. This ubiquitin regulation specifically restricts the infection of DNA virus HSV-1 but not the RNA virus IAV [73]. cGAS is also found to interact with RNF185 in HSV-1 infection and can be polyubiquitinated with the K27 linkage by RNF185 [74]. Whether there is a regulatory switch in the two cGAS ubiquitination reactions during HSV-1 infection is not known, neither is how the two different reactions may affect the HSV-1 pathogenesis.

The downstream effector STING has shown a tremendously complex ubiquitin regulation. For example, TRIM56, TRIM32 and mitochondrial E3 ligase MUL1 can all add K63 polyubiquitin to STING, but at different lysine residues, to stimulate TBK1-mediated IFN expression [75–77]. Meanwhile, an E3 ubiquitin ligase complex of autocrine motility factor receptor (AMFR) is regulated by insulin-induced gene 1 (INSIG1) to add K27 polyubiquitin to STING to stimulate the TBK1 [78]. However, RNF5 can add K48 polyubiquitin to K150 to trigger STING degradation [79], whereas RNF26 adding K11 polyubiquitin at K150 can prevent the formation of K48 chain to protect STING integrity [80]. Based on such complexity in the STING ubiquitin code, host cells may have the ability to distinguish the various forms of viral DNA with great specificity. Indeed, evidence of specific virus counteractions against STING has already been reported from human T lymphotropic virus 1 (HTLV-1) and hepatitis B virus (HBV), both of which are reverse transcribing viruses, with the former having RNA genome and latter having DNA genome. They produce different viral proteins, Tax for HTLV and polymerase for HBV, to reduce the K63 polyubiquitin chain and block the IFN production [81,82]. Whether the various ubiquitin linkages of STING have additional roles in viral replication needs further studies.

5.3.2. Ubiquitin Specificity in RNA Induced IFN Production and Viral Counteractions

dsRNA is a characteristic feature of virus infection very distinguishable from the uninfected cells. Retinoic acid-inducible gene I (RIG-I) is an RNA helicase that contains two caspase recruitment domains (CARD) [83]. The RIG-I-dsRNA complex interacts the mitochondrial antiviral signaling proteins (MAVS) at the mitochondria membrane, which in turn induces the MAVS interaction to IKKε and TBK1. As a result, IRF3, IRF7 and NFκB get phosphorylated and translocated to the nucleus where they activate IFN expression [70].

TRIM25 is the first E3 ubiquitin ligase identified that catalyzes K63 polyubiquitination at the CARD domains of RIG-I to stabilize the RIG-I dimerization and dsRNA binding, which play a critical role in the RNA-sensor mediated IFN production [84]. Subsequent studies show that other E3s such as Riplet, MEX3C and TRIM4 also add K63-linked polyubiquitin to RIG-I [85]. These results indicate the E3 redundancy and the importance of regulating RIG-I K63 polyubiquitination. Naturally, in order to evade the RIG-I induced IFN production to promote virus replication, many RNA viruses have evolved specific means to inhibit the RIG-I pathway. For example, although IAV, West Nile Virus (WNV) and respiratory syncytial virus (RSV) belong to different RNA virus families, they all express their own NS1 protein to interact with TRIM25 and inhibit the K63-linked ubiquitination on RIG-I [86–88]. Other RNA viruses can directly code for ubiquitin enzymes to interfere the RIG-I mediated IFN production. For example, severe acute respiratory syndrome coronavirus (SARS-CoV) expresses a papain-like viral protease to mimic DUB and remove the polyubiquitin chain from RIG-I, while Toscana virus (TOSV) expresses a RBR-type E3 ubiquitin ligase to add K48-linked ubiquitin chain to RIG-I and target it for proteasomal degradation [89,90]. Besides RNA viruses, DNA viruses also generate dsRNA during infection due to the overlapped coding sequences located in both DNA strands. Therefore DNA viruses such as Epstein Barr virus (EBV) and human papillomavirus (HPV) are also found to use their specific viral proteins, BPLF1 and E6, respectively, to manipulate TRIM25 and block this critical event [91,92].

For the downstream MAVS protein, it is recently found that TRIM31 can add K63-linked ubiquitin chain to MAVS and TRIM21 can add the K27-linked chain, whereas YOD1 serves as a DUB to remove the K63 chain [93–95]. It is not yet clear how viruses may be modulating the MAVS ubiquitination in the infection process. A recent report showed that Zika virus (ZIKV) NS3 protein can interact with MAVS and trigger K48-linked polyubiquitination of MAVS to target it for degradation. Whether a cellular or viral E3 is responsible for the process remains to be investigated [96].

5.3.3. Ubiquitin Specificity of Transcription Factors Promoting IFN Production and Viral Counteractions

The promoter of type I IFN are activated by a few transcription factors, such as IRF3 and NFκB. IRF3 can be activated by both the cGAS-STING and RIG-I pathways. Examples from both DNA and RNA viruses have been reported, in which viral proteins manipulate cellular E3s to add K48-linked polyubiquitin to IRF3 for degradation [97,98]. It is a common viral counteraction to target IRF3 and prevent IFN production.

NF-κB is a family of key transcription factors regulating the production of IFN and many other cytokines [99]. In IFN production, upstream effectors such as RIG-I activate the IKK complex to phosphorylate IκB, which leads to the K48-linked ubiquitination and degradation of IκB. Subsequently, NF-κB, a transcription factor for multiple innate pathways, is released and translocated to the nucleus to activate the expression of IFN, along with other cytokines [100]. Since IκB degradation is the key step in NFκB activation, both DNA and RNA viruses frequently target this step to control the NFκB activity. Some viruses express proteins, such as NSP1 of rotavirus and ORF61 of Varicella-Zoster Virus, to directly interact with the E3 ubiquitin ligase and block the IκB ubiquitination, [101,102]. Other viruses, such as HSV-1, use viral DUB to directly deubiquitinate IκB to prevent its degradation [103]. Interestingly, not all viruses inhibit IκB degradation. HTLV-1 Tax has been reported to interact with LUBAC to add M1 ubiquitin and generate the M1/K63 hybrid linkage on IKK, which constantly activates IKK and prevents NF-κB from being inhibited by IκB [104]. As aforementioned, Tax also modulates STING ubiquitination to counteract IFN production [81], the seemingly contradictory reports may reflect a complex coordination between IFN and inflammatory cytokine regulation in the HTLV-1 infection.

5.3.4. Ubiquitin in IFN Responses and Viral Counteractions

IFN works in both autocrine and paracrine manners to amplify the anti-viral effects by stimulating the expression of many interferon stimulated genes (ISGs). In IFN responsive pathways, individual IFN family members are recognized by specific membrane receptors to initiate signaling transduction, by which the IFN-receptor complexes activate members of the Janus kinase (JAK) family to phosphorylate the signal transducers and activators of transcription (STAT) family of transcription factors, which in turn promote the transcription of ISGs [105,106].

To evade the massive interference of IFN, viruses have also evolved ways to inhibit the IFN responses, in which ubiquitination again plays an important role. On the IFN receptor level, hemagglutinin (HA) protein of IAV is found to induce a K48-linked ubiquitination of IFNα receptor 1 (IFNAR1) and downregulate its membrane abundancy [107]. IFNAR1 ubiquitination is also observed in HBV infection, in which a cellular protein matrix metalloproteinase 9 (MMP9) is activated to mediate the IFNAR1 ubiquitination in order to reduce its membrane abundancy [108].

Because of the extensive cross-talking of JAK/STAT signaling in the overall cytokine responses, the details of viral specific regulation via the JAK and STAT proteins are not clearly understood. For example, NS1 protein of RSV, NS5 protein of Dengue virus and Zika virus, and V protein of mumps virus are all found to trigger proteasomal degradation of STATs. Some of the viral proteins interact with components in E3 ubiquitin ligase complexes, while others may act as E3 ubiquitin ligases by themselves [109–112]. At this point, details in the viral specific regulation are still missing. Further investigations are needed.

5.3.5. Ubiquitin Specificity in Intrinsic Responses and Viral Counteractions

To create an environment that supports viral transcription and replication, viruses heavily manipulate cell cycle and cause cell stress, which frequently trigger the cell level intrinsic anti-viral responses, involving pathways such as apoptosis, chromatin remodeling and stress responses.

In virus infection, both the host and virus attempt to manipulate the apoptotic pathways to their own advantage. From the host, apoptosis helps to shut down cell machineries to constrain the infection, whereas the virus exploits apoptosis for selective viral expression and virus dissemination [113]. In this process, virus also manipulates the ubiquitin system to achieve a fine balance between the proapoptotic and antiapoptotic activities. For example, tumor suppressor p53 plays a pivotal role in DNA damage response and apoptosis.

Many viruses have evolved various strategies to up- or down-regulate the p53 activity to capitalize on the pro/anti-apoptotic molecules according to their needs. For example, the E6 protein of HPV is famous for hijacking a cellular E3 ubiquitin ligase E6AP to degrade p53 by polyubiquitination [114]. Recent results showed that R175 residue of p53 is essential for the p53-E6 interaction which bridges for an E6/E6AP/p53 complex. Interestingly, E6 itself is also ubiquitinated by E6AP, so decrease in E6 stability or blockage in the p53-E6 interaction upregulate the p53 level [115]. It is plausible to hypothesize that the fine tuning of the trimeric interactions to achieve a balance in E6 and p53 levels may be associated with the persistent HPV infection in differentiating epithelium, the cause of HPV induced cancer. Caspase family proteins are also important modulators for apoptosis. A20 is an E3 ubiquitin ligase that catalyzes the K63-linked polyubiquitination of caspase 8. In HBV infection, HBx protein promotes apoptosis in the infected hepatocytes by inducing the expression of miR-125a, a microRNA for A20 [116]. Whether a balanced ubiquitination of caspase 8 plays a role in HBV tumorigenesis has not yet been investigated (Figure 3).

Nuclear domain 10 (ND10) is a dynamic nuclear structure that has promiscuous functions in many cellular events such as DNA repair, gene regulation, cell cycle regulation, and antiviral defenses. This nuclear structure has over 150 components assembled together through the interaction of SUMOylated proteins [117]. ND10 is frequently found to colocalize at the incoming viral DNA, and some of its major component proteins restrict viral transcription and replication via chromatin repression [118]. Many viruses express specific viral proteins to disrupt or modify ND10 structure while establishing the replication, such as ICP0 of HSV-1, pp71 of human cytomegalovirus (HCMV), and E4-ORF3 of AdV [119]. Among them, ICP0 is the best studied viral RING-type E3 ubiquitin ligase that ubiquitinates the organizer of ND10, promyelocytic leukemia (PML) protein, and targets it for proteasomal degradation (Figure 3) [120]. The unique phenomenon in this ubiquitin regulation is that ICP0 has the ability to distinguish the different isoforms of PML, which share the same N-terminal exons 1-6 and differ only at their C-terminus [121,122]. Whether the specific recognition of these similar substrates by ICP0 can lead to distinct ubiquitin linkages or differential roles in viral infection remains unclear. RTA protein of KSHV also has E3 ubiquitin ligase activity to target the SUMOylated PML for degradation [123]. It is not yet clear whether the ability of ICP0 to differentially recognize similar substrates is conserved in other members of the herpesvirus family.

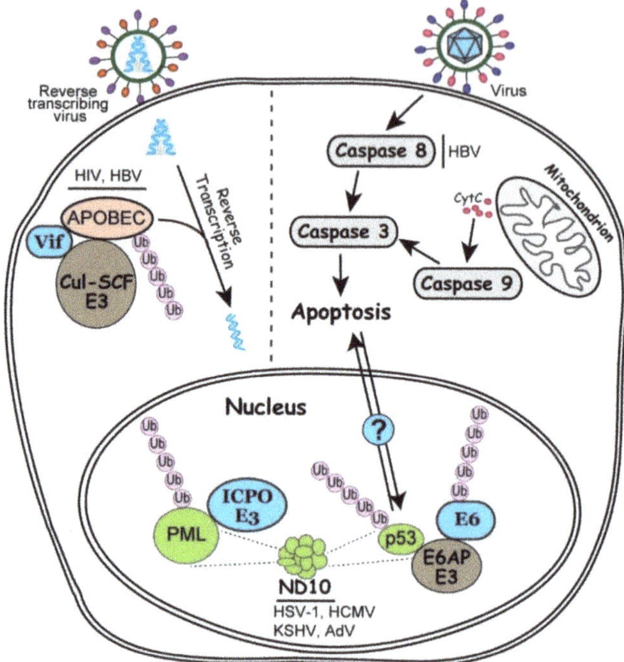

Figure 3. Involvement of ubiquitin in the intrinsic antiviral responses including apoptosis, ND10 nuclear bodies and APOBEC enzymes. In the nucleus, ND10 associated PML and p53 regulations are enlarged to show detailed interactions. On the right side of the grey dashed line, infection by a generic virus (as described in Figure 1) can trigger exogenous or endogenous apoptosis via caspase activation. Potential communication of the p53 and caspase induced apoptosis is indicated by a question mark. On the left side of the grey dashed line, a reverse transcribing virus associated with APOBEC editing is illustrated. Steps involving ubiquitination and viral regulation of the ubiquitination process are discussed in the text.

APOBEC (apolipoprotein B mRNA editing enzyme catalytic polypeptide) is a family of cytosine deaminases that are capable of the C to U RNA editing. Therefore, they act as a layer of intrinsic immunity to restrict the replication of retroviruses and some DNA viruses through the induction of hyper mutations [124]. A classic example of viral counteraction of APOBEC is the Vif protein of HIV-1, which induces the degradation of APOBEC3G mediated by the Cul5-SCF E3 ubiquitin ligase [125]. In the absence of Vif, APOBEC3G is packaged into the virions, which induces hyper mutations in the cDNA synthesis upon infection and therefore impair viral genome integrity and virus production (Figure 3) [125,126]. The same virus-host interaction also exists in the reverse transcribing HBV infection, in which viral protein HBx upregulates the MSL2 E3 ubiquitin ligase to trigger the APOBEC3B degradation [127].

5.3.6. Ubiquitin Specificity in Viral Replication Enzymes

Besides relying on host transcription/translation machineries, many viruses encode their own DNA/RNA polymerases to carry out viral transcription and genome replication. Ubiquitin regulation on these viral encoded replication enzymes has also been reported. For example, all three subunits

(PA, PB1 and PB2) of the IAV RNA polymerase are ubiquitinated to enhance the genome replication [128]. More detailed study now have revealed that Cul4 E3 ubiquitin ligase can add K29-linked polyubiquitin to PB2 to improve replication [129]. The significance of the ubiquitination on other subunits and how these ubiquitin codes coordinate to regulate the IAV genome replication remain to be seen. Similarly, Vp35 of Ebola virus (EBOV) is a cofactor for its RNA polymerase. In EBOV infection, Vp5 interacts with the cellular E3 ubiquitin ligase TRIM6 for itself to be ubiquitinated in order to enhance the genome replication [130].

Ubiquitination is also important in modulating the host DNA-dependent RNA polymerase in viral transcription. The Tat protein of HIV is well known for substantially improving HIV gene transcription [131]. Recent results revealed that Tat does so by hijacking a cellular E3 ligase UBE2O to monoubiquitinate HEXIM1, an inhibitory protein that binds to the positive transcription elongation factor b (P-TEFb). The non-degradative monoubiquitination of HEXIM1 leads to the release of P-TEFb and drives RNA pol II into the elongation mode [132]. Presumably, Tat may promote the transcription elongation in a viral specific manner to tilt the reaction towards HIV RNA production, but evidence still remains to be seen.

5.4. Role of Ubiquitination in Virus Egress

Virus assembly and release is the last but not least step in a virus infection cycle. Without this step, a continuous viral propagation via shedding and transmission cannot be ensured. Most of the molecular foundation for viral egress comes from studying the budding process of enveloped viruses, a process depending on the endosomal sorting complex required for transport (ESCRT). Similar to the role of ubiquitination in the receptor internalization during the virus entry, ubiquitination of cargo membrane proteins that marks them for ESCRT recognition is again important for the viral egress [133].

Several families of enveloped viruses contain late assembly domain (L domain) in their Gag proteins, including HIV, EBOV and RSV. One conserved feature for these viruses is the presence of PPxY motif in the L domain that can be recognized by the NEDD4 family of E3s for K63-linked ubiquitination, a key to the efficient budding of these viruses [134–136]. In HIV-1 variants that lack the PPxY motif, the virus can circumvent the defect by using a cellular protein, angiomotin, to bridge the interaction of Gag to a NEDD4-2 isoform [137,138]. The extensive efforts in Gag ubiquitination indicate its importance in the HIV egress, likely to compensate for the high mutation rate and slow productivity of HIV.

6. Conclusions

Virus replication relies on its abilities to overpower the host defenses as well as to harness the cell machinery. To achieve both goals simultaneously, it is essential for the virus to walk a fine line in sorting out friends and foes in the host cell proteome. Ubiquitin modification has enormous power to modulate the protein stability, specificity, and affinity to fine tune its position in an interaction network. Therefore the ubiquitin system is an ideal cell machinery for the virus to adopt and manipulate to its own advantage. Delineation of virus specificity in controlling the ubiquitin system to write viral ubiquitin code will help us to identify the beneficial and inimical factors in a virus infection. This will become the cornerstone in developing novel treatment strategies for viral diseases.

Author Contributions: All authors have read and agreed to the published version of the manuscript. Writing, H.G.; Literature search, B.J.F.

Funding: This research was funded by National Institute of Allergy and Infectious Diseases, grant number RO1AI118992.

Abbreviations

AdV	Adenovirus
AMFR	Autocrine motility factor receptor
APC/C	Cullin-containing anaphase-promoting complex
APOBEC	Apolipoprotein B mRNA editing enzyme catalytic polypeptide
CARD	Caspase recruitment domains
cGAS	Cyclic GMP-AMP synthase
DUB	Deubiquitinase
EBOV	Ebola virus
EBV	Epstein Barr virus
ESCRT	Endosomal sorting complex required for transport
HBV	Hepatitis B virus
HCMV	Human cytomegalovirus
HECT	Homologous to the E6AP carboxyl terminus
HPV	Human papillomavirus
HSV-1	Herpes simplex virus 1
HTLV-1	Human T lymphotropic virus 1
IAV	Influenza A virus
IFN	Interferon
INSIG1	Insulin-induced gene 1
IKK	Inhibitor of nuclear factor kappa-B kinase
ISG	Interferon stimulated genes
JAK	Janus kinase
KSHV	Kaposi's sarcoma herpesvirus
LUBAC	Linear ubiquitin chain assembly complex
MAVS	Mitochondrial antiviral signaling proteins
ND10	Nuclear domain 10
PAMPs	Pathogen associated molecular patterns
PML	Promyelocytic leukemia
P-TEFb	Positive transcription elongation factor b
PTM	Post-translational modification
RIG-I	Retinoic acid-inducible gene I
RING	Really interesting new gene
RBR	RING-between-RING
SARS-CoV	Severe acute respiratory syndrome coronavirus
STAT	Signal transducers and activators of transcription
STING	Stimulator of interferon genes
TBK1	TANK-binding kinase 1
TIM	Immunoglobulin and mucin
TOSV	Toscana virus
TRIM	Tripartite motif
UBD	Ubiquitin binding domain
ZIKV	Zika virus

References

1. Goldknopf, I.L.; Busch, H. Isopeptide linkage between nonhistone and histone 2A polypeptides of chromosomal conjugate-protein A24. *Proc. Natl. Acad. Sci. USA* **1977**, *74*, 864–868. [CrossRef] [PubMed]
2. Hunt, L.T.; Dayhoff, M.O. Amino-terminal sequence identity of ubiquitin and the nonhistone component of nuclear protein A24. *Biochem. Biophys. Res. Commun.* **1977**, *74*, 650–655. [CrossRef]
3. Hershko, A.; Ciechanover, A. The ubiquitin system. *Annu. Rev. Biochem.* **1998**, *67*, 425–479. [CrossRef]
4. Swatek, K.N.; Komander, D. Ubiquitin modifications. *Cell Res.* **2016**, *26*, 399–422. [CrossRef]
5. Haakonsen, D.L.; Rape, M. Branching Out: Improved Signaling by Heterotypic Ubiquitin Chains. *Trends Cell Biol.* **2019**, *29*, 704–716. [CrossRef]

6. Pohl, C.; Dikic, I. Cellular quality control by the ubiquitin-proteasome system and autophagy. *Science* **2019**, *366*, 818–822. [CrossRef]

7. Zheng, N.; Shabek, N. Ubiquitin Ligases: Structure, Function, and Regulation. *Annu. Rev. Biochem.* **2017**, *86*, 129–157. [CrossRef]

8. Schulman, B.A.; Harper, J.W. Ubiquitin-like protein activation by E1 enzymes: The apex for downstream signalling pathways. *Nat. Rev.. Mol. Cell Biol.* **2009**, *10*, 319–331. [CrossRef]

9. Morreale, F.E.; Walden, H. Types of Ubiquitin Ligases. *Cell* **2016**, *165*, 248. [CrossRef]

10. Clague, M.J.; Urbe, S.; Komander, D. Breaking the chains: Deubiquitylating enzyme specificity begets function. *Nat. Rev.. Mol. Cell Biol.* **2019**, *20*, 338–352. [CrossRef]

11. Mevissen, T.E.T.; Komander, D. Mechanisms of Deubiquitinase Specificity and Regulation. *Annu. Rev. Biochem.* **2017**, *86*, 159–192. [CrossRef] [PubMed]

12. Komander, D.; Rape, M. The ubiquitin code. *Annu. Rev. Biochem.* **2012**, *81*, 203–229. [CrossRef] [PubMed]

13. George, A.J.; Hoffiz, Y.C.; Charles, A.J.; Zhu, Y.; Mabb, A.M. A Comprehensive Atlas of E3 Ubiquitin Ligase Mutations in Neurological Disorders. *Front. Genet.* **2018**, *9*, 29. [CrossRef] [PubMed]

14. Pinto-Fernandez, A.; Kessler, B.M. DUBbing Cancer: Deubiquitylating Enzymes Involved in Epigenetics, DNA Damage and the Cell Cycle As Therapeutic Targets. *Front. Genet.* **2016**, *7*, 133. [CrossRef] [PubMed]

15. Akutsu, M.; Dikic, I.; Bremm, A. Ubiquitin chain diversity at a glance. *J. Cell Sci.* **2016**, *129*, 875–880. [CrossRef] [PubMed]

16. Kliza, K.; Husnjak, K. Resolving the Complexity of Ubiquitin Networks. *Front. Mol. Biosci.* **2020**, *7*, 21. [CrossRef] [PubMed]

17. Ohtake, F.; Tsuchiya, H. The emerging complexity of ubiquitin architecture. *J. Biochem.* **2017**, *161*, 125–133. [CrossRef] [PubMed]

18. McClellan, A.J.; Laugesen, S.H.; Ellgaard, L. Cellular functions and molecular mechanisms of non-lysine ubiquitination. *Open Biol.* **2019**, *9*, 190147. [CrossRef]

19. Husnjak, K.; Dikic, I. Ubiquitin-binding proteins: Decoders of ubiquitin-mediated cellular functions. *Annu. Rev. Biochem.* **2012**, *81*, 291–322. [CrossRef]

20. Xu, P.; Duong, D.M.; Seyfried, N.T.; Cheng, D.; Xie, Y.; Robert, J.; Rush, J.; Hochstrasser, M.; Finley, D.; Peng, J. Quantitative proteomics reveals the function of unconventional ubiquitin chains in proteasomal degradation. *Cell* **2009**, *137*, 133–145. [CrossRef]

21. Kim, W.; Bennett, E.J.; Huttlin, E.L.; Guo, A.; Li, J.; Possemato, A.; Sowa, M.E.; Rad, R.; Rush, J.; Comb, M.J.; et al. Systematic and quantitative assessment of the ubiquitin-modified proteome. *Mol. Cell* **2011**, *44*, 325–340. [CrossRef] [PubMed]

22. Thrower, J.S.; Hoffman, L.; Rechsteiner, M.; Pickart, C.M. Recognition of the polyubiquitin proteolytic signal. *Embo, J.* **2000**, *19*, 94–102. [CrossRef] [PubMed]

23. Jacobson, A.D.; Zhang, N.Y.; Xu, P.; Han, K.J.; Noone, S.; Peng, J.; Liu, C.W. The lysine 48 and lysine 63 ubiquitin conjugates are processed differently by the 26 s proteasome. *J. Biol. Chem.* **2009**, *284*, 35485–35494. [CrossRef] [PubMed]

24. Ohtake, F.; Tsuchiya, H.; Saeki, Y.; Tanaka, K. K63 ubiquitylation triggers proteasomal degradation by seeding branched ubiquitin chains. *Proc. Natl. Acad. Sci. USA* **2018**, *115*, E1401–e1408. [CrossRef] [PubMed]

25. Rajsbaum, R.; Versteeg, G.A.; Schmid, S.; Maestre, A.M.; Belicha-Villanueva, A.; Martínez-Romero, C.; Patel, J.R.; Morrison, J.; Pisanelli, G.; Miorin, L.; et al. Unanchored K48-linked polyubiquitin synthesized by the E3-ubiquitin ligase TRIM6 stimulates the interferon-IKKε kinase-mediated antiviral response. *Immunity* **2014**, *40*, 880–895. [CrossRef]

26. Zeng, W.; Sun, L.; Jiang, X.; Chen, X.; Hou, F.; Adhikari, A.; Xu, M.; Chen, Z.J. Reconstitution of the RIG-I pathway reveals a signaling role of unanchored polyubiquitin chains in innate immunity. *Cell* **2010**, *141*, 315–330. [CrossRef]

27. Pertel, T.; Hausmann, S.; Morger, D.; Zuger, S.; Guerra, J.; Lascano, J.; Reinhard, C.; Santoni, F.A.; Uchil, P.D.; Chatel, L.; et al. TRIM5 is an innate immune sensor for the retrovirus capsid lattice. *Nature* **2011**, *472*, 361–365. [CrossRef]

28. Hao, R.; Nanduri, P.; Rao, Y.; Panichelli, R.S.; Ito, A.; Yoshida, M.; Yao, T.P. Proteasomes activate aggresome disassembly and clearance by producing unanchored ubiquitin chains. *Mol. Cell* **2013**, *51*, 819–828. [CrossRef]

29. Matsumoto, M.L.; Wickliffe, K.E.; Dong, K.C.; Yu, C.; Bosanac, I.; Bustos, D.; Phu, L.; Kirkpatrick, D.S.; Hymowitz, S.G.; Rape, M.; et al. K11-linked polyubiquitination in cell cycle control revealed by a K11 linkage-specific antibody. *Mol. Cell* **2010**, *39*, 477–484. [CrossRef] [PubMed]

30. Brown, N.G.; VanderLinden, R.; Watson, E.R.; Weissmann, F.; Ordureau, A.; Wu, K.P.; Zhang, W.; Yu, S.; Mercredi, P.Y.; Harrison, J.S.; et al. Dual RING E3 Architectures Regulate Multiubiquitination and Ubiquitin Chain Elongation by APC/C. *Cell* **2016**, *165*, 1440–1453. [CrossRef]

31. Li, Z.; Wang, Y.; Li, Y.; Yin, W.; Mo, L.; Qian, X.; Zhang, Y.; Wang, G.; Bu, F.; Zhang, Z.; et al. Ube2s stabilizes beta-Catenin through K11-linked polyubiquitination to promote mesendoderm specification and colorectal cancer development. *Cell Death Dis.* **2018**, *9*, 456. [CrossRef] [PubMed]

32. Michel, M.A.; Swatek, K.N.; Hospenthal, M.K.; Komander, D. Ubiquitin Linkage-Specific Affimers Reveal Insights into K6-Linked Ubiquitin Signaling. *Mol. Cell* **2017**, *68*, 233–246. [CrossRef]

33. Ordureau, A.; Heo, J.M.; Duda, D.M.; Paulo, J.A.; Olszewski, J.L.; Yanishevski, D.; Rinehart, J.; Schulman, B.A.; Harper, J.W. Defining roles of PARKIN and ubiquitin phosphorylation by PINK1 in mitochondrial quality control using a ubiquitin replacement strategy. *Proc. Natl. Acad. Sci. USA* **2015**, *112*, 6637–6642. [CrossRef] [PubMed]

34. Kirisako, T.; Kamei, K.; Murata, S.; Kato, M.; Fukumoto, H.; Kanie, M.; Sano, S.; Tokunaga, F.; Tanaka, K.; Iwai, K. A ubiquitin ligase complex assembles linear polyubiquitin chains. *Embo J.* **2006**, *25*, 4877–4887. [CrossRef]

35. Tokunaga, F.; Sakata, S.; Saeki, Y.; Satomi, Y.; Kirisako, T.; Kamei, K.; Nakagawa, T.; Kato, M.; Murata, S.; Yamaoka, S.; et al. Involvement of linear polyubiquitylation of NEMO in NF-kappaB activation. *Nat. Cell Biol.* **2009**, *11*, 123–132. [CrossRef] [PubMed]

36. Gerlach, B.; Cordier, S.M.; Schmukle, A.C.; Emmerich, C.H.; Rieser, E.; Haas, T.L.; Webb, A.I.; Rickard, J.A.; Anderton, H.; Wong, W.W.; et al. Linear ubiquitination prevents inflammation and regulates immune signalling. *Nature* **2011**, *471*, 591–596. [CrossRef]

37. Spit, M.; Rieser, E.; Walczak, H. Linear ubiquitination at a glance. *J. Cell Sci.* **2019**, *132*. [CrossRef]

38. Dittmar, G.; Winklhofer, K.F. Linear Ubiquitin Chains: Cellular Functions and Strategies for Detection and Quantification. *Front. Chem.* **2019**, *7*, 915. [CrossRef]

39. Helenius, A. Virus Entry: Looking Back and Moving Forward. *J. Mol. Biol.* **2018**, *430*, 1853–1862. [CrossRef]

40. Yamauchi, Y.; Helenius, A. Virus entry at a glance. *J. Cell Sci.* **2013**, *126*, 1289–1295. [CrossRef]

41. Foot, N.; Henshall, T.; Kumar, S. Ubiquitination and the Regulation of Membrane Proteins. *Physiol. Rev.* **2017**, *97*, 253–281. [CrossRef] [PubMed]

42. Simhadri, V.R.; Andersen, J.F.; Calvo, E.; Choi, S.C.; Coligan, J.E.; Borrego, F. Human CD300a binds to phosphatidylethanolamine and phosphatidylserine, and modulates the phagocytosis of dead cells. *Blood* **2012**, *119*, 2799–2809. [CrossRef] [PubMed]

43. Freeman, G.J.; Casasnovas, J.M.; Umetsu, D.T.; DeKruyff, R.H. TIM genes: A family of cell surface phosphatidylserine receptors that regulate innate and adaptive immunity. *Immunol. Rev.* **2010**, *235*, 172–189. [CrossRef]

44. Brouillette, R.B.; Phillips, E.K.; Patel, R.; Mahauad-Fernandez, W.; Moller-Tank, S.; Rogers, K.J.; Dillard, J.A.; Cooney, A.L.; Martinez-Sobrido, L.; Okeoma, C.; et al. TIM-1 Mediates Dystroglycan-Independent Entry of Lassa Virus. *J. Virol.* **2018**, *92*. [CrossRef]

45. Niu, J.; Jiang, Y.; Xu, H.; Zhao, C.; Zhou, G.; Chen, P.; Cao, R. TIM-1 Promotes Japanese Encephalitis Virus Entry and Infection. *Viruses* **2018**, *10*. [CrossRef] [PubMed]

46. Hu, M.; Wang, F.; Li, W.; Zhang, X.; Zhang, Z.; Zhang, X.E.; Cui, Z. Ebola Virus Uptake into Polarized Cells from the Apical Surface. *Viruses* **2019**, *11*. [CrossRef]

47. Dejarnac, O.; Hafirassou, M.L.; Chazal, M.; Versapuech, M.; Gaillard, J.; Perera-Lecoin, M.; Umana-Diaz, C.; Bonnet-Madin, L.; Carnec, X.; Tinevez, J.Y.; et al. TIM-1 Ubiquitination Mediates Dengue Virus Entry. *Cell Rep.* **2018**, *23*, 1779–1793. [CrossRef]

48. Levkowitz, G.; Waterman, H.; Zamir, E.; Kam, Z.; Oved, S.; Langdon, W.Y.; Beguinot, L.; Geiger, B.; Yarden, Y. c-Cbl/Sli-1 regulates endocytic sorting and ubiquitination of the epidermal growth factor receptor. *Genes Dev.* **1998**, *12*, 3663–3674. [CrossRef]

49. Duval, M.; Bedard-Goulet, S.; Delisle, C.; Gratton, J.P. Vascular endothelial growth factor-dependent down-regulation of Flk-1/KDR involves Cbl-mediated ubiquitination. Consequences on nitric oxide production from endothelial cells. *J. Biol. Chem.* **2003**, *278*, 20091–20097. [CrossRef]

50. Cooper, J.A.; Kaneko, T.; Li, S.S. Cell regulation by phosphotyrosine-targeted ubiquitin ligases. *Mol. Cell. Biol.* **2015**, *35*, 1886–1897. [CrossRef]

51. Chakraborty, S.; ValiyaVeettil, M.; Sadagopan, S.; Paudel, N.; Chandran, B. c-Cbl-mediated selective virus-receptor translocations into lipid rafts regulate productive Kaposi's sarcoma-associated herpesvirus infection in endothelial cells. *J. Virol.* **2011**, *85*, 12410–12430. [CrossRef]

52. Valiya Veettil, M.; Sadagopan, S.; Kerur, N.; Chakraborty, S.; Chandran, B. Interaction of c-Cbl with myosin IIA regulates Bleb associated macropinocytosis of Kaposi's sarcoma-associated herpesvirus. *Plos Pathog.* **2010**, *6*, e1001238. [CrossRef]

53. Deschamps, T.; Dogrammatzis, C.; Mullick, R.; Kalamvoki, M. Cbl E3 Ligase Mediates the Removal of Nectin-1 from the Surface of Herpes Simplex Virus 1-Infected Cells. *J. Virol.* **2017**, *91*. [CrossRef] [PubMed]

54. Liang, Y.; Kurakin, A.; Roizman, B. Herpes simplex virus 1 infected cell protein 0 forms a complex with CIN85 and Cbl and mediates the degradation of EGF receptor from cell surfaces. *Proc. Natl. Acad. Sci. USA* **2005**, *102*, 5838–5843. [CrossRef] [PubMed]

55. Gu, H. Infected cell protein 0 functional domains and their coordination in herpes simplex virus replication. *World J. Virol.* **2016**, *5*, 1–13. [CrossRef] [PubMed]

56. Greber, U.F.; Willetts, M.; Webster, P.; Helenius, A. Stepwise dismantling of adenovirus 2 during entry into cells. *Cell* **1993**, *75*, 477–486. [CrossRef]

57. Wiethoff, C.M.; Wodrich, H.; Gerace, L.; Nemerow, G.R. Adenovirus protein VI mediates membrane disruption following capsid disassembly. *J. Virol.* **2005**, *79*, 1992–2000. [CrossRef]

58. Wodrich, H.; Henaff, D.; Jammart, B.; Segura-Morales, C.; Seelmeir, S.; Coux, O.; Ruzsics, Z.; Wiethoff, C.M.; Kremer, E.J. A capsid-encoded PPxY-motif facilitates adenovirus entry. *PLoS Pathog.* **2010**, *6*, e1000808. [CrossRef]

59. Montespan, C.; Marvin, S.A.; Austin, S.; Burrage, A.M.; Roger, B.; Rayne, F.; Faure, M.; Campell, E.M.; Schneider, C.; Reimer, R.; et al. Multi-layered control of Galectin-8 mediated autophagy during adenovirus cell entry through a conserved PPxY motif in the viral capsid. *PLoS Pathog.* **2017**, *13*, e1006217. [CrossRef]

60. Bauer, M.; Flatt, J.W.; Seiler, D.; Cardel, B.; Emmenlauer, M.; Boucke, K.; Suomalainen, M.; Hemmi, S.; Greber, U.F. The E3 Ubiquitin Ligase Mind Bomb 1 Controls Adenovirus Genome Release at the Nuclear Pore Complex. *Cell Rep.* **2019**, *29*, 3785–3795. [CrossRef]

61. Banerjee, I.; Miyake, Y.; Nobs, S.P.; Schneider, C.; Horvath, P.; Kopf, M.; Matthias, P.; Helenius, A.; Yamauchi, Y. Influenza A virus uses the aggresome processing machinery for host cell entry. *Science* **2014**, *346*, 473–477. [CrossRef] [PubMed]

62. Yan, J. Interplay between HDAC6 and its interacting partners: Essential roles in the aggresome-autophagy pathway and neurodegenerative diseases. *DNA Cell Biol.* **2014**, *33*, 567–580. [CrossRef] [PubMed]

63. Kattenhorn, L.M.; Korbel, G.A.; Kessler, B.M.; Spooner, E.; Ploegh, H.L. A deubiquitinating enzyme encoded by HSV-1 belongs to a family of cysteine proteases that is conserved across the family Herpesviridae. *Mol. Cell* **2005**, *19*, 547–557. [CrossRef] [PubMed]

64. Schipke, J.; Pohlmann, A.; Diestel, R.; Binz, A.; Rudolph, K.; Nagel, C.H.; Bauerfeind, R.; Sodeik, B. The C terminus of the large tegument protein pUL36 contains multiple capsid binding sites that function differently during assembly and cell entry of herpes simplex virus. *J. Virol.* **2012**, *86*, 3682–3700. [CrossRef]

65. Schneider, S.M.; Pritchard, S.M.; Wudiri, G.A.; Trammell, C.E.; Nicola, A.V. Early Steps in Herpes Simplex Virus Infection Blocked by a Proteasome Inhibitor. *mBio* **2019**, *10*. [CrossRef]

66. Isaacs, A.; Lindenmann, J.; Valentine, R.C. Virus interference. II. Some properties of interferon. *Proc. R. Soc. Lond.. Ser. BBiol. Sci.* **1957**, *147*, 268–273. [CrossRef]

67. Isaacs, A.; Lindenmann, J. Virus interference. I. The interferon. *Proc. R. Soc. Lond.* **1957**, *147*, 258–267. [CrossRef]

68. Samuel, C.E. Antiviral actions of interferons. *Clin. Microbiol. Rev.* **2001**, *14*, 778–809. [CrossRef]

69. Cao, X. Self-regulation and cross-regulation of pattern-recognition receptor signalling in health and disease. *Nat. Rev. Immunol.* **2016**, *16*, 35–50. [CrossRef]

70. Ablasser, A.; Hur, S. Regulation of cGAS- and RLR-mediated immunity to nucleic acids. *Nat. Immunol.* **2020**, *21*, 17–29. [CrossRef]

71. Sun, L.; Wu, J.; Du, F.; Chen, X.; Chen, Z.J. Cyclic GMP-AMP synthase is a cytosolic DNA sensor that activates the type I interferon pathway. *Science* **2013**, *339*, 786–791. [CrossRef] [PubMed]

72. Van Gent, M.; Sparrer, K.M.J.; Gack, M.U. TRIM Proteins and Their Roles in Antiviral Host Defenses. *Annu. Rev. Virol.* **2018**, *5*, 385–405. [CrossRef] [PubMed]

73. Seo, G.J.; Kim, C.; Shin, W.J.; Sklan, E.H.; Eoh, H.; Jung, J.U. TRIM56-mediated monoubiquitination of cGAS for cytosolic DNA sensing. *Nat. Commun.* **2018**, *9*, 613. [CrossRef] [PubMed]

74. Wang, Q.; Huang, L.; Hong, Z.; Lv, Z.; Mao, Z.; Tang, Y.; Kong, X.; Li, S.; Cui, Y.; Liu, H.; et al. The E3 ubiquitin ligase RNF185 facilitates the cGAS-mediated innate immune response. *Plos Pathog.* **2017**, *13*, e10006264. [CrossRef] [PubMed]

75. Tsuchida, T.; Zou, J.; Saitoh, T.; Kumar, H.; Abe, T.; Matsuura, Y.; Kawai, T.; Akira, S. The ubiquitin ligase TRIM56 regulates innate immune responses to intracellular double-stranded DNA. *Immunity* **2010**, *33*, 765–776. [CrossRef]

76. Zhang, J.; Hu, M.M.; Wang, Y.Y.; Shu, H.B. TRIM32 protein modulates type I interferon induction and cellular antiviral response by targeting MITA/STING protein for K63-linked ubiquitination. *J. Biol. Chem.* **2012**, *287*, 28646–28655. [CrossRef]

77. Ni, G.; Konno, H.; Barber, G.N. Ubiquitination of STING at lysine 224 controls IRF3 activation. *Sci. Immunol.* **2017**, *2*. [CrossRef]

78. Wang, Q.; Liu, X.; Cui, Y.; Tang, Y.; Chen, W.; Li, S.; Yu, H.; Pan, Y.; Wang, C. The E3 ubiquitin ligase AMFR and INSIG1 bridge the activation of TBK1 kinase by modifying the adaptor STING. *Immunity* **2014**, *41*, 919–933. [CrossRef]

79. Zhong, B.; Zhang, L.; Lei, C.; Li, Y.; Mao, A.P.; Yang, Y.; Wang, Y.Y.; Zhang, X.L.; Shu, H.B. The ubiquitin ligase RNF5 regulates antiviral responses by mediating degradation of the adaptor protein MITA. *Immunity* **2009**, *30*, 397–407. [CrossRef]

80. Qin, Y.; Zhou, M.T.; Hu, M.M.; Hu, Y.H.; Zhang, J.; Guo, L.; Zhong, B.; Shu, H.B. RNF26 temporally regulates virus-triggered type I interferon induction by two distinct mechanisms. *PLoS Pathog.* **2014**, *10*, e1004358. [CrossRef]

81. Wang, J.; Yang, S.; Liu, L.; Wang, H.; Yang, B. HTLV-1 Tax impairs K63-linked ubiquitination of STING to evade host innate immunity. *Virus Res.* **2017**, *232*, 13–21. [CrossRef] [PubMed]

82. Liu, Y.; Li, J.; Chen, J.; Li, Y.; Wang, W.; Du, X.; Song, W.; Zhang, W.; Lin, L.; Yuan, Z. Hepatitis B virus polymerase disrupts K63-linked ubiquitination of STING to block innate cytosolic DNA-sensing pathways. *J. Virol.* **2015**, *89*, 2287–2300. [CrossRef] [PubMed]

83. Yoneyama, M.; Kikuchi, M.; Natsukawa, T.; Shinobu, N.; Imaizumi, T.; Miyagishi, M.; Taira, K.; Akira, S.; Fujita, T. The RNA helicase RIG-I has an essential function in double-stranded RNA-induced innate antiviral responses. *Nat. Immunol.* **2004**, *5*, 730–737. [CrossRef]

84. Gack, M.U.; Shin, Y.C.; Joo, C.H.; Urano, T.; Liang, C.; Sun, L.; Takeuchi, O.; Akira, S.; Chen, Z.; Inoue, S.; et al. TRIM25 RING-finger E3 ubiquitin ligase is essential for RIG-I-mediated antiviral activity. *Nature* **2007**, *446*, 916–920. [CrossRef] [PubMed]

85. Okamoto, M.; Kouwaki, T.; Fukushima, Y.; Oshiumi, H. Regulation of RIG-I Activation by K63-Linked Polyubiquitination. *Front. Immunol.* **2017**, *8*, 1942. [CrossRef] [PubMed]

86. Rajsbaum, R.; Albrecht, R.A.; Wang, M.K.; Maharaj, N.P.; Versteeg, G.A.; Nistal-Villan, E.; Garcia-Sastre, A.; Gack, M.U. Species-specific inhibition of RIG-I ubiquitination and IFN induction by the influenza A virus NS1 protein. *Plos Pathog.* **2012**, *8*, e1003059. [CrossRef] [PubMed]

87. Zhang, H.L.; Ye, H.Q.; Liu, S.Q.; Deng, C.L.; Li, X.D.; Shi, P.Y.; Zhang, B. West Nile Virus NS1 Antagonizes Interferon Beta Production by Targeting RIG-I and MDA5. *J. Virol.* **2017**, *91*. [CrossRef] [PubMed]

88. Ban, J.; Lee, N.R.; Lee, N.J.; Lee, J.K.; Quan, F.S.; Inn, K.S. Human Respiratory Syncytial Virus NS 1 Targets TRIM25 to Suppress RIG-I Ubiquitination and Subsequent RIG-I-Mediated Antiviral Signaling. *Viruses* **2018**, *10*. [CrossRef]

89. Sun, L.; Xing, Y.; Chen, X.; Zheng, Y.; Yang, Y.; Nichols, D.B.; Clementz, M.A.; Banach, B.S.; Li, K.; Baker, S.C.; et al. Coronavirus papain-like proteases negatively regulate antiviral innate immune response through disruption of STING-mediated signaling. *PLoS ONE* **2012**, *7*, e30802. [CrossRef]

90. Gori Savellini, G.; Anichini, G.; Gandolfo, C.; Prathyumnan, S.; Cusi, M.G. Toscana virus non-structural protein NSs acts as E3 ubiquitin ligase promoting RIG-I degradation. *Plos Pathog.* **2019**, *15*, e1008186. [CrossRef]

91. Gupta, S.; Yla-Anttila, P.; Callegari, S.; Tsai, M.H.; Delecluse, H.J.; Masucci, M.G. Herpesvirus deconjugases inhibit the IFN response by promoting TRIM25 autoubiquitination and functional inactivation of the RIG-I signalosome. *Plos Pathog.* **2018**, *14*, e1006852. [CrossRef] [PubMed]

92. Chiang, C.; Pauli, E.K.; Biryukov, J.; Feister, K.F.; Meng, M.; White, E.A.; Munger, K.; Howley, P.M.; Meyers, C.; Gack, M.U. The Human Papillomavirus E6 Oncoprotein Targets USP15 and TRIM25 To Suppress RIG-I-Mediated Innate Immune Signaling. *J. Virol.* **2018**, *92*. [CrossRef] [PubMed]

93. Liu, B.; Zhang, M.; Chu, H.; Zhang, H.; Wu, H.; Song, G.; Wang, P.; Zhao, K.; Hou, J.; Wang, X.; et al. The ubiquitin E3 ligase TRIM31 promotes aggregation and activation of the signaling adaptor MAVS through Lys63-linked polyubiquitination. *Nat. Immunol.* **2017**, *18*, 214–224. [CrossRef] [PubMed]

94. Xue, B.; Li, H.; Guo, M.; Wang, J.; Xu, Y.; Zou, X.; Deng, R.; Li, G.; Zhu, H. TRIM21 Promotes Innate Immune Response to RNA Viral Infection through Lys27-Linked Polyubiquitination of MAVS. *J. Virol.* **2018**, *92*. [CrossRef] [PubMed]

95. Liu, C.; Huang, S.; Wang, X.; Wen, M.; Zheng, J.; Wang, W.; Fu, Y.; Tian, S.; Li, L.; Li, Z.; et al. The Otubain YOD1 Suppresses Aggregation and Activation of the Signaling Adaptor MAVS through Lys63-Linked Deubiquitination. *J Immunol* **2019**, *202*, 2957–2970. [CrossRef]

96. Li, W.; Li, N.; Dai, S.; Hou, G.; Guo, K.; Chen, X.; Yi, C.; Liu, W.; Deng, F.; Wu, Y.; et al. Zika virus circumvents host innate immunity by targeting the adaptor proteins MAVS and MITA. *Faseb J. Off. Publ. Fed. Am. Soc. Exp. Biol.* **2019**, *33*, 9929–9944. [CrossRef]

97. Yu, Y.; Hayward, G.S. The ubiquitin E3 ligase RAUL negatively regulates type i interferon through ubiquitination of the transcription factors IRF7 and IRF3. *Immunity* **2010**, *33*, 863–877. [CrossRef] [PubMed]

98. Wang, P.; Zhao, W.; Zhao, K.; Zhang, L.; Gao, C. TRIM26 negatively regulates interferon-beta production and antiviral response through polyubiquitination and degradation of nuclear IRF3. *Plos Pathog.* **2015**, *11*, e1004726. [CrossRef] [PubMed]

99. Ghosh, S.; Dass, J.F.P. Study of pathway cross-talk interactions with NF-kappaB leading to its activation via ubiquitination or phosphorylation: A brief review. *Gene* **2016**, *584*, 97–109. [CrossRef] [PubMed]

100. Courtois, G.; Fauvarque, M.O. The Many Roles of Ubiquitin in NF-kappaB Signaling. *Biomedicines* **2018**, *6*. [CrossRef]

101. Davis, K.A.; Morelli, M.; Patton, J.T. Rotavirus NSP1 Requires Casein Kinase II-Mediated Phosphorylation for Hijacking of Cullin-RING Ligases. *mBio* **2017**, *8*. [CrossRef] [PubMed]

102. Whitmer, T.; Malouli, D.; Uebelhoer, L.S.; DeFilippis, V.R.; Fruh, K.; Verweij, M.C. The ORF61 Protein Encoded by Simian Varicella Virus and Varicella-Zoster Virus Inhibits NF-kappaB Signaling by Interfering with IkappaBalpha Degradation. *J. Virol.* **2015**, *89*, 8687–8700. [CrossRef] [PubMed]

103. Ye, R.; Su, C.; Xu, H.; Zheng, C. Herpes Simplex Virus 1 Ubiquitin-Specific Protease UL36 Abrogates NF-kappaB Activation in DNA Sensing Signal Pathway. *J. Virol.* **2017**, *91*. [CrossRef]

104. Shibata, Y.; Tokunaga, F.; Goto, E.; Komatsu, G.; Gohda, J.; Saeki, Y.; Tanaka, K.; Takahashi, H.; Sawasaki, T.; Inoue, S.; et al. HTLV-1 Tax Induces Formation of the Active Macromolecular IKK Complex by Generating Lys63- and Met1-Linked Hybrid Polyubiquitin Chains. *Plos Pathog.* **2017**, *13*, e1006162. [CrossRef]

105. Morris, R.; Kershaw, N.J.; Babon, J.J. The molecular details of cytokine signaling via the JAK/STAT pathway. *Protein Sci. A Publ. Protein Soc.* **2018**, *27*, 1984–2009. [CrossRef]

106. Goraya, M.U.; Zaighum, F.; Sajjad, N.; Anjum, F.R.; Sakhawat, I.; Rahman, S.U. Web of interferon stimulated antiviral factors to control the influenza A viruses replication. *Microb. Pathog.* **2020**, *139*, 103919. [CrossRef]

107. Xia, C.; Vijayan, M.; Pritzl, C.J.; Fuchs, S.Y.; McDermott, A.B.; Hahm, B. Hemagglutinin of Influenza A Virus Antagonizes Type I Interferon (IFN) Responses by Inducing Degradation of Type I IFN Receptor 1. *J. Virol.* **2015**, *90*, 2403–2417. [CrossRef]

108. Chen, J.; Xu, W.; Chen, Y.; Xie, X.; Zhang, Y.; Ma, C.; Yang, Q.; Han, Y.; Zhu, C.; Xiong, Y.; et al. Matrix Metalloproteinase 9 Facilitates Hepatitis B Virus Replication through Binding with Type I Interferon (IFN) Receptor 1 To Repress IFN/JAK/STAT Signaling. *J. Virol.* **2017**, *91*. [CrossRef]

109. Elliott, J.; Lynch, O.T.; Suessmuth, Y.; Qian, P.; Boyd, C.R.; Burrows, J.F.; Buick, R.; Stevenson, N.J.; Touzelet, O.; Gadina, M.; et al. Respiratory syncytial virus NS1 protein degrades STAT2 by using the Elongin-Cullin E3 ligase. *J. Virol.* **2007**, *81*, 3428–3436. [CrossRef]

110. Morrison, J.; Laurent-Rolle, M.; Maestre, A.M.; Rajsbaum, R.; Pisanelli, G.; Simon, V.; Mulder, L.C.; Fernandez-Sesma, A.; Garcia-Sastre, A. Dengue virus co-opts UBR4 to degrade STAT2 and antagonize type I interferon signaling. *Plos Pathog.* **2013**, *9*, e1003265. [CrossRef]

111. Grant, A.; Ponia, S.S.; Tripathi, S.; Balasubramaniam, V.; Miorin, L.; Sourisseau, M.; Schwarz, M.C.; Sanchez-Seco, M.P.; Evans, M.J.; Best, S.M.; et al. Zika Virus Targets Human STAT2 to Inhibit Type I Interferon Signaling. *Cell Host Microbe* **2016**, *19*, 882–890. [CrossRef] [PubMed]

112. Ulane, C.M.; Kentsis, A.; Cruz, C.D.; Parisien, J.P.; Schneider, K.L.; Horvath, C.M. Composition and assembly of STAT-targeting ubiquitin ligase complexes: Paramyxovirus V protein carboxyl terminus is an oligomerization domain. *J. Virol.* **2005**, *79*, 10180–10189. [CrossRef] [PubMed]

113. Zhou, X.; Jiang, W.; Liu, Z.; Liu, S.; Liang, X. Virus Infection and Death Receptor-Mediated Apoptosis. *Viruses* **2017**, *9*. [CrossRef] [PubMed]

114. Scheffner, M.; Huibregtse, J.M.; Vierstra, R.D.; Howley, P.M. The HPV-16 E6 and E6-AP complex functions as a ubiquitin-protein ligase in the ubiquitination of p53. *Cell* **1993**, *75*, 495–505. [CrossRef]

115. Li, S.; Hong, X.; Wei, Z.; Xie, M.; Li, W.; Liu, G.; Guo, H.; Yang, J.; Wei, W.; Zhang, S. Ubiquitination of the HPV Oncoprotein E6 Is Critical for E6/E6AP-Mediated p53 Degradation. *Front. Microbiol.* **2019**, *10*, 2483. [CrossRef]

116. Zhang, H.; Huang, C.; Wang, Y.; Lu, Z.; Zhuang, N.; Zhao, D.; He, J.; Shi, L. Hepatitis B Virus X Protein Sensitizes TRAIL-Induced Hepatocyte Apoptosis by Inhibiting the E3 Ubiquitin Ligase A20. *PLoS ONE* **2015**, *10*, e0127329. [CrossRef]

117. Lallemand-Breitenbach, V.; de The, H. PML nuclear bodies: From architecture to function. *Curr. Opin. Cell Biol.* **2018**, *52*, 154–161. [CrossRef]

118. Gu, H.; Zheng, Y. Role of ND10 nuclear bodies in the chromatin repression of HSV-1. *Virol. J.* **2016**, *13*, 62. [CrossRef]

119. Scherer, M.; Stamminger, T. Emerging Role of PML Nuclear Bodies in Innate Immune Signaling. *J. Virol.* **2016**, *90*, 5850–5854. [CrossRef]

120. Chelbi-Alix, M.K.; de The, H. Herpes virus induced proteasome-dependent degradation of the nuclear bodies-associated PML and Sp100 proteins. *Oncogene* **1999**, *18*, 935–941. [CrossRef]

121. Zheng, Y.; Samrat, S.K.; Gu, H. A Tale of Two PMLs: Elements Regulating a Differential Substrate Recognition by the ICP0 E3 Ubiquitin Ligase of Herpes Simplex Virus 1. *J. Virol.* **2016**, *90*, 10875–10885. [CrossRef] [PubMed]

122. Jan Fada, B.; Kaadi, E.; Samrat, S.; Zheng, Y.; Gu, H. Effect of SUMO-SIM interaction on the ICP0-mediated degradation of PML isoform II and its associated proteins in HSV-1 infection. *J. Virol.* **2020**, *94*, e00470-20. [CrossRef] [PubMed]

123. Izumiya, Y.; Kobayashi, K.; Kim, K.Y.; Pochampalli, M.; Izumiya, C.; Shevchenko, B.; Wang, D.H.; Huerta, S.B.; Martinez, A.; Campbell, M.; et al. Kaposi's sarcoma-associated herpesvirus K-Rta exhibits SUMO-targeting ubiquitin ligase (STUbL) like activity and is essential for viral reactivation. *Plos Pathog.* **2013**, *9*, e1003506. [CrossRef] [PubMed]

124. Harris, R.S.; Dudley, J.P. APOBECs and virus restriction. *Virology* **2015**, *479–480*, 131–145. [CrossRef] [PubMed]

125. Yu, X.; Yu, Y.; Liu, B.; Luo, K.; Kong, W.; Mao, P.; Yu, X.F. Induction of APOBEC3G ubiquitination and degradation by an HIV-1 Vif-Cul5-SCF complex. *Science* **2003**, *302*, 1056–1060. [CrossRef]

126. Sheehy, A.M.; Gaddis, N.C.; Choi, J.D.; Malim, M.H. Isolation of a human gene that inhibits HIV-1 infection and is suppressed by the viral Vif protein. *Nature* **2002**, *418*, 646–650. [CrossRef]

127. Gao, Y.; Feng, J.; Yang, G.; Zhang, S.; Liu, Y.; Bu, Y.; Sun, M.; Zhao, M.; Chen, F.; Zhang, W.; et al. Hepatitis B virus X protein-elevated MSL2 modulates hepatitis B virus covalently closed circular DNA by inducing degradation of APOBEC3B to enhance hepatocarcinogenesis. *Hepatology* **2017**, *66*, 1413–1429. [CrossRef]

128. Kirui, J.; Mondal, A.; Mehle, A. Ubiquitination Upregulates Influenza Virus Polymerase Function. *J. Virol.* **2016**, *90*, 10906–10914. [CrossRef]

129. Karim, M.; Biquand, E.; Declercq, M.; Jacob, Y.; van der Werf, S.; Demeret, C. Nonproteolytic K29-Linked Ubiquitination of the PB2 Replication Protein of Influenza A Viruses by Proviral Cullin 4-Based E3 Ligases. *mBio* **2020**, *11*. [CrossRef]

130. Bharaj, P.; Atkins, C.; Luthra, P.; Giraldo, M.I.; Dawes, B.E.; Miorin, L.; Johnson, J.R.; Krogan, N.J.; Basler, C.F.; Freiberg, A.N.; et al. The Host E3-Ubiquitin Ligase TRIM6 Ubiquitinates the Ebola Virus VP35 Protein and Promotes Virus Replication. *J. Virol.* **2017**, *91*. [CrossRef]

131. Laspia, M.F.; Rice, A.P.; Mathews, M.B. HIV-1 Tat protein increases transcriptional initiation and stabilizes elongation. *Cell* **1989**, *59*, 283–292. [CrossRef]

132. Faust, T.B.; Li, Y.; Bacon, C.W.; Jang, G.M.; Weiss, A.; Jayaraman, B.; Newton, B.W.; Krogan, N.J.; D'Orso, I.; Frankel, A.D. The HIV-1 Tat protein recruits a ubiquitin ligase to reorganize the 7SK snRNP for transcriptional activation. *eLife* **2018**, *7*. [CrossRef]
133. Ahmed, I.; Akram, Z.; Iqbal, H.M.N.; Munn, A.L. The regulation of Endosomal Sorting Complex Required for Transport and accessory proteins in multivesicular body sorting and enveloped viral budding-An overview. *Int. J. Biol. Macromol.* **2019**, *127*, 1–11. [CrossRef] [PubMed]
134. Strack, B.; Calistri, A.; Accola, M.A.; Palu, G.; Gottlinger, H.G. A role for ubiquitin ligase recruitment in retrovirus release. *Proc. Natl. Acad. Sci. USA* **2000**, *97*, 13063–13068. [CrossRef]
135. Harty, R.N.; Brown, M.E.; Wang, G.; Huibregtse, J.; Hayes, F.P. A PPxY motif within the VP40 protein of Ebola virus interacts physically and functionally with a ubiquitin ligase: Implications for filovirus budding. *Proc. Natl. Acad. Sci. USA* **2000**, *97*, 13871–13876. [CrossRef]
136. Kikonyogo, A.; Bouamr, F.; Vana, M.L.; Xiang, Y.; Aiyar, A.; Carter, C.; Leis, J. Proteins related to the Nedd4 family of ubiquitin protein ligases interact with the L domain of Rous sarcoma virus and are required for gag budding from cells. *Proc. Natl. Acad. Sci. USA* **2001**, *98*, 11199–11204. [CrossRef] [PubMed]
137. Weiss, E.R.; Popova, E.; Yamanaka, H.; Kim, H.C.; Huibregtse, J.M.; Gottlinger, H. Rescue of HIV-1 release by targeting widely divergent NEDD4-type ubiquitin ligases and isolated catalytic HECT domains to Gag. *Plos Pathog.* **2010**, *6*, e1001107. [CrossRef]
138. Mercenne, G.; Alam, S.L.; Arii, J.; Lalonde, M.S.; Sundquist, W.I. Angiomotin functions in HIV-1 assembly and budding. *eLife* **2015**, *4*. [CrossRef]

International Journal of
Molecular Sciences

Review

The Role of Tissue-Specific Ubiquitin Ligases, RNF183, RNF186, RNF182 and RNF152, in Disease and Biological Function

Takumi Okamoto [†], Kazunori Imaizumi and Masayuki Kaneko [*,†]

Department of Biochemistry, Graduate School of Biomedical and Health Sciences, Hiroshima University, 1-2-3 Kasumi, Minami-ku, Hiroshima 734-8553, Japan; toka@hiroshima-u.ac.jp (T.O.); imaizumi@hiroshima-u.ac.jp (K.I.)
* Correspondence: mkaneko@hiroshima-u.ac.jp; Tel.: +81-82-257-5131
† These authors contributed equally to this work.

Received: 8 May 2020; Accepted: 28 May 2020; Published: 30 May 2020

Abstract: Ubiquitylation plays multiple roles not only in proteasome-mediated protein degradation but also in various other cellular processes including DNA repair, signal transduction, and endocytosis. Ubiquitylation is mediated by ubiquitin ligases, which are predicted to be encoded by more than 600 genes in humans. RING finger (RNF) proteins form the majority of these ubiquitin ligases. It has also been predicted that there are 49 RNF proteins containing transmembrane regions in humans, several of which are specifically localized to membrane compartments in the secretory and endocytic pathways. Of these, *RNF183*, *RNF186*, *RNF182*, and *RNF152* are closely related genes with high homology. These genes share a unique common feature of exhibiting tissue-specific expression patterns, such as in the kidney, nervous system, and colon. The products of these genes are also reported to be involved in various diseases such as cancers, inflammatory bowel disease, Alzheimer's disease, and chronic kidney disease, and in various biological functions such as apoptosis, endoplasmic reticulum stress, osmotic stress, nuclear factor-kappa B (NF-κB), mammalian target of rapamycin (mTOR), and Notch signaling. This review summarizes the current knowledge of these tissue-specific ubiquitin ligases, focusing on their physiological roles and significance in diseases.

Keywords: RNF183; RNF186; RNF182; RNF152; ubiquitin ligase; RING finger; mTOR; NF-κB; endoplasmic reticulum stress; osmotic stress

---

1. Introduction

Ubiquitin (Ub) is a 76-amino-acid protein that is highly conserved among all eukaryotes. Ubiquitylation, which involves the conjugation of ubiquitin to the lysine residues of various cellular proteins, is one of the most prevalent post-translational modifications of proteins, and is usually catalyzed by a three-enzyme cascade consisting of Ub-activating enzymes (E1s), Ub-conjugating enzymes (E2s), and Ub ligases (E3s). In mammals, there are 10 or fewer E1 activating enzymes, dozens of E2 conjugating enzymes, and hundreds of E3 ligases; these enzymes regulate the ubiquitylation of numerous proteins [1]. In ubiquitylation, E1 initially activates ubiquitin by adenylating it at the C-terminal glycine residue in an adenosine triphosphate-dependent process; this activated ubiquitin is then captured by the catalytic cysteine of the E1, forming a thioester intermediate. Then, the thioester ubiquitin is transferred from the enzyme active site of E1 to the catalytic center cysteine residue of E2 via a trans-thioesterification reaction [2]. Finally, E3 mediates the transfer of ubiquitin from E2 to a substrate protein. Both the efficiency and the substrate specificity of the ubiquitylation reaction depend on E3 ligases. Depending on the mechanism by which ubiquitin is transferred from E2 to the substrate, E3 is classified into three broad families: Really Interesting New Gene (RING) finger domain-,

Homologous to E6-associated protein C Terminus (HECT) domain-, or RING Between RING (RBR) domain-containing ubiquitin ligases. While RING E3 ligases, the major family among them, facilitate the direct transfer of ubiquitin from E2–ubiquitin intermediates to the substrate protein [3], HECT and RBR E3 ligases contain an active-site cysteine that forms a thioester with ubiquitin before transferring it to the substrate protein [4,5]. The selective pairing between E2 and the multiple cognate E3s confers the specificity necessary for the regulation by ubiquitylation of various biological pathways.

Ubiquitin is conjugated with a lysine residue of a substrate via its C-terminal carboxyl group and can also attach itself via the N-terminal methionine (M1) and seven lysine residues (K6, K11, K27, K29, K33, K48, and K63). As a result, a substrate protein is modified with a single monoubiquitin, multiple monoubiquitins, or polyubiquitin chain. Such ubiquitin attachments can be reversed in the process of deubiquitylation by deubiquitinase (DUB) [6]. Owing to the reversible nature of this modification, the ubiquitin pool of cells is divided into different fractions, including free monoubiquitins, covalently linked mono- and polyubiquitin–protein complexes, and unanchored polyubiquitin. These different linkage types and lengths affect substrate proteins in different biological and biochemical ways and play an essential role in regulating a considerable number of significant cellular functions (e.g., protein degradation, endocytosis of membrane proteins, transcriptional control, DNA repair, and cell cycle regulation) [7]. The principal and abundant forms are K48-linked and K63-linked polyubiquitin chains. K48-linked polyubiquitin functions as a signal of proteasomal degradation, whereas K63-linked polyubiquitin chains have non-degradative roles in cellular signaling, intracellular trafficking, the DNA damage response, and other contexts [8,9].

In this review, we focus on E3 ligases, *RNF183*, *RNF186*, *RNF182*, and *RNF152*, which are closely related genes encoding a RING-finger domain (C3HC4) at its N-terminus and one or two predicted transmembrane domains at its C-terminus with high homology (Figure 1) [10]. As common features, these E3s are expressed in specific tissues, such as the kidney, nervous system, and colon, and are localized in the lysosome (Table 1) [10]. Hereinafter, these ubiquitin ligases are referred to as the RNF183 family. In this review, we summarize our current understanding of the molecular mechanisms underlying the functions and regulation of these E3s in diseases.

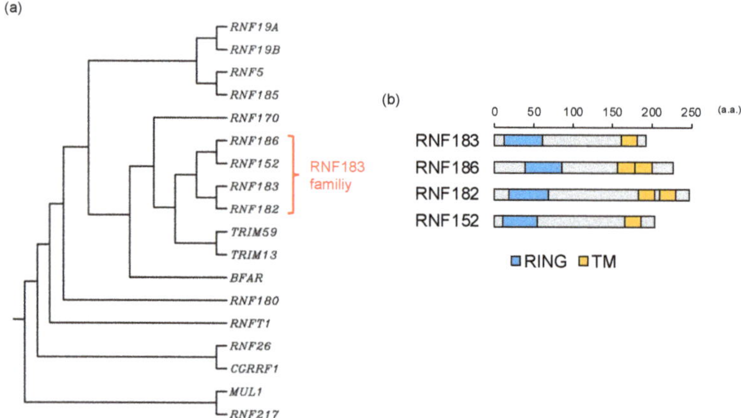

Figure 1. RNF183 family: (**a**) The phylogenetic tree for C3H2C3-RING E3s with transmembrane. Protein sequences for E3s were aligned with a multiple sequence alignment using the CLUSTALW (http://www.genome.jp/tools/clustalw); (**b**) The comparison of the domain structures of the RNF183 family. Information on the domain structure of RNF183 family protein was obtained from UniProt (https://www.uniprot.org) for RNF183 (Q96D59), RNF186 (Q9NXI6), RNF182 (Q8N6D2), and RNF152 (Q8N8N0). RING, C3H2C3-RING domain; TM, transmembrane domain; a.a., amino acids.

Table 1. Features of RNF183 family.

Gene	Cellular Localization	Expressing Tissue	Induction Mechanism	Substrate Protein	Associated Signaling Pathway	Types of Ubiquitin Chain	E2	Associated Disease/Biological Function
RNF183	ER, Golgi, lysosome [11–13]	kidney, testis [10]; renal medullary collecting duct [13]	prolonged ER stress [11]; NFAT5 [13,14]	BIK [15]; BNIP3L [16]; Bcl-xL [11]; IκBα [17]; DR5 [18]; Na,K-ATPase β1 subunit [19]	apoptosis [11,14–16,18]; NF-κB [17,20,21]	K48 [11,15–17]; K63 [18,19]	Ubc5c (in vitro) [11]; UbcH5c (in vitro) [12]	IBD [17,18]; endometrial carcinoma [22]; colorectal cancer [15,16,20]; Ewing Sarcoma [21]
RNF186	ER [23]; lysosome? [24]	lower gastrointestinal tract, kidney [10]		BNip1 [23]; Occludin [25]; Sestrin-2 [24]	apoptosis [23,26]; mTORC1 [24]	K29 [23]; K48 [24,25]; K63 [23,24]		IBD [25,27–33]; CKD [34]
RNF182	lysosome [35]	nervous system (cortex, hippocampus, cerebellum, spinal cord) [10,35]	oxygen and glucose deprivation [35]; MeCP2 mutation [36]; ischemia-reperfusion injury [37]; TLR stimuli [38]	ATP6V0C [35]; NF-κB p65 subunit [38]	apoptosis [35]; mTORC1 [37]; NF-κB [38]	K48 [35,38]	Ubc5a (in vitro) [35]	AD [35]; Rett syndrome [36]; colorectal cancer [39]; myocardial ischemia [37]
RNF152	lysosome [40]	kidney [10]; eyes, neural tube [41]; floor plate [42]	FoxA2 [42]	RagA [43]; Rheb [44]	apoptosis [40]; mTORC1 [42–44]; Notch [41]	K48 [40]; K63 [43]; mono [44]	UBC13 (in vivo) [43]; Ubc5a (in vitro) [40]	breast and prostate cancer [45]; colorectal cancer [44,46–48]; development of the eyes, midbrain and hindbrain (zebrafish) [41]; proliferation of floor plate cells [42]

2. RNF183

The E3 ubiquitin ligase RNF183 has been identified as a new biomarker of endometrial carcinoma (EC) via gene expression screening and protein level experiments on carcinoma samples. Furthermore, the differential expression of RNF183 in primary endometrial tumors has been shown to be correlated with its expression level in corresponding uterine fluid samples and it exhibits an analogous value in the initial stage of EC [22]. EC is the most common invasive tumor of the female genital tract, which is usually detected in its initial stages. However, 20% of patients are at an advanced stage at the time of detection. Because molecular markers for the diagnosis of EC have yet to be validated, new methods for the medical prognostication and classification of EC are needed to combat this deadly disease. RNF183 could be helpful as a precise molecular tool to diagnose EC and reduce unnecessary biopsies.

Another study has indicated that RNF183 interacts with fetal and adult testis-expressed 1 (FATE1) in tumors and negatively regulates the apoptosis effector Bcl-2-interacting killer (BIK), leading to increased viability of tumor cells [15]. FATE1 is one of the cancer/testis antigens whose expression is biased to the testes but is also activated in cancer [49,50]. Depletion of FATE1 reduces the viability of cancer cells. Large-scale proteomic studies have revealed that BIK is a FATE1-interacting partner. At the same time, RNF183 has also been revealed to be a FATE1-interacting partner. BIK associates with both FATE1 and RNF183, and both RNF183 depletion and the mutant RNF183, which exhibits the loss of enzyme activity, increase BIK protein accumulation. Thus, FATE1 and RNF183 collaborate to suppress BIK protein levels and escape from BIK-related apoptotic signaling [15]. However, in Ewing sarcoma cells, no appreciable levels of BIK protein are detectable even in the presence of the proteasome inhibitor MG132, and FATE1 depletion does not induce BIK accumulation.

There is another context-selective mechanism in Ewing sarcoma. FATE1 is most robustly induced by the Ewing sarcoma breakpoint region 1-Friend Leukemia Integration 1 (EWSR1-FLI1) chimeric transcription factor caused by a pathognomonic chromosomal translocation of Ewing sarcoma and interacts with Bcl-2/adenovirus E1B 19 kDa protein-interacting protein 3-like (BNIP3L). Then, BNIP3L is degraded in the presence of RNF183 [16]. Because BNIP3L is a tumor suppressor [51], its depletion increases tumorigenesis in vivo [16].

Moreover, *RNF183* has been identified as a gene conferring resistance to trametinib. Trametinib is one of the anticancer drugs inhibiting MEK1/2 [52]. RNF183 expression is increased after trametinib treatment, which in turn activates the NF-κB pathway. Then, the activated NF-κB increases the expression of the pro-inflammatory cytokine interleukin-8 (IL-8), which is a downstream target of NF-κB [20]. IL-8 signaling increases the proliferation and survival of cancer cells and potentiates their migration [53]. Thus, RNF183 confers resistance to trametinib on colorectal cancer (CRC) cells and promotes their proliferation and metastasis [20].

Under physiological conditions, RNF183 is not expressed in the large intestine, but is specifically expressed in the kidney [10]. The abnormal expression of RNF183 is thought to be involved in several diseases, not only tumorigenesis but also inflammatory conditions such as inflammatory bowel disease (IBD), including Crohn's disease (CD) and ulcerative colitis (UC), which is a chronic, idiopathic, inflammatory, gastrointestinal disease, the molecular mechanism underlying the development and pathophysiology of which have not been fully elucidated. [15–17,20–22]. However, FATE1 is not expressed in the intestine. Therefore, there is a FATE1-independent inflammatory mechanism involving RNF183 in the large intestine. In fact, some studies have shown that RNF183 is upregulated in colon samples of the intestinal tissues of IBD patients [17] and the colons of mice with colitis treated with trinitrobenzene sulfonic acid (TNBS) or dextran sulfate sodium (DSS) [17,18].

It has been reported that RNF183 is largely involved in executing apoptosis in response to prolonged ER stress. It is considered that the mechanism of apoptosis involving RNF183 features the ubiquitylation and degradation of B-cell lymphoma extra-large (Bcl-xL), which functions as an inhibitor of apoptosis by preventing cytochrome c release [11]. Bcl-xL is usually localized to the mitochondria [54], whereas RNF183 is predominantly localized to the ER, Golgi, and lysosome [12]. Some Bcl-xL may be targeted to the ER [55], where it is in the vicinity of RNF183. Then, since their cytosolic domains can interact with each other, they interact directly and RNF183 ubiquitylates Bcl-xL [11]. The detailed mechanism behind this involves inositol requiring 1α (IRE1α) being activated by prolonged ER stress and readily decreasing microRNA-7 (miR-7) and microRNA-96 (miR-96), presumably by the digestion of miR precursors through the IRE1-dependent decay of mRNA [56,57]. Since miR-7 and miR-96 negatively regulate RNF183 by directly interacting with its 3'-UTR [17], their decrease eventually stabilizes the RNF183 mRNA and leads to increased protein levels. This increase in RNF183 in turn promotes its binding to Bcl-xL, polyubiquitylation, and subsequent degradation. The gradual decrease in Bcl-xL levels eventually triggers the intrinsic apoptotic pathway [11]. It has also been reported that increased RNF183 due to decreased miR-7 may contribute to the pathogenesis of IBD by recognizing NF-κB inhibitor α (IκBα), not Bcl-xL, as a substrate and degrading ubiquitylated IκBα [17]. Because IκBα is a suppressor of NF-κB, the reduction of IκBα by ubiquitylation and degradation induces NF-κB activation.

Recently, another mechanism of RNF183-related IBD pathogenesis has also been reported. Specifically, RNF183 recognizes DR5 as a substrate protein and K63-ubiquitylated DR5 is transported to lysosomes for degradation. In addition, RNF183 promotes TRAIL-induced caspase activation and apoptosis, providing new insights into the potential roles of RNF183 in DR5-mediated caspase activation in the pathogenesis of IBD [18]. RNF183-mediated ubiquitylation of substrates, Bcl-xL, IκBα, and DR5, and the negative regulation of RNF183 by miR-7 may be important novel epigenetic mechanisms in the pathogenesis of IBD.

In human and mouse tissues, RNF183 is specifically expressed in the kidney [10]. In particular, high Rnf183 expression in the renal medullary collecting duct has been reported from a tissue analysis using GFP-knock-in mice [13]. The kidney is the only tissue that is continuously under hypertonic conditions, and this hypertonicity gradually increases from the outer medulla down to the inner medulla. Nuclear factor of activated T cells 5 (NFAT5)/tonicity-responsive enhancer-binding protein is a transcription factor essential for the adaptation to hypertonic conditions, under which it stimulates the transcription of some genes [58]. The Rnf183 gene is also downstream of NFAT5 [14]. Indeed, the expression of Rnf183 in the renal medulla is dramatically decreased upon treatment with the loop diuretic furosemide, which can downregulate NFAT5 levels by inhibiting the Na-K-Cl cotransporter type 2 (NKCC2) and inducing hypotonicity in the medulla [13]. This is consistent with the decrease in NFAT5 protein and the mRNA expression of its target gene. Additionally, Rnf183 expression increases markedly in mouse inner-medullary collecting duct (mIMCD-3) cells treated with hypertonic NaCl. Rnf183, as well as several NFAT5 downstream genes, protects renal medullary cells from hypertonicity-induced apoptosis. mIMCD-3 cells transfected with siRNA targeting Rnf183 exhibit significant increases in cleaved caspase-3 protein levels. Therefore, Rnf183 expression is involved in the osmotic tolerance of mIMCD-3 cells [14].

In terms of its subcellular localization, RNF183 is predominantly localized to the endoplasmic reticulum (ER), Golgi apparatus and lysosome. Its stability depends on its interaction with Sec16A, which is involved in the formation of coat protein complex II (COPII) vesicles. However, Sec16A is not ubiquitylated by RNF183 [12]. Recently, it has been identified that the Na,K-ATPase β1 subunit, which forms a complex with Na,K-ATPase α1 subunit on the plasma membrane, is one of the substrates for RNF183. Na,K-ATPase contributes to the regulation of cell volume and solute absorption by the active transport of Na$^+$ and K$^+$ across the plasma membrane [59], and the expression of both α1 and β1 subunits is increased for adapting hypertonic condition in human renal cells [60]. RNF183 ubiquitylates only the β1 subunit, not the α1 subunit. Then, a complex with α1 and β1 subunits translocates from the plasma membrane to the lysosome, where it is degraded [19]. Therefore, RNF183 may play an important role in the kidney for the adaptation to hyperosmotic stress by regulating the level of Na,K-ATPase.

As described above, the kidney-specific ubiquitin ligase RNF183 protects cells from apoptosis induced by hypertonic stress in the kidney, whereas its aberrant expression such as in the colon induces inflammatory and tumorigenesis. Understanding the function of RNF183 could lead to new therapeutic strategies for patients with IBD and various types of cancer.

3. RNF186

A study of the distribution of RNF186 has indicated that its expression is highest in the lower gastrointestinal tract in both human and mouse tissues [10]. Initially, the RNF186 gene was identified as a locus associated with the risk of UC by several genome-wide association studies (GWAS) [27–30]. However, its function has only recently been clarified. Recently, deep resequencing of GWAS loci identified a causal variant in RNF186 that encodes an alanine-to-threonine substitution at position 64 [31]. This variant confers risk for the development of UC. The A64T variant is located in the RING-finger domain of RNF186. Since the RING-finger domain possesses E3 ubiquitin-protein ligase activity, there is the possibility that the A64T variant exhibits a reduction or complete loss of enzyme activity. A report indicates that RNF186 is thus a candidate for an association with the development of chronic inflammation in the intestine of UC. In addition, a recent study focusing on protein-truncating variants identified a novel genetic variant in RNF186, R179X [32]. This variant contributes to protection against UC. The R179X truncation is expressed at reduced levels and is diffusely localized, not in the ER, but preferentially in the plasma membrane [32]. Studying the specific effects of this variant, which has not yet been performed, should be useful for understanding the mechanism by which it protects against UC. In humans, RNF186 is also expressed in the kidney [10]. Interestingly, R179X truncation, which confers protection against UC, has also been identified as a factor associated with an increased risk of chronic kidney disease (CKD) [34].

Several RNF186 substrates have been identified, for example, Bcl2/adenovirus E1B 19-kDa interacting protein 1 (BNip1), occludin, and Sestrin-2 [23–25]. BNip1, a Bcl-2 family protein, co-localizes with and binds to RNF186 in the ER. RNF186 conjugates polyubiquitin through K29 and K63 linkage to BNip1 and the ubiquitylated BNip1 is transferred from the ER to the mitochondria [23]. Because BNip1 can induce a moderate level of apoptosis [61], RNF186 may regulate ER stress-induced apoptosis by the ubiquitylation of BNip1. The overexpression of RNF186 upregulates the critical regulators in the unfolded protein response (UPR), such as chaperone protein BiP/GRP78 and pro-apoptotic transcription factor CHOP/GADD153. Additionally, RNF186 triggers caspase-12 activation, which plays a role in inducing apoptosis in ER stress. Ca^{2+} is a secondary messenger of ER stress, and RNF186 actually promotes Ca^{2+} leakage from the ER [23]. Moreover, the expression of Rnf186 has been shown to be induced in the livers of mice with diabetes, obesity, and diet-induced obesity. In mouse primary hepatocytes, the overexpression of Rnf186 increases the protein levels of the ER stress markers, IRE1 and CHOP, as well as the level of eIF2α phosphorylation, and the treatment with tauroursodeoxycholic acid, the inhibitor of ER stress, decreases the expression of ER stress markers. It has been indicated that the overexpression of Rnf186 induces ER stress. These effects interfere with insulin action through the phosphorylation of insulin receptor substrate 1 (IRS1) by c-Jun N-terminal kinase (JNK). Furthermore, it has also been shown that the expression levels of proinflammatory cytokines, including TNF-α, IL-6, and MCP1, are increased by overexpressing Rnf186. Thus, Rnf186 may be a novel therapeutic target for the treatment of metabolic diseases associated with insulin resistance [26].

Occludin is one of the tight junction proteins, which control gastrointestinal tract permeability. In the colonic epithelia of mice with Rnf186 knockout, the expression of occludin protein is increased and its distribution is changed, resulting in increased permeability of small organic molecules. Rnf186 ubiquitylates occludin through K48-linked chains. Because K48-linked polyubiquitylation contributes to regulating protein turnover through proteasomal degradation, it is consistent with the increased expression of occludin in colonic epithelia of Rnf186 knockout mice. Disordered protein homeostasis in Rnf186 knockout mice correlates with enhanced ER stress in colonic epithelium and increased susceptibility to intestinal inflammation by DSS treatment. Therefore, it has been suggested that RNF186 is an E3 ubiquitin ligase controlling the homeostasis of occludin and ER stress in the colon [25].

Sestrin-2 is one of the intracellular amino acid sensors and functions as an inhibitor of mTORC1 signaling by interacting with GAP activity toward Rags 2 (GATOR2) [62]. This function occurs in a cytosolic leucine-dependent manner. When leucine is abundant in the cytosol, interference of the interaction between Sestrin-2 and GATOR2 occurs, allowing the activation of GATOR2. Because GATOR2 negatively regulates the GAP (GTPase-activating protein) activity of GATOR1, which inhibits mTORC1 activity, the decrease of Sestrin-2 indirectly activates mTOR1. Although Sestrin-2 has been shown to be transcriptionally controlled by several mechanisms, the regulation of the post-translational degradation of Sestrin-2 remains unclear. A recent study has shown that RNF186 and Sestrin-2 bind to each other via C-terminal motifs and RNF186 polyubiquitylated lysine residue at position 13 in Sestrin-2 with K48 linkage [24]. The ubiquitylated Sestrin-2 is degraded through the proteasome system. This is a new mechanism regulating mTORC1 activity through E3 ligases. In addition, Sestrin-2 also contains polyubiquitin chains with K63 linkages. Although this type has been shown to be involved in various processes, such as cellular trafficking, the role of the K63-linked ubiquitylation of Sestrin-2 remains unknown.

4. RNF182

RNF182 is the member of the RNF183 family on which the fewest reports have been published. It was initially identified as an upregulated gene in postmortem brains of patients with Alzheimer's disease (AD) [35]. RNF182 is selectively expressed in the nervous system, such as the cortex, hippocampus, cerebellum, and spinal cord, but not in the heart, liver, kidney, and skeletal muscle. RNF182 is expressed in differentiated Ntera2 (NT2) neurons and was shown to be upregulated by oxygen and glucose deprivation. RNF182 overexpression can initiate the death of neuronal cells. In addition, yeast two-hybrid screening revealed that V-type proton ATPase 16 kDa proteolipid subunit (ATP6V0C) is a substrate of RNF182. RNF182 promotes the degradation of ATP6V0C via the proteasome pathway. Because ATP6V0C is a key component of gap junctions and neurotransmitter release channels, the upregulation of RNF182 in AD brains might induce ATP6V0C degradation and contribute to the pathophysiology of AD.

Next, RNF182 was identified as a gene associated with Rett syndrome. Patients with Rett syndrome often exhibit heterozygous mutations in the methyl-CpG-binding protein 2 (MECP2) gene encoding a transcriptional modulator [63]. Gene expression profiles obtained by microarray analysis revealed that RNF182 is upregulated in mutant MeCP2 fibroblasts obtained from patients with Rett syndrome [36]. Since, as described above [35], RNF182 possibly compromises brain function, it is speculated to be involved in postnatal neurodevelopmental abnormalities associated with Rett syndrome.

Although RNF182 appears to be specifically expressed in the nervous system, its expression and function in other tissues were reported. Rnf182 was also shown to be upregulated in a rat model of myocardial ischemia–reperfusion injury (MIRI) [37]. Its silencing by shRNA was shown to reduce myocardial infarct size and myocardial cell apoptosis in this rat model. Intriguingly, RNF182, as well as RNF152 [43,44] and RNF186 [24], is associated with the mTOR signaling pathway, since Rnf182 silencing and treatment with phosphoesterase, an inhibitor of the mTOR signaling pathway, reverse the effect of Rnf182 silencing in the MIRI rat. Thus, it is likely that Rnf182 silencing activates the mTOR signaling pathway, resulting in improved ventricular remodeling after MIRI. However, the target substrate of RNF182 was not identified in this study.

The expression of RNF182 has been observed in immune tissues such as lymph nodes and spleen and in immune cells such as macrophages and dendritic cells [38], although the levels of expression were not compared with that in nerve tissues. Lipopolysaccharide, poly(I:C) and CpG, which is involved in Toll-like receptor (TLR) signaling, induced RNF182 expression. Moreover, RNF182 silencing can specifically suppress TLR-induced activation of NF-κB and the NF-κB-mediated production of proinflammatory cytokines. RNF182 mediates K48-linked polyubiquitylation of the subunit p65 of NF-κB and its degradation via the proteasome pathway in macrophages. Therefore, RNF182 may be upregulated by immune responses and involved in the feedback of inflammatory responses.

5. RNF152

In terms of studies on RNF152, it was first reported as having decreased expression in breast and prostate cancer cell lines [45]. Next, RNF152 was cloned and characterized as a ubiquitin ligase [40]. It is localized in lysosomes and its overexpression can mediate self-polyubiquitylation through K48 linkage and induce apoptosis. However, its substrates and function initially remained unknown.

For the first time among members of the RNF183 family, RNF152 has been reported to be involved in mTOR signaling [43]. The ubiquitylation of RagA, which belongs to the Rag family of small GTPases and recruits mTORC1 to the lysosomal membrane [62], is markedly increased in response to amino acid starvation. RNF152 was identified among some lysosomal ubiquitin ligases and ubiquitylated RagA in a K63-linked manner. Increased amino acid levels facilitate the interaction of RNF152 with RagA and RNF152-mediated RagA ubiquitylation. RNF152 selectively ubiquitylates the inactive form of RagA (RagA-GDP/RagC-GTP), but not the active form (RagA-GTP/RagC-GDP), suggesting that the interaction between RNF152 and RagA is regulated by the nucleotide-bound status of RagA. Furthermore, the authors carefully demonstrated that RNF152 acts as a negative regulator of amino-acid-induced mTORC1 activation. RNF152-mediated RagA ubiquitylation promotes the binding between RagA and GATOR1, a GAP complex for Rag GTPases [62], resulting in the inactivation of RagA. USP17L2/DUB3 (ubiquitin carboxyl-terminal hydrolase 17), a deubiquitylating enzyme, was identified for RagA ubiquitylated by RNF152. Moreover, RNF152 knockout revealed that the deficiency of RNF152 causes the hyperactivation of mTORC1 and inhibits amino-acid-starvation-induced autophagy.

Four years later, the same group demonstrated that RNF152 is involved in the ubiquitylation of Rheb, another mTORC1 signal molecule, in growth factor-induced mTORC1 activation, not amino acids [62]. Rheb acts as a small GTPase like RagA and activates mTORC1 in its GTP-bound form. RNF152 induces Rheb monoubiquitylation that enhances binding to TSC2, which is a core subunit of the TSC complex and functions as a GAP [62], resulting in mTORC1 inactivation. On the other hand, USP4 was identified as a DUB for Rheb [64]. USP4 promotes the activation of Rheb by removing ubiquitin from it. USP4 activity is regulated by the EGF-Akt-mediated phosphorylation at S445. Therefore, RNF152 and USP4 can regulate Rheb activity negatively and positively, respectively, downstream of the EGF pathway. Since Rheb ubiquitylation negatively regulates mTORC1 activation, RNF152 can regulate cellular autophagy positively and cell proliferation negatively. Studies of USP4 knockout mice revealed that USP4 upregulates tumor growth in an mTORC1-dependent manner. In contrast, TCGA database indicates that RNF152 expression is downregulated in various types of cancer, including colon, lung, kidney, and liver cancers.

RNF152 in addition to RNF183 is also associated with CRC. 1,2-Dimethylhydrazine (DMH) is a potent carcinogen that acts as a DNA methylating agent. Microarray gene expression analysis showed that Rnf152 was upregulated in DMH-injected CRC model mice provided with high-calcium feed, compared with that in those provided with normal feed [46]. An additional relationship of RNF152 with CRC was reported in mTORC1 signaling [47]. RNF152 expression is significantly reduced in CRC tissues compared with that in normal tissues. The expression levels of RNF152 were reported to be correlated with prognosis in patients with CRC. Using cell lines and xenografts, it was demonstrated that RNF152 overexpression significantly decreased CRC cell growth in vitro and in vivo. RNF152 inhibits CRC cell proliferation by suppressing mTORC1, resulting in the induction of autophagy and apoptotic cell death. Although the ubiquitylation of RagA by RNF152 was not demonstrated in CRC cells, this finding is consistent with RNF152-mediated mTORC1 downregulation.

Rnf152 was first revealed to play a physiological role using zebrafish embryos [41]. The rnf152 transcript is now known to be expressed from the one-cell stage (maternally) to 48 h post-fertilization (hpf) (zygotically). Rnf152 is ubiquitously expressed in the brain at 24 hpf, whereas its expression is restricted to the eyes, midbrain–hindbrain boundary (MHB), and rhombomeres at 48 hpf. Since Rnf152 knockdown in zebrafish embryos leads to morphological defects in the eyes, MHB, and rhombomeres at 24 hpf, Rnf152 is required for appropriate development of the eyes and neural tube later than 18 hpf during embryogenesis. NeuroD is a marker for the inner and outer layers of the eyes at 48 hpf and plays a crucial role in regulating cell cycle exit and cell fate determination in mitotic cells [65,66]. NeuroD expression in rnf152-deficient embryos disappears in the marginal zone, outer nuclear layer (ONL), inner nuclear layer (INL), and ganglion cell layer (GCL) of the eyes at 27 hpf. Furthermore, the expression of deltaD and notch1a in rnf152 morphants is remarkably reduced in the ONL, INL, subpallium, tectum, and cerebellum [67]. Knockdown of rnf152 was found to cause decreases of the expression of her4 and ascl1a, Notch target genes [68], in specific regions at 24 hpf. Taking these findings together, Rnf152 may play essential roles in the development of the eyes, midbrain, and hindbrain, as well as the activation of Delta-Notch signaling. However, Deng et al. have already reported that Rnf152 knockout mice exhibited no difference in birth rates from the expected Mendelian ratios, suggesting that RNF152 is not required for embryonic development, at least in mouse [43]. Although these findings suggest that ubiquitin ligases other than Rnf152 can work in the neural development of mammals, it remains unknown whether Rnf182, which is also expressed in the nervous system, works compensatorily.

One study reported on the relationship between the RNF152–mTORC1 axis and neural development [42]. The cell proliferation rate of the floor plate in the ventral region of the neural tube remains low [69]. Forkhead-type transcription factor FoxA2 was demonstrated to be a negative regulator of mTORC1 signaling in the floor plate [70]. Furthermore, RNF152 was identified as a target gene for FoxA2 among mTORC1 signaling genes. The silencing of RNF152 causes aberrant mTOR activation and aberrant cell division in the floor plate. Therefore, RNF152 may function as a negative regulator for the mTORC1 signaling in the floor plate downstream of FoxA2.

6. mTORC1 in Cancer

A common feature of members of the RNF183 family, with the exception of RNF183, is the ability to regulate mTORC1 signaling in the lysosomal membrane, albeit in different manners. RNF152 [43,44] and RNF182 [37] negatively regulate it, whereas RNF186 [24] positively regulates it (Figure 2). Since mTORC1 signaling is important for cell growth [62], these E3s may be involved in cancer progression. Indeed, members of the RNF183 family are known to be commonly associated with cancer (Table 1). As demonstrated by the findings for RNF152 [44,47], the expression levels of these E3s may affect cell proliferation by regulating mTORC1. Moreover, members of the RNF183 family, except RNF186, are frequently associated with CRC. The expression of RNF183 and RNF182 is upregulated in CRC [21,39], whereas that of RNF152 is downregulated [44,47]. This tendency is largely consistent with the findings for other cancers in TCGA database. Thus, the expression levels of these E3s may also be associated with the progression of other cancers besides CRC. Further investigations are needed to determine whether members of the RNF183 family are involved in other cancers, which are regulated by mTORC1. The mTORC1 signal is currently attracting attention as a target for drug discovery, and mTOR inhibitors such as everolimus are actually in use as molecular-targeted drugs for various cancers [62]. Therefore, if the mTORC1 signal of the RNF183 family and its involvement in cancer are clarified, they have potential as targets for new anticancer agents.

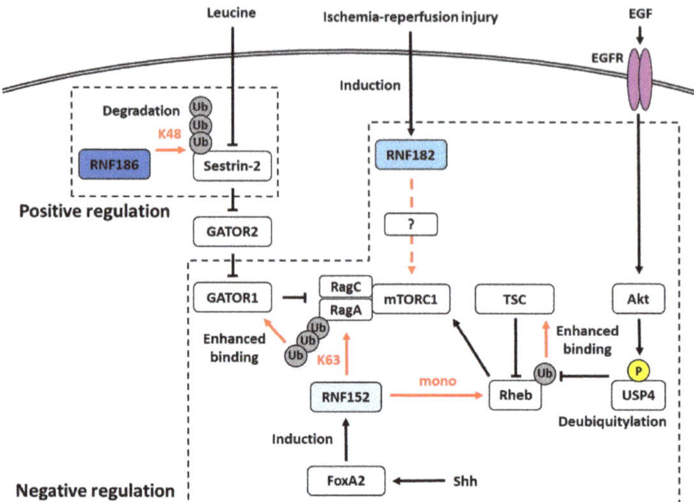

Figure 2. RNF186, RNF182, and RNF152 are involved in the regulation of the mTORC1 pathway. RNF186 regulates positively by degrading Sestrin-2 ubiquitylated with K48-linked chain, whereas RNF182 and RNF152 are native regulators. RNF152 involves two pathways: the K63-linked ubiquitylation of RagA and the mono ubiquitylation of Rheb. The ubiquitylated RagA and Rheb enhance the binding performance with GATOR1 and TSC, respectively. Phosphorylated USP4, which is identified as a DUB for Rheb, deubiquitylates the ubiquitylated Rheb and cancels the negative regulation by mono ubiquitylation. USP4 is phosphorylated by the EGF-Akt pathway. Shh-dependent FoxA2-transcriptional activity induces RNF152 expression. The expression of RNF182 is induced by ischemia-reperfusion injury; however, the mechanism of mTORC1 suppression by RNF182 remains unknown. The arrows indicate activation, and the T-bars indicate inhibition. The red arrows indicate the effects of ubiquitylation by RNF186 or RNF152. The red characters indicate the types of ubiquitylation by RNF186 or RNF152. The dotted arrow indicates that the detailed mechanism is unknown.

7. NF-κB and NFAT5 in Inflammation and Osmotic Stress

RNF183 and RNF182 have the opposite effects on the NF-κB pathway in the degradation of IκBα [17] and NF-κB p65 subunit [38], respectively. RNF183 and RNF182 are induced by miR-7 reduction [11,17] and TLR signaling [38], respectively (Figure 3). Therefore, these E3s are not only induced by inflammatory stimuli but also regulate inflammatory responses as a feedback mechanism. Furthermore, RNF183 expression is upregulated by NFAT5 under hypertonic conditions [14]. NFAT5 belongs to the NF-κB/Rel family and is involved in inflammatory responses like NF-κB [58]. Since hyperosmotic response genes include a number of inflammatory genes, which are upregulated by NFAT5, RNF183 may play important roles in inflammation as well as hypertonic stress response. Recently, the relationship between osmotic stress and inflammation was suggested [71]. Thus, RNF183 may be a key factor that links osmotic stress to inflammation. In fact, since it has been reported that osmotic stress exacerbates IBD [72], the osmotic stress response associated with RNF183 may play an important role in the pathology of IBD. Therefore, it is anticipated that substrates of RNF183 related to osmotic stress in IBD will be identified.

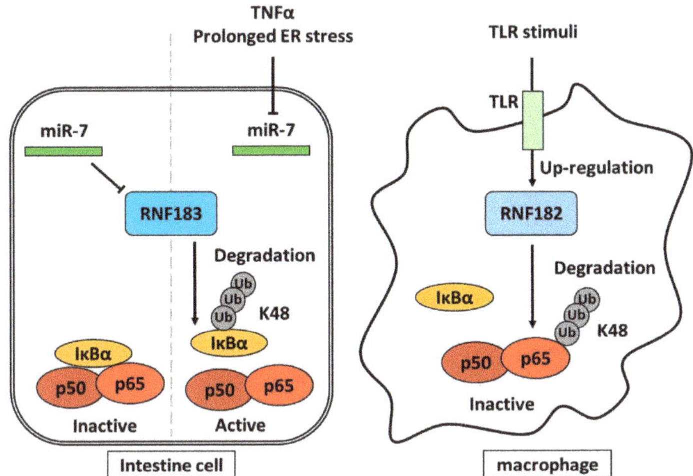

Figure 3. RNF183 and RNF182 are involved in the regulation of the NF-κB pathway. First, RNF183 ubiquitylates IκBα. Then, the ubiquitylated IκBα is degraded, and next, the NF-κB pathway is activated. The miR-7 suppresses the transcription of RNF183. TNFα and prolonged ER stress decrease miR-7, resulting in an increased expression of RNF183. RNF182 ubiquitylates and degrades the subunit p65 of NF-κB, and the TLR stimuli induces the expression of RNF182. Both IκBα and the subunit p65 of NF-κB are ubiquitylated with K48 chain. The arrows indicate activation and the T-bars indicate inhibition.

8. Conclusions

Members of the RNF183 family are widely expressed in the ER, Golgi apparatus, endosomes, and lysosomes. Although their substrates do not overlap with each other, they are involved in some common signaling pathways, such as mTORC1 and NF-κB (Table 1). The tissue expression pattern of these E3s is restricted, but overlaps in some tissues, such as the kidney (RNF183, RNF152, and RNF186), the nervous system (RNF152 and RNF182), and the colon (RNF186 and RNF183) [10]. Thus, it is possible that these E3s could function competitively or cooperatively in the regulation of common pathways. The tissue- and development-specific expression of the RNF183 family suggests that it plays an essential role in specific biological functions. To address the problem of dissociated phenotype in the development of knockout mice [43] and species other than mammals [41,42], it is necessary to establish mice with the knockout of two or more genes. It is also important to investigate whether the RNF183 family works in a coordinated and compensatory manner. On the other hand, it is very difficult to identify the substrates of E3, and experiments involving overexpression alone cannot reveal the true physiological function of a substrate. Therefore, the development of technology for identifying in vivo substrates will be important.

Author Contributions: T.O. and M.K. wrote the original draft; K.I. and M.K. reviewed and edited the manuscript. All authors have read and agreed to the published version of the manuscript.

Funding: This study was supported by Grants-in-Aid for Scientific Research (KAKENHI: 19K16373, 18H06105, 18K06685, 17H06416, and 17H01424) from the Japan Society for the Promotion of Science, Japan.

Conflicts of Interest: The authors declare no conflict of interest.

Abbreviations

AD	Alzheimer's disease
ATP6V0C	V-type proton ATPase 16 kDa proteolipid subunit
Bcl-xL	B-cell lymphoma extra large
BIK	Bcl-2-interacting killer
BNip1	Bcl2/adenovirus E1B 19-kDa interacting protein 1
BNIP3L	cl-2/adenovirus E1B 19 kDa protein-interacting protein 3-like
CD	Crohn's disease
CKD	chronic kidney disease
CRC	colorectal cancer
DR5	Death receptor 5
DSS	dextran sulfate sodium
DUB	deubiquitinase
E1	ubiquitin-activating enzyme
E2	ubiquitin-conjugating enzyme
E3	ubiquitin ligase
EC	endometrial carcinoma
ER	endoplasmic reticulum
EWSR1-FLI1	Ewing sarcoma breakpoint region 1-Friend Leukemia Integration 1
FATE1	fetal and adult testis-expressed 1
GAP	GTPase-activating protein
GATOR	GAP activity toward RAGs
GCL	ganglion cell layer
GWAS	genome-wide association study
hpf	hours post-fertilization
IBD	inflammatory bowel disease
IκB	Inhibitor of NF-κB
INL	inner nuclear layer
IRE1α	Inositol requiring 1α
IRS1	insulin receptor substrate 1
JNK	c-Jun N-terminal kinase
MeCP2	methyl-CpG-binding protein 2
MHB	midbrain–hindbrain boundary
mIMCD	mouse inner medullary collecting duct
MIRI	myocardial ischemia–reperfusion injury
mTORC1	mammalian target of rapamycin complex 1
NFAT5	Nuclear factor of activated T cells 5
NF-κB	nuclear factor-kappa B
NT2	Ntera2
ONL	outer nuclear layer
Shh	Sonic hedgehog
TLR	Toll-like receptor
TNBS	trinitrobenzene sulfonic acid
TNF-α	Tumor necrosis factor-α
UC	ulcerative colitis
UPR	unfolded protein response
Ub	Ubiquitin

References

1. Rennie, M.L.; Chaugule, V.K.; Walden, H. Modes of allosteric regulation of the ubiquitination machinery. *Curr. Opin. Struct. Biol.* **2020**, *62*, 189–196. [CrossRef]
2. Hann, Z.S.; Ji, C.; Olsen, S.K.; Lu, X.; Lux, M.C.; Tan, D.S.; Lima, C.D. Structural basis for adenylation and thioester bond formation in the ubiquitin E1. *Proc. Natl. Acad. Sci. USA* **2019**, *116*, 15475–15484. [CrossRef]

3. Buetow, L.; Huang, D.T. Structural insights into the catalysis and regulation of E3 ubiquitin ligases. *Nat. Rev. Mol. Cell Biol.* **2016**, *17*, 626–642. [CrossRef]
4. Walden, H.; Rittinger, K. RBR ligase-mediated ubiquitin transfer: A tale with many twists and turns. *Nat. Struct. Mol. Biol.* **2018**, *25*, 440–445. [CrossRef]
5. Bernassola, F.; Chillemi, G.; Melino, G. HECT-Type E3 Ubiquitin Ligases in Cancer. *Trends Biochem. Sci.* **2019**, *44*, 1057–1075. [CrossRef]
6. Ruan, J.; Schlüter, D.; Wang, X. Deubiquitinating enzymes (DUBs): DoUBle-edged swords in CNS autoimmunity. *J. Neuroinflamm.* **2020**, *17*, 102. [CrossRef]
7. Kwon, Y.T.; Ciechanover, A. The Ubiquitin Code in the Ubiquitin-Proteasome System and Autophagy. *Trends Biochem. Sci.* **2017**, *42*, 873–886. [CrossRef]
8. Davis, M.E.; Gack, M.U. Ubiquitination in the antiviral immune response. *Virology* **2015**, *479*, 52–65. [CrossRef]
9. Grice, G.L.; Nathan, J.A. The recognition of ubiquitinated proteins by the proteasome. *Cell. Mol. Life Sci.* **2016**, *73*, 3497–3506. [CrossRef]
10. Kaneko, M.; Iwase, I.; Yamasaki, Y.; Takai, T.; Wu, Y.; Kanemoto, S.; Matsuhisa, K.; Asada, R.; Okuma, Y.; Watanabe, T.; et al. Genome-wide identification and gene expression profiling of ubiquitin ligases for endoplasmic reticulum protein degradation. *Sci. Rep.* **2016**, *6*, 30955. [CrossRef]
11. Wu, Y.; Li, X.; Jia, J.; Zhang, Y.; Li, J.; Zhu, Z.; Wang, H.; Tang, J.; Hu, J. Transmembrane E3 ligase RNF183 mediates ER stress-induced apoptosis by degrading Bcl-xL. *Proc. Natl. Acad. Sci. USA* **2018**, *115*, E2762–E2771. [CrossRef]
12. Wu, Y.; Guo, X.P.; Kanemoto, S.; Maeoka, Y.; Saito, A.; Asada, R.; Matsuhisa, K.; Ohtake, Y.; Imaizumi, K.; Kaneko, M. Sec16A, a key protein in COPII vesicle formation, regulates the stability and localization of the novel ubiquitin ligase RNF183. *PLoS ONE* **2018**, *13*, e0190407. [CrossRef]
13. Maeoka, Y.; Okamoto, T.; Wu, Y.; Saito, A.; Asada, R.; Matsuhisa, K.; Terao, M.; Takada, S.; Masaki, T.; Imaizumi, K.; et al. Renal medullary tonicity regulates RNF183 expression in the collecting ducts via NFAT5. *Biochem. Biophys. Res. Commun.* **2019**, *514*, 436–442. [CrossRef]
14. Maeoka, Y.; Wu, Y.; Okamoto, T.; Kanemoto, S.; Guo, X.P.; Saito, A.; Asada, R.; Matsuhisa, K.; Masaki, T.; Imaizumi, K.; et al. NFAT5 up-regulates expression of the kidney-specific ubiquitin ligase gene Rnf183 under hypertonic conditions in inner-medullary collecting duct cells. *J. Biol. Chem.* **2019**, *294*, 101–115. [CrossRef]
15. Maxfield, K.E.; Taus, P.J.; Corcoran, K.; Wooten, J.; Macion, J.; Zhou, Y.; Borromeo, M.; Kollipara, R.K.; Yan, J.; Xie, Y.; et al. Comprehensive functional characterization of cancer-testis antigens defines obligate participation in multiple hallmarks of cancer. *Nat. Commun.* **2015**, *6*, 8840. [CrossRef]
16. Gallegos, Z.R.; Taus, P.; Gibbs, Z.A.; McGlynn, K.; Gomez, N.C.; Davis, I.; Whitehurst, A.W. EWSR1-FLI1 Activation of the Cancer/Testis Antigen FATE1 Promotes Ewing Sarcoma Survival. *Mol. Cell Biol.* **2019**, *39*. [CrossRef]
17. Yu, Q.; Zhang, S.; Chao, K.; Feng, R.; Wang, H.; Li, M.; Chen, B.; He, Y.; Zeng, Z.; Chen, M. E3 Ubiquitin ligase RNF183 Is a Novel Regulator in Inflammatory Bowel Disease. *J. Crohn's Colitis* **2016**, *10*, 713–725. [CrossRef]
18. Wu, Y.; Kimura, Y.; Okamoto, T.; Matsuhisa, K.; Asada, R.; Saito, A.; Sakaue, F.; Imaizumi, K.; Kaneko, M. Inflammatory bowel disease-associated ubiquitin ligase RNF183 promotes lysosomal degradation of DR5 and TRAIL-induced caspase activation. *Sci. Rep.* **2019**, *9*, 20301. [CrossRef]
19. Okamoto, T.; Wu, Y.; Matsuhisa, K.; Saito, A.; Sakaue, F.; Imaizumi, K.; Kaneko, M. Hypertonicity-responsive ubiquitin ligase RNF183 promotes Na, K-ATPase lysosomal degradation through ubiquitination of its β1 subunit. *Biochem. Biophys. Res. Commun.* **2020**, *521*, 1030–1035. [CrossRef]
20. Geng, R.; Tan, X.; Zuo, Z.; Wu, J.; Pan, Z.; Shi, W.; Liu, R.; Yao, C.; Wang, G.; Lin, J.; et al. Synthetic lethal short hairpin RNA screening reveals that ring finger protein 183 confers resistance to trametinib in colorectal cancer cells. *Chin. J. Cancer* **2017**, *36*, 63. [CrossRef]
21. Geng, R.; Tan, X.; Wu, J.; Pan, Z.; Yi, M.; Shi, W.; Liu, R.; Yao, C.; Wang, G.; Lin, J.; et al. RNF183 promotes proliferation and metastasis of colorectal cancer cells via activation of NF-κB-IL-8 axis. *Cell Death Dis.* **2017**, *8*, e2994. [CrossRef] [PubMed]
22. Colas, E.; Perez, C.; Cabrera, S.; Pedrola, N.; Monge, M.; Castellvi, J.; Eyzaguirre, F.; Gregorio, J.; Ruiz, A.; Llaurado, M.; et al. Molecular markers of endometrial carcinoma detected in uterine aspirates. *Int. J. Cancer* **2011**, *129*, 2435–2444. [CrossRef] [PubMed]

23. Wang, P.; Wu, Y.; Li, Y.; Zheng, J.; Tang, J. A novel RING finger E3 ligase RNF186 regulate ER stress-mediated apoptosis through interaction with BNip1. *Cell Signal.* **2013**, *25*, 2320–2333. [CrossRef] [PubMed]

24. Lear, T.B.; Lockwood, K.C.; Ouyang, Y.; Evankovich, J.W.; Larsen, M.B.; Lin, B.; Liu, Y.; Chen, B.B. The RING-type E3 ligase RNF186 ubiquitinates Sestrin-2 and thereby controls nutrient sensing. *J. Biol. Chem.* **2019**, *294*, 16527–16534. [CrossRef] [PubMed]

25. Fujimoto, K.; Kinoshita, M.; Tanaka, H.; Okuzaki, D.; Shimada, Y.; Kayama, H.; Okumura, R.; Furuta, Y.; Narazaki, M.; Tamura, A.; et al. Regulation of intestinal homeostasis by the ulcerative colitis-associated gene RNF186. *Mucosal. Immunol.* **2017**, *10*, 446–459. [CrossRef] [PubMed]

26. Tong, X.; Zhang, Q.; Wang, L.; Ji, Y.; Zhang, L.; Xie, L.; Chen, W.; Zhang, H. RNF186 impairs insulin sensitivity by inducing ER stress in mouse primary hepatocytes. *Cell Signal.* **2018**, *52*, 155–162. [CrossRef] [PubMed]

27. Anderson, C.A.; Boucher, G.; Lees, C.W.; Franke, A.; D'Amato, M.; Taylor, K.D.; Lee, J.C.; Goyette, P.; Imielinski, M.; Latiano, A.; et al. Meta-analysis identifies 29 additional ulcerative colitis risk loci, increasing the number of confirmed associations to 47. *Nat. Genet.* **2011**, *43*, 246–252. [CrossRef]

28. Juyal, G.; Prasad, P.; Senapati, S.; Midha, V.; Sood, A.; Amre, D.; Juyal, R.C.; Bk, T. An investigation of genome-wide studies reported susceptibility loci for ulcerative colitis shows limited replication in north Indians. *PLoS ONE* **2011**, *6*, e16565. [CrossRef]

29. Uniken Venema, W.T.; Voskuil, M.D.; Dijkstra, G.; Weersma, R.K.; Festen, E.A. The genetic background of inflammatory bowel disease: From correlation to causality. *J. Pathol.* **2017**, *241*, 146–158. [CrossRef]

30. Venkataraman, G.R.; Rivas, M.A. Rare and common variant discovery in complex disease: The IBD case study. *Hum. Mol. Genet.* **2019**, *28*, R162–R169. [CrossRef]

31. Beaudoin, M.; Goyette, P.; Boucher, G.; Lo, K.S.; Rivas, M.A.; Stevens, C.; Alikashani, A.; Ladouceur, M.; Ellinghaus, D.; Torkvist, L.; et al. Deep resequencing of GWAS loci identifies rare variants in CARD9, IL23R and RNF186 that are associated with ulcerative colitis. *PLoS Genet.* **2013**, *9*, e1003723. [CrossRef] [PubMed]

32. Rivas, M.A.; Graham, D.; Sulem, P.; Stevens, C.; Desch, A.N.; Goyette, P.; Gudbjartsson, D.; Jonsdottir, I.; Thorsteinsdottir, U.; Degenhardt, F.; et al. A protein-truncating R179X variant in RNF186 confers protection against ulcerative colitis. *Nat. Commun.* **2016**, *7*, 12342. [CrossRef] [PubMed]

33. Yang, S.K.; Hong, M.; Zhao, W.; Jung, Y.; Tayebi, N.; Ye, B.D.; Kim, K.J.; Park, S.H.; Lee, I.; Shin, H.D.; et al. Genome-wide association study of ulcerative colitis in Koreans suggests extensive overlapping of genetic susceptibility with Caucasians. *Inflamm. Bowel Dis.* **2013**, *19*, 954–966. [CrossRef] [PubMed]

34. Sveinbjornsson, G.; Mikaelsdottir, E.; Palsson, R.; Indridason, O.S.; Holm, H.; Jonasdottir, A.; Helgason, A.; Sigurdsson, S.; Jonasdottir, A.; Sigurdsson, A.; et al. Rare mutations associating with serum creatinine and chronic kidney disease. *Hum. Mol. Genet.* **2014**, *23*, 6935–6943. [CrossRef]

35. Liu, Q.Y.; Lei, J.X.; Sikorska, M.; Liu, R. A novel brain-enriched E3 ubiquitin ligase RNF182 is up regulated in the brains of Alzheimer's patients and targets ATP6V0C for degradation. *Mol. Neurodegener.* **2008**, *3*, 4. [CrossRef]

36. Nectoux, J.; Fichou, Y.; Rosas-Vargas, H.; Cagnard, N.; Bahi-Buisson, N.; Nusbaum, P.; Letourneur, F.; Chelly, J.; Bienvenu, T. Cell cloning-based transcriptome analysis in Rett patients: Relevance to the pathogenesis of Rett syndrome of new human MeCP2 target genes. *J. Cell Mol. Med.* **2010**, *14*, 1962–1974. [CrossRef]

37. Wang, J.H.; Wei, Z.F.; Gao, Y.L.; Liu, C.C.; Sun, J.H. Activation of the mammalian target of rapamycin signaling pathway underlies a novel inhibitory role of ring finger protein 182 in ventricular remodeling after myocardial ischemia-reperfusion injury. *J. Cell Biochem.* **2018**. [CrossRef]

38. Cao, Y.; Sun, Y.; Chang, H.; Sun, X.; Yang, S. The E3 ubiquitin ligase RNF182 inhibits TLR-triggered cytokine production through promoting p65 ubiquitination and degradation. *FEBS Lett.* **2019**, *593*, 3210–3219. [CrossRef]

39. Brim, H.; Abu-Asab, M.S.; Nouraie, M.; Salazar, J.; Deleo, J.; Razjouyan, H.; Mokarram, P.; Schaffer, A.A.; Naghibhossaini, F.; Ashktorab, H. An integrative CGH, MSI and candidate genes methylation analysis of colorectal tumors. *PLoS ONE* **2014**, *9*, e82185. [CrossRef]

40. Zhang, S.; Wu, W.; Wu, Y.; Zheng, J.; Suo, T.; Tang, H.; Tang, J. RNF152, a novel lysosome localized E3 ligase with pro-apoptotic activities. *Protein Cell* **2010**, *1*, 656–663. [CrossRef]

41. Kumar, A.; Huh, T.L.; Choe, J.; Rhee, M. Rnf152 Is Essential for NeuroD Expression and Delta-Notch Signaling in the Zebrafish Embryos. *Mol. Cells* **2017**, *40*, 945–953. [CrossRef] [PubMed]

42. Kadoya, M.; Sasai, N. Negative Regulation of mTOR Signaling Restricts Cell Proliferation in the Floor Plate. *Front. Neurosci.* **2019**, *13*, 1022. [CrossRef] [PubMed]

43. Deng, L.; Jiang, C.; Chen, L.; Jin, J.; Wei, J.; Zhao, L.; Chen, M.; Pan, W.; Xu, Y.; Chu, H.; et al. The ubiquitination of rag A GTPase by RNF152 negatively regulates mTORC1 activation. *Mol. Cell* **2015**, *58*, 804–818. [CrossRef] [PubMed]

44. Deng, L.; Chen, L.; Zhao, L.; Xu, Y.; Peng, X.; Wang, X.; Ding, L.; Jin, J.; Teng, H.; Wang, Y.; et al. Ubiquitination of Rheb governs growth factor-induced mTORC1 activation. *Cell Res.* **2019**, *29*, 136–150. [CrossRef] [PubMed]

45. Yamamoto, F.; Yamamoto, M. Scanning copy number and gene expression on the 18q21-qter chromosomal region by the systematic multiplex PCR and reverse transcription-PCR methods. *Electrophoresis* **2007**, *28*, 1882–1895. [CrossRef] [PubMed]

46. Wang, J.L.; Lin, Y.W.; Chen, H.M.; Kong, X.; Xiong, H.; Shen, N.; Hong, J.; Fang, J.Y. Calcium prevents tumorigenesis in a mouse model of colorectal cancer. *PLoS ONE* **2011**, *6*, e22566. [CrossRef] [PubMed]

47. Cui, X.; Shen, W.; Wang, G.; Huang, Z.; Wen, D.; Yang, Y.; Liu, Y.; Cui, L. Ring finger protein 152 inhibits colorectal cancer cell growth and is a novel prognostic biomarker. *Am. J. Trans. Res.* **2018**, *10*, 3701–3712.

48. Fu, J.; Tang, W.; Du, P.; Wang, G.; Chen, W.; Li, J.; Zhu, Y.; Gao, J.; Cui, L. Identifying microRNA-mRNA regulatory network in colorectal cancer by a combination of expression profile and bioinformatics analysis. *BMC Syst. Biol.* **2012**, *6*, 68. [CrossRef]

49. Olesen, C.; Larsen, N.J.; Byskov, A.G.; Harboe, T.L.; Tommerup, N. Human FATE is a novel X-linked gene expressed in fetal and adult testis. *Mol. Cell. Endocrinol.* **2001**, *184*, 25–32. [CrossRef]

50. Dong, X.Y.; Su, Y.R.; Qian, X.P.; Yang, X.A.; Pang, X.W.; Wu, H.Y.; Chen, W.F. Identification of two novel CT antigens and their capacity to elicit antibody response in hepatocellular carcinoma patients. *Br. J. Cancer* **2003**, *89*, 291–297. [CrossRef]

51. Fei, P.; Wang, W.; Kim, S.H.; Wang, S.; Burns, T.F.; Sax, J.K.; Buzzai, M.; Dicker, D.T.; McKenna, W.G.; Bernhard, E.J.; et al. Bnip3L is induced by p53 under hypoxia, and its knockdown promotes tumor growth. *Cancer Cell* **2004**, *6*, 597–609. [CrossRef] [PubMed]

52. Infante, J.R.; Fecher, L.A.; Falchook, G.S.; Nallapareddy, S.; Gordon, M.S.; Becerra, C.; DeMarini, D.J.; Cox, D.S.; Xu, Y.; Morris, S.R.; et al. Safety, pharmacokinetic, pharmacodynamic, and efficacy data for the oral MEK inhibitor trametinib: A phase 1 dose-escalation trial. *Lancet Oncol.* **2012**, *13*, 773–781. [CrossRef]

53. Waugh, D.J.; Wilson, C. The interleukin-8 pathway in cancer. *Clin. Cancer Res.* **2008**, *14*, 6735–6741. [CrossRef] [PubMed]

54. González-García, M.; Pérez-Ballestero, R.; Ding, L.; Duan, L.; Boise, L.H.; Thompson, C.B.; Núñez, G. bcl-XL is the major bcl-x mRNA form expressed during murine development and its product localizes to mitochondria. *Development* **1994**, *120*, 3033–3042.

55. Eno, C.O.; Eckenrode, E.F.; Olberding, K.E.; Zhao, G.; White, C.; Li, C. Distinct roles of mitochondria- and ER-localized Bcl-xL in apoptosis resistance and Ca2+ homeostasis. *Mol. Biol. Cell* **2012**, *23*, 2605–2618. [CrossRef]

56. Lerner, A.G.; Upton, J.P.; Praveen, P.V.; Ghosh, R.; Nakagawa, Y.; Igbaria, A.; Shen, S.; Nguyen, V.; Backes, B.J.; Heiman, M.; et al. IRE1α induces thioredoxin-interacting protein to activate the NLRP3 inflammasome and promote programmed cell death under irremediable ER stress. *Cell Metab.* **2012**, *16*, 250–264. [CrossRef]

57. Upton, J.P.; Wang, L.; Han, D.; Wang, E.S.; Huskey, N.E.; Lim, L.; Truitt, M.; McManus, M.T.; Ruggero, D.; Goga, A.; et al. IRE1α cleaves select microRNAs during ER stress to derepress translation of proapoptotic Caspase-2. *Science* **2012**, *338*, 818–822. [CrossRef]

58. Halterman, J.A.; Kwon, H.M.; Wamhoff, B.R. Tonicity-independent regulation of the osmosensitive transcription factor TonEBP (NFAT5). *Am. J. Physiol. Cell Physiol.* **2012**, *302*, C1–C8. [CrossRef]

59. Kaplan, J.H. Biochemistry of Na,K-ATPase. *Annu. Rev. Biochem.* **2002**, *71*, 511–535. [CrossRef]

60. Yordy, M.R.; Bowen, J.W. Na,K-ATPase expression and cell volume during hypertonic stress in human renal cells. *Kidney Int.* **1993**, *43*, 940–948. [CrossRef]

61. Yasuda, M.; Chinnadurai, G. Functional identification of the apoptosis effector BH3 domain in cellular protein BNIP1. *Oncogene* **2000**, *19*, 2363–2367. [CrossRef] [PubMed]

62. Saxton, R.A.; Sabatini, D.M. mTOR Signaling in Growth, Metabolism, and Disease. *Cell* **2017**, *168*, 960–976. [CrossRef] [PubMed]

63. Amir, R.E.; Van den Veyver, I.B.; Wan, M.; Tran, C.Q.; Francke, U.; Zoghbi, H.Y. Rett syndrome is caused by mutations in X-linked MECP2, encoding methyl-CpG-binding protein 2. *Nat. Genet.* **1999**, *23*, 185–188. [CrossRef] [PubMed]

64. McCann, A.P.; Scott, C.J.; Van Schaeybroeck, S.; Burrows, J.F. Deubiquitylating enzymes in receptor endocytosis and trafficking. *Biochem. J.* **2016**, *473*, 4507–4525. [CrossRef]

65. Thomas, J.L.; Ochocinska, M.J.; Hitchcock, P.F.; Thummel, R. Using the Tg(nrd:egfp)/albino zebrafish line to characterize in vivo expression of neurod. *PLoS ONE* **2012**, *7*, e29128. [CrossRef]

66. Ochocinska, M.J.; Hitchcock, P.F. Dynamic expression of the basic helix-loop-helix transcription factor neuroD in the rod and cone photoreceptor lineages in the retina of the embryonic and larval zebrafish. *J. Comp. Neurol.* **2007**, *501*, 1–12. [CrossRef]

67. Raymond, P.A.; Barthel, L.K.; Bernardos, R.L.; Perkowski, J.J. Molecular characterization of retinal stem cells and their niches in adult zebrafish. *BMC Dev. Biol.* **2006**, *6*, 36. [CrossRef]

68. Taylor, S.M.; Alvarez-Delfin, K.; Saade, C.J.; Thomas, J.L.; Thummel, R.; Fadool, J.M.; Hitchcock, P.F. The bHLH Transcription Factor NeuroD Governs Photoreceptor Genesis and Regeneration Through Delta-Notch Signaling. *Investig. Ophthalmol. Vis. Sci.* **2015**, *56*, 7496–7515. [CrossRef]

69. Le Dreau, G.; Marti, E. Dorsal-ventral patterning of the neural tube: A tale of three signals. *Dev. Neurobiol.* **2012**, *72*, 1471–1481. [CrossRef]

70. Kutejova, E.; Sasai, N.; Shah, A.; Gouti, M.; Briscoe, J. Neural Progenitors Adopt Specific Identities by Directly Repressing All Alternative Progenitor Transcriptional Programs. *Dev. Cell* **2016**, *36*, 639–653. [CrossRef]

71. Berry, M.R.; Mathews, R.J.; Ferdinand, J.R.; Jing, C.; Loudon, K.W.; Wlodek, E.; Dennison, T.W.; Kuper, C.; Neuhofer, W.; Clatworthy, M.R. Renal Sodium Gradient Orchestrates a Dynamic Antibacterial Defense Zone. *Cell* **2017**, *170*, 860–874. [CrossRef] [PubMed]

72. Neuhofer, W. Role of NFAT5 in inflammatory disorders associated with osmotic stress. *Curr. Genom.* **2010**, *11*, 584–590. [CrossRef] [PubMed]

International Journal of
Molecular Sciences

Review

Overview of Mitochondrial E3 Ubiquitin Ligase MITOL/MARCH5 from Molecular Mechanisms to Diseases

Isshin Shiiba [1,2], Keisuke Takeda [1], Shun Nagashima [1] and Shigeru Yanagi [1,2,*]

[1] Laboratory of Molecular Biochemistry, School of Life Sciences, Tokyo University of Pharmacy and Life Sciences, Hachioji, Tokyo 192-0392, Japan; isshin.shiiba@gakushuin.ac.jp (I.S.); stakeda@toyaku.ac.jp (K.T.); nagashi@toyaku.ac.jp (S.N.)

[2] Laboratory of Molecular Biochemistry, Department of Life Science, Faculty of Science, Gakushuin University, 1-5-1 Mejiro, Toshima-ku, Tokyo 171-8588, Japan

* Correspondence: shigeru.yanagi@gakushuin.ac.jp

Received: 3 May 2020; Accepted: 26 May 2020; Published: 27 May 2020

Abstract: The molecular pathology of diseases seen from the mitochondrial axis has become more complex with the progression of research. A variety of factors, including the failure of mitochondrial dynamics and quality control, have made it extremely difficult to narrow down drug discovery targets. We have identified MITOL (mitochondrial ubiquitin ligase: also known as MARCH5) localized on the mitochondrial outer membrane and previously reported that it is an important regulator of mitochondrial dynamics and mitochondrial quality control. In this review, we describe the pathological aspects of MITOL revealed through functional analysis and its potential as a drug discovery target.

Keywords: mitochondria; E3 ubiquitin ligase; MITOL/MARCH5

1. Introduction

There are many functional proteins on the mitochondrial outer membrane. Not only mitochondrial quality control, but also various types of signal transduction including in the innate immune response, is performed on mitochondria as a scaffold. Mitochondria are dynamic organelles that form diverse networks through morphological changes via their fusion and fission, their intracellular movement along microtubules, and their interaction with other organelles. Since mitochondrial function is tightly regulated by mitochondrial dynamics, its molecular mechanisms and their association with disease have attracted substantial attention. In the ubiquitylation system, E3 ubiquitin ligases play a key role in determining substrate specificity and catalyzing the transfer of ubiquitin from E2 enzymes to the substrate. Growing evidence has shown that E3 ubiquitin ligases are involved in the regulation of mitochondrial functions. HECT domain type E3 ligases have one more transthiolation reaction to transfer the ubiquitin onto the E3, whereas the much more common RING finger domain type ligases transfer ubiquitin directly from E2 to the substrate. MITOL (mitochondrial ubiquitin ligase: also known as MARCH5) is a mitochondrial membrane-associated RING finger E3 ubiquitin ligase and was initially identified as a regulator of mitochondrial dynamics, which involves two different aspects: the regulation of protein activation by K63 ubiquitin chain attachment and the regulation of proteasome-dependent degradation by K48 ubiquitin chain attachment via ubiquitin signaling on the outer mitochondrial membrane. A number of MITOL substrates have been identified so far and their association with disease has been suggested. In this review, we present the latest findings on MITOL as a potential drug target and its relevance to disease and aging.

2. Mitochondria and Intracellular Quality Control by MITOL

2.1. Neurodegenerative Diseases

Various quality control systems enable the production of high-quality proteins in the cell. The endoplasmic reticulum is equipped with a protein quality control mechanism termed endoplasmic reticulum-associated degradation (ERAD), in which the ubiquitin–proteasome system (UPS) plays a central role [1]. The UPS selectively and rapidly degrades and removes abnormal proteins by adding ubiquitin chains [2]. Recently, a new quality control mechanism called mitochondrial protein translocation-associated degradation (mitoTAD) has been discovered in yeast [3]. It would be interesting to determine whether such a sophisticated quality control mechanism is preserved in mammals. The accumulation of dysfunctional proteins has been observed in many neurodegenerative diseases, and it has become clear that it is deeply involved in their pathogenesis. The increase in dysfunctional proteins due to mutations in the causative gene and the decrease in proteasome activity with aging are thought to be the main pathogenic factors. The accumulation of dysfunctional proteins in cells leads to a decrease in proteasome activity, suggesting the presence of a downward spiral that further promotes the accumulation of aggregated proteins [4]. It has been reported that such highly aggregated dysfunctional proteins can also accumulate in mitochondria and impair their function [5]. As an example, mutant superoxide dismutase-1 (mSOD1), one of the gene products responsible for amyotrophic lateral sclerosis (ALS), aggregates due to its conformational changes, accumulates in mitochondria, and inhibits mitochondrial protein transport, thereby impairing mitochondrial function. In addition, in spinal cerebellar degeneration, such as polyglutamine disease, the abnormal elongation of the polyglutamine chain resulting in conformational changes leading to accumulation of the protein PolyQ in the nucleus and mitochondria, and reduces mitochondrial function, which is thought to be one of the mechanisms of neuropathy. We reported that MITOL specifically ubiquitinates and promotes the proteasomal degradation of the misfolded proteins mSOD1 and PolyQ as substrates [6,7]. MITOL has a disordered domain in its C-terminal region, which may recognize denatured proteins. This suggests that MITOL is involved in mitochondrial quality control by eliminating dysfunctional proteins from the mitochondria. However, it is controversial whether MITOL recognizes target proteins in the process of aggregation or after their aggregation.

2.2. Innate Immune Response

In mammals, the innate immune system against RNA viruses is regulated by two different signaling pathways: the TLR (Toll-like receptor) pathway and the RLR (RIG-I-like receptor) pathway. The membrane-associated RING-CH (MARCH) family of E3 ubiquitin ligases is thought to perform immunoregulatory functions by controlling the localization and abundance of immune receptors [8]. In the RLR pathway, RIG-I/MDA-5 interacts with mitochondrial antiviral signaling (MAVS), also known as VISA, on mitochondria as an adaptor and transmits signals downstream. MITOL, the only member of the MARCH family localized in mitochondria, prevents excessive immune responses by ubiquitinating phosphorylated MAVs and activated RIG-I for degradation [9–11]. In addition, MITOL acts as a positive modulator in the TLR pathway, particularly through the ubiquitination of TANK, a negative regulator of TLR7 signaling [12]. Furthermore, MITOL preserves mitochondrial function by ubiquitinating HBx, which is localized in mitochondria for degradation in response to hepatitis B virus [13]. These findings suggest that MITOL could regulate immunosignaling on mitochondria (Figure 1A).

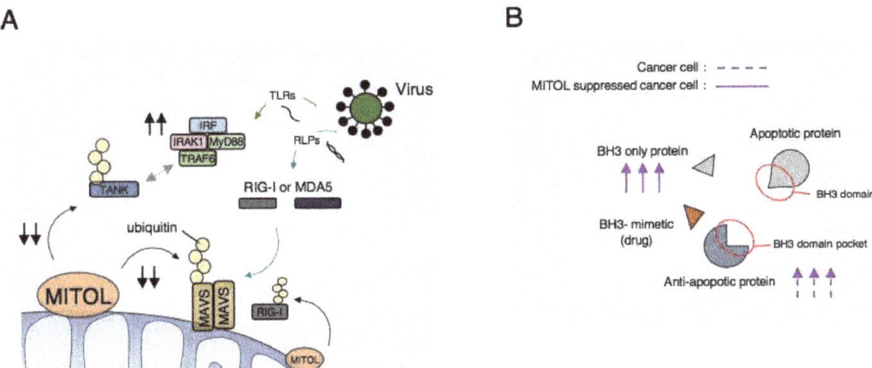

Figure 1. (**A**) MITOL regulates both RIG-I-like receptor (RLR) pathway and Toll-like receptor (TLR) pathway; (**B**) Relativity between MITOL and cell death signaling in cancer cells.

2.3. Cell Death in Cancer Cells

Mitochondria store apoptotic factors such as cytochrome c, Apoptosis-inducing factor (AIF), and caspase-9 in the intermembrane space and release them from the mitochondria to the cytoplasm in response to extracellular death signals, which initiates a cascade of apoptosis. The mitochondrial membrane permeability of cytochrome c is regulated by the B-cell leukemia gene-2 (Bcl-2) family, with pro-apoptotic proteins (Bax and Bak with BH3 domain) and anti-apoptotic proteins (Bcl-2, Bcl-xL, and Mcl-1 with BH3 binding site). There are also BH3-only proteins (Bim, Bid, Bad, Noxa, and Puma) that have only a BH3 domain among the Bcl-2 family members. These BH3-only proteins induce apoptosis by binding to the anti-apoptotic member bcl-2 and counteracting its apoptotic inhibitory effect. Mechanistically, NOXA/Mcl-1 complex was basically ubiquitinated by MITOL for degradation, cooperating with MTCH2 in a steady state [14]. Because many cancer cells escape cell death by activating the anti-apoptotic pathway such as via the upregulation of Mcl-1, BH3 mimetics are used for cancer treatment to inhibit anti-apoptotic activity and promote cell death [15].

At first glance, the activation of MITOL appears to be effective for cancer treatment. However, the quantity of NOXA is important for the promotion of cell death in cancer cells treated with ABT737 (BH3 mimetic), suggesting that the suppression of MITOL is effective to promote cancer cell death, resulting in the upregulation of NOXA [16,17]. As it was also shown that the overexpression of MITOL in breast cancer cells promotes tumor growth and metastasis [18], the suppression of MITOL might be effective for cancer treatment (Figure 1B).

2.4. Mitophagy

Mitophagy, a mitochondrion-specific type of autophagy, is responsible for removing damaged mitochondria and protecting cells from injury. PTEN-induced putative protein kinase 1 (PINK1) and Parkin are key players in the regulation of ubiquitin-dependent mitophagy [19]. In healthy mitochondria, PINK1 is transported into the inner mitochondrial membrane (IMM), where it is processed and cleaved by several proteases [20,21]. The truncated form of PINK1 is degraded by the ubiquitin–proteasome system [20,21]. Following the loss of the mitochondrial membrane potential, mitochondrial import is prevented, facilitating PINK1 stabilization on the outer mitochondrial membrane (OMM) [20–22]. PINK1 is activated by auto-phosphorylation leading to the translocation of Parkin to the mitochondrial surface. In this process, MITOL plays a role in the initial step in Parkin recruitment to mitochondria, resulting in ubiquitination of its substrate [23]. Although the physiological significance of this is still unclear, it has also been shown that MITOL is transferred to other organelles during mitophagy [24]. We have actually observed translocation to other organelles as well. Several mitochondrial proteins, including Nix/BNIP3L, BNIP3, FUNDC1, Optineurin, NDP52, and Bcl2-L-13, have been reported as mitophagy receptors in mammalian cells [19]. In particular, FUNDC1 is thought to be involved in mitophagy through its interaction with LC3 (yeast Atg8 ortholog) and ULK1 (yeast Atg1 ortholog) through hypoxic stress and inhibitory treatment of oxidative phosphorylation [25]. MITOL has been shown to regulate mitophagy by ubiquitinating FUNDC1 for degradation during hypoxic stress [26], suggesting that MITOL also acts as a regulator of such dynamic mitochondrial quality control. These observations suggest that MITOL may determine mitochondrial fate in response to a variety of mitochondrial stresses.

2.5. Aging

Mitochondrial dysfunction has been shown to lead to oxidative stress due to the generation of excessive reactive oxygen species, triggering senescence and age-related diseases. Since cell lines with MITOL deficiency induces cellular senescence [27], we also focused on the association between reduced MITOL function and senescence and age-related diseases in vivo. In our in vivo analysis, mice lacking MITOL in a skin-specific manner showed significant signs of aging, including gray hair and hair loss. These results and MITOL downregulation during aging suggest that MITOL may block the induction of physiological aging. We hoped that drugs that upregulate MITOL expression could inhibit or delay aging, and in a collaborative study with pharmaceutical companies, we screened drugs that upregulate MITOL mRNA from a Chinese medicine library using cultured human keratinocytes. We found that extracts of *Coptis japonica* and *Phellodendron amurense* in a Chinese medicine library upregulated MITOL mRNA approximately threefold (Patent No: P2019-52145A). When berberine, common compound of *Coptis japonica* and *Phellodendron amurense*, was mixed with drinking water, the expression of mRNA and protein of MITOL increased in various organs and tissues, including skin. To investigate the anti-aging activity of berberine, mice treated with it were irradiated with ultraviolet light to induce the formation of wrinkles in the skin, and a significant inhibitory effect on this was observed compared with that in control mice. In addition, with collaboration research, we found that the mice that took berberine for more than 1 year were significantly less likely to show signs of aging, such as skin hair loss and thickening of the epidermis, than control mice. Because the anti-aging effect of berberine on mice deficient of MITOL specifically in the skin was not confirmed, anti-aging effects are induced especially via the upregulation of MITOL expression levels. These results strongly indicate that MITOL has the potential to be a target for anti-aging drugs. In the future, it is anticipated that new MITOL activators other than berberine will be identified and applied to various aging-related diseases such as neurodegenerative diseases and heart failure. Interestingly, the expression of MITOL has been observed to gradually decrease with aging in many organs such as the brain and heart of mice. Moreover, Alzbase, a database for gene dysregulation in Alzheimer's disease (AD), suggested that the expression level of the mitol/march5 gene was significantly reduced in the brain of AD patients. If the worsening of AD is linked to the reduced MITOL expression, the compounds we have identified may exert a good therapeutic effect on AD. Therefore, we speculate that the decreased degradation of denatured proteins caused by decreased expression of MITOL may be one of the factors exacerbating the development of neurodegenerative diseases with aging (Figure 2).

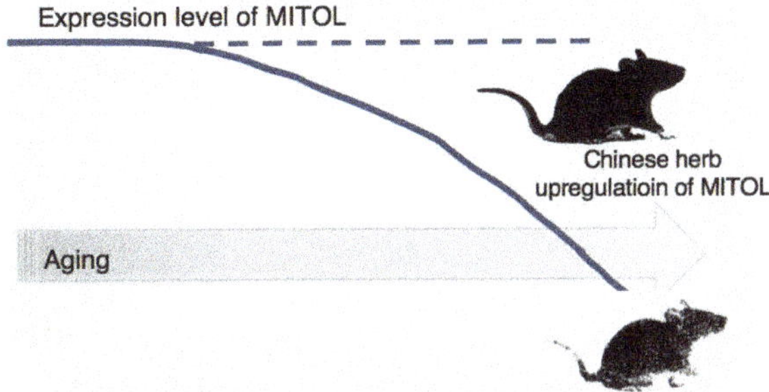

Figure 2. Relativity between MITOL and aging. Since MITOL gradually decrease following aging, upregulation of MITOL by a Chinese herb might be able to effect anti-aging.

3. Regulation of Mitochondrial Dynamics by MITOL

3.1. Mitochondrial Fission

Drp1, a dynamin-related GTPase, is the central player in mitochondrial fission [28–30]. Drp1 basically exists in the cytosol and can be recruited to the surface of mitochondria to wrap around mitochondrial tubules. Increasing Drp1 GTP hydrolysis activity leads to conformational changes [31,32] that increase the tightness of mitochondrial constriction and the scission of mitochondrial tubules. In terms of the chemical modifications of Drp1, phosphorylation, ubiquitination, and SUMOylation by several kinases and E3 ligase, including ubiquitin and SUMO ligase, have been particularly well studied [33–35]. Drp1 is ubiquitylated by APC/CCdh1 E3 ubiquitin ligase complex for degradation, the regulator of the M- to G1-phase transition, for regulating mitochondrial morphology during the G1/S phase [36]. Parkin, an E3 ligase mainly localized in the cytosol in a steady state, also induces the proteasomal degradation of Drp1 [37,38], suggesting that Parkin targets Drp1 as a substrate and plays an inhibitory role in mitochondrial fission. Controversial results have been reported on functions of MITOL in regulating mitochondrial dynamics. We and others showed that MITOL interacts with Drp1 and leads to its ubiquitylation and proteasome-dependent degradation to inhibit excessive mitochondrial fission [39,40]. However, the opposite role of MITOL in mitochondrial fission, namely, its positive regulation of Drp1, was observed in another study [41]. In a later study, Karbowski's group reported that MITOL ubiquitinates a Mid49, a Drp1 receptor, for degradation to control mitochondrial fission [42]. These findings suggest that there may be some specificity regarding regulation of MITOL for its substrates, especially Drp1, such as relating to modifications, structural changes, and interactions with or without Drp1 receptors (Figure 3A).

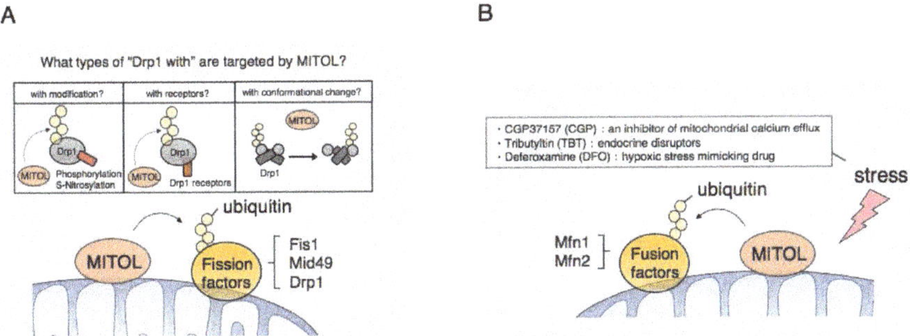

Figure 3. Regulation of mitochondrial fission and fusion factors by MITOL. (**A**) MITOL might have some specificity for its target; e.g., especially Drp1, such as relating to modifications, structural changes, and interactions with or without Drp1 receptors; (**B**) Alteration of substrate specificity for mitochondrial fusion factors in response to stress.

Recently, denitrosylase *S*-nitrosoglutathione reductase (GSNOR)-deficient mice were generated and it was reported that excessive *S*-nitrosylation resulted in the disruption of their mitochondrial dynamics and presented a typical senescence phenotype [43]. At the molecular level, it has also been reported that Drp1 can be in active form when subjected to *S*-nitrosylation by NO, leading to excessive mitochondrial division and the induction of neurotoxicity [44]. Similarly, MAP1B-LC1, a microtubule-associated factor that regulates retrograde mitochondrial transport, is also activated by *S*-nitrosylation, changing its steric structure [45], which in turn causes mitochondrial aggregation and cytotoxicity [46]. Thus, NO disrupts mitochondrial dynamics via the *S*-nitrosylation of Drp1 and MAP1B-LC1, even at the molecular level, and is thought to induce cellular damage as a result. We identified MAP1B-LC1 as a binding protein of MITOL using the yeast two-hybrid method and found that MITOL specifically recognizes and induces the degradation of *S*-nitrosylation-modified MAP1B-LC1 [47]. We also confirmed similar results in Drp1, suggesting that MITOL has selectivity for recognizing its substrates and protecting against the collapse of mitochondrial dynamics partly by degrading active Drp1 and MAP1B-LC1.

Mitochondrion–organelle contacts are involved in mitochondrial fission [48–51]. In mammals, Fis1 is a mitochondrial recruitment factor for TBC1D15, which drives lysosomal Rab7 GTP hydrolysis [52,53] and indirectly regulates mitochondrial fission at the mitochondrion–lysosome contact sites [50]. It has been reported that MITOL ubiquitinates Fis1 and promotes its degradation [39]. These findings suggest that MITOL also regulates mitochondrial fission at the mitochondrion–lysosome contact sites via Fis1. Further work examining the mechanistic role of MITOL in mitochondrion–lysosome contacts is required.

Although MITOL is clearly involved in regulating the fission machinery, we need to comprehensively consider the involvement of multiple signaling platforms explained by the diverse roles in cellular biology, variations in tissue-specific expression, and the activity of fission members for the regulation of fission by MITOL.

3.2. Mitochondrial Fusion

Mitochondria are fused through a few steps, termed mitochondrial fusion. First, mitochondria tether together via mitofusin (MFN), a large GTPase called Fzo1 in yeast, and MFN1/2 in mammals. Then, following MFN activation and conformational changes induced by GTP hydrolysis, each OMM attaches and fuses [54–56]. In recent years, various E3 ligases have been shown to regulate mitochondrial fusion via the modulation of one or both MFNs in response to various physiological or stress-induced conditions. Glycoprotein 78 (Gp78), an ER membrane-anchored E3 ubiquitin ligase, interacts with both MFNs, and Gp78 overexpression induces mitochondrial fragmentation [57]. Moreover, autocrine motility factor (AMF) prevents the Gp78-induced degradation of both MFNs [58]. MGRN1, an E3 ligase located in the cytoplasm, plasma membrane, endosomes, and nucleus, was reported to promote mitochondrial fusion via the non-degradative ubiquitylation of MFN1, consistent with previous observations [59,60]. The OMM E3 ligase MAPL/Mul1 leads to the specific ubiquitylation and degradation of MFN2 to regulate mitochondrial morphology [61]. HUWE1, a cytoplasmic E3 HECT family ubiquitin ligase also termed Mule/ARF-BP1/HectH9/E3Histone/Lasu12, ubiquitinated MFN2 associated with genotoxic stress to regulate mitochondrial fusion [62]. MITOL-mediated ubiquitylation and degradation of MFN1, but not of MFN2, lead to mitochondrial fragmentation in various inducible stresses and situations. In prostate cancer cells, the induction of cell death with CGP, an inhibitor of mitochondrial calcium efflux, led to ubiquitylation and degradation of MFN1 by MITOL [63]. It was also shown that MITOL ubiquitylates and degrades MFN1 at G2/M, the notable phase of mitochondrial fragmentation before cellular division [64]. However, under hypoxic stress induced by deferoxamine (DFO), MITOL interacts with MFN2 and is responsible for the ubiquitylation and degradation of MFN2 in cells lacking HDAC6 [65]. In conclusion, although mitochondrial fission and fusion were clearly regulated by ubiquitylation, further studies are required to understand how MITOL divides substrate-specificity according to its surroundings in order to regulate mitochondrial dynamics

(Figure 3B). In addition, mitochondrial dysfunction is suspected to be one of the causes of many neurodegenerative diseases such as Alzheimer's disease and Parkinson's disease, so investigating the involvement of MITOL in these diseases may lead to future therapies.

4. Relation between MITOL and Membrane Bontact Site with the Endoplasmic Reticulum

4.1. Membrane Contact Site with the Endoplasmic Reticulum

The mitochondrial surface also represents the signal hub where a host of metabolic systems cross-talk through inter-organelle communication. Mitochondria indeed have a unique microdomain physically and functionally connecting to other organelles such as the endoplasmic reticulum (ER). The membrane contact site (MCS) between the ER and mitochondria is maintained by some tethering or spacer proteins such as PDZD8, Fis1-BAP31, VDAC-IP3R, PTPIP51-VAPB, and MFN2 (the tethering function of Mfn2 appears to still be controversial) in mammals [66,67]. Mitochondria and the ER can exchange lipids and calcium ion through their MCS [68,69]. The proximal domain between the ER and mitochondria is also available as a membrane scaffold for signal transmission including autophagy, inflammation, and the unfolded protein response (UPR) due to the raft-like membrane structure [70]. Taking these findings together, the ER and mitochondria complement each other through inter-organelle communication such as membrane contacts.

4.2. MCS Formation by MITOL

MITOL can control signals contributing to various features outside mitochondria through the MCS between the ER and mitochondria. A primary finding that MITOL is enriched in the MCS between the ER and mitochondria led us to investigate MITOL with a focus on MCS.

In basal and physiological conditions, MITOL interacts with and ubiquitinates the mitochondrial GTPase MFN2 [71]. MFN2 acts as a tethering factor between the ER and mitochondria, as well as a factor for inter-mitochondrial fusion via its GTPase activity [72]. The MFN2 ubiquitination by MITOL contributes to its function related to ER–mitochondrion contacts through the GTPase activation of MFN2 (Figure 4A). Thus, MFN2 mutated at K192, the lysine specific for MITOL-mediated ubiquitination, leads to the inability to connect the membrane between the ER and mitochondria, resulting in a failure of calcium ion transfer from the ER to mitochondria in the cell. The disruption of MCS between the ER and mitochondria is not fatal for the cell but affects setting the threshold for the complementation of mitochondrial or ER defects triggered by pathological conditions. Therefore, abnormal formation of the MCS might initiate or aggravate progressive, not early-developmental, diseases in humans. Actually, mutations in the MFN2 gene lead to its catalytic inactivation and subsequently trigger Charcot-Marie-Tooth disease type 2A (CMT2A) in the peripheral nervous system [73]. Similarly, a primary mutation in Sig1R or SOD1, a cause of inherited juvenile ALS, was shown to result in the disruption of MCS between the ER and mitochondria [74].

What is the in vivo and physiological contribution of MCS between the ER and mitochondria? What is triggered by the perturbation of MCS between the ER and mitochondria in vivo remains poorly characterized. There are also serious concerns regarding the accuracy of the methods for analyzing the MCS structure in vivo. To obtain an understanding of the morphology of MCS between the ER and mitochondria in vivo, electron microscopy is mostly adopted (sometimes potentially being the only method available in vivo). However, both organelles, mitochondria and the ER, exhibit complex and diverse morphology. A single image obtained from an electron microscope was limited and restricts us to evaluating a whole picture of mitochondrial states. It is also difficult to judge whether the membrane structure in a single image is part of the continuous ER or an independent part. Therefore, there is an urgent need to accurately investigate the structure of MCS between continuous ER and mitochondria in the brain.

To obtain a precise understanding of the morphology of MCS between mitochondria and the ER in neurons, we recently performed three-dimensional (3D) reconstructions from serial electron microscopy images of mitochondria using serial block—face scanning electron microscopy (SBF-SEM) [75]. Interestingly, over 95% of mitochondria had at least one contact site with the continuous ER in the brain regardless of the individual morphology of mitochondria, suggesting that MCS with the ER is pivotal for almost all mitochondria, at least in neurons (in this analysis, only neurons were selected morphologically from serial images). Larger mitochondria required more MCS with continuous ER. However, each MCS with continuous ER displayed morphological differences, such as large and small types. Contacts between the ER and mitochondria with distinct sizes might involve distinct roles. We also examined the physiological contribution of the MITOL-MFN2 axis to MCS using mice with nerve-specific ablation of MITOL. The MITOL-deleted brain showed the formation of fewer and smaller MCS between continuous ER and mitochondria. However, the phenotype of MITOL-deleted neurons regarding MCS appeared to be restricted to only larger mitochondria. When taking the findings as a whole, mitochondrial defects in the brain with MITOL deletion were mild, leading to slight developmental abnormalities in the brain. Currently, we are examining the pathological contribution of disrupted mitochondrion-ER connections during disease development by performing crossing with murine models for aging-related diseases.

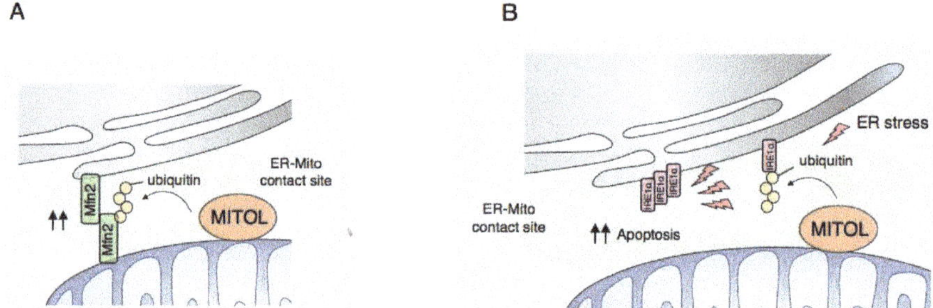

Figure 4. (**A**) Membrane contact site (MCS) formation by MITOL-MFN2 axis. MITOL ubiquitinates mitochondrial MFN2. The ubiquitinated MFN2 enhances its GTPase activity, triggering trans-oligomerization between mitochondrial MFN2 and ER-localized MFN2 for tethering both organelles; mitochondria and the ER; (**B**) UPR regulation by MITOL-IRE1α axis. MITOL ubiquitinates IRE1α at ER-mitochondria contact site in unstressed and low-stressed conditions regarding the ER. The ubiquitination of IRE1α does not perturbate the oligomerization itself in response to ER stress, however, leading to short-term stabilization and smaller oligomerization. In contract, the ubiquitination by MITOL is reduced under severe or chronic ER stress, resulting in continuous oligomerization of IRE1α and apoptotic switching of IRE1α signaling.

4.3. UPR Regulation by MITOL

In addition to the basic role of MITOL in unstressed conditions, recent evidence has implied that MITOL serves as a unique signaling regulator in several specific conditions. We recently identified a novel substrate, IRE1α, for MITOL at the MCS between the ER and mitochondria [76]. IRE1α is an ER membrane-integrated protein that possesses bifunctional activity as a kinase and endoribonuclease. The luminal domain of IRE1α contributes to monitoring the stress level of the ER by sensing the emergence of unfolded proteins. Meanwhile, the cytosolic domain of IRE1α, containing both catalytic domains, contributes to signal transduction from the ER to outside it in order to recover from ER damage during ER stress conditions. The ubiquitination of IRE1α by MITOL determines the persistence of IRE1α activation upon ER stress (Figure 4B). MITOL-catalyzed ubiquitination of IRE1α is accompanied by binding to the mediator BIM for a smaller state of (or less stable) IRE1α oligomerization, allowing the catalytic activation of IRE1α only for the short term. This regulation of IRE1α is pivotal with regard to termination of the UPR after the recovery of ER homeostasis. Unresolved IRE1α activation indeed triggers an alternative outcome of IRE1α signaling, namely, cell death. Importantly, severe ER stress attenuates the ubiquitination levels of IRE1α by MITOL via unclear mechanisms. Therefore, the unlimited IRE1α oligomerization and activation promote the induction of cell death during severe and irremediable ER stress. Taking these findings together, MITOL prevents the signal switching of IRE1α from cell survival to cell death via direct ubiquitination of IRE1α at the MCS between the ER and mitochondria. The ER network accumulates severe abnormalities, both functional and morphological, during irremediable ER stress. Thus, it might be reasonable to set the monitoring system, related to the signal switch of the UPR, outside the ER, such as on the mitochondrial surface. Several mitochondrial molecules therefore hold therapeutic potential in diseases initiated or aggravated by ER stress, not only mitochondrial stress.

5. Discussion

It has become clear that ubiquitin not only regulates proteasome-dependent proteolysis, but also partly regulates various cellular functions such as signal transduction, membrane protein transport, selective autophagy, and aging. The multiple functions of ubiquitin are derived from the structural diversity of ubiquitin modifications, eight different linking modes, chain length, branching, and a wide variety of higher-order structures resulting from combinations of post-translational modifications of ubiquitin. Globally, the success of proteasome inhibitors in the treatment of cancer has led to substantial progress in the development of ubiquitin drugs. In particular, targeted proteolysis and induction techniques using small molecules such as PROTACs and SNIPERs are attracting substantial attention as a new generation of drug discovery methods, and there is growing momentum for the formation of groups based on the fusion of ubiquitin research and chemical biology. We expect MITOL, an E3 ubiquitin ligase, to be one of the successful examples of academic drug discovery, as it is becoming a potential target for promising seeds.

Author Contributions: Conceptualization, I.S. and Y.S.; writing—I.S., K.T., S.N. and S.Y.; writing—review and editing—I.S. and S.Y. All authors have read and agreed to the published version of the manuscript.

Funding: This research received no external funding.

Conflicts of Interest: The authors declare no conflict of interest.

References

1. Amm, I.; Sommer, T.; Wolf, D.H. Protein quality control and elimination of protein waste: The role of the ubiquitin-proteasome system. *Biochim. Biophys. Acta Mol. Cell Res.* **2014**, *1843*, 182–196. [CrossRef] [PubMed]
2. Saeki, Y. Ubiquitin recognition by the proteasome. *J. Biochem.* **2017**, *161*, 113–124. [CrossRef] [PubMed]
3. Martensson, C.U.; Priesnitz, C.; Song, J.; Ellenrieder, L.; Doan, K.N.; Boos, F.; Floerchinger, A.; Zufall, N.; Oeljeklaus, S.; Warscheid, B.; et al. Mitochondrial protein translocation-associated degradation. *Nature* **2019**, *569*, 679–683. [CrossRef] [PubMed]
4. Vilchez, D.; Saez, I.; Dillin, A. The role of protein clearance mechanisms in organismal ageing and age-related diseases. *Nat. Commun.* **2014**, *5*, 5659. [CrossRef] [PubMed]
5. Deng, H.X.; Shi, Y.; Furukawa, Y.; Zhai, H.; Fu, R.; Liu, E.; Gorrie, G.H.; Khan, M.S.; Hung, W.Y.; Bigio, E.H.; et al. Conversion to the amyotrophic lateral sclerosis phenotype is associated with intermolecular linked insoluble aggregates of sod1 in mitochondria. *Proc. Natl. Acad. Sci. USA* **2006**, *103*, 7142–7147. [CrossRef] [PubMed]
6. Sugiura, A.; Yonashiro, R.; Fukuda, T.; Matsushita, N.; Nagashima, S.; Inatome, R.; Yanagi, S. A mitochondrial ubiquitin ligase mitol controls cell toxicity of polyglutamine-expanded protein. *Mitochondrion* **2011**, *11*, 139–146. [CrossRef]
7. Yonashiro, R.; Sugiura, A.; Miyachi, M.; Fukuda, T.; Matsushita, N.; Inatome, R.; Ogata, Y.; Suzuki, T.; Dohmae, N.; Yanagi, S. Mitochondrial ubiquitin ligase mitol ubiquitinates mutant sod1 and attenuates mutant sod1-induced ros generation. *Mol. Biol. Cell* **2009**, *20*, 4254–4530. [CrossRef]
8. Lin, H.; Li, S.; Shu, H.B. The membrane-associated march e3 ligase family: Emerging roles in immune regulation. *Front. Immunol.* **2019**, *10*, 1751. [CrossRef]

9. Park, Y.J.; Oanh, N.T.K.; Heo, J.; Kim, S.G.; Lee, H.S.; Lee, H.; Lee, J.H.; Kang, H.C.; Lim, W.; Yoo, Y.S.; et al. Dual targeting of rig-i and mavs by march5 mitochondria ubiquitin ligase in innate immunity. *Cell. Signal.* **2020**, *67*, 109520. [CrossRef] [PubMed]

10. Yan, B.R.; Zhou, L.; Hu, M.M.; Li, M.; Lin, H.; Yang, Y.; Wang, Y.Y.; Shu, H.B. Pkacs attenuate innate antiviral response by phosphorylating visa and priming it for march5-mediated degradation. *PLoS Pathog.* **2017**, *13*, e1006648. [CrossRef] [PubMed]

11. Yoo, Y.S.; Park, Y.Y.; Kim, J.H.; Cho, H.; Kim, S.H.; Lee, H.S.; Kim, T.H.; Sun Kim, Y.; Lee, Y.; Kim, C.J.; et al. The mitochondrial ubiquitin ligase march5 resolves mavs aggregates during antiviral signalling. *Nat. Commun.* **2015**, *6*, 7910. [CrossRef] [PubMed]

12. Shi, H.X.; Liu, X.; Wang, Q.; Tang, P.P.; Liu, X.Y.; Shan, Y.F.; Wang, C. Mitochondrial ubiquitin ligase march5 promotes tlr7 signaling by attenuating tank action. *PLoS Pathog.* **2011**, *7*, e1002057. [CrossRef] [PubMed]

13. Yoo, Y.S.; Park, Y.J.; Lee, H.S.; Oanh, N.T.K.; Cho, M.Y.; Heo, J.; Lee, E.S.; Cho, H.; Park, Y.Y.; Cho, H. Mitochondria ubiquitin ligase, march5 resolves hepatitis b virus x protein aggregates in the liver pathogenesis. *Cell Death Dis.* **2019**, *10*, 938. [CrossRef] [PubMed]

14. Djajawi, T.M.; Liu, L.; Gong, J.N.; Huang, A.S.; Luo, M.J.; Xu, Z.; Okamoto, T.; Call, M.J.; Huang, D.C.S.; van Delft, M.F. March5 requires mtch2 to coordinate proteasomal turnover of the mcl1:Noxa complex. *Cell Death Differ.* **2020**, *10*, 1–13. [CrossRef] [PubMed]

15. Timucin, A.C.; Basaga, H.; Kutuk, O. Selective targeting of antiapoptotic bcl-2 proteins in cancer. *Med. Res. Rev.* **2019**, *39*, 146–175. [CrossRef] [PubMed]

16. Subramanian, A.; Andronache, A.; Li, Y.C.; Wade, M. Inhibition of march5 ubiquitin ligase abrogates mcl1-dependent resistance to bh3 mimetics via noxa. *Oncotarget* **2016**, *7*, 15986–16002. [CrossRef]

17. Haschka, M.D.; Karbon, G.; Soratroi, C.; O'Neill, K.L.; Luo, X.; Villunger, A. March5-dependent degradation of mcl1/noxa complexes defines susceptibility to antimitotic drug treatment. *Cell Death Differ.* **2020**. [CrossRef]

18. Tang, H.; Peng, S.; Dong, Y.; Yang, X.; Yang, P.; Yang, L.; Yang, B.; Bao, G. March5 overexpression contributes to tumor growth and metastasis and associates with poor survival in breast cancer. *Cancer Manag. Res.* **2019**, *11*, 201–215. [CrossRef]

19. Pickles, S.; Vigie, P.; Youle, R.J. Mitophagy and quality control mechanisms in mitochondrial maintenance. *Curr. Biol.* **2018**, *28*, R170–R185. [CrossRef]

20. Sekine, S.; Youle, R.J. Pink1 import regulation; a fine system to convey mitochondrial stress to the cytosol. *BMC Biol.* **2018**, *16*, 2. [CrossRef]

21. Harper, J.W.; Ordureau, A.; Heo, J.M. Building and decoding ubiquitin chains for mitophagy. *Nat. Rev. Mol. Cell Biol.* **2018**, *19*, 93–108. [CrossRef] [PubMed]

22. Hasson, S.A.; Kane, L.A.; Yamano, K.; Huang, C.H.; Sliter, D.A.; Buehler, E.; Wang, C.; Heman-Ackah, S.M.; Hessa, T.; Guha, R.; et al. High-content genome-wide rnai screens identify regulators of parkin upstream of mitophagy. *Nature* **2013**, *504*, 291–295. [CrossRef] [PubMed]

23. Koyano, F.; Yamano, K.; Kosako, H.; Tanaka, K.; Matsuda, N. Parkin recruitment to impaired mitochondria for nonselective ubiquitylation is facilitated by mitol. *J. Biol. Chem.* **2019**, *294*, 10300–10314. [CrossRef] [PubMed]

24. Koyano, F.; Yamano, K.; Kosako, H.; Kimura, Y.; Kimura, M.; Fujiki, Y.; Tanaka, K.; Matsuda, N. Parkin-mediated ubiquitylation redistributes mitol/march5 from mitochondria to peroxisomes. *EMBO Rep.* **2019**, *20*, e47728. [CrossRef] [PubMed]

25. Wei, H.; Liu, L.; Chen, Q. Selective removal of mitochondria via mitophagy: Distinct pathways for different mitochondrial stresses. *Biochim. Biophys. Acta Mol. Cell Res.* **2015**, *1853*, 2784–2790. [CrossRef]

26. Chen, Z.; Liu, L.; Cheng, Q.; Li, Y.; Wu, H.; Zhang, W.; Wang, Y.; Sehgal, S.A.; Siraj, S.; Wang, X.; et al. Mitochondrial e3 ligase march5 regulates fundc1 to fine-tune hypoxic mitophagy. *EMBO Rep.* **2017**, *18*, 495–509. [CrossRef] [PubMed]

27. Park, Y.Y.; Lee, S.; Karbowski, M.; Neutzner, A.; Youle, R.J.; Cho, H. Loss of march5 mitochondrial e3 ubiquitin ligase induces cellular senescence through dynamin-related protein 1 and mitofusin 1. *J. Cell Sci.* **2010**, *123*, 619–626. [CrossRef] [PubMed]

28. Smirnova, E.; Griparic, L.; Shurland, D.L.; van der Bliek, A.M. Dynamin-related protein drp1 is required for mitochondrial division in mammalian cells. *Mol. Biol. Cell* **2001**, *12*, 2245–2256. [CrossRef] [PubMed]

29. Kraus, F.; Ryan, M.T. The constriction and scission machineries involved in mitochondrial fission. *J. Cell Sci.* **2017**, *130*, 2953–2960. [CrossRef] [PubMed]

30. Tilokani, L.; Nagashima, S.; Paupe, V.; Prudent, J. Mitochondrial dynamics: Overview of molecular mechanisms. *Essays Biochem.* **2018**, *62*, 341–360. [PubMed]

31. Mears, J.A.; Lackner, L.L.; Fang, S.; Ingerman, E.; Nunnari, J.; Hinshaw, J.E. Conformational changes in dnm1 support a contractile mechanism for mitochondrial fission. *Nat. Struct. Mol. Biol.* **2011**, *18*, 20–26. [CrossRef] [PubMed]

32. Koirala, S.; Guo, Q.; Kalia, R.; Bui, H.T.; Eckert, D.M.; Frost, A.; Shaw, J.M. Interchangeable adaptors regulate mitochondrial dynamin assembly for membrane scission. *Proc. Natl. Acad. Sci. USA* **2013**, *110*, E1342–E1351. [CrossRef] [PubMed]

33. Elgass, K.; Pakay, J.; Ryan, M.T.; Palmer, C.S. Recent advances into the understanding of mitochondrial fission. *Biochim. Biophys. Acta Mol. Cell Res.* **2013**, *1833*, 150–161. [CrossRef] [PubMed]

34. Prudent, J.; McBride, H.M. The mitochondria-endoplasmic reticulum contact sites: A signalling platform for cell death. *Curr. Opin. Cell Biol.* **2017**, *47*, 52–63. [CrossRef] [PubMed]

35. Braschi, E.; Zunino, R.; McBride, H.M. Mapl is a new mitochondrial sumo e3 ligase that regulates mitochondrial fission. *EMBO Rep.* **2009**, *10*, 748–754. [CrossRef] [PubMed]

36. Horn, S.R.; Thomenius, M.J.; Johnson, E.S.; Freel, C.D.; Wu, J.Q.; Coloff, J.L.; Yang, C.S.; Tang, W.; An, J.; Ilkayeva, O.R.; et al. Regulation of mitochondrial morphology by apc/ccdh1-mediated control of drp1 stability. *Mol. Biol. Cell* **2011**, *22*, 1207–1216. [CrossRef] [PubMed]

37. Cui, M.; Tang, X.; Christian, W.V.; Yoon, Y.; Tieu, K. Perturbations in mitochondrial dynamics induced by human mutant pink1 can be rescued by the mitochondrial division inhibitor mdivi-1. *J. Biol. Chem.* **2010**, *285*, 11740–11752. [CrossRef]

38. Wang, H.; Song, P.; Du, L.; Tian, W.; Yue, W.; Liu, M.; Li, D.; Wang, B.; Zhu, Y.; Cao, C.; et al. Parkin ubiquitinates drp1 for proteasome-dependent degradation: Implication of dysregulated mitochondrial dynamics in parkinson disease. *J. Biol. Chem.* **2011**, *286*, 11649–11658. [CrossRef]

39. Yonashiro, R.; Ishido, S.; Kyo, S.; Fukuda, T.; Goto, E.; Matsuki, Y.; Ohmura-Hoshino, M.; Sada, K.; Hotta, H.; Yamamura, H.; et al. A novel mitochondrial ubiquitin ligase plays a critical role in mitochondrial dynamics. *EMBO J.* **2006**, *25*, 3618–3626. [CrossRef]

40. Nakamura, N.; Kimura, Y.; Tokuda, M.; Honda, S.; Hirose, S. March-v is a novel mitofusin 2- and drp1-binding protein able to change mitochondrial morphology. *EMBO Rep.* **2006**, *7*, 1019–1022. [CrossRef]

41. Karbowski, M.; Neutzner, A.; Youle, R.J. The mitochondrial e3 ubiquitin ligase march5 is required for drp1 dependent mitochondrial division. *J. Cell Biol.* **2007**, *178*, 71–84. [CrossRef] [PubMed]

42. Xu, S.; Cherok, E.; Das, S.; Li, S.; Roelofs, B.A.; Ge, S.X.; Polster, B.M.; Boyman, L.; Lederer, W.J.; Wang, C.; et al. Mitochondrial e3 ubiquitin ligase march5 controls mitochondrial fission and cell sensitivity to stress-induced apoptosis through regulation of mid49 protein. *Mol. Biol. Cell* **2016**, *27*, 349–359. [CrossRef] [PubMed]

43. Rizza, S.; Cardaci, S.; Montagna, C.; Di Giacomo, G.; De Zio, D.; Bordi, M.; Maiani, E.; Campello, S.; Borreca, A.; Puca, A.A.; et al. S-nitrosylation drives cell senescence and aging in mammals by controlling mitochondrial dynamics and mitophagy. *Proc. Natl. Acad. Sci. USA* **2018**, *115*, E3388–E3397. [CrossRef] [PubMed]

44. Cho, D.H.; Nakamura, T.; Fang, J.; Cieplak, P.; Godzik, A.; Gu, Z.; Lipton, S.A. S-nitrosylation of drp1 mediates beta-amyloid-related mitochondrial fission and neuronal injury. *Science* **2009**, *324*, 102–105. [CrossRef] [PubMed]

45. Stroissnigg, H.; Trancikova, A.; Descovich, L.; Fuhrmann, J.; Kutschera, W.; Kostan, J.; Meixner, A.; Nothias, F.; Propst, F. S-nitrosylation of microtubule-associated protein 1b mediates nitric-oxide-induced axon retraction. *Nat. Cell Biol.* **2007**, *9*, 1035–1045. [CrossRef] [PubMed]

46. Liu, L.; Vo, A.; Liu, G.; McKeehan, W.L. Distinct structural domains within c19orf5 support association with stabilized microtubules and mitochondrial aggregation and genome destruction. *Cancer Res.* **2005**, *65*, 4191–4201. [CrossRef]

47. Yonashiro, R.; Kimijima, Y.; Shimura, T.; Kawaguchi, K.; Fukuda, T.; Inatome, R.; Yanagi, S. Mitochondrial ubiquitin ligase mitol blocks s-nitrosylated map1b-light chain 1-mediated mitochondrial dysfunction and neuronal cell death. *Proc. Natl. Acad. Sci. USA* **2012**, *109*, 2382–2387. [CrossRef]

48. Friedman, J.R.; Lackner, L.L.; West, M.; DiBenedetto, J.R.; Nunnari, J.; Voeltz, G.K. Er tubules mark sites of mitochondrial division. *Science* **2011**, *334*, 358–362. [CrossRef]

49. Murley, A.; Nunnari, J. The emerging network of mitochondria-organelle contacts. *Mol. Cell* **2016**, *61*, 648–653. [CrossRef]

50. Wong, Y.C.; Ysselstein, D.; Krainc, D. Mitochondria-lysosome contacts regulate mitochondrial fission via rab7 gtp hydrolysis. *Nature* **2018**, *554*, 382–386. [CrossRef]

51. Nagashima, S.; Tabara, L.C.; Tilokani, L.; Paupe, V.; Anand, H.; Pogson, J.H.; Zunino, R.; McBride, H.M.; Prudent, J. Golgi-derived pi(4)p-containing vesicles drive late steps of mitochondrial division. *Science* **2020**, *367*, 1366–1371. [CrossRef] [PubMed]

52. Onoue, K.; Jofuku, A.; Ban-Ishihara, R.; Ishihara, T.; Maeda, M.; Koshiba, T.; Itoh, T.; Fukuda, M.; Otera, H.; Oka, T.; et al. Fis1 acts as a mitochondrial recruitment factor for tbc1d15 that is involved in regulation of mitochondrial morphology. *J. Cell Sci.* **2013**, *126*, 176–185. [CrossRef] [PubMed]

53. Yamano, K.; Fogel, A.I.; Wang, C.; van der Bliek, A.M.; Youle, R.J. Mitochondrial rab gaps govern autophagosome biogenesis during mitophagy. *eLife* **2014**, *3*, e01612. [CrossRef] [PubMed]

54. Legros, F.; Lombes, A.; Frachon, P.; Rojo, M. Mitochondrial fusion in human cells is efficient, requires the inner membrane potential, and is mediated by mitofusins. *Mol. Biol. Cell* **2002**, *13*, 4343–4354. [CrossRef] [PubMed]

55. Chen, H.; Detmer, S.A.; Ewald, A.J.; Griffin, E.E.; Fraser, S.E.; Chan, D.C. Mitofusins mfn1 and mfn2 coordinately regulate mitochondrial fusion and are essential for embryonic development. *J. Cell Biol.* **2003**, *160*, 189–200. [CrossRef]

56. Ishihara, N.; Eura, Y.; Mihara, K. Mitofusin 1 and 2 play distinct roles in mitochondrial fusion reactions via gtpase activity. *J. Cell Sci.* **2004**, *117*, 6535–6546. [CrossRef] [PubMed]

57. Fu, M.; St-Pierre, P.; Shankar, J.; Wang, P.T.; Joshi, B.; Nabi, I.R. Regulation of mitophagy by the gp78 e3 ubiquitin ligase. *Mol. Biol. Cell* **2013**, *24*, 1153–1162. [CrossRef]

58. Shankar, J.; Kojic, L.D.; St-Pierre, P.; Wang, P.T.; Fu, M.; Joshi, B.; Nabi, I.R. Raft endocytosis of amf regulates mitochondrial dynamics through rac1 signaling and the gp78 ubiquitin ligase. *J. Cell Sci.* **2013**, *126*, 3295–3304. [CrossRef]

59. Bagher, P.; Jiao, J.; Owen Smith, C.; Cota, C.D.; Gunn, T.M. Characterization of mahogunin ring finger-1 expression in mice. *Pigment Cell Res.* **2006**, *19*, 635–643. [CrossRef]

60. Yue, W.; Chen, Z.; Liu, H.; Yan, C.; Chen, M.; Feng, D.; Yan, C.; Wu, H.; Du, L.; Wang, Y.; et al. A small natural molecule promotes mitochondrial fusion through inhibition of the deubiquitinase usp30. *Cell Res.* **2014**, *24*, 482–496. [CrossRef]

61. Tang, F.L.; Liu, W.; Hu, J.X.; Erion, J.R.; Ye, J.; Mei, L.; Xiong, W.C. Vps35 deficiency or mutation causes dopaminergic neuronal loss by impairing mitochondrial fusion and function. *Cell Rep.* **2015**, *12*, 1631–1643. [CrossRef] [PubMed]

62. Leboucher, G.P.; Tsai, Y.C.; Yang, M.; Shaw, K.C.; Zhou, M.; Veenstra, T.D.; Glickman, M.H.; Weissman, A.M. Stress-induced phosphorylation and proteasomal degradation of mitofusin 2 facilitates mitochondrial fragmentation and apoptosis. *Mol. Cell* **2012**, *47*, 547–557. [CrossRef] [PubMed]

63. Choudhary, V.; Kaddour-Djebbar, I.; Alaisami, R.; Kumar, M.V.; Bollag, W.B. Mitofusin 1 degradation is induced by a disruptor of mitochondrial calcium homeostasis, cgp37157: A role in apoptosis in prostate cancer cells. *Int. J. Oncol.* **2014**, *44*, 1767–1773. [CrossRef] [PubMed]

64. Park, Y.Y.; Cho, H. Mitofusin 1 is degraded at g2/m phase through ubiquitylation by march5. *Cell Div.* **2012**, *7*, 25. [CrossRef] [PubMed]

65. Kim, H.J.; Nagano, Y.; Choi, S.J.; Park, S.Y.; Kim, H.; Yao, T.P.; Lee, J.Y. Hdac6 maintains mitochondrial connectivity under hypoxic stress by suppressing march5/mitol dependent mfn2 degradation. *Biochem. Biophys. Res. Commun.* **2015**, *464*, 1235–1240. [CrossRef] [PubMed]

66. Csordas, G.; Weaver, D.; Hajnoczky, G. Endoplasmic reticulum-mitochondrial contactology: Structure and signaling functions. *Trends Cell Biol.* **2018**, *28*, 523–540. [CrossRef]

67. Hirabayashi, Y.; Kwon, S.K.; Paek, H.; Pernice, W.M.; Paul, M.A.; Lee, J.; Erfani, P.; Raczkowski, A.; Petrey, D.S.; Pon, L.A.; et al. Er-mitochondria tethering by pdzd8 regulates ca(2+) dynamics in mammalian neurons. *Science* **2017**, *358*, 623–630. [CrossRef]

68. Tamura, Y.; Kawano, S.; Endo, T. Organelle contact zones as sites for lipid transfer. *J. Biochem.* **2019**, *165*, 115–123. [CrossRef]

69. Raffaello, A.; Mammucari, C.; Gherardi, G.; Rizzuto, R. Calcium at the center of cell signaling: Interplay between endoplasmic reticulum, mitochondria, and lysosomes. *Trends Biochem. Sci.* **2016**, *41*, 1035–1049. [CrossRef]

70. Scorrano, L.; De Matteis, M.A.; Emr, S.; Giordano, F.; Hajnoczky, G.; Kornmann, B.; Lackner, L.L.; Levine, T.P.; Pellegrini, L.; Reinisch, K.; et al. Coming together to define membrane contact sites. *Nat. Commun.* **2019**, *10*, 1287. [CrossRef]

71. Sugiura, A.; Nagashima, S.; Tokuyama, T.; Amo, T.; Matsuki, Y.; Ishido, S.; Kudo, Y.; McBride, H.M.; Fukuda, T.; Matsushita, N.; et al. Mitol regulates endoplasmic reticulum-mitochondria contacts via mitofusin2. *Mol. Cell* **2013**, *51*, 20–34. [CrossRef] [PubMed]

72. Filadi, R.; Pendin, D.; Pizzo, P. Mitofusin 2: From functions to disease. *Cell Death Dis.* **2018**, *9*, 330. [CrossRef] [PubMed]

73. Zuchner, S.; Mersiyanova, I.V.; Muglia, M.; Bissar-Tadmouri, N.; Rochelle, J.; Dadali, E.L.; Zappia, M.; Nelis, E.; Patitucci, A.; Senderek, J.; et al. Mutations in the mitochondrial gtpase mitofusin 2 cause charcot-marie-tooth neuropathy type 2a. *Nat. Genet.* **2004**, *36*, 449–451. [CrossRef] [PubMed]

74. Watanabe, S.; Ilieva, H.; Tamada, H.; Nomura, H.; Komine, O.; Endo, F.; Jin, S.; Mancias, P.; Kiyama, H.; Yamanaka, K. Mitochondria-associated membrane collapse is a common pathomechanism in sigmar1- and sod1-linked als. *EMBO Mol. Med.* **2016**, *8*, 1421–1437. [CrossRef] [PubMed]

75. Nagashima, S.; Takeda, K.; Ohno, N.; Ishido, S.; Aoki, M.; Saitoh, Y.; Takada, T.; Tokuyama, T.; Sugiura, A.; Fukuda, T.; et al. Mitol deletion in the brain impairs mitochondrial structure and er tethering leading to oxidative stress. *Life Sci. Alliance* **2019**, *2*. [CrossRef] [PubMed]

76. Takeda, K.; Nagashima, S.; Shiiba, I.; Uda, A.; Tokuyama, T.; Ito, N.; Fukuda, T.; Matsushita, N.; Ishido, S.; Iwawaki, T.; et al. Mitol prevents er stress-induced apoptosis by ire1alpha ubiquitylation at er-mitochondria contact sites. *EMBO J.* **2019**, *38*, e100999. [CrossRef]

International Journal of
Molecular Sciences

Review

Degradation of Tyrosine Hydroxylase by the Ubiquitin-Proteasome System in the Pathogenesis of Parkinson's Disease and Dopa-Responsive Dystonia

Ichiro Kawahata * and Kohji Fukunaga *

Department of Pharmacology, Graduate School of Pharmaceutical Sciences, Tohoku University, Sendai 980-8578, Japan
* Correspondence: kawahata@tohoku.ac.jp (I.K.); kfukunaga@tohoku.ac.jp (K.F.); Tel.: +81-22-795-6838 (I.K.); +81-22-795-6836 (K.F.); Fax: +81-22-795-6835 (I.K. & K.F.)

Received: 7 April 2020; Accepted: 25 May 2020; Published: 27 May 2020

Abstract: Nigrostriatal dopaminergic systems govern physiological functions related to locomotion, and their dysfunction leads to movement disorders, such as Parkinson's disease and dopa-responsive dystonia (Segawa disease). Previous studies revealed that expression of the gene encoding nigrostriatal tyrosine hydroxylase (TH), a rate-limiting enzyme of dopamine biosynthesis, is reduced in Parkinson's disease and dopa-responsive dystonia; however, the mechanism of TH depletion in these disorders remains unclear. In this article, we review the molecular mechanism underlying the neurodegeneration process in dopamine-containing neurons and focus on the novel degradation pathway of TH through the ubiquitin-proteasome system to advance our understanding of the etiology of Parkinson's disease and dopa-responsive dystonia. We also introduce the relation of α-synuclein propagation with the loss of TH protein in Parkinson's disease as well as anticipate therapeutic targets and early diagnosis of these diseases.

Keywords: Parkinson's disease; dopa-responsive dystonia; tyrosine hydroxylase; α-synuclein; fatty acid-binding protein 3; ubiquitination; proteasomal degradation; ubiquitin-proteasome system

---

1. Introduction

Parkinson's disease (PD) is a common disease whose prevalence is increasing owing to the aging society. PD is clinically characterized by movement disabilities, such as resting tremor, rigidity, and bradykinesia [1]. PD is also defined pathologically by the selective degeneration of dopaminergic neurons in the substantia nigra pars compacta (SNpc) and by the cytoplasmic accumulation of proteinaceous inclusions, termed Lewy bodies [2,3]. Dopa-responsive dystonia (DRD), also termed as Segawa disease, is a disorder that involves involuntary muscle contractions, tremors, and other uncontrolled movements, which usually appear during childhood [4]. DRD patients present with reduced nigrostriatal dopaminergic function [5,6]. As widely known, PD and DRD are neurodegenerative disorders that predominately affect midbrain dopamine-producing neurons. Though dysfunctions of the dopaminergic system are involved in neurological disorders, such as Tourette's syndrome [7], schizophrenia [8,9], pituitary tumors [10], PD [11–15], and DRD [4,5,16], the loss of nigrostriatal tyrosine hydroxylase (TH) protein is distinctive in PD and DRD. The etiology of PD and DRD has been studied in the past quarter-century; however, the molecular mechanism of the onset of the disorders has not been completely elucidated. In particular, the reason why the TH protein, which is a rate-limiting enzyme of dopamine biosynthesis, is lost in mesencephalic dopaminergic neurons in PD and DRD, and is not entirely understood. In this review, we focus on the molecular mechanism of the loss of TH protein in the neurodegeneration process in PD and DRD by introducing the degradation of phosphorylated TH protein through the ubiquitin-proteasome system. We also introduce the relation between the loss of TH protein and the propagation of α-synuclein, which is a well-known protein in PD pathology, to clarify the mechanism underlying the reduction of nigrostriatal dopamine function and the loss of TH protein in these movement disorders.

2. Pathology of Parkinson's Disease and Dopa-Responsive Dystonia

PD was first diagnosed and described in detail by James Parkinson in 1817 [1]. PD affects over 10 million worldwide, particularly 1%–3% of the global population aged over 60 years and up to 50% of individuals aged over 85 [17]. The clinical features of PD are resting tremor, rigidity, bradykinesia, gait disturbances, postural instability [1], and dementia, which becomes common in the advanced stage of the disease [18]. Pathologically, PD is characterized by the loss of dopamine-biosynthesizing neurons in the substantia nigra pars compacta (SNpc), and by the abnormal deposition of α-synuclein in the cell body (called Lewy body) and in neuronal processes (called Lewy neurites). The risk of developing PD is twice as high in men than in women; particularly, women have a higher mortality rate and faster progression of the disease [19]. Moreover, 90% of PD are sporadic, and hereditary and environmental factors are thought to be involved in the etiology of PD. Currently, over 20 causative or putative genes of hereditary PD have been identified by genetic linkage analysis [20]; for example, *SNCA* (*PARK1*, *PARK4*), *Parkin* (*PARK2*), *DJ-1* (*PARK7*), and *LRRK2* (*PARK8*) [21–27], which encode α-synuclein, Parkin, protein/nucleic acid deglycase DJ-1, and leucine-rich repeat kinase 2 (LRRK2) protein, respectively. These different gene mutations in familial PD point to the possibility that an alteration in protein conformation and/or degradation could be a key to the degenerative process.

Another dopaminergic disorder, dystonia, is a heterogenous, neurological disorder characterized by abnormal involuntary sustained muscle contractions, frequently causing twisting and repetitive movements or abnormal postures [28]. It is believed that approximately 70% of all patients with dystonia have idiopathic rather than symptomatic dystonia. The mechanisms of dystonia pathogenesis include abnormalities in the regulation of dopaminergic transcription, nigrostriatal dopamine signaling, and loss of inhibition at neuronal circuits. There are at least 11 different genes involved in autosomal dominant inherited dystonia, one in autosomal recessive inherited dystonia, and another in X-linked recessive inherited dystonia [29]. One of the most common genetic dystonia, dopa-responsive dystonia (DRD, *DYT5*), is mainly caused by the mutation of *GCH1* [4,30], which encodes GTP cyclohydrolase 1 (GCH1). Women are more commonly affected, with men showing a lower penetrance of mutations [31,32]; this disease develops in early childhood at approximately age 5–8 [4].

In common, PD and DRD are associated with impaired nigrostriatal dopaminergic function [33]. Nigrostriatal dopaminergic projections play a central role in the control of voluntary movements, and their degeneration has been implicated in Parkinsonian clinical symptoms. In addition, the dopaminergic system, originating in the SNpc and the ventral tegmental area (VTA), which mainly projects to the striatum (mesostriatal pathway) and the prefrontal context (mesocortical pathway), plays a major motivational role in behavioral actions [34–36]. Consistently, lesions in nigral neurons lead to simultaneous dysfunction of agonist and antagonist muscle pairs in animal models of parkinsonism [37] and idiopathic PD [15]. The dopaminergic function is regulated by dopamine, which is biosynthesized from L-tyrosine by TH and aromatic L-amino acid decarboxylase (AADC). TH requires tetrahydrobiopterin, which is biosynthesized by GCH1, to perform its enzymatic activity. Because the enzymatic activity of TH protein strictly controls the rate-limiting step of dopamine biosynthesis, unlike those of other dopamine biosynthesizing enzymes, the expression level and activity of TH directly affect intracellular dopamine amount. Thus, we next focus on the physiological features of TH protein and its implications in PD and DRD pathogenesis.

3. Physiology of Tyrosine Hydroxylase Phosphorylation

TH is a rate-limiting enzyme for dopamine biosynthesis [38] and is selectively expressed in monoaminergic neurons in the central nervous system. In humans, TH protein has four isoforms with different molecular weight, which are derived from the same gene through alternative splicing of mRNA [39,40]. This protein also has two isoforms in monkeys and only a single isoform in all nonprimate mammals [41,42]. The catalytic domain of TH is located within the C-terminal area, whereas the region that controls enzyme activity (the regulatory domain) is located at the N-terminal end [43]. Four phosphorylation sites, namely Ser8, Ser19, Ser31, and Ser40, have been identified in the N-terminal region of TH [44], whereas the catalytic domain is in 188–456 amino acid residue [45]. TH is a homotetramer consisting of four subunits, and the C-terminal domain forms this homotetramer structure [46].

Two mechanisms can modulate the activity of TH: one is a medium- to long-term regulation of gene expression, such as enzyme stability, transcriptional regulation, RNA stability, alternative RNA splicing, and translational regulation. The regulation of TH is well known; its expression level depends on transcription driven by cyclic adenosine monophosphate (cAMP)-dependent responsive element (in promoter) [47] in a manner dependent on activator protein 1 (AP-1) [48,49], serum-responsive factor (SRF) [50], and nuclear receptor related-1 (Nurr1) [51]. The other is a short-term regulation of enzyme activity, such as feedback inhibition, allosteric regulation, and phosphorylation [47,52,53]. Many factors strictly regulate the activity of TH to control dopamine biosynthesis. Upon depolarization, cyclic AMP-dependent protein kinase (PKA) and calcium-calmodulin-dependent protein kinase II (CaMKII) are activated [54–56]. PKA phosphorylates TH at Ser40 and CaMKII phosphorylates TH at Ser19 [57,58]. Phosphorylation of Ser19 increases Ser40 phosphorylation, indicating that the phosphorylation of Ser19 can potentiate the phosphorylation of Ser40 and subsequent activation of TH [59]. Other stress-related protein kinases can also phosphorylate TH at Ser40 [52,53]. Phosphorylation at Ser40 leads to the liberation of dopamine from the active site of TH and changes the conformation to the high specific activity form [60]. Cytosolic free dopamine can bind to the active site of TH and deactivate the enzyme to suppress dopamine overproduction [61,62]. It has been reported that the phosphorylated form of TH is highly labile, whereas the dopamine-bound form is stable [63]. TH phosphorylated at Ser40 (pSer40-TH) is dephosphorylated by a protein phosphatase, such as protein phosphatase 2A (PP2A), because inhibition of PP2A with okadaic acid or microcystin induces an increase in pSer40-TH level [64–66]. Ser31 phosphorylation is mediated by extracellular signal-regulated kinase 1 (ERK1) and ERK2 [42,67], and its dephosphorylation is mediated by PP2A [66]. Because ERK signals are usually activated as part of the mitogen-activated protein kinase (MAPK) cascade for cell survival, dephosphorylation of TH phosphorylated at Ser31 (pSer31-TH) is very rare in living cells. Phosphorylation of TH at Ser8 has been shown in cultured rat pheochromocytoma PC12 cells and permeabilized bovine chromaffin cells after treatment with okadaic acid [57,66]. In contrast, no significant phenomena have been reported in cultured dopaminergic neurons and in vivo. These data suggest that TH regulation by Ser8 phosphorylation is not critical in the central nervous system.

4. Linkage of Tyrosine Hydroxylase Phosphorylation to Dopaminergic Pathology

As mentioned above, nigrostriatal TH protein is lost in PD and DRD. Ichinose et al. previously showed that Parkinsonian brains had very low levels of TH mRNAs in the substantia nigra compared with control brains, but no significant differences were found between schizophrenic and normal brains [68]. In addition, DRD patients have severely reduced (<3%) TH protein levels in the putamen [5,6]. These results suggest that TH protein levels in the nigrostriatal dopaminergic neurons are markedly decreased in both PD and DRD, but not in schizophrenia. Furthermore, Mogi et al. found that a decrease in total TH protein level in the striatum was greater than that in the total enzyme activity, as assessed by enzyme immunoassay [69]. This result suggests that upregulation of TH phosphorylation, which compensates decreased dopamine level, is linked to the reduction of nigrostriatal TH protein in PD. Intriguingly, we previously found that proteasomal inhibition leads to accumulation of pSer40-TH, which is ubiquitin-immunopositive, in nerve growth factor (NGF)-differentiated PC12D cells [70]. Moreover, Lewy bodies and Lewy neurites are pSer40-TH-immunopositive in PD [71]. TH protein, particularly phosphorylated TH, apparently forms intracellular aggregates easily [70,72]. In contrast, the dopamine- or biopterin-deficient state, which corresponds to PD or DRD, respectively, facilitates TH phosphorylation and leads to reduction of the total TH level in cultured cells [73,74]. The reduction of TH immunoreactivity can be observed in the midbrain and striatum of 6-pyruvoyl-tetrahydrobiopterin synthase-null and sepiapterin reductase-null mice, which are mouse models of tetrahydrobiopterin biosynthesis dysfunction [75,76]. Importantly, there is a difference in pathological features between PD and DRD, namely the presence or absence of abnormal protein accumulation. Here, a question arises. By which mechanism is nigral TH protein depleted, and does TH protein accumulate to form inclusions? Before we discuss the possible mechanism underlying the decrease in TH protein, let us take a brief look at protein degradation pathways.

5. Protein Degradation Pathways: Lysosome and Proteasome

The autophagy-lysosome and ubiquitin-proteasome pathways are the two main routes of protein and organelle clearance in eukaryotic cells [77] (Figure 1). Autophagy is a phenomenon in which cytoplasmic components are transported to lysosomes and degrade substrates, such as protein complexes and organelles, using lysosomal enzymes. There are various types of autophagy, namely selective and nonselective autophagy. The bulk degradation of cytoplasmic proteins or organelles is largely mediated by nonselective macroautophagy; a process generally referred to as autophagy. Selective macroautophagy employs the same core machinery used for nonselective macroautophagy. A small number of additional cargo-ligand-receptor-proteins serve to make the process selective [78–83]. Another well-known selective autophagy is chaperone-mediated. In chaperone-mediated autophagy, substrate proteins are selectively recognized by a cytosolic chaperone, the heat shock cognate protein of 70 kDa (hsc70) [84]. The interaction between the chaperone and the substrate in the cytosol targets the complex to the lysosomal membrane, where it binds to the lysosome-associated membrane protein type 2A (LAMP-2A), which acts as a receptor for this pathway [84,85]. In contrast, chaperone-unmediated autophagy is thought to function in the degradation of mitochondria.

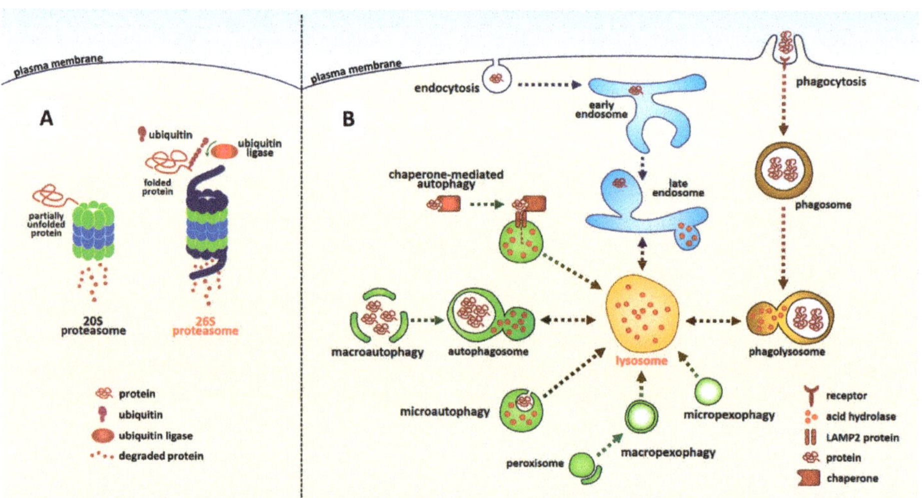

Figure 1. The representative protein degradation pathways in eukaryotic cells. Proteasome-dependent (**A**) and lysosome-dependent (**B**) pathways are shown. Note that not all the degradation pathways are illustrated in the figure.

Proteasomes are multiprotein complexes that predominantly degrade nuclear and cytosolic proteins. Most proteins are targeted for proteasomal degradation after being covalently modified with ubiquitin, which is conjugated through its carboxy terminus [86–90]. This reaction is called ubiquitination. Ubiquitin-protein conjugates are subsequently recognized and degraded by 26S proteasomes, which are multisubunit proteases found in the cytosol, perinuclear regions, and nucleus of eukaryotic cells [91]. The degradation products of 26S proteasomal catalysis are short peptide fragments and amino acids that can be recycled to produce new proteins. Simultaneously, polyubiquitin chains are released from targeted proteins and then disassembled by ubiquitin carboxy-terminal hydrolases to produce monomeric ubiquitin molecules that re-enter the ubiquitin-proteasome system, from which point they can contribute to the clearance of other abnormal proteins [92,93]. Failure of the ubiquitin-proteasome system is implicated in the pathogenesis of both sporadic and familial PD [22–24,94–96].

6. Ubiquitination and Proteasomal Degradation of Phosphorylated Tyrosine Hydroxylase

Here, we introduce an evidence of the ubiquitination of phosphorylated TH and its proteasomal degradation by the ubiquitin-proteasome system, and discuss its possible physiological significance in PD and DRD (Table 1). First, Lazar et al. revealed that activated TH purified from bovine striatum showed decreased half-life at 50 °C [97]. They suggested that phosphorylation of TH could greatly increase the degradation rate of the enzyme in vivo. Several years later, Døskeland and Flatmark reported that human recombinant TH protein is ubiquitinated and degraded in the reticulocyte lysate system [98]. Subsequently, Urano et al. reported that recombinant human TH protein forms insoluble aggregates in the presence of tetrahydrobiopterin in vitro [99]. Recombinant TH is free from dopamine and presumably similar to phosphorylated TH [99]. We further revealed that 26S proteasomal inhibition leads to accumulation of TH phosphorylated at Ser40 (pSer40-TH), which are ubiquitin-positive, as well as formation of its insoluble inclusions in NGF-differentiated PC12D cells [70]. These observations support the novel pathway of proteasomal degradation of TH protein. The phenomenon of intracellular pSer40-TH insolubility unveiled the characteristics of pSer40-TH that it easily forms aggregates in living cells (Figure 2). Insight into the reduction of proteasomal activity in PD [94–96] further supports the evidence of the accumulation of pSer40-TH to form inclusion bodies in PD patients [71]. A publication by Nakashima et al. also showed the proteasomal degradation of the TH protein and evidence that phosphorylation of the N-terminal TH domain causes proteasomal degradation [100,101]. Carbajosa et al. also reported that short-term inhibition of proteasome increases the accumulation of ubiquitinated TH protein in PC12 cells and brainstem neurons [102], indicating that TH, especially phosphorylated TH, is ubiquitinated, resulting in its degradation by the ubiquitin-proteasome system.

What effect does the reduction of dopamine and biopterin levels have on the proteasomal degradation of phosphorylated TH in PD and DRD? Interestingly, dopamine and biopterin deficiencies lead to reduced total TH protein, which is caused by the degradation of pSer40-TH [74]. This pSer40-TH degradation was sensitive to MG-132, a 26S proteasome inhibitor [74], indicating a ubiquitin-proteasome system-mediated degradation. Salvatore et al. further revealed that knockout of dopamine transporter decreased dopamine content in the terminals of dopaminergic neurons, and this phenomenon was accompanied by the elevation of pSer40-TH and reduction of total TH protein [103]. Altogether, these data strongly suggest that phosphorylated TH protein is ubiquitinated to be degraded by the ubiquitin-proteasome system (Figure 2). Moreover, the lack of dopamine accelerates the proteasomal degradation of TH and its phosphorylation through PKA activation, resulting in the loss of TH protein and the negative spiral of TH depletion (Figure 3).

Table 1. Advances of the study for the ubiquitination and proteasomal degradation of phosphorylated tyrosine hydroxylase protein (original articles).

Evidence	Year	Reference
Activated tyrosine hydroxylase purified from bovine striatum decreases its thermal stability	1981	[97]
Human recombinant TH protein is ubiquitinated and degraded in the reticulocyte lysate system	2002	[98]
Recombinant human TH forms insoluble aggregates in the presence of tetrahydrobiopterin	2006	[99]
Proteasomal inhibition accumulates ubiquitinated TH protein phosphorylated at ^{40}Ser to form insoluble aggregates in NGF-differentiated PC12D cells	2009	[70]
Phosphorylation of the N-terminal domain of tyrosine hydroxylase triggers proteasomal digestion	2011	[100]
Short-term inhibition of proteasome increases the accumulation of ubiquitinated TH protein in PC12 cell and brainstem neurons	2015	[102]
Dopamine or biopterin deficiency potentiates phosphorylation at ^{40}Ser and ubiquitination of TH protein to be degraded by the ubiquitin proteasome system	2015	[74]
Inhibition of USP14 to activate proteasome decreases TH protein phosphorylated at ^{19}Ser	2016	[101]
Dopamine transporter-deficiency increases TH phosphorylation and decreases TH protein in striatum and nucleus accumbens	2016	[103]

TH, tyrosine hydroxylase; NGF, nerve growth factor; USP14, Ubiquitin-specific protease 14.

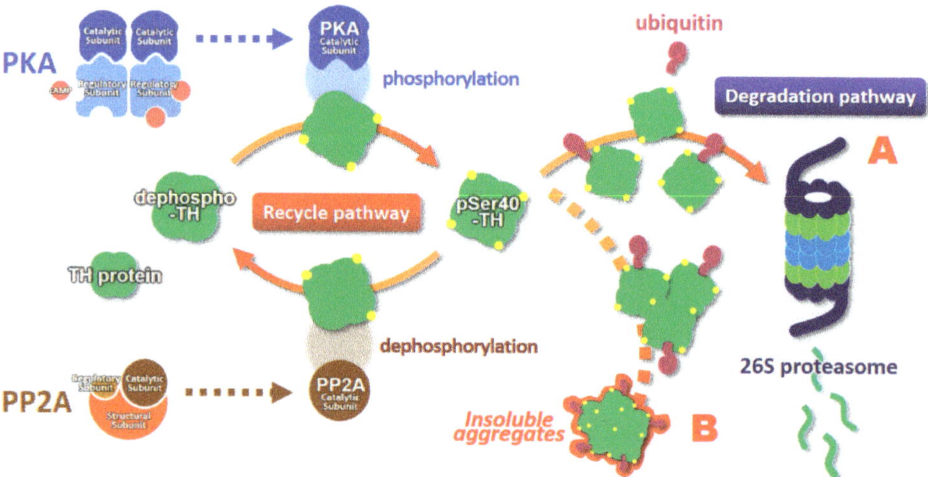

Figure 2. Schematic model of the proteasomal degradation of phosphorylated tyrosine hydroxylase (pSer40-TH). The degradation pathway is indicated in A, and the accumulation pathway to form insoluble aggregates is shown in B. pSer40-TH, tyrosine hydroxylase phosphorylated at serine 40 residue; cAMP, cyclic adenosine monophosphate; PKA, cAMP-dependent protein kinase; PP2A, protein phosphatase 2a.

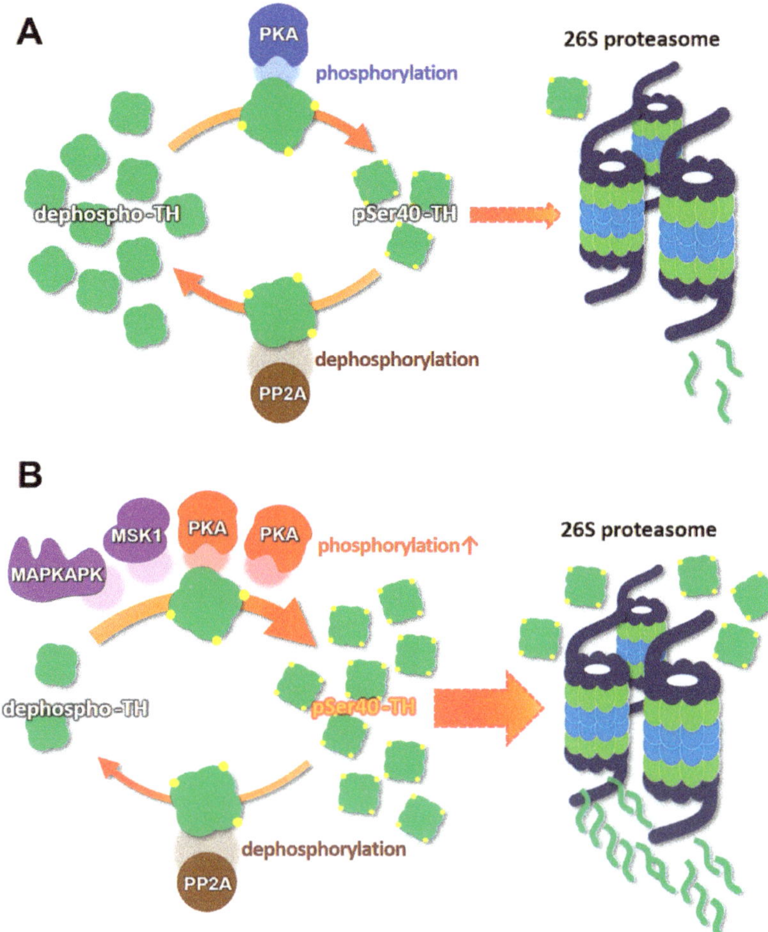

Figure 3. Schematic model of the molecular mechanism of the loss of tyrosine hydroxylase (TH) protein in the dopaminergic neurons. (**A**) A balanced state between the recycling TH protein and degrading TH protein. (**B**) Dopamine/biopterin deficient state activates PKA (red), and α-Synuclein aggregation presumably activates MAPKAPK and MSK1 (purple). Both activations accelerate TH phosphorylation (pSer40-TH), which is accompanied by proteasomal degradation. PKA, cAMP-dependent protein kinase; PP2A, protein phosphatase 2a. MAPKAPK, mitogen-activated protein kinase activated protein kinase; MSK1, mitogen- and stress-activated kinase 1.

7. Modification of Tyrosine Hydroxylase Phosphorylation by α-Synuclein

α-Synuclein is a major component of Lewy bodies, and its deposition is a subset hallmark of neurodegenerative disorders, including PD, dementia with Lewy bodies (DLB), and multiple system atrophy, collectively referred to as synucleopathies. α-Synuclein was found in filamentous aggregates of Lewy bodies and Lewy neuritis [2,3], and the protein itself was first identified in 1993 as a nonamyloid β component of Alzheimer's disease (AD) [104]. α-Synuclein isolated from DLB patients was phosphorylated [105]. α-Synuclein is degraded by proteasomes [106,107], and phosphorylated α-synuclein is ubiquitinated in α-synucleinopathy lesions [108], indicating that the ubiquitin-proteasome system degrades phosphorylated synuclein. Chaperone-mediated autophagy, which contributes to the degradation of intracellular proteins in lysosomes (Figure 1), also degrades α-synuclein [109–111].

α-Synuclein itself seems to contribute to the maintenance of presynaptic function by participating in the assembly of the SNARE protein complex [112,113]. Furthermore, α-synuclein in the soluble form physically interacts with TH and maintains the level of phosphorylated TH in a PP2A-dependent manner [114–116], which suggests the possibility that α-synuclein monomer prevents excessive phosphorylation of TH by activating PP2A. Because the overexpression of wild-type or mutant human α-synuclein caused by the TH promoter did not result in the formation of pathological inclusions nor alter the behavior and sensitivity to 1-methyl-4-phenyl-1,2,3,6- tetrahydropyridine (MPTP) in C57BL/6 mice [117–119], factors other than α-synuclein itself may be associated with the neuronal degeneration of dopaminergic neurons. Thus, we hypothesize that not soluble α-synuclein monomers themselves, but oligomerized filaments and aggregates are associated with neurodegeneration. For instance, the failure of the ubiquitin-proteasome system in the substantia nigra in PD [94] presumably impairs the degradation of α-synuclein, which facilitates the formation of filamentous inclusions. Furthermore, dopamine-modified α-synuclein blocks chaperone-mediated autophagy [109,110], which induces abnormal intracellular accumulation α-synuclein in PD [120]. Plasma α-synuclein level in PD is higher than that in healthy controls [121], indicating possible reduction of protein degradation rate. Such aggregation of α-synuclein presumably potentiates TH phosphorylation and reduces total TH protein [116,122,123]. Indeed, we revealed that the formation of intracellular aggregations of filamentous α-synuclein led to a decrease in the total TH protein levels with increased pSer40-TH in cultured dopaminergic neurons (Figure 4). α-Synuclein activates stress-related protein kinases to potentiate TH phosphorylation at serine 40, suggesting the possible mechanism of pSer40-TH elevation by α-synuclein aggregation [52,124,125]. The α-synuclein-induced abnormal upregulation of TH phosphorylation, combined with the reduction of gene transcription by aging and aging-related disorders [126–129], results in the acceleration of pSer40-TH degradation to reduce total TH protein (Figure 3).

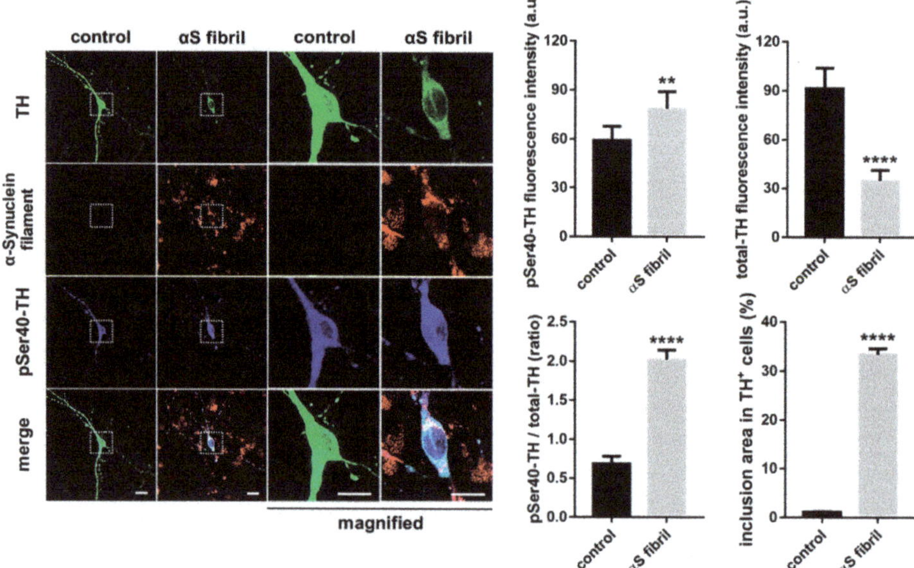

Figure 4. Exposure to α-Synuclein fibrils formed intracellular filamentous inclusions, which was accompanied by the acceleration of TH phosphorylation and the reduction of total TH protein in the cultured dopaminergic neurons in the presence of cycloheximide. Scale bars indicate 10 μm. The right columns show the quantified data (Student's *t*-test, **** $p < 0.0001$, ** $p < 0.01$, $n > 20$). αS indicate α-Synuclein.

8. Novel Therapeutic Targets for α-Synuclein Propagation

Previously, when the molecular mechanism of PD pathogenesis was not well understood, PD patients have been prescribed a dopamine precursor, L-3,4-dihydroxyphenylalanine (L-DOPA) [130]. Oral administration of L-DOPA led to partial improvement of PD symptoms; however, L-DOPA exerts side effects, such as nausea and vomiting, which had been able to be attenuated by slowing the increases in the daily dose [130,131]. Second, after prolonged treatment with L-DOPA, as many as 72% of Parkinsonism patients will suffer from movement disorders. These disorders consist of uncontrollable facial movements, namely grimacing, tongue protrusion, and chewing motions [131,132]. A third side effect is a loss of blood pressure upon standing; approximately 33% of patients have shown this effect [131,132]. This problem tends to disappear in patients receiving the drug for a sufficiently long period. Although L-DOPA has such uncomfortable side effects [133], it is still useful for treating PD and DRD [134] and often used in combination with carbidopa, which inhibits peripheral metabolism of L-DOPA. Therefore, L-DOPA is expected to be used in combination with novel therapeutic agents.

Recently, the propagation of α-synuclein is focused on PD pathogenesis [135,136]. Accumulation of propagated α-synuclein results in synucleinopathies, including PD, DLB, and multiple system atrophy [137]. As introduced in Section 7, the aggregation of propagated α-synuclein alters TH phosphorylation, which is accompanied by the proteasomal degradation of pSer40-TH to decrease total TH protein (Figure 4). Furthermore, α-synuclein contributes to the fibrilization of amyloid-β and tau [138], which are two critical proteins in AD, suggesting a central role of α-synuclein toxicity in neurodegeneration. Thus, α-synuclein uptake into living neuronal cells is critical for the pathogenesis of synucleinopathies. Then, how can we prevent α-synuclein propagation and its uptake into dopaminergic neurons?

Various molecular mechanisms are expected to be involved in α-synuclein uptake; for example, mechanisms related to the α3-subunit of Na^+/K^+-ATPase [139], neurexin [140,141], flotillin [142], and particular endocytic pathways [143]. Very recently, we showed that fatty acid-binding protein 3 (FABP3) is critical for α-synuclein uptake into dopaminergic neurons [144] and enhancement of α-synuclein spreading [145]. FABP3 is also essential in 1-methyl-4-phenylpyridinium (MPP^+)-induced morphological abnormality, mitochondrial dysfunction and neurotoxicity [144]. The injury induced by MPTP or its metabolite, MPP^+-, to dopaminergic neurons of the nigrostriatal pathways of nonhuman primates has been an important model for parkinsonism as well as dystonia [146–149]. These data suggest that FABP3 is a potential therapeutic target in synucleinopathies that can act by preventing α-synuclein uptake into dopaminergic neurons. Intriguingly, FABP3 ligand, which we have recently synthesized, inhibits α-synuclein oligomerization in PD mouse models [150,151]. These data suggest that FABP3-targeting ligands are potential therapeutic candidates for synucleinopathies.

Intriguingly, serum FABP3 level is increased in PD [152]. Although cerebrospinal fluid (CSF) is the nearest body fluid to the cerebral parenchyma as a biomarker of the central nervous system, the method of obtaining CSF is invasive and painful. Serum or plasma derived from blood is an ideal body fluid that can be used for screening of biomarker levels, as it is easily obtainable, and its collection process causes minimal discomfort. Previously, plasma levels of phosphorylated tau [153], amyloid-β (1-40/1-42) [154–159], and α-synuclein [121] have been studied for their potential to predict or diagnose AD and PD. The average value of each biomarker changes significantly; however, it is not sufficient to accurately predict specific disorders because some patients with AD or PD show lower plasma amyloid-β and α-synuclein levels than those of healthy controls. Therefore, novel diagnostic tools will be required. When we can predict PD at the very early stage and prevent the interaction of α-synuclein and FABP3 before the onset of PD, accumulation of the protein and its-induced neurotoxicity will be abolished. We will further study the pharmacologic action and molecular mechanism of FABP3-targeted compounds to prevent dopaminergic neurons from α-synuclein propagation and to promote neuronal survival [50,160–164], and we will develop a diagnostic method for predicting PD at the very early stages.

9. Conclusions

It is unclear why dopaminergic neurons preferentially degenerate in PD and DRD. Many factors may contribute to this, including mitochondrial dysfunction [165,166], oxidative stress, decreased glutathione content [167], increased iron levels [168], and production of oxygen radicals through the combination of dopamine and tetrahydrobiopterin [169,170]. Here, we present clues to understanding this selective degeneration of dopamine-containing neurons, which are sensitive to dopamine/biopterin deficiency and α-synuclein invasion. The consequent elevation of TH phosphorylation is followed by the degradation of pSer40-TH by the ubiquitin-proteasome system. Interestingly, proteasomal inhibition results in TH aggregation, whereas choline acetyltransferase does not show such aggregations [70]. Owing to such characteristics of TH protein to aggregate and be degraded easily, especially pSer40-TH, handling phosphorylated TH is somehow tricky for dopaminergic neurons. The formation of insoluble inclusions of pSer40-TH further reduces cytoplasmic operable TH. We suggest that the negative spiral mechanism of TH phosphorylation-induced degradation is involved in the loss of nigrostriatal TH protein in PD and DRD (Figure 5).

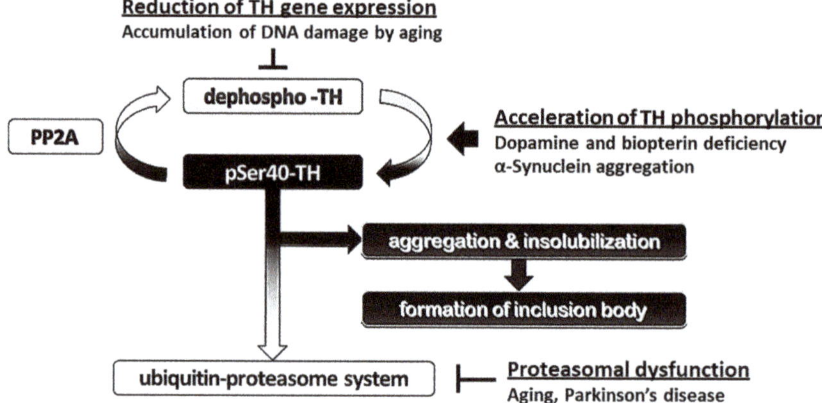

Figure 5. Schematic model for the pathways of degradation and accumulation of tyrosine hydroxylase (TH) protein. The ubiquitin-proteasome system degrades phosphorylated TH at serine 40 (pSer40-TH). Otherwise, pSer40-TH is accumulated in the proteasomal deficient state and forms insoluble aggregates.

In the present article, we have reviewed the molecular mechanism of the loss of mesencephalic TH protein in PD and DRD. We conclude that the ubiquitin-proteasome system participates in the degradation of phosphorylated TH. The mechanism of ubiquitin-proteasome-linked dopaminergic pathogenesis might help explain the dopaminergic neuron-selective loss of TH protein in PD and DRD. These insights may lead to more focused efforts to develop therapeutics and strategies to prevent the onset of neurodegeneration in PD and DRD.

Author Contributions: Conceptualization, I.K. and K.F.; writing—original draft preparation, I.K.; writing—review and editing, I.K. and K.F.; funding acquisition, K.F. and I.K. All authors have read and agreed to the published version of the manuscript.

Funding: This research was funded in part by the Strategic Research Program for Brain Sciences of the Japan Agency for Medical Research and Development, AMED (JP18dm0107071 and JP19dm0107071) to K.F., as well as by the Japan Society for the Promotion of Science, KAKENHI (19K07097), Kobayashi Foundation, and Intelligent Cosmos Academic Foundation to I.K.

Acknowledgments: We appreciate Hiroshi Ichinose (Tokyo Institute of Technology) for helpful discussion on the proteasomal degradation of tyrosine hydroxylase protein. We also acknowledge Emeritus Toshiharu Nagatsu at Fujita Health University for valuable discussion on the characteristics of tyrosine hydroxylase.

Conflicts of Interest: The authors declare no conflicts of interest.

Abbreviations

AD	Alzheimer's disease
AP-1	Activator protein 1
CaMKII	Calcium/calmodulin-dependent protein kinase II
cAMP	Cyclic adenosine monophosphate
CMA	Chaperone-mediated autophagy
CSF	Cerebrospinal fluid
DLB	Dementia with Lewy bodies
DRD	Dopa-responsive dystonia (Segawa disease)
ERK	Extracellular signal-regulated kinase
FABP	Fatty acid-binding protein
GTP	guanosine triphosphate
GCH1	GTP cyclohydrolase 1
Hsc70	Heat shock cognate protein of 70 kDa
L-DOPA	L-3,4-dihydroxyphenylalanine
LAMP	Lysosome-associated membrane protein
LRRK2	Leucine-rich repeat kinase 2
MAPK	mitogen-activated protein kinase
MAPKAPK	Mitogen-activated protein kinase activated protein kinase
MPP+	1-methyl-4-phenylpyridinium
MPTP	1-methyl-4-phenyl-1,2,3,6-tetrahydropyridine
MSK1	Mitogen- and stress-activated kinase 1
NGF	Nerve growth factor
Nurr1	Nuclear receptor related-1
PD	Parkinson's disease
PP2A	Protein phosphatase 2a
pSer40-TH	Tyrosine hydroxylase phosphorylated at Ser40
SNpc	Substantia nigra pars compacta
SRF	Serum-responsive factor
TH	Tyrosine hydroxylase
VTA	Ventral tegmental area

References

1. Parkinson, J. An essay on the shaking palsy. 1817. *J. Neuropsychiatry Clin. Neurosci.* **2002**, *14*, 223–236. [CrossRef] [PubMed]
2. Spillantini, M.G.; Crowther, R.A.; Jakes, R.; Hasegawa, M.; Goedert, M. alpha-Synuclein in filamentous inclusions of Lewy bodies from Parkinson's disease and dementia with lewy bodies. *Proc. Natl. Acad. Sci. USA* **1998**, *95*, 6469–6473. [CrossRef]
3. Spillantini, M.G.; Crowther, R.A.; Jakes, R.; Cairns, N.J.; Lantos, P.L.; Goedert, M. Filamentous alpha-synuclein inclusions link multiple system atrophy with Parkinson's disease and dementia with Lewy bodies. *Neurosci. Lett.* **1998**, *251*, 205–208. [CrossRef]
4. Segawa, M. Dopa-responsive dystonia. *Handb. Clin. Neurol.* **2011**, *100*, 539–557. [CrossRef] [PubMed]
5. Furukawa, Y.; Nygaard, T.G.; Gutlich, M.; Rajput, A.H.; Pifl, C.; DiStefano, L.; Chang, L.J.; Price, K.; Shimadzu, M.; Hornykiewicz, O.; et al. Striatal biopterin and tyrosine hydroxylase protein reduction in dopa-responsive dystonia. *Neurology* **1999**, *53*, 1032–1041. [CrossRef] [PubMed]
6. Furukawa, Y.; Kapatos, G.; Haycock, J.W.; Worsley, J.; Wong, H.; Kish, S.J.; Nygaard, T.G. Brain biopterin and tyrosine hydroxylase in asymptomatic dopa-responsive dystonia. *Ann. Neurol.* **2002**, *51*, 637–641. [CrossRef]
7. Leckman, J.F.; Bloch, M.H.; Smith, M.E.; Larabi, D.; Hampson, M. Neurobiological substrates of Tourette's disorder. *J. Child Adolesc. Psychopharmacol.* **2010**, *20*, 237–247. [CrossRef]
8. Carlsson, M.; Svensson, A. Interfering with glutamatergic neurotransmission by means of NMDA antagonist administration discloses the locomotor stimulatory potential of other transmitter systems. *Pharmacol. Biochem. Behav.* **1990**, *36*, 45–50. [CrossRef]

9.   David, C.; Ewert, M.; Seeburg, P.H.; Fuchs, S. Antipeptide antibodies differentiate between long and short isoforms of the D2 dopamine receptor. *Biochem. Biophys. Res. Commun.* **1991**, *179*, 824–829. [CrossRef]

10.   Cooper, O.; Greenman, Y. Dopamine Agonists for Pituitary Adenomas. *Front. Endocrinol. (Lausanne)* **2018**, *9*, 469. [CrossRef]

11.   Albanese, A. Extrapyramidal system, motor Ganglia and movement disorders. *Rev. Neurosci.* **1990**, *2*, 145–164. [CrossRef] [PubMed]

12.   Robertson, H.A. Synergistic interactions of D1- and D2-selective dopamine agonists in animal models for Parkinson's disease: Sites of action and implications for the pathogenesis of dyskinesias. *Can. J. Neurol. Sci.* **1992**, *19*, 147–152. [CrossRef] [PubMed]

13.   Haavik, J.; Toska, K. Tyrosine hydroxylase and Parkinson's disease. *Mol. Neurobiol.* **1998**, *16*, 285–309. [CrossRef] [PubMed]

14.   Dauer, W.; Przedborski, S. Parkinson's disease: Mechanisms and models. *Neuron* **2003**, *39*, 889–909. [CrossRef]

15.   Hoefer, P.F.A. Action Potentials of Muscles in Rigidity and Tremor. *Arch. Neurol. Psychiatry* **1940**, *43*, 704–725. [CrossRef]

16.   Muller, K.; Homberg, V.; Lenard, H.G. Motor control in childhood onset dopa-responsive dystonia (Segawa syndrome). *Neuropediatrics* **1989**, *20*, 185–191. [CrossRef]

17.   Hirsch, L.; Jette, N.; Frolkis, A.; Steeves, T.; Pringsheim, T. The Incidence of Parkinson's Disease: A Systematic Review and Meta-Analysis. *Neuroepidemiology* **2016**, *46*, 292–300. [CrossRef]

18.   Sveinbjornsdottir, S. The clinical symptoms of Parkinson's disease. *J. Neurochem.* **2016**, *139*, 318–324. [CrossRef]

19.   Cerri, S.; Mus, L.; Blandini, F. Parkinson's Disease in Women and Men: What's the Difference? *J. Parkinsons Dis.* **2019**, *9*, 501–515. [CrossRef]

20.   Bandres-Ciga, S.; Diez-Fairen, M.; Kim, J.J.; Singleton, A.B. Genetics of Parkinson's disease: An introspection of its journey towards precision medicine. *Neurobiol. Dis.* **2020**, *137*, 104782. [CrossRef]

21.   Polymeropoulos, M.H.; Lavedan, C.; Leroy, E.; Ide, S.E.; Dehejia, A.; Dutra, A.; Pike, B.; Root, H.; Rubenstein, J.; Boyer, R.; et al. Mutation in the alpha-synuclein gene identified in families with Parkinson's disease. *Science* **1997**, *276*, 2045–2047. [CrossRef] [PubMed]

22.   Kitada, T.; Asakawa, S.; Hattori, N.; Matsumine, H.; Yamamura, Y.; Minoshima, S.; Yokochi, M.; Mizuno, Y.; Shimizu, N. Mutations in the parkin gene cause autosomal recessive juvenile parkinsonism. *Nature* **1998**, *392*, 605–608. [CrossRef] [PubMed]

23.   Hattori, N.; Matsumine, H.; Asakawa, S.; Kitada, T.; Yoshino, H.; Elibol, B.; Brookes, A.J.; Yamamura, Y.; Kobayashi, T.; Wang, M.; et al. Point mutations (Thr240Arg and Gln311Stop) [correction of Thr240Arg and Ala311Stop] in the Parkin gene. *Biochem. Biophys. Res. Commun.* **1998**, *249*, 754–758. [CrossRef] [PubMed]

24.   Shimura, H.; Hattori, N.; Kubo, S.; Mizuno, Y.; Asakawa, S.; Minoshima, S.; Shimizu, N.; Iwai, K.; Chiba, T.; Tanaka, K.; et al. Familial Parkinson disease gene product, parkin, is a ubiquitin-protein ligase. *Nat. Genet.* **2000**, *25*, 302–305. [CrossRef]

25.   Bonifati, V.; Rizzu, P.; Van Baren, M.J.; Schaap, O.; Breedveld, G.J.; Krieger, E.; Dekker, M.C.; Squitieri, F.; Ibanez, P.; Joosse, M.; et al. Mutations in the DJ-1 gene associated with autosomal recessive early-onset parkinsonism. *Science* **2003**, *299*, 256–259. [CrossRef]

26.   Ohta, E.; Hasegawa, K.; Gasser, T.; Obata, F. Independent occurrence of I2020T mutation in the kinase domain of the leucine rich repeat kinase 2 gene in Japanese and German Parkinson's disease families. *Neurosci. Lett.* **2007**, *417*, 21–23. [CrossRef]

27.   Funayama, M.; Hasegawa, K.; Ohta, E.; Kawashima, N.; Komiyama, M.; Kowa, H.; Tsuji, S.; Obata, F. An LRRK2 mutation as a cause for the parkinsonism in the original PARK8 family. *Ann. Neurol.* **2005**, *57*, 918–921. [CrossRef]

28.   Balint, B.; Mencacci, N.E.; Valente, E.M.; Pisani, A.; Rothwell, J.; Jankovic, J.; Vidailhet, M.; Bhatia, K.P. Dystonia. *Nat. Rev. Dis. Primers* **2018**, *4*, 25. [CrossRef]

29.   Camargo, C.H.F.; Camargos, S.T.; Cardoso, F.E.C.; Teive, H.A.G. The genetics of the dystonias a review based on the new classification of the dystonias. *Arq. Neuropsiquiatr.* **2015**, *73*, 350–358. [CrossRef]

30.   Ichinose, H.; Ohye, T.; Takahashi, E.; Seki, N.; Hori, T.; Segawa, M.; Nomura, Y.; Endo, K.; Tanaka, H.; Tsuji, S.; et al. Hereditary progressive dystonia with marked diurnal fluctuation caused by mutations in the GTP cyclohydrolase I gene. *Nat. Genet.* **1994**, *8*, 236–242. [CrossRef]

31. Ceravolo, R.; Nicoletti, V.; Garavaglia, B.; Reale, C.; Kiferle, L.; Bonuccelli, U. Expanding the clinical phenotype of DYT5 mutations: Is multiple system atrophy a possible one? *Neurology* **2013**, *81*, 301–302. [CrossRef] [PubMed]

32. Bernal-Pacheco, O.; Oyama, G.; Briton, A.; Singleton, A.B.; Fernandez, H.H.; Rodriguez, R.L.; Malaty, I.A.; Okun, M.S. A Novel DYT-5 Mutation with Phenotypic Variability within a Colombian Family. *Tremor Other Hyperkinet Mov. (N Y)* **2013**, *3*. [CrossRef]

33. Shetty, A.S.; Bhatia, K.P.; Lang, A.E. Dystonia and Parkinson's disease: What is the relationship? *Neurobiol. Dis.* **2019**, *132*, 104462. [CrossRef]

34. Montague, P.R.; Hyman, S.E.; Cohen, J.D. Computational roles for dopamine in behavioural control. *Nature* **2004**, *431*, 760–767. [CrossRef]

35. Schultz, W. Behavioral theories and the neurophysiology of reward. *Annu. Rev. Psychol.* **2006**, *57*, 87–115. [CrossRef]

36. Berridge, K.C. From prediction error to incentive salience: Mesolimbic computation of reward motivation. *Eur. J. Neurosci.* **2012**, *35*, 1124–1143. [CrossRef] [PubMed]

37. Stern, G. The effects of lesions in the substantia nigra. *Brain* **1966**, *89*, 449–478. [CrossRef] [PubMed]

38. Nagatsu, T.; Levitt, M.; Udenfriend, S. Tyrosine Hydroxylase. The initial step in norepinephrine biosynthesis. *J. Biol. Chem.* **1964**, *239*, 2910–2917. [PubMed]

39. Kaneda, N.; Kobayashi, K.; Ichinose, H.; Kishi, F.; Nakazawa, A.; Kurosawa, Y.; Fujita, K.; Nagatsu, T. Isolation of a novel cDNA clone for human tyrosine hydroxylase: Alternative RNA splicing produces four kinds of mRNA from a single gene. *Biochem. Biophys. Res. Commun.* **1987**, *146*, 971–975. [CrossRef]

40. Grima, B.; Lamouroux, A.; Boni, C.; Julien, J.F.; Javoy-Agid, F.; Mallet, J. A single human gene encoding multiple tyrosine hydroxylases with different predicted functional characteristics. *Nature* **1987**, *326*, 707–711. [CrossRef]

41. Ichinose, H.; Ohye, T.; Fujita, K.; Yoshida, M.; Ueda, S.; Nagatsu, T. Increased heterogeneity of tyrosine hydroxylase in humans. *Biochem. Biophys. Res. Commun.* **1993**, *195*, 158–165. [CrossRef] [PubMed]

42. Haycock, J.W. Species differences in the expression of multiple tyrosine hydroxylase protein isoforms. *J. Neurochem.* **2002**, *81*, 947–953. [CrossRef] [PubMed]

43. Abate, C.; Joh, T.H. Limited proteolysis of rat brain tyrosine hydroxylase defines an N-terminal region required for regulation of cofactor binding and directing substrate specificity. *J. Mol. Neurosci.* **1991**, *2*, 203–215. [PubMed]

44. Campbell, D.G.; Hardie, D.G.; Vulliet, P.R. Identification of four phosphorylation sites in the N-terminal region of tyrosine hydroxylase. *J. Biol. Chem.* **1986**, *261*, 10489–10492.

45. Goodwill, K.E.; Sabatier, C.; Marks, C.; Raag, R.; Fitzpatrick, P.F.; Stevens, R.C. Crystal structure of tyrosine hydroxylase at 2.3 A and its implications for inherited neurodegenerative diseases. *Nat. Struct. Biol.* **1997**, *4*, 578–585. [CrossRef]

46. Mitchell, J.P.; Hardie, D.G.; Vulliet, P.R. Site-specific phosphorylation of tyrosine hydroxylase after KCl depolarization and nerve growth factor treatment of PC12 cells. *J. Biol. Chem.* **1990**, *265*, 22358–22364.

47. Kumer, S.C.; Vrana, K.E. Intricate regulation of tyrosine hydroxylase activity and gene expression. *J. Neurochem.* **1996**, *67*, 443–462. [CrossRef]

48. Chae, H.D.; Suh, B.C.; Joh, T.H.; Kim, K.T. AP1-mediated transcriptional enhancement of the rat tyrosine hydroxylase gene by muscarinic stimulation. *J. Neurochem.* **1996**, *66*, 1264–1272. [CrossRef]

49. Guo, Z.; Du, X.; Iacovitti, L. Regulation of tyrosine hydroxylase gene expression during transdifferentiation of striatal neurons: Changes in transcription factors binding the AP-1 site. *J. Neurosci.* **1998**, *18*, 8163–8174. [CrossRef]

50. Kawahata, I.; Lai, Y.; Morita, J.; Kato, S.; Ohtaku, S.; Tomioka, Y.; Tabuchi, A.; Tsuda, M.; Sumi-Ichinose, C.; Kondo, K.; et al. V-1/CP complex formation is required for genetic co-regulation of adult nigrostriatal dopaminergic function via the RHO/MAL/SRF pathway in vitro and in vivo. *J. Neurol. Sci.* **2017**, *381*, 359–360. [CrossRef]

51. Kadkhodaei, B.; Ito, T.; Joodmardi, E.; Mattsson, B.; Rouillard, C.; Carta, M.; Muramatsu, S.; Sumi-Ichinose, C.; Nomura, T.; Metzger, D.; et al. Nurr1 is required for maintenance of maturing and adult midbrain dopamine neurons. *J. Neurosci.* **2009**, *29*, 15923–15932. [CrossRef] [PubMed]

52. Dunkley, P.R.; Bobrovskaya, L.; Graham, M.E.; Von Nagy-Felsobuki, E.I.; Dickson, P.W. Tyrosine hydroxylase phosphorylation: Regulation and consequences. *J. Neurochem.* **2004**, *91*, 1025–1043. [CrossRef] [PubMed]

53. Dunkley, P.R.; Dickson, P.W. Tyrosine hydroxylase phosphorylation in vivo. *J. Neurochem.* **2019**, *149*, 706–728. [CrossRef] [PubMed]

54. Fukunaga, K.; Rich, D.P.; Soderling, T.R. Generation of the Ca2(+)-independent form of Ca2+/calmodulin-dependent protein kinase II in cerebellar granule cells. *J. Biol. Chem.* **1989**, *264*, 21830–21836.

55. Fukunaga, K.; Miyamoto, E.; Soderling, T.R. Regulation of Ca2+/calmodulin-dependent protein kinase II by brain gangliosides. *J. Neurochem.* **1990**, *54*, 103–109. [CrossRef]

56. Soderling, T.R.; Fukunaga, K.; Rich, D.P.; Fong, Y.L.; Smith, K.; Colbran, R.J. Regulation of brain Ca2+/calmodulin-dependent protein kinase II. *Adv. Second Messenger Phosphoprot. Res.* **1990**, *24*, 206–211.

57. Haycock, J.W. Phosphorylation of tyrosine hydroxylase in situ at serine 8, 19, 31, and 40. *J. Biol. Chem.* **1990**, *265*, 11682–11691.

58. Haycock, J.W.; Haycock, D.A. Tyrosine hydroxylase in rat brain dopaminergic nerve terminals. Multiple-site phosphorylation in vivo and in synaptosomes. *J. Biol. Chem.* **1991**, *266*, 5650–5657.

59. Bobrovskaya, L.; Dunkley, P.R.; Dickson, P.W. Phosphorylation of Ser19 increases both Ser40 phosphorylation and enzyme activity of tyrosine hydroxylase in intact cells. *J. Neurochem.* **2004**, *90*, 857–864. [CrossRef]

60. Daubner, S.C.; Lauriano, C.; Haycock, J.W.; Fitzpatrick, P.F. Site-directed mutagenesis of serine 40 of rat tyrosine hydroxylase. Effects of dopamine and cAMP-dependent phosphorylation on enzyme activity. *J. Biol. Chem.* **1992**, *267*, 12639–12646.

61. Okuno, S.; Fujisawa, H. A new mechanism for regulation of tyrosine 3-monooxygenase by end product and cyclic AMP-dependent protein kinase. *J. Biol. Chem.* **1985**, *260*, 2633–2635. [PubMed]

62. Fujisawa, H.; Okuno, S. Regulatory mechanism of tyrosine hydroxylase activity. *Biochem. Biophys. Res. Commun.* **2005**, *338*, 271–276. [CrossRef] [PubMed]

63. Royo, M.; Fitzpatrick, P.F.; Daubner, S.C. Mutation of regulatory serines of rat tyrosine hydroxylase to glutamate: Effects on enzyme stability and activity. *Arch. Biochem. Biophys.* **2005**, *434*, 266–274. [CrossRef] [PubMed]

64. Haavik, J.; Schelling, D.L.; Campbell, D.G.; Andersson, K.K.; Flatmark, T.; Cohen, P. Identification of protein phosphatase 2A as the major tyrosine hydroxylase phosphatase in adrenal medulla and corpus striatum: Evidence from the effects of okadaic acid. *FEBS Lett.* **1989**, *251*, 36–42. [CrossRef]

65. Goncalves, C.A.; Hall, A.; Sim, A.T.; Bunn, S.J.; Marley, P.D.; Cheah, T.B.; Dunkley, P.R. Tyrosine hydroxylase phosphorylation in digitonin-permeabilized bovine adrenal chromaffin cells: The effect of protein kinase and phosphatase inhibitors on Ser19 and Ser40 phosphorylation. *J. Neurochem.* **1997**, *69*, 2387–2396. [CrossRef] [PubMed]

66. Leal, R.B.; Sim, A.T.; Goncalves, C.A.; Dunkley, P.R. Tyrosine hydroxylase dephosphorylation by protein phosphatase 2A in bovine adrenal chromaffin cells. *Neurochem. Res.* **2002**, *27*, 207–213. [CrossRef]

67. Haycock, J.W. Peptide substrates for ERK1/2: Structure-function studies of serine 31 in tyrosine hydroxylase. *J. Neurosci. Methods* **2002**, *116*, 29–34. [CrossRef]

68. Ichinose, H.; Ohye, T.; Fujita, K.; Pantucek, F.; Lange, K.; Riederer, P.; Nagatsu, T. Quantification of mRNA of tyrosine hydroxylase and aromatic L-amino acid decarboxylase in the substantia nigra in Parkinson's disease and schizophrenia. *J. Neural Transm. Park. Dis. Dement. Sect.* **1994**, *8*, 149–158. [CrossRef]

69. Mogi, M.; Harada, M.; Kiuchi, K.; Kojima, K.; Kondo, T.; Narabayashi, H.; Rausch, D.; Riederer, P.; Jellinger, K.; Nagatsu, T. Homospecific activity (activity per enzyme protein) of tyrosine hydroxylase increases in parkinsonian brain. *J. Neural Transm.* **1988**, *72*, 77–82. [CrossRef]

70. Kawahata, I.; Tokuoka, H.; Parvez, H.; Ichinose, H. Accumulation of phosphorylated tyrosine hydroxylase into insoluble protein aggregates by inhibition of an ubiquitin-proteasome system in PC12D cells. *J. Neural Transm.* **2009**, *116*, 1571–1578. [CrossRef]

71. Kawahata, I.; Yagishita, S.; Hasegawa, K.; Nagatsu, I.; Nagatsu, T.; Ichinose, H. Immunohistochemical analyses of the postmortem human brains from patients with Parkinson's disease with anti-tyrosine hydroxylase antibodies. *Biog. Amines* **2009**, *23*, 1–7.

72. Baumann, A.; Jorge-Finnigan, A.; Jung-Kc, K.; Sauter, A.; Horvath, I.; Morozova-Roche, L.A.; Martinez, A. Tyrosine Hydroxylase Binding to Phospholipid Membranes Prompts Its Amyloid Aggregation and Compromises Bilayer Integrity. *Sci. Rep.* **2016**, *6*, 39488. [CrossRef] [PubMed]

73. Kawahata, I.; Ichinose, H. Long-term up-regulation of the level of phosphorylated tyrosine hydroxylase by depletion of dopamine or biopterin. *Neurosci. Res.* **2009**, *65*, S66. [CrossRef]

74. Kawahata, I.; Ohtaku, S.; Tomioka, Y.; Ichinose, H.; Yamakuni, T. Dopamine or biopterin deficiency potentiates phosphorylation at (40)Ser and ubiquitination of tyrosine hydroxylase to be degraded by the ubiquitin proteasome system. *Biochem. Biophys. Res. Commun.* **2015**, *465*, 53–58. [CrossRef] [PubMed]
75. Sumi-Ichinose, C.; Urano, F.; Kuroda, R.; Ohye, T.; Kojima, M.; Tazawa, M.; Shiraishi, H.; Hagino, Y.; Nagatsu, T.; Nomura, T.; et al. Catecholamines and serotonin are differently regulated by tetrahydrobiopterin. A study from 6-pyruvoyltetrahydropterin synthase knockout mice. *J. Biol. Chem.* **2001**, *276*, 41150–41160. [CrossRef]
76. Takazawa, C.; Fujimoto, K.; Homma, D.; Sumi-Ichinose, C.; Nomura, T.; Ichinose, H.; Katoh, S. A brain-specific decrease of the tyrosine hydroxylase protein in sepiapterin reductase-null mice–as a mouse model for Parkinson's disease. *Biochem. Biophys. Res. Commun.* **2008**, *367*, 787–792. [CrossRef]
77. Rubinsztein, D.C. The roles of intracellular protein-degradation pathways in neurodegeneration. *Nature* **2006**, *443*, 780–786. [CrossRef]
78. Stern, S.T.; Adiseshaiah, P.P.; Crist, R.M. Autophagy and lysosomal dysfunction as emerging mechanisms of nanomaterial toxicity. *Part. Fibre Toxicol.* **2012**, *9*, 20. [CrossRef]
79. Cohignac, V.; Landry, M.J.; Boczkowski, J.; Lanone, S. Autophagy as a Possible Underlying Mechanism of Nanomaterial Toxicity. *Nanomaterials* **2014**, *4*, 548–582. [CrossRef]
80. Kenney, D.L.; Benarroch, E.E. The autophagy-lysosomal pathway: General concepts and clinical implications. *Neurology* **2015**, *85*, 634–645. [CrossRef]
81. Kaushik, S.; Cuervo, A.M. The coming of age of chaperone-mediated autophagy. *Nat. Rev. Mol. Cell Biol.* **2018**, *19*, 365–381. [CrossRef] [PubMed]
82. Dikic, I.; Elazar, Z. Mechanism and medical implications of mammalian autophagy. *Nat. Rev. Mol. Cell Biol.* **2018**, *19*, 349–364. [CrossRef] [PubMed]
83. Bingol, B. Autophagy and lysosomal pathways in nervous system disorders. *Mol Cell Neurosci* **2018**, *91*, 167–208. [CrossRef] [PubMed]
84. Dice, J.F.; Terlecky, S.R.; Chiang, H.L.; Olson, T.S.; Isenman, L.D.; Short-Russell, S.R.; Freundlieb, S.; Terlecky, L.J. A selective pathway for degradation of cytosolic proteins by lysosomes. *Semin. Cell Biol.* **1990**, *1*, 449–455. [PubMed]
85. Chiang, H.L.; Terlecky, S.R.; Plant, C.P.; Dice, J.F. A role for a 70-kilodalton heat shock protein in lysosomal degradation of intracellular proteins. *Science* **1989**, *246*, 382–385. [CrossRef]
86. Hershko, A.; Leshinsky, E.; Ganoth, D.; Heller, H. ATP-dependent degradation of ubiquitin-protein conjugates. *Proc. Natl. Acad. Sci. USA* **1984**, *81*, 1619–1623. [CrossRef] [PubMed]
87. Hershko, A.; Heller, H.; Elias, S.; Ciechanover, A. Components of ubiquitin-protein ligase system. Resolution, affinity purification, and role in protein breakdown. *J. Biol. Chem.* **1983**, *258*, 8206–8214. [PubMed]
88. Ciechanover, A.; Elias, S.; Heller, H.; Hershko, A. "Covalent affinity" purification of ubiquitin-activating enzyme. *J. Biol. Chem.* **1982**, *257*, 2537–2542.
89. Hershko, A.; Ciechanover, A.; Heller, H.; Haas, A.L.; Rose, I.A. Proposed role of ATP in protein breakdown: Conjugation of protein with multiple chains of the polypeptide of ATP-dependent proteolysis. *Proc. Natl. Acad. Sci. USA* **1980**, *77*, 1783–1786. [CrossRef]
90. Ciehanover, A.; Hod, Y.; Hershko, A. A heat-stable polypeptide component of an ATP-dependent proteolytic system from reticulocytes. *Biochem. Biophys. Res. Commun.* **1978**, *81*, 1100–1105. [CrossRef]
91. DeMartino, G.N.; Slaughter, C.A. The proteasome, a novel protease regulated by multiple mechanisms. *J. Biol. Chem.* **1999**, *274*, 22123–22126. [CrossRef] [PubMed]
92. Pickart, C.M. Ubiquitin biology: An old dog learns an old trick. *Nat. Cell Biol.* **2000**, *2*, E139–E141. [CrossRef] [PubMed]
93. Pickart, C.M. Ubiquitin in chains. *Trends Biochem. Sci.* **2000**, *25*, 544–548. [CrossRef]
94. McNaught, K.S.; Jenner, P. Proteasomal function is impaired in substantia nigra in Parkinson's disease. *Neurosci. Lett.* **2001**, *297*, 191–194. [CrossRef]
95. McNaught, K.S.; Olanow, C.W.; Halliwell, B.; Isacson, O.; Jenner, P. Failure of the ubiquitin-proteasome system in Parkinson's disease. *Nat. Rev. Neurosci.* **2001**, *2*, 589–594. [CrossRef] [PubMed]
96. McNaught, K.S.; Olanow, C.W. Proteolytic stress: A unifying concept for the etiopathogenesis of Parkinson's disease. *Ann. Neurol.* **2003**, *53*, S73–S84. [CrossRef]

97. Lazar, M.A.; Truscott, R.J.; Raese, J.D.; Barchas, J.D. Thermal denaturation of native striatal tyrosine hydroxylase: Increased thermolability of the phosphorylated form of the enzyme. *J. Neurochem.* **1981**, *36*, 677–682. [CrossRef] [PubMed]

98. Døskeland, A.P.; Flatmark, T. Ubiquitination of soluble and membrane-bound tyrosine hydroxylase and degradation of the soluble form. *Eur. J. Biochem.* **2002**, *269*, 1561–1569. [CrossRef]

99. Urano, F.; Hayashi, N.; Arisaka, F.; Kurita, H.; Murata, S.; Ichinose, H. Molecular mechanism for pterin-mediated inactivation of tyrosine hydroxylase: Formation of insoluble aggregates of tyrosine hydroxylase. *J. Biochem.* **2006**, *139*, 625–635. [CrossRef]

100. Nakashima, A.; Mori, K.; Kaneko, Y.S.; Hayashi, N.; Nagatsu, T.; Ota, A. Phosphorylation of the N-terminal portion of tyrosine hydroxylase triggers proteasomal digestion of the enzyme. *Biochem. Biophys. Res. Commun.* **2011**, *407*, 343–347. [CrossRef]

101. Nakashima, A.; Ohnuma, S.; Kodani, Y.; Kaneko, Y.S.; Nagasaki, H.; Nagatsu, T.; Ota, A. Inhibition of deubiquitinating activity of USP14 decreases tyrosine hydroxylase phosphorylated at Ser19 in PC12D cells. *Biochem. Biophys. Res. Commun.* **2016**, *472*, 598–602. [CrossRef] [PubMed]

102. Carbajosa, N.A.; Corradi, G.; Verrilli, M.A.; Guil, M.J.; Vatta, M.S.; Gironacci, M.M. Tyrosine hydroxylase is short-term regulated by the ubiquitin-proteasome system in PC12 cells and hypothalamic and brainstem neurons from spontaneously hypertensive rats: Possible implications in hypertension. *PLoS ONE* **2015**, *10*, e0116597. [CrossRef] [PubMed]

103. Salvatore, M.F.; Calipari, E.S.; Jones, S.R. Regulation of Tyrosine Hydroxylase Expression and Phosphorylation in Dopamine Transporter-Deficient Mice. *ACS Chem. Neurosci.* **2016**, *7*, 941–951. [CrossRef]

104. Ueda, K.; Fukushima, H.; Masliah, E.; Xia, Y.; Iwai, A.; Yoshimoto, M.; Otero, D.A.; Kondo, J.; Ihara, Y.; Saitoh, T. Molecular cloning of cDNA encoding an unrecognized component of amyloid in Alzheimer disease. *Proc. Natl. Acad. Sci. USA* **1993**, *90*, 11282–11286. [CrossRef] [PubMed]

105. Fujiwara, H.; Hasegawa, M.; Dohmae, N.; Kawashima, A.; Masliah, E.; Goldberg, M.S.; Shen, J.; Takio, K.; Iwatsubo, T. alpha-Synuclein is phosphorylated in synucleinopathy lesions. *Nat. Cell Biol.* **2002**, *4*, 160–164. [CrossRef]

106. Bennett, M.C.; Bishop, J.F.; Leng, Y.; Chock, P.B.; Chase, T.N.; Mouradian, M.M. Degradation of alpha-synuclein by proteasome. *J. Biol. Chem.* **1999**, *274*, 33855–33858. [CrossRef]

107. Giasson, B.I.; Lee, V.M. Are ubiquitination pathways central to Parkinson's disease? *Cell* **2003**, *114*, 1–8. [CrossRef]

108. Hasegawa, M.; Fujiwara, H.; Nonaka, T.; Wakabayashi, K.; Takahashi, H.; Lee, V.M.; Trojanowski, J.Q.; Mann, D.; Iwatsubo, T. Phosphorylated alpha-synuclein is ubiquitinated in alpha-synucleinopathy lesions. *J. Biol. Chem.* **2002**, *277*, 49071–49076. [CrossRef]

109. Cuervo, A.M.; Stefanis, L.; Fredenburg, R.; Lansbury, P.T.; Sulzer, D. Impaired degradation of mutant alpha-synuclein by chaperone-mediated autophagy. *Science* **2004**, *305*, 1292–1295. [CrossRef]

110. Vogiatzi, T.; Xilouri, M.; Vekrellis, K.; Stefanis, L. Wild type alpha-synuclein is degraded by chaperone-mediated autophagy and macroautophagy in neuronal cells. *J. Biol. Chem.* **2008**, *283*, 23542–23556. [CrossRef]

111. Cuervo, A.M.; Wong, E. Chaperone-mediated autophagy: Roles in disease and aging. *Cell Res.* **2014**, *24*, 92–104. [CrossRef] [PubMed]

112. Chandra, S.; Fornai, F.; Kwon, H.B.; Yazdani, U.; Atasoy, D.; Liu, X.; Hammer, R.E.; Battaglia, G.; German, D.C.; Castillo, P.E.; et al. Double-knockout mice for alpha- and beta-synucleins: Effect on synaptic functions. *Proc. Natl. Acad. Sci. USA* **2004**, *101*, 14966–14971. [CrossRef]

113. Burre, J.; Sharma, M.; Tsetsenis, T.; Buchman, V.; Etherton, M.R.; Sudhof, T.C. Alpha-synuclein promotes SNARE-complex assembly in vivo and in vitro. *Science* **2010**, *329*, 1663–1667. [CrossRef]

114. Peng, X.; Tehranian, R.; Dietrich, P.; Stefanis, L.; Perez, R.G. Alpha-synuclein activation of protein phosphatase 2A reduces tyrosine hydroxylase phosphorylation in dopaminergic cells. *J. Cell Sci.* **2005**, *118*, 3523–3530. [CrossRef] [PubMed]

115. Lou, H.; Montoya, S.E.; Alerte, T.N.; Wang, J.; Wu, J.; Peng, X.; Hong, C.S.; Friedrich, E.E.; Mader, S.A.; Pedersen, C.J.; et al. Serine 129 phosphorylation reduces the ability of alpha-synuclein to regulate tyrosine hydroxylase and protein phosphatase 2A in vitro and in vivo. *J. Biol. Chem.* **2010**, *285*, 17648–17661. [CrossRef] [PubMed]

116. Perez, R.G.; Waymire, J.C.; Lin, E.; Liu, J.J.; Guo, F.; Zigmond, M.J. A role for alpha-synuclein in the regulation of dopamine biosynthesis. *J. Neurosci.* **2002**, *22*, 3090–3099. [CrossRef] [PubMed]
117. Matsuoka, Y.; Vila, M.; Lincoln, S.; McCormack, A.; Picciano, M.; LaFrancois, J.; Yu, X.; Dickson, D.; Langston, W.J.; McGowan, E.; et al. Lack of nigral pathology in transgenic mice expressing human alpha-synuclein driven by the tyrosine hydroxylase promoter. *Neurobiol. Dis.* **2001**, *8*, 535–539. [CrossRef]
118. Rathke-Hartlieb, S.; Kahle, P.J.; Neumann, M.; Ozmen, L.; Haid, S.; Okochi, M.; Haass, C.; Schulz, J.B. Sensitivity to MPTP is not increased in Parkinson's disease-associated mutant alpha-synuclein transgenic mice. *J. Neurochem.* **2001**, *77*, 1181–1184. [CrossRef]
119. Richfield, E.K.; Thiruchelvam, M.J.; Cory-Slechta, D.A.; Wuertzer, C.; Gainetdinov, R.R.; Caron, M.G.; Di Monte, D.A.; Federoff, H.J. Behavioral and neurochemical effects of wild-type and mutated human alpha-synuclein in transgenic mice. *Exp. Neurol.* **2002**, *175*, 35–48. [CrossRef]
120. Martinez-Vicente, M.; Talloczy, Z.; Kaushik, S.; Massey, A.C.; Mazzulli, J.; Mosharov, E.V.; Hodara, R.; Fredenburg, R.; Wu, D.C.; Follenzi, A.; et al. Dopamine-modified alpha-synuclein blocks chaperone-mediated autophagy. *J. Clin. Investig.* **2008**, *118*, 777–788. [CrossRef]
121. Ng, A.S.L.; Tan, Y.J.; Lu, Z.; Ng, E.Y.L.; Ng, S.Y.E.; Chia, N.S.Y.; Setiawan, F.; Xu, Z.; Tay, K.Y.; Prakash, K.M.; et al. Plasma alpha-synuclein detected by single molecule array is increased in PD. *Ann. Clin. Transl. Neurol.* **2019**, *6*, 615–619. [CrossRef] [PubMed]
122. Perez, R.G.; Hastings, T.G. Could a loss of alpha-synuclein function put dopaminergic neurons at risk? *J. Neurochem.* **2004**, *89*, 1318–1324. [CrossRef] [PubMed]
123. Alerte, T.N.; Akinfolarin, A.A.; Friedrich, E.E.; Mader, S.A.; Hong, C.S.; Perez, R.G. Alpha-synuclein aggregation alters tyrosine hydroxylase phosphorylation and immunoreactivity: Lessons from viral transduction of knockout mice. *Neurosci. Lett.* **2008**, *435*, 24–29. [CrossRef] [PubMed]
124. Toska, K.; Kleppe, R.; Armstrong, C.G.; Morrice, N.A.; Cohen, P.; Haavik, J. Regulation of tyrosine hydroxylase by stress-activated protein kinases. *J. Neurochem.* **2002**, *83*, 775–783. [CrossRef]
125. Klegeris, A.; Pelech, S.; Giasson, B.I.; Maguire, J.; Zhang, H.; McGeer, E.G.; McGeer, P.L. Alpha-synuclein activates stress signaling protein kinases in THP-1 cells and microglia. *Neurobiol. Aging* **2008**, *29*, 739–752. [CrossRef]
126. Collier, T.J.; Lipton, J.; Daley, B.F.; Palfi, S.; Chu, Y.; Sortwell, C.; Bakay, R.A.; Sladek, J.R., Jr.; Kordower, J.H. Aging-related changes in the nigrostriatal dopamine system and the response to MPTP in nonhuman primates: Diminished compensatory mechanisms as a prelude to parkinsonism. *Neurobiol. Dis.* **2007**, *26*, 56–65. [CrossRef]
127. Burhans, W.C.; Weinberger, M. DNA replication stress, genome instability and aging. *Nucleic Acids Res.* **2007**, *35*, 7545–7556. [CrossRef]
128. Anisimova, A.S.; Alexandrov, A.I.; Makarova, N.E.; Gladyshev, V.N.; Dmitriev, S.E. Protein synthesis and quality control in aging. *Aging (Albany NY)* **2018**, *10*, 4269–4288. [CrossRef]
129. Rodriguez, M.; Rodriguez-Sabate, C.; Morales, I.; Sanchez, A.; Sabate, M. Parkinson's disease as a result of aging. *Aging Cell* **2015**, *14*, 293–308. [CrossRef]
130. Cotzias, G.C.; Papavasiliou, P.S.; Gellene, R. Modification of Parkinsonism—chronic treatment with L-dopa. *N. Engl. J. Med.* **1969**, *280*, 337–345. [CrossRef]
131. Cotzias, G.C.; Miller, S.T.; Tang, L.C.; Papavasiliou, P.S. Levodopa, fertility, and longevity. *Science* **1977**, *196*, 549–551. [CrossRef] [PubMed]
132. Papavasiliou, P.S.; Cotzias, G.C.; Rosal, V.L.; Miller, S.T. Treatment of parkinsonism with N-n-propyl norapomorphine and levodopa (with or without carbidopa). *Arch. Neurol.* **1978**, *35*, 787–791. [CrossRef] [PubMed]
133. Pandey, S.; Srivanitchapoom, P. Levodopa-induced Dyskinesia: Clinical Features, Pathophysiology, and Medical Management. *Ann. Indian Acad. Neurol.* **2017**, *20*, 190–198. [CrossRef] [PubMed]
134. Poewe, W.; Antonini, A.; Zijlmans, J.C.; Burkhard, P.R.; Vingerhoets, F. Levodopa in the treatment of Parkinson's disease: An old drug still going strong. *Clin. Interv. Aging* **2010**, *5*, 229–238. [CrossRef] [PubMed]
135. Recasens, A.; Ulusoy, A.; Kahle, P.J.; Di Monte, D.A.; Dehay, B. In vivo models of alpha-synuclein transmission and propagation. *Cell Tissue Res.* **2018**, *373*, 183–193. [CrossRef] [PubMed]
136. Breen, D.P.; Halliday, G.M.; Lang, A.E. Gut-brain axis and the spread of alpha-synuclein pathology: Vagal highway or dead end? *Mov. Disord.* **2019**, *34*, 307–316. [CrossRef] [PubMed]

137. Wong, Y.C.; Krainc, D. alpha-synuclein toxicity in neurodegeneration: Mechanism and therapeutic strategies. *Nat. Med.* **2017**, *23*, 1–13. [CrossRef]

138. Clinton, L.K.; Blurton-Jones, M.; Myczek, K.; Trojanowski, J.Q.; LaFerla, F.M. Synergistic Interactions between Abeta, tau, and alpha-synuclein: Acceleration of neuropathology and cognitive decline. *J. Neurosci.* **2010**, *30*, 7281–7289. [CrossRef]

139. Shrivastava, A.N.; Redeker, V.; Fritz, N.; Pieri, L.; Almeida, L.G.; Spolidoro, M.; Liebmann, T.; Bousset, L.; Renner, M.; Lena, C.; et al. alpha-synuclein assemblies sequester neuronal alpha3-Na+/K+-ATPase and impair Na+ gradient. *EMBO J.* **2015**, *34*, 2408–2423. [CrossRef] [PubMed]

140. Rodriguez, L.; Marano, M.M.; Tandon, A. Import and Export of Misfolded alpha-Synuclein. *Front. Neurosci.* **2018**, *12*, 344. [CrossRef] [PubMed]

141. Mao, X.; Ou, M.T.; Karuppagounder, S.S.; Kam, T.I.; Yin, X.; Xiong, Y.; Ge, P.; Umanah, G.E.; Brahmachari, S.; Shin, J.H.; et al. Pathological alpha-synuclein transmission initiated by binding lymphocyte-activation gene 3. *Science* **2016**, *353*. [CrossRef] [PubMed]

142. Kobayashi, J.; Hasegawa, T.; Sugeno, N.; Yoshida, S.; Akiyama, T.; Fujimori, K.; Hatakeyama, H.; Miki, Y.; Tomiyama, A.; Kawata, Y.; et al. Extracellular alpha-synuclein enters dopaminergic cells by modulating flotillin-1-assisted dopamine transporter endocytosis. *FASEB J.* **2019**, *33*, 10240–10256. [CrossRef] [PubMed]

143. Delenclos, M.; Trendafilova, T.; Mahesh, D.; Baine, A.M.; Moussaud, S.; Yan, I.K.; Patel, T.; McLean, P.J. Investigation of Endocytic Pathways for the Internalization of Exosome-Associated Oligomeric Alpha-Synuclein. *Front. Neurosci.* **2017**, *11*, 172. [CrossRef] [PubMed]

144. Kawahata, I.; Bousset, L.; Melki, R.; Fukunaga, K. Fatty Acid-Binding Protein 3 is Critical for alpha-Synuclein Uptake and MPP(+)-Induced Mitochondrial Dysfunction in Cultured Dopaminergic Neurons. *Int. J. Mol. Sci.* **2019**, *20*, 5358. [CrossRef] [PubMed]

145. Yabuki, Y.; Matsuo, K.; Kawahata, I.; Fukui, N.; Mizobata, T.; Kawata, Y.; Owada, Y.; Shioda, N.; Fukunaga, K. Fatty Acid Binding Protein 3 Enhances the Spreading and Toxicity of alpha-Synuclein in Mouse Brain. *Int. J. Mol. Sci.* **2020**, *21*, 2230. [CrossRef]

146. Perlmutter, J.S.; Tempel, L.W.; Black, K.J.; Parkinson, D.; Todd, R.D. MPTP induces dystonia and parkinsonism. Clues to the pathophysiology of dystonia. *Neurology* **1997**, *49*, 1432–1438. [CrossRef]

147. Tabbal, S.D.; Mink, J.W.; Antenor, J.A.; Carl, J.L.; Moerlein, S.M.; Perlmutter, J.S. 1-Methyl-4-phenyl-1,2,3,6-tetrahydropyridine-induced acute transient dystonia in monkeys associated with low striatal dopamine. *Neuroscience* **2006**, *141*, 1281–1287. [CrossRef]

148. Langston, J.W. The MPTP Story. *J. Parkinsons Dis.* **2017**, *7*, S11–S19. [CrossRef]

149. Lai, F.; Jiang, R.; Xie, W.; Liu, X.; Tang, Y.; Xiao, H.; Gao, J.; Jia, Y.; Bai, Q. Intestinal Pathology and Gut Microbiota Alterations in a Methyl-4-phenyl-1,2,3,6-tetrahydropyridine (MPTP) Mouse Model of Parkinson's Disease. *Neurochem. Res.* **2018**, *43*, 1986–1999. [CrossRef]

150. Matsuo, K.; Cheng, A.; Yabuki, Y.; Takahata, I.; Miyachi, H.; Fukunaga, K. Inhibition of MPTP-induced alpha-synuclein oligomerization by fatty acid-binding protein 3 ligand in MPTP-treated mice. *Neuropharmacology* **2019**, *150*, 164–174. [CrossRef]

151. Cheng, A.; Shinoda, Y.; Yamamoto, T.; Miyachi, H.; Fukunaga, K. Development of FABP3 ligands that inhibit arachidonic acid-induced alpha-synuclein oligomerization. *Brain Res.* **2019**, *1707*, 190–197. [CrossRef]

152. Teunissen, C.E.; Veerhuis, R.; De Vente, J.; Verhey, F.R.; Vreeling, F.; Van Boxtel, M.P.; Glatz, J.F.; Pelsers, M.A. Brain-specific fatty acid-binding protein is elevated in serum of patients with dementia-related diseases. *Eur. J. Neurol.* **2011**, *18*, 865–871. [CrossRef] [PubMed]

153. Tatebe, H.; Kasai, T.; Ohmichi, T.; Kishi, Y.; Kakeya, T.; Waragai, M.; Kondo, M.; Allsop, D.; Tokuda, T. Quantification of plasma phosphorylated tau to use as a biomarker for brain Alzheimer pathology: Pilot case-control studies including patients with Alzheimer's disease and down syndrome. *Mol. Neurodegener.* **2017**, *12*, 63. [CrossRef] [PubMed]

154. Xia, W.; Yang, T.; Shankar, G.; Smith, I.M.; Shen, Y.; Walsh, D.M.; Selkoe, D.J. A specific enzyme-linked immunosorbent assay for measuring beta-amyloid protein oligomers in human plasma and brain tissue of patients with Alzheimer disease. *Arch. Neurol.* **2009**, *66*, 190–199. [CrossRef] [PubMed]

155. Lewczuk, P.; Kornhuber, J.; Vanmechelen, E.; Peters, O.; Heuser, I.; Maier, W.; Jessen, F.; Burger, K.; Hampel, H.; Frolich, L.; et al. Amyloid beta peptides in plasma in early diagnosis of Alzheimer's disease: A multicenter study with multiplexing. *Exp. Neurol.* **2010**, *223*, 366–370. [CrossRef] [PubMed]

156. Scheuner, D.; Eckman, C.; Jensen, M.; Song, X.; Citron, M.; Suzuki, N.; Bird, T.D.; Hardy, J.; Hutton, M.; Kukull, W.; et al. Secreted amyloid beta-protein similar to that in the senile plaques of Alzheimer's disease is increased in vivo by the presenilin 1 and 2 and APP mutations linked to familial Alzheimer's disease. *Nat. Med.* **1996**, *2*, 864–870. [CrossRef] [PubMed]

157. Van Oijen, M.; Hofman, A.; Soares, H.D.; Koudstaal, P.J.; Breteler, M.M. Plasma Abeta(1-40) and Abeta(1-42) and the risk of dementia: A prospective case-cohort study. *Lancet Neurol.* **2006**, *5*, 655–660. [CrossRef]

158. Graff-Radford, N.R.; Crook, J.E.; Lucas, J.; Boeve, B.F.; Knopman, D.S.; Ivnik, R.J.; Smith, G.E.; Younkin, L.H.; Petersen, R.C.; Younkin, S.G. Association of low plasma Abeta42/Abeta40 ratios with increased imminent risk for mild cognitive impairment and Alzheimer disease. *Arch. Neurol.* **2007**, *64*, 354–362. [CrossRef]

159. Schupf, N.; Patel, B.; Pang, D.; Zigman, W.B.; Silverman, W.; Mehta, P.D.; Mayeux, R. Elevated plasma beta-amyloid peptide Abeta(42) levels, incident dementia, and mortality in Down syndrome. *Arch. Neurol.* **2007**, *64*, 1007–1013. [CrossRef]

160. Haga, H.; Yamada, R.; Izumi, H.; Shinoda, Y.; Kawahata, I.; Miyachi, H.; Fukunaga, K. Novel fatty acid-binding protein 3 ligand inhibits dopaminergic neuronal death and improves motor and cognitive impairments in Parkinson's disease model mice. *Pharmacol. Biochem. Behav.* **2020**, *191*, 172891. [CrossRef]

161. Kawahata, I.; Yoshida, M.; Sun, W.; Nakajima, A.; Lai, Y.; Osaka, N.; Matsuzaki, K.; Yokosuka, A.; Mimaki, Y.; Naganuma, A.; et al. Potent activity of nobiletin-rich Citrus reticulata peel extract to facilitate cAMP/PKA/ERK/CREB signaling associated with learning and memory in cultured hippocampal neurons: Identification of the substances responsible for the pharmacological action. *J. Neural Transm.* **2013**, *120*, 1397–1409. [CrossRef] [PubMed]

162. Kawahata, I.; Suzuki, T.; Rico, E.G.; Kusano, S.; Tamura, H.; Mimaki, Y.; Yamakuni, T. Fermented Citrus reticulata (ponkan) fruit squeezed draff that contains a large amount of 4'-demethylnobiletin prevents MK801-induced memory impairment. *J. Nat. Med.* **2017**, *71*, 617–631. [CrossRef] [PubMed]

163. Kawahata, I.; Xu, H.; Takahashi, M.; Murata, K.; Han, W.; Yamaguchi, Y.; Fujii, A.; Yamaguchi, K.; Yamakuni, T. Royal jelly coordinately enhances hippocampal neuronal expression of somatostatin and neprilysin genes conferring neuronal protection against toxic soluble amyloid-β oligomers implicated in Alzheimer's disease pathogenesis. *J. Funct. Foods* **2018**, *51*, 28–38. [CrossRef]

164. Kawahata, I. Drug discovery study of fundamental therapy for dopamine related diseases targeting new V-1 / CP complex. *Impact* **2019**, *2019*, 49–51. [CrossRef]

165. Schapira, A.H.; Cooper, J.M.; Dexter, D.; Jenner, P.; Clark, J.B.; Marsden, C.D. Mitochondrial complex I deficiency in Parkinson's disease. *Lancet* **1989**, *1*, 1269. [CrossRef]

166. Schapira, A.H.; Cooper, J.M.; Dexter, D.; Clark, J.B.; Jenner, P.; Marsden, C.D. Mitochondrial complex I deficiency in Parkinson's disease. *J. Neurochem.* **1990**, *54*, 823–827. [CrossRef] [PubMed]

167. Sian, J.; Dexter, D.T.; Lees, A.J.; Daniel, S.; Agid, Y.; Javoy-Agid, F.; Jenner, P.; Marsden, C.D. Alterations in glutathione levels in Parkinson's disease and other neurodegenerative disorders affecting basal ganglia. *Ann. Neurol.* **1994**, *36*, 348–355. [CrossRef]

168. Dexter, D.T.; Carayon, A.; Javoy-Agid, F.; Agid, Y.; Wells, F.R.; Daniel, S.E.; Lees, A.J.; Jenner, P.; Marsden, C.D. Alterations in the levels of iron, ferritin and other trace metals in Parkinson's disease and other neurodegenerative diseases affecting the basal ganglia. *Brain* **1991**, *114*, 1953–1975. [CrossRef]

169. Choi, H.J.; Jang, Y.J.; Kim, H.J.; Hwang, O. Tetrahydrobiopterin is released from and causes preferential death of catecholaminergic cells by oxidative stress. *Mol. Pharmacol.* **2000**, *58*, 633–640. [CrossRef]

170. Kim, S.T.; Choi, J.H.; Chang, J.W.; Kim, S.W.; Hwang, O. Immobilization stress causes increases in tetrahydrobiopterin, dopamine, and neuromelanin and oxidative damage in the nigrostriatal system. *J. Neurochem.* **2005**, *95*, 89–98. [CrossRef]

International Journal of
Molecular Sciences

Review

Linear Ubiquitin Code: Its Writer, Erasers, Decoders, Inhibitors, and Implications in Disorders

Daisuke Oikawa [1], Yusuke Sato [2], Hidefumi Ito [3] and Fuminori Tokunaga [1,*]

[1] Department of Pathobiochemistry, Graduate School of Medicine, Osaka City University, Osaka 545-8585, Japan; oikawa.daisuke@med.osaka-cu.ac.jp
[2] Center for Research on Green Sustainable Chemistry, Tottori University, Tottori 680-8552, Japan; yusato@iam.u-tokyo.ac.jp
[3] Department of Neurology, Faculty of Medicine, Wakayama Medical University, Wakayama 641-8510, Japan; itohid@kuhp.kyoto-u.ac.jp
* Correspondence: ftokunaga@med.osaka-cu.ac.jp; Tel.: +81-6-6645-3720

Received: 31 March 2020; Accepted: 6 May 2020; Published: 11 May 2020

Abstract: The linear ubiquitin chain assembly complex (LUBAC) is a ubiquitin ligase composed of the Heme-oxidized IRP2 ubiquitin ligase-1L (HOIL-1L), HOIL-1L-interacting protein (HOIP), and Shank-associated RH domain interactor (SHARPIN) subunits. LUBAC specifically generates the N-terminal Met1-linked linear ubiquitin chain and regulates acquired and innate immune responses, such as the canonical nuclear factor-κB (NF-κB) and interferon antiviral pathways. Deubiquitinating enzymes, OTULIN and CYLD, physiologically bind to HOIP and control its function by hydrolyzing the linear ubiquitin chain. Moreover, proteins containing linear ubiquitin-specific binding domains, such as NF-κB-essential modulator (NEMO), optineurin, A20-binding inhibitors of NF-κB (ABINs), and A20, modulate the functions of LUBAC, and the dysregulation of the LUBAC-mediated linear ubiquitination pathway induces cancer and inflammatory, autoimmune, and neurodegenerative diseases. Therefore, inhibitors of LUBAC would be valuable to facilitate investigations of the molecular and cellular bases for LUBAC-mediated linear ubiquitination and signal transduction, and for potential therapeutic purposes. We identified and characterized α,β-unsaturated carbonyl-containing chemicals, named HOIPINs (HOIP inhibitors), as LUBAC inhibitors. We summarize recent advances in elucidations of the pathophysiological functions of LUBAC-mediated linear ubiquitination and identifications of its regulators, toward the development of LUBAC inhibitors.

Keywords: inflammation; inhibitor; innate immune; interferon; LUBAC; NF-κB; ubiquitin

---

1. Introduction

Ubiquitin, a 76-residue (8.6 kDa) small globular protein, is evolutionarily conserved in most eukaryotes. Ubiquitin functions as a spatiotemporal-specific post-translational modifier. In most cases, ubiquitin is covalently conjugated to the ε-NH_2 group of Lys in the targeted proteins via an isopeptide bond [1]. In particular cases, ubiquitin is conjugated to the N-terminal α-NH_2-group of Lys-less proteins and internal Ser, Thr, and Cys residues [2,3]. The human ubiquitin system is composed of two ubiquitin-activating enzymes (E1), ~40 ubiquitin-conjugating enzymes (E2), >600 ubiquitin ligases (E3), and ~100 deubiquitinating enzymes (DUBs). Among them, the E3s play crucial roles in recognizing and conjugating one or more ubiquitins to substrates, and are classified into the homologous to the E6AP carboxyl terminus (HECT)-type, the really interesting new gene (RING)-type, and the RING-in between-RING (RBR)-type [4]. In the human genome, most of the E3s are RING-type [5], and limited numbers of HECT-type (28 members) and RBR-type (14 members) E3s have been identified.

Protein ubiquitination regulates numerous cellular functions, including proteasomal degradation, membrane trafficking, DNA repair, and signal transduction, through the conjugation of one or more ubiquitins to substrates [1]. The conjugation of monoubiquitin and multiple-monoubiquitins to substrates is principally involved in membrane trafficking and endocytosis. The isopeptide bond-linked ubiquitination is mediated via seven internal Lys residues (K6, K11, K27, K29, K33, K48, and K63). Among them, the K48-linked polyubiquitin chain, which is the most common, serves as a typical proteasomal degradation signal, and the K63-linked polyubiquitin chain, the second most predominant linkage, is involved in non-proteasomal functions, such as signal transduction and DNA repair [1,6]. In addition to these Lys-linked polyubiquitin chains, the N-terminal Met1(M1)-linked linear polyubiquitination is specifically generated by the E3 complex, named the linear ubiquitin chain assembly complex (LUBAC). LUBAC functions in the regulation of the innate and acquired immune responses and anti-apoptosis [7]. In addition to the mammalian LUBACs, an ortholog of a HOIP subunit of LUBAC, named LUBEL, was identified in *Drosophila*, and it also catalyzes linear ubiquitination upon heat shock [8]. These findings indicated that linear ubiquitination is evolutionarily conserved to maintain cellular homeostasis. In addition to the homotypic polyubiquitin chains, heterogeneous complex-types of polyubiquitin chains, such as mixed, hybrid, and branched ubiquitin chains, have also been identified. Furthermore, specific residues of ubiquitin are chemically modified, by phosphorylation, acetylation, and ADP-ribosylation, and these modifications regulate the pathophysiological functions of ubiquitination. These diverse ubiquitin linkages exhibit multiple functions in a system called the "ubiquitin code" [6], in which E1, E2, and E3 function as "writers", DUBs are "erasers", and ubiquitin-binding proteins serve as "decoders". In this review, we focus on the structures, catalytic mechanisms, inhibitors, and pathophysiological functions of the LUBAC-mediated "linear ubiquitin code", revealed by studies using human cell lines, diseases, and phenotypes of genetically deficient mice.

2. LUBAC: The Only Writer of the Linear Ubiquitin Code

2.1. Structure and Catalytic Mechanism of LUBAC

Mammalian LUBAC, a ~600 kDa complex composed of the HOIL-1L (also known as RBCK1) [9], HOIP (RNF31, ZIBRA, and PAUL) [10], and SHARPIN [11–13] subunits, is the sole E3 that generates the N-terminal M1-linked linear polyubiquitin chain, using the E2s UBE2L3 (UbcH7) and UbcH5s [14,15]. LUBAC subunits contain multiple domains (Figure 1). Although the detailed architecture of LUBAC has not been solved, the ubiquitin-like (UBL) domains in SHARPIN and HOIL-1L bind to the ubiquitin-associated (UBA)1 and UBA2 domains, respectively, in HOIP [16]. Moreover, the LUBAC-tethering motifs (LTMs) in HOIL-1L and SHARPIN associate with each other to form a globular domain [16]. HOIL-1L and HOIP possess RBR-type E3 motifs. The RBR-type E3 family members reportedly generate polyubiquitin chains through a unique RING-HECT-hybrid reaction [17,18]. During the course of linear ubiquitination, the RING1 domain in HOIP binds a ubiquitin-charged E2. Subsequently, the donor ubiquitin is transiently transferred to the active Cys885 in the RING2 domain of HOIP via a thioester-linkage. Finally, the donor ubiquitin is conjugated to an acceptor ubiquitin, which is captured in the C-terminal linear ubiquitin chain determining domain (LDD) of HOIP, to specifically generate a linear ubiquitin chain [19–22]. In contrast to HOIP, the RBR domain in HOIL-1L uniquely catalyzes the oxyester-bond monoubiquitination of Ser/Thr residues through the active Cys458 [23].

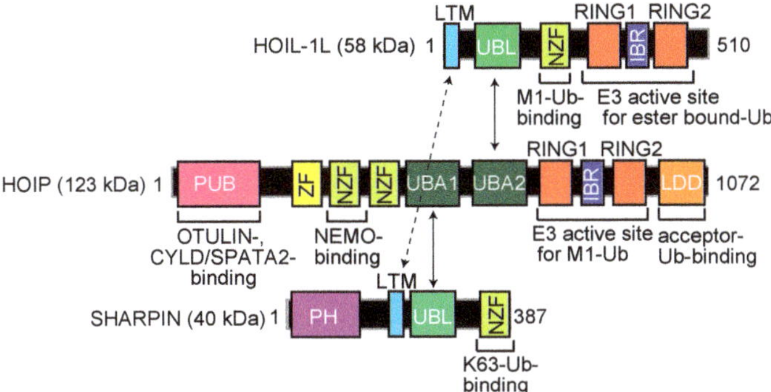

Figure 1. Domain structure and functional regions of the LUBAC subunits, HOIL-1L, HOIP, and SHARPIN. LTM, LUBAC-tethering motif; UBL, ubiquitin-like; NZF, Npl4-type zinc finger; RING, really interesting new gene; IBR, in-between RING; PUB, PNGase/UBA or UBX; ZF, zinc finger; UBA, ubiquitin-associated; LDD, linear ubiquitin chain determining domain; PH, Pleckstrin-homology.

The N-terminal portion of HOIP contains a PNGase/UBA or UBX (PUB) domain (Figure 1), which is reportedly an AAA-ATPase p97-interacting domain [24] that plays an important role to recruit linear ubiquitin-editing DUBs, such as OTULIN [25] and the CYLD-SPATA2 complex [26–29]. Thus, LUBAC forms complexes with negative regulators through the PUB domain. Furthermore, LUBAC includes several zinc finger domains (ZFs). Among them, the Npl4-type zinc finger (NZF) domain in HOIL-1L specifically binds linear ubiquitin [30], whereas the NZF domain in SHARPIN binds K63-ubiquitin to regulate the cell death pathway [31]. There are two NZF domains in HOIP, and NZF1 binds NEMO, a LUBAC substrate, during linear ubiquitination [32]. Collectively, these findings indicate that LUBAC consists of multiple functional domains for the regulation of linear ubiquitination and participates in various physiological phenomena.

2.2. Cellular Functions of LUBAC

2.2.1. LUBAC in the Inflammatory Cytokine-Induced canonical NF-κB Activation Pathway

LUBAC principally participates in the regulation of the canonical NF-κB signaling pathway in various mammalian cells. NF-κB is a master transcription factor for the biological defense system, and is composed of homo- or hetero-dimers of Rel-homology domain-containing proteins, such as p65 (RelA), RelB, c-Rel, p105/p50, and p100/p52. NF-κB expression leads to the transcription of target genes in the inflammatory and immune responses [33]. LUBAC regulates the NF-κB activation pathways induced by proinflammatory cytokines, such as TNF-α and IL-1β, various pathogen-associated molecular patterns (PAMPs), T cell receptor (TCR) agonists, genotoxic stress, and NOD2-mediated inflammasome activation [14,15]. However, LUBAC is not involved in either the B cell receptor (BCR)-mediated pathway or the noncanonical NF-κB pathway [12,34].

Upon stimulation of cells with TNF-α, LUBAC is recruited to the TNF receptor (TNFR) through binding to K63-linked polyubiquitin chains, which are antecedently generated by c-IAP-1/2, TRAF2, and TRAF5, and functions as a member of the TNFR signaling complex I [35–37]. LUBAC then conjugates linear ubiquitin chains to NEMO, RIP1, and FADD (Figure 2) [11,38,39]. The linear ubiquitin chain functions as a scaffold to recruit canonical IκB kinase (IKK) molecules, which are composed of the kinase subunits of IKKα and IKKβ, and a regulatory subunit of NEMO. Importantly, NEMO contains a high-affinity linear ubiquitin binding site that accumulates multiple IKK molecules on the linear ubiquitin chain. The *trans*-phosphorylation of the IKK molecules principally leads to the activation of IKKβ, which subsequently phosphorylates the inhibitory protein of NF-κB, IκBα. Interestingly, the conjugation of two linearly linked molecules of ubiquitin (linear diubiquitin) to NEMO sufficiently induces IKK activation [40]. The phosphorylated IκBα is ubiquitinated by the E3 complex SCF$^{β\text{-TrCP}}$ for the K48-ubiquitination-mediated proteasomal degradation of IκBα. After liberation from IκBα, the canonical NF-κB transcription factors, predominantly composed of homo- or hetero-dimers of p65 (RelA) and/or p50, translocate into the nucleus and activate NF-κB target genes (Figure 2) [32]. Upon TNF-α stimulation, mammalian Ste20-like kinase (MST1, also called STK4) is recruited to TNFR in a TRAF2-dependent manner and phosphorylates Ser1066 in the LDD domain of HOIP, which attenuates the E3 activity of LUBAC [41]. Recently, *Parkin-coregulated gene* (PACRG) was identified as a functional replacement of SHARPIN in TNF signaling in human and mouse cells [42]. Therefore, multiple factors regulate the LUBAC-mediated NF-κB activation pathway.

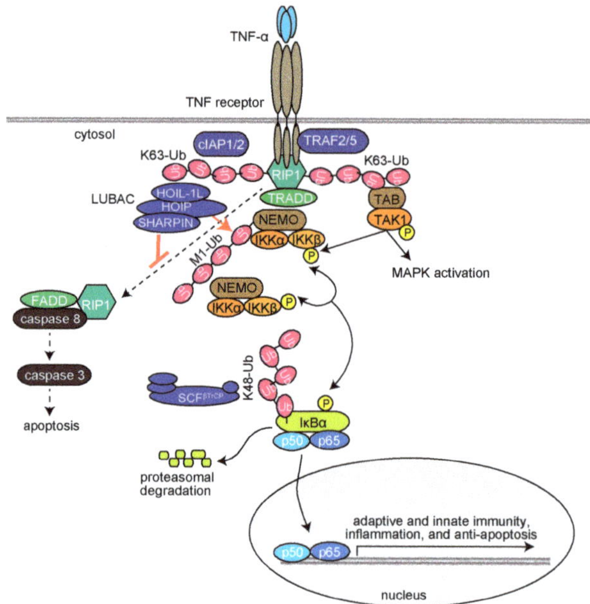

Figure 2. LUBAC-mediated regulation of the TNF-α-induced canonical NF-κB activation pathway and extrinsic apoptosis pathway.

Although IL-1β is another prominent proinflammatory cytokine that activates the canonical NF-κB activation pathway, both K63- and M1-linked ubiquitinations are required for the formation of the NEMO-containing punctate structure upon IL-1β stimulation [43]. Importantly, the K63/M1-hybrid ubiquitin chain can become conjugated to interleukin 1 receptor-associated kinase 1 (IRAK1) and IRAK4 [44]. Furthermore, HOIL-1L conjugates oxyester-bond monoubiquitin to its own Ser/Thr residues, as well as those in SHARPIN, IRAK1/2, and MyD88 in human keratinocyte HaCaT cells and mouse bone marrow-derived macrophages [23]. Thus, the E3 activity of HOIL-1L regulates the Myddosome components upon innate immune responses. These results indicate the differences in the LUBAC functions between the TNF-α- and IL-1β-mediated canonical NF-κB activation pathways.

2.2.2. LUBAC in Acquired Immune Responses

The NF-κB activity plays important roles in lymphocyte development and antigen receptor-mediated acquired immune responses in mammals [33]. Characteristically, a protein complex composed of CARMA1, BCL10, and MALT1 (CBM complex) is critical to activate the B cell receptor (BCR)- and T cell receptor (TCR)-mediated NF-κB activation pathways [45]. In mice B cells, LUBAC has no influence on the IgM-induced BCR pathway, whereas the LUBAC activity is critical for the

CD40-mediated NF-κB activation pathway and B1 cell development [34]. In contrast, in T cells, LUBAC is involved in the TCR-mediated NF-κB activation pathway, FOXP3$^+$ regulatory T cell (Treg) development, and homeostasis [46]. In the course of the TCR pathway, HOIL-1L is cleaved at Arg165-Gly166 by MALT1, a paracaspase [47]. Moreover, BCL10 is linearly ubiquitinated by LUBAC [48]. However, the importance of the E3 activity of LUBAC in the antigen receptor-mediated NF-κB activation pathway remains to be established [49]. Therefore, further studies are necessary to clarify the function of LUBAC in the antigen receptor-mediated NF-κB activation pathways in lymphocytes.

2.2.3. LUBAC in the Genotoxic Stress Response and Inflammasome Activation

DNA damaging anti-cancer agents, such as camptothecin, etoposide, and doxorubicin, stimulate the NF-κB pathway through the activation of ataxia telangiectasia mutated (ATM) kinase and various post-translational modifications of NEMO, such as phosphorylation, SUMOylation, and ubiquitination [50]. In the genotoxic stress-induced NF-κB activation pathway, X-linked inhibitor of apoptosis (XIAP) conjugates K63-ubiquitin chains to ELKS, which then induces the LUBAC-mediated linear ubiquitination of NEMO in the cytosol [51]. Similarly, the XIAP-mediated K63-linked ubiquitination of RIP2 recruits LUBAC to activate the NOD2-mediated NF-κB activation pathway [52], which plays an important role in the bacterial peptidoglycan-mediated innate immune response.

The inflammasome is a protein complex that activates pro-inflammatory cytokines, such as pro-IL-1β and pro-IL-18. Upon stimulation through Toll-like receptors (TLRs) by damage-associated molecular patterns (DAMPs) and PAMPs, inflammasomes become oligomerized and activate caspase 1. The ubiquitin system functions as both a negative and positive regulator of inflammasomes [53]. The nucleotide binding and leucine-rich repeat-containing protein 3 (NLRP3) is one of the best characterized inflammasomes. LUBAC conjugates a linear ubiquitin chain to the caspase-recruit domain (CARD) of the ASC component, and activates the NLRP3 inflammasome in macrophages [54].

2.2.4. LUBAC-Mediated Regulation of Cell Death

The TNF-α-induced expression of NF-κB-target genes basically functions in anti-apoptosis. However, under conditions where the expression of NF-κB-target genes is suppressed, such as by the protein synthesis inhibitor cycloheximide, TNF-α stimulation extensively induces apoptosis through the generation of TNFR complex IIa, which is composed of RIP1, FADD, and caspase 8 [55] (Figure 2). Subsequently, caspase 8 activates caspase 3 to induce extrinsic apoptosis, and the activated caspases cleave the N-terminal portion of HOIP [39,56]. A genetic deficiency of LUBAC subunits causes reduced expression of NF-κB genes, and thus efficiently induces TNF-α-mediated apoptosis in mice [11–13,38,57]. Characteristically, spontaneous *Sharpin*-deficient mice (*cpdm* mice) exhibit severe chronic proliferative dermatitis, and lack secondary lymphoid organs [58–60]. The combined genetic deletions of *Tnf* [11] or *Tnfr* [61] with *Sharpin$^{cpdm/cpdm}$* mice prevented the skin lesions, indicating that TNF-α-mediated apoptosis plays a critical role in dermatitis in *Sharpin*-deficient mice. Importantly, the lack of secondary lymphoid organs, such as Peyer's patches, was not alleviated by the attenuation of TNF signaling. However, the genetic deletions of caspase 8 and Rip3k in mice (*Sharpin$^{cpd/cpdm}$/Casp8$^{+/−}$/Rip3k$^{−/−}$*) completely alleviated the phenotype [62]. Recently, Sharma and coworkers showed that the genetic ablation of *MyD88* in *Sharpin*-deficient *cpdm* mice (*Sharpin$^{cpd/cpdm}$/Myd88$^{−/−}$*) completely and partially rescued the skin lesions and systemic inflammation, respectively [63]. Interestingly, they proposed that gut microbiota may play a role in inflammation induction in *cpdm* mice. Therefore, LUBAC seems to be involved in the regulation of not only apoptosis, but also necroptosis.

2.2.5. LUBAC-Mediated Regulation of Interferon Signaling

In innate immune responses, PAMPs are recognized by host pattern-recognition receptors, and then activate the NF-κB and type I interferon (IFN) production pathways. In the course of the IFN antiviral pathway, the phosphorylation of transcription factors, such as interferon regulatory factor

3 (IRF3), by TANK-binding kinase 1 (TBK1) and IKKε causes the transcription of IRF3-target genes, including *IFNβ*, *ISG15*, and *ISG56*. [64]. Although various E3s are known to regulate the antiviral pathway [65], whether LUBAC activates or suppresses the type I IFN production pathway remains controversial. LUBAC and the linear ubiquitination activity reportedly down-regulate the type I IFN production pathway [66,67]. STAT1 is linearly ubiquitinated at the K511 and K652 residues by LUBAC, which inhibits binding to the type I IFN receptor, IFNAR2, and the linear ubiquitination of STAT1 is removed by OTULIN [68]. By contrast, virus infection induces the enhanced LUBAC-mediated NF-κB activation, which in turn, inhibits IFN-STAT1 signaling.

However, some studies suggested that the LUBAC activity is necessary for the TLR-mediated IRF3 activation [69–71], and LUBAC is reportedly indispensable for the TNF-induced TBK1 and IKKε activation and prevention of cell death [72]. We showed that the LPS-, poly(I:C)-, and SeV-mediated IFN production pathway was impaired in *HOIP*-deficient mouse embryonic fibroblasts and human Jurkat T-lymphoblasts, and LUBAC inhibitors, HOIPINs, suppressed the antiviral pathway through the reduced activation of TBK1 and IRF3, supporting the positive function of LUBAC in this pathway [73]. Further studies are required to clarify the function of LUBAC in the IFN antiviral pathway.

2.2.6. Involvement of Linear Ubiquitination in Selective Autophagy

Damaged organelles and invading pathogens are selectively degraded through the autophagy pathway [74]. The selective autophagy against intracellular bacteria is referred to as xenophagy. For instance, invaded *Salmonella* replicates within the host-derived membrane vacuole, and the rupture of the vacuole causes ubiquitination and autophagy [75]. The ubiquitinome analysis showed that *Salmonella* infection promotes CDC42 and LUBAC activities with the enhanced NF-κB activity through linear ubiquitination in human colon cancer HCT116 cells and HeLa cells [76]. In addition to several E3s, such as Parkin, Smurf1, RNF166, ARIH1, and LRSAM1, LUBAC indeed restricts *Salmonella* proliferation through linear ubiquitination, which functions to recruit optineurin (OPTN) and NEMO to induce xenophagy and activate NF-κB, respectively [77,78]. Moreover, a linear ubiquitin-specific DUB, OTULIN, antagonizes LUBAC function in the xenophagy of *Salmonella* [79]. These results indicate that linear (de)-ubiquitination is indispensable for the regulation of bacterial pathogenesis and selective autophagy.

3. Erasers of the Linear Ubiquitin Code

DUBs function in the biosynthesis, recycling, editing, and cleavage of ubiquitin linkages, and consequently maintain the intracellular free ubiquitin pool [80,81]. The human DUBs are classified into seven subfamilies: ubiquitin-specific protease (USP; 54 members), ubiquitin C-terminal hydrolase (UCH; 4 members), ovarian tumor protease (OTU; 16 members), Josephins (4 members), motif interacting with ubiquitin (MIU)-containing novel DUB (MINDY; 4 members) [82], zinc finger with UFM1-specific peptidase domain protein (ZUFSP; 2 members) [83], and JAB1/MPN/MOV34 metalloenzymes (JAMM/MPN+; 16 members) [80,81]. The USP, UCH, OTU, Josephin, MINDY, and ZFUBP DUBs belong to the Cys protease family, whereas the JAMM/MPN+ family proteins are zinc metalloproteases. Among the DUBs, OTULIN (OTU family), CYLD (USP family), and A20 (OTU family) (Figure 3) are well known DUBs that regulate the LUBAC-mediated NF-κB signaling pathway [84]. In this chapter, we introduce OTULIN and CYLD as "erasers" of the linear ubiquitin code, and classify A20 as a "decoder" (see Section 4.2).

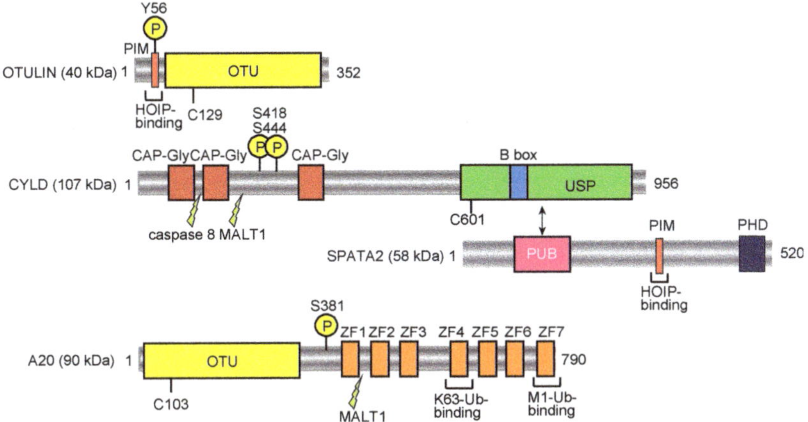

Figure 3. DUBs involved in the regulation of the linear ubiquitin code. Domain structures and functions of LUBAC-regulating factors, such as OTULIN, CYLD-SPATA2, and A20, are shown. PIM, PUB domain-interacting motif; OTU, ovarian tumor protease; CAP-Gly, cytoskeleton-associated protein Gly-rich domain; B box, B-box-type zinc finger domain; PUB, PNGase/UBA or UBX; PHD, plant homeodomain; ZF, zinc finger. Phosphorylation sites are denoted by encircled Ps, and caspase 8 and MALT1 cleavage sites are also indicated.

3.1. OTULIN

OTULIN (also called FAM105B and Gumby) is an OTU-family DUB with the catalytic Cys129 residue (Figure 3). OTULIN exclusively hydrolyzes the M1-linked peptide bond between ubiquitins, but none of the K-linked isopeptide linkages [85–87]. Therefore, OTULIN downregulates the LUBAC-mediated innate immune responses. Interestingly, OTULIN binds to the N-terminal PUB domain of HOIP through the PUB domain-interacting motif (PIM), and the phosphorylation of Tyr56 in the PIM of OTULIN abrogated the HOIP binding [25,88,89]. In humans, genetic mutations in *OTULIN* cause multiple symptoms, such as recurrent fevers, autoantibodies, diarrhea, panniculitis, and arthritis, which are collectively referred to as OTULIN-related autoinflammatory syndrome (ORAS). These mutations deregulate the LUBAC-mediated linear ubiquitination signal [90–92], and the *OTULIN*-deficiency causes cell-type-specific LUBAC degradation [85,93]. Interestingly, knock-in mice expressing the active site mutant of Otulin(C129A) showed enhanced cell death through apoptosis and necroptosis, and the increased production of type I IFN, due to the reduced LUBAC activity [94]. Therefore, under physiological conditions, OTULIN prevents the auto-linear ubiquitination of LUBAC, and functions to maintain the LUBAC activity. During cell death, OTULIN regulates the linear ubiquitination of RIP1, which is cleaved at Asp31 by caspase 3. Moreover, the phosphorylation of Tyr56 in OTULIN is increased, and this is counteracted by the dual-specificity phosphatase 14 (DUSP14) during necroptosis [95]. The OTULIN-mediated suppression of hepatocyte apoptosis plays a crucial role in liver pathogeneses, such as hepatitis, fibrosis, and hepatocellular carcinoma [96]. TRIM32, a RING-type E3, conjugates the K63-linked ubiquitin chain to OTULIN to interfere with its interaction with HOIP in human embryonic kidney (HEK) 293 cells [97], and the interaction between TRIM32 and SNX27, which regulates endosome-to-plasma membrane trafficking, has been confirmed [98]. Thus, OTULIN is a crucial pathophysiological regulator in the linear ubiquitin code.

3.2. CYLD

CYLD, a USP family DUB (Figure 3), was initially identified as a cylindromatosis tumor suppressor gene in humans [99]. CYLD downregulates the NF-κB activation pathway by hydrolyzing K63-linked ubiquitin chains [100,101]. The genetic deficiency in *CYLD* causes trichoepithelioma. Importantly,

CYLD efficiently hydrolyzes the K63-linked ubiquitin chain and the linear ubiquitin chain, but not the K48-chain [102], and thus regulates innate immune signaling [103]. We and another group showed that the catalytic activity of CYLD is indispensable for the downregulation of LUBAC-mediated NF-κB activation [104,105]. Furthermore, we revealed the structural bases of the CYLD USP domain recognition of either K63 or linear diubiquitin [106], and the interaction of a potential CYLD inhibitor, subquinocin [107]. We further identified that mind bomb homolog 2 (MIB2) is an E3 that leads to the proteasomal degradation of CYLD, and therefore MIB2 affects LUBAC-mediated NF-κB activation [108]. Importantly, the USP domain of CYLD binds to the PUB domain of SPATA2, and the PIM in SPATA2 associates with the PUB domain of HOIP (Figure 3) [26–29]. Recently, a familial variant of *CYLD* with the M719V missense mutation was identified in the induction of frontotemporal dementia (FTD) and amyotrophic lateral sclerosis (ALS), with enhanced K63-deubiquitinating activity and NF-κB suppression [109]. Moreover, this M719V variant of CYLD impairs autophagosome maturation and increases the cytosolic localization of TDP-43. Thus, the up-regulation of CYLD activity is associated with neurodegenerative diseases, whereas the down-regulation of *CYLD* causes cancers.

4. Decoders of the Linear Ubiquitin Code

In the ubiquitin code, various types of ubiquitin chains serve as scaffolds to recruit their specific binding proteins, and subsequently, these locally concentrated proteins are responsible for the cellular functions of the ubiquitin code. Therefore, the ubiquitin chain-specific binding proteins are referred to as "decoders". In the linear ubiquitin code, several protein motifs, such as the UBD in ABIN proteins and the NEMO (UBAN) domain [110], the NZF domain in HOIL-1L [30], and the 7th zinc finger domain in A20 (A20 ZF7) [104,111], have been identified as linear ubiquitin chain-specific binding domains [112]. In this chapter, we summarize the structures and functions of linear ubiquitin-binding proteins.

4.1. UBAN Domain-Containing Proteins: NEMO, OPTN, and ABINs

The UBAN domain forms a homodimeric coiled-coil structure, and is present in proteins such as NEMO, OPTN, and A20-binding inhibitors of NF-κB (ABIN), including ABIN-1, ABIN-2, and ABIN-3 [110]. Although the K63-linked and linear diubiquitins adopt similar conformations, the NEMO UBAN domain shows approximately 100-fold higher affinity to the linear ubiquitin chain than the K63-chain. The hydrophobic patches centered at Ile44 and Phe4 of the distal and proximal parts of linear ubiquitin, respectively, are crucial for the interactions with the UBAN domain [113,114]. The NEMO-UBAN domain is critical for the recruitment of the IKK complex onto the linear ubiquitin chain, which induces the canonical NF-κB activation [32]. Missense mutations in the NEMO-UBAN domain in humans cause X-linked anhidrotic ectodermal dysplasia with immunodeficiency (EDA-ID) [115,116]. Therefore, the linear ubiquitin-binding function of NEMO is indispensable for homeostasis.

OPTN, initially identified as a gene responsible for primary open-angle glaucoma (POAG) [117], has a similar domain organization to that of NEMO, although it does not interact with IKKα/β. OPTN is a multifunctional protein that participates in the regulation of the NF-κB and antiviral signaling pathways, vesicular transport, and selective autophagy, such as mitophagy and xenophagy [118]. In 2010, reports demonstrated that genetic mutations in *OPTN* are associated with ALS [119,120], and include an E478G missense mutation in the UBAN domain of human OPTN. We analyzed the effects of the POAG- and ALS-associated mutants of OPTN on the LUBAC- and TNF-α-mediated NF-κB activation in HEK293T cells, and showed that the ALS-associated OPTN mutants lost their ability to suppress NF-κB activation, mainly due to the dysfunction in the UBAN domain of OPTN [121]. The OPTN-UBAN domain strongly bound linear ubiquitin, with a Kd value of 1.0 μM, and the crystal structure of the OPTN-UBAN domain complexed with linear ubiquitin revealed that it shares a similar architecture with that of NEMO [121]. Thus, linear ubiquitin binding by OPTN regulates NF-κB activation and apoptosis, and consequently suppresses ALS.

ABINs were originally identified as inhibitors of NF-κB signaling, and they regulate multiple signal transduction pathways, apoptosis, virus replication, and cancer progression [122]. At present, the crystal structures of the UBAN domains in ABIN-1 and -2 with linear ubiquitin have been solved, and they basically adopt conformations similar to those in NEMO and OPTN [123,124]. Therefore, the ABINs are potential decoders of the linear ubiquitin code, through linear ubiquitin-binding by their UBAN domains.

4.2. A20

A20 (also called TNFAIP3) consists of an OTU family DUB domain at the N-terminus, followed by seven zinc finger (ZF) domains (Figure 3) [125]. A20 is an anti-inflammatory protein strongly induced by TNF-α stimulation. Moreover, dysfunctions and polymorphisms of A20 are correlated with various disorders, such as B cell lymphoma, systemic lupus erythematosus (SLE), inflammatory bowel disease, rheumatoid arthritis, and psoriasis. A20 reportedly removes the K63-linked ubiquitin chain from RIP1 by the DUB activity through the OTU domain, and conjugates K48-linked ubiquitin to RIP1 by the E3 activity in the ZF4 domain, leading to the proteasomal degradation of RIP1 [126]. We showed that A20 strongly inhibits the LUBAC-mediated NF-κB activation in a DUB-activity independent manner [104]. Indeed, although A20 hydrolyzes K48- and K63-linked ubiquitin chains, it does not cleave a linear ubiquitin chain. In contrast, the ZF7 domain of A20 is indispensable for the inhibition of LUBAC-mediated NF-κB activation, through specific binding to the linear ubiquitin chain, with a Kd value of 9 μM [104,111]. We solved the crystal structure of human A20 ZF7 with linear ubiquitin, and found that the B cell lymphoma-inducible missense mutations within ZF7 caused the lack of linear ubiquitin-binding. Furthermore, the DUB activity and ZF4 were not necessary for A20-potentiated RIP1-dependent apoptosis, whereas ZF7 is critical in A20 dimerization and intestinal epithelial cell death [127]. Recent studies using knock-in mice revealed that those carrying mutations in ZF7 spontaneously developed arthritis, although mice with mutations of the OTU or ZF4 domain showed no overt inflammatory phenotype [128,129]. Thus, A20 physiologically functions as a decoder, but not as an eraser, in the linear ubiquitin code through the linear ubiquitin-specific binding by ZF7.

5. LUBAC-Related Disorders

NF-κB signaling plays pivotal roles in the innate and adaptive immune responses, and anti-apoptosis. Therefore, the impaired NF-κB activation is closely associated with various disorders, such as cancers, inflammatory, autoimmune, and neurodegenerative diseases, and metabolic syndrome [33,130]. In this chapter, we introduce the disorders closely associated with the dysregulation of the linear ubiquitin code.

5.1. Genetic Deficiency of LUBAC Subunits and Related Diseases

The *Hoip* [34] or *Hoil-1l* [131] knockout mice are embryonic lethal, while the spontaneous deficiency of *Sharpin* reportedly induces severe dermatitis [60], indicating that the LUBAC activity is crucial for development and homeostasis. In humans, inherited *HOIL-1L* mutations reportedly induce polyglucosan body myopathy with or without immunodeficiency and autoinflammation (Table 1) [132–134]. Moreover, the L72P missense mutation in the PUB domain of *HOIP* in a patient with multiorgan autoinflammation, immunodeficiency, amylopectinosis, and systemic lymphangiectasia [135], and another case of *HOIP* deficiency with early-onset immunodeficiency and autoinflammation, but not amylopectinosis and lymphangiectasia [136], were recently identified. These *HOIP* mutations affect the expression of type I IFN regulated genes. Taken together, these results indicate that the LUBAC activity is required to suppress myopathy, systemic inflammation, and immunodeficiency in humans.

Table 1. Genetic deficiencies of human *HOIL-1L* and *HOIP*, and their symptoms.

Gene	Mutations	Symptoms	References
HOIL-1L	p.Q185X (c.553C > T), p.L41fsX7 (c.121_122delCT), c.ex1_ex4del	polyglucosan myopathy with immunodeficiency, sepsis, chronic autoinflammation	[132]
	p.Q222X (c.90C > T) p.E190fs (c.68_69insAGGAGCG) c.456+1G > C	polyglucosan body myopathy, cardiomyopathy, muscle atrophy	[133]
	p.E243X (c.727G > T), p.N387S (c.1160A > G), p.E299VfsX18 (c.896_899delAGTG), p.A241GfsX34 (c.722delC), p.A18P (c.52G > C), p.E243GfsX114 (c.727_728ins GGCG), c.ex1_ex4del p.R352X (c.1054C > T) p.R298RfsX40 (c.917+3_917+4insG), p.R165RfsX111 (c.494delG)	polyglucosan myopathy without immunodeficiency	[134]
HOIP	p.L72P (c.215T > C)	multiorgan autoinflammation with immunodeficiency, amylopectinosis, systemic lymphangiectasia	[135]
	p.Q399H (c.1197G > C) c.1737+3A > G	eczematous dermatitis, splenomegaly, clubbing of fingers	[136]

5.2. Enhanced LUBAC Expression and Cancers

Diffuse large B cell lymphoma (DLBCL) is the most common type of non-Hodgkin's lymphoma, and is subclassified into the germinal center B cell-like (GCB) and activated B cell-like (ABC) subtypes [137]. Patients with ABC-DLBCL usually have a worse prognosis than those with GCB-DLBCL, and oncogenic mutations in NF-κB signaling proteins, such as in the *CD79B, CARMA1, MYD88*, and *A20* genes, are associated with ABC-DLBCL. Staudt's group identified single nucleotide polymorphisms (SNPs) in human *HOIP* that cause the Q622L and Q584H substitutions, which are significantly associated with ABC-DLBCL [138]. Unexpectedly, these replacements, located in the UBA domain of HOIP, enhance the interactions with HOIL-1L and HOIP, resulting in the increased NF-κB activity. The authors developed stapled peptides, an α-helical short peptide with a hydrocarbon cross-link, targeting the HOIP-HOIL-1L interface, and demonstrated the suppressed viability of ABC-DLBCL cells. Indeed, the silencing of *HOIP* also reportedly reduces the viability of ABC-DLBCL cells [49]. Moreover, the E3s of c-IAP-1/2 are involved in ABC-DLBCL via the K63 ubiquitination of BCL10, which results in the recruitment of LUBAC and IKK to the CBM complex, thus inducing BCR-mediated NF-κB activation [139]. Therefore, SMAC mimetics, which lead to the autodegradation of c-IAP-1/2, block the growth of ABC-DLBCL cells. These results indicate that the enhanced LUBAC activity is associated with a poor prognosis in B cell lymphoma (ABC-DLBCL), and thus the inhibition of the LUBAC activity may be a valid therapeutic target.

Recently, Ruiz et al. showed that the expression of LUBAC subunits is enhanced in human and murine lung squamous cell carcinoma (LSCC) cells, but not adenocarcinoma cells, which results in the increased linear ubiquitination, NF-κB activation, and cisplatin-resistance in LSCC [140]. Moreover, the suppression of LUBAC activity, in addition to the suppression of TAK1, ameliorated the chemotherapy resistance of LSCC. Thus, LUBAC inhibitors seem to be therapeutic drug seeds to treat LSCC.

5.3. Linear Ubiquitination in Neurodegenerative Diseases

Neurodegenerative diseases, such as Alzheimer's disease, Parkinson's disease, Huntington's disease, ALS, FTD, and related tauopathies, are characterized by the progressive degeneration of the central or peripheral nervous system [141], and at present, no disease-modifying therapies or medicines have been developed. The accumulation of misfolded, aggregated, and ubiquitinated proteins is a common mechanism underlying neurodegenerative diseases. Disease-specific misfolded and aggregated proteins have been identified, such as amyloid-β and tau in Alzheimer's disease, α-synuclein in Parkinson's disease, huntingtin in Huntington's disease, and superoxide dismutase 1 and TAR DNA-binding protein 43 (TDP-43) in ALS [142]. Although the aggregates or inclusions of these proteins are ubiquitin-positive, the types of ubiquitin linkages within these protein aggregates have not been identified.

ALS is a fatal progressive neurodegenerative disease caused by the loss of motor neurons. Although most ALS cases are sporadic, around 10% are familial, and mutations in approximately 20 genes encoding proteins involved in protein/RNA aggregation (*SOD1*, *TDP-43*, *hnRNPA1/2*, and *FUS*), neuroinflammation (*TBK1*), the ubiquitin-proteasome pathway (*UBQLN2*), and autophagy (*C9orf72*, *OPTN*, *SQSTM1/p62*, and *VCP*) have been identified [143]. Although OPTN reportedly functions as an autophagy receptor, we determined that the ALS-associated *OPTN* mutations, *E478G* and *Q398X*, abrogated the inhibitory effects of OPTN on LUBAC-mediated NF-κB activation, and accelerated TNF-induced cell death in HEK293T and HeLa cells [121]. Importantly, the intracytoplasmic inclusions in neurons from patients with the heterozygous *E478G* and homozygous *Q398X* mutations reacted with an anti-linear ubiquitin antibody, and they were co-localized with TDP-43 and phosphorylated p65, which is an activated form of NF-κB [121]. Moreover, these spinal anterior cells are positive to anti-cleaved caspase 8 and 3 antibodies, suggesting that OPTN regulates neuroinflammation and cell death. We recently showed that the linear ubiquitination of not only the *OPTN*-associated ALS, but also TDP-43-containing inclusions was detected in the cases of sporadic ALS [144]. Interestingly, all of the neuronal cytoplasmic inclusions (NCIs) in spinal cords, including fine "wisps", were immunolabeled by the anti-K48-linked ubiquitin antibody (Figure 4A,C), whereas the linear ubiquitin was mainly detected in the intermediate and thick bundles of TDP-43-positive inclusions (Figure 4B,E). Similarly, we showed that K63-linked ubiquitin was predominantly detected in intermediate and thick bundles, and linear- and K63-linked ubiquitin immunoreactants were not always co-localized [144]. Moreover, OPTN and phosphorylated p65 were co-localized with these inclusions derived from sporadic ALS patients. Furthermore, we showed that K48-linked ubiquitination is present in all of the tau neurofibrillary tangles, including small ones, whereas a subset of thick neurofibrillary tangles, dystrophic neurites of senile plaques, and neuropil threads were immunopositive for linear ubiquitin in Alzheimer's disease [145]. These results suggested that various ubiquitinations, including linear ubiquitin, is involved in protein aggregates of neurodegenerative diseases.

Importantly, Winklhofer's group reported that linear ubiquitin is indispensable for protein quality control [146]. They showed that LUBAC is recruited to the aggregates derived from overexpressed huntingtin-derived polyglutamine (Htt-polyQ) in human neuroblastoma SH-SY5Y cells, and linear ubiquitin also co-localized with the Htt-polyQ aggregates. During the recruitment of LUBAC, AAA-ATPase p97/VCP, which binds to the PUB domain of HOIP, plays an important role to suppress the proteotoxicity, since the linear ubiquitination of Htt-polyQ modulates the aberrant interaction of Htt-polyQ with the transcription factor Sp1. They further indicated that linear ubiquitination is involved in various disease-associated aggregable proteins, such as ataxin-3 (Machado-Joseph disease), and SOD1, TDP-43, and OPTN (ALS). Thus LUBAC, together with its linear ubiquitination activity, seems to be a crucial regulator of various neurodegenerative diseases.

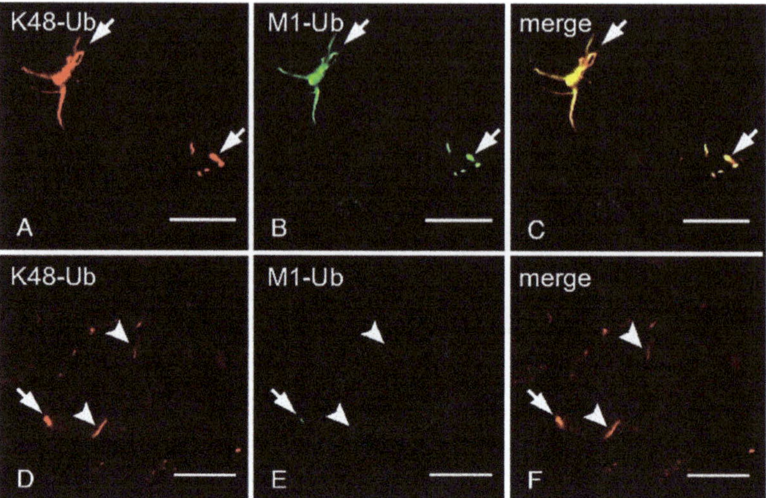

Figure 4. Thick bundles of inclusions are immunopositive for anti-K48 and anti-linear ubiquitin antibodies, whereas fine wisps exclusively reacted with anti-K48 ubiquitin, but not anti-linear ubiquitin. The double immunofluorescence staining using anti-K48-linked ubiquitin antibody (**A,D**), anti-linear ubiquitin antibody (**B,E**), and their merged images (**C,F**) are shown. Arrows; neuronal cytoplasmic inclusions co-localized with K48- and linear-ubiquitins. Arrowheads; K48-ubiquitin-positive inclusions, which lack immunoreactivity for anti-linear ubiquitin antibody (Bars = 20 µm, modified from Ref. [144] with permission).

6. LUBAC Inhibitors

LUBAC is the sole E3 that can generate a linear ubiquitin chain to regulate acquired and innate immune responses. Therefore, LUBAC inhibitors will facilitate investigations of its enzymatic mechanisms and the cellular bases for immune responses, and serve as potential therapeutics for various LUBAC-related disorders. As described above, stapled peptides targeting the HOIP-HOIL-1L interface suppressed the viability of ABC-DLBCL cells [138]. In this chapter, we summarize the recent advances in the development of small-molecule chemical inhibitors of LUBAC.

6.1. Chemical Inhibitors of LUBAC

To date, BAY11-7082 [147], gliotoxin [148], and bendamustine [149] have been reported as chemical inhibitors of LUBAC. However, using non-toxic concentrations, we recently showed that these inhibitors lack the selectivity and inhibitory effects on LUBAC-induced linear ubiquitination and NF-κB activation in HEK293T cells [73].

Recently, Rittinger's group developed α,β-unsaturated methyl ester-containing compounds, such as compound [5], as LUBAC inhibitors [150,151]. Among them, compound [11a] reportedly inhibited the overexpressed LUBAC-induced NF-κB activity in HEK293T cells (IC_{50} = 37 µM). Importantly, they showed that compound [5] was covalently attached to the catalytic Cys885 of HOIP via Michael addition, by accommodation in a hydrophobic pocket formed by Tyr878, Leu880, and Phe888, and stabilization by hydrogen bonds with the main-chain CO and NH groups of His889 and the Oγ atom of Ser899. These compound [5]-interacting residues of HOIP are located in the RING2 domain [150,151], but are not conserved in other RBR E3s. These sequence variations of RING2 may be beneficial for the HOIP specificity of compound [5].

6.2. HOIPINs

To search for novel inhibitors of LUBAC, we constructed a FRET-based assay system to quantify the linear ubiquitination level, using a recombinant LUBAC fragment [152]. After screening 250,000 small molecular chemicals, we identified a thiol-reactive, α,β-unsaturated carbonyl-containing compound, sodium (*E*)-2-(3-(2-methoxyphenyl)-3-oxoprop-1-en-1-yl)benzoate, named HOIPIN-1 from HOIP inhibitor-1, as a LUBAC inhibitor (Figure 5). We developed derivatives of HOIPIN-1, and found that sodium (*E*)-2-(3-(2,6-difluoro-4-(1*H*-pyrazol-4-yl)phenyl)-3-oxoprop-1-en-1-yl)-4-(1-methyl-1*H*-pyrazol-4-yl)benzoate, designated as HOIPIN-8, is the most potent LUBAC inhibitor (Figure 5) [153]. HOIPIN-1 and HOIPIN-8 inhibited the in vitro linear ubiquitination activity of recombinant LUBAC, with IC$_{50}$ values of 2.8 μM and 11 nM, respectively. Furthermore, HOIPIN-1 and HOIPIN-8 suppressed the overexpressed LUBAC-induced NF-κB activity in HEK293T cells, with IC$_{50}$ values of 4.0 μM and 0.42 μM, respectively, indicating that HOIPIN-8 is the most potent among the reported LUBAC inhibitors.

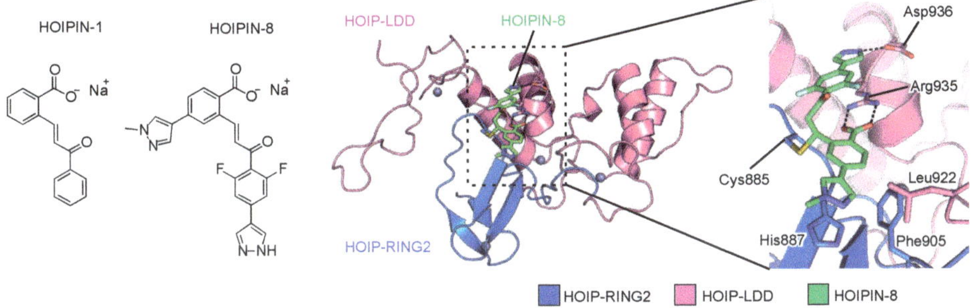

Figure 5. LUBAC inhibitors, HOIPIN-1 and HOIPIN-8. Chemical structures of HOIPIN-1 and -8, and crystal structure of the HOIPIN-8-bound RING2-LDD domain of HOIP are shown.

We recently determined that HOIPINs are conjugated to the active site Cys885 in the RING2 domain of HOIP through Michael addition, and interrupt the RING-HECT-hybrid reaction in HOIP [73]. The benzoate, 2,6-difluorophenyl, and 1*H*-pyrazol-4-yl moieties of HOIPIN-8 interact with Arg935 and Asp936, which are located in the LDD domain of human HOIP (Figure 5). Arg935 and Asp936 in the LDD domain reportedly interact with Glu16 and Thr14 in the acceptor ubiquitin, and the Ala mutations of these residues abrogated the linear ubiquitination activity [19–22]. Thus, HOIPIN-8 not only interacts with the active Cys885, but also masks the critical residues for acceptor ubiquitin-binding by the benzoate, 2,6-difluorophenyl, and 1*H*-pyrazol-4-yl moieties. In addition, the 1-methyl-1*H*-pyrazol-4-yl moiety of HOIPIN-8 interacts with His887, Phe905, and Leu922 of HOIP (Figure 5). These sequence variations of RING2 may be beneficial for the HOIP specificity of compound [5]. On the other hand, the LDD domain of HOIP may be a key determinant for the HOIP specificity of HOIPINs. Therefore, the mechanisms underlying the HOIP specificity are completely different between HOIPINs and compound [5].

Although the α,β-unsaturated carbonyl-containing chemicals seemed to react with various SH-groups, HOIPINs did not inhibit the E1-mediated ubiquitin transfer to E2, or the activities of the HECT-, RING-, and other RBR-type E3s, and specifically suppressed the intracellular linear ubiquitin level induced by TNF-α, IL-1β, and poly(I:C) [73]. As shown in Section 5.2, the enhanced LUBAC activity is associated with the progression of ABC-DLBCL, and we found that HOIPINs effectively suppressed the cell viability of human ABC-DLBCL cells, but not GCB-DLBCL cells, by inhibiting the linear ubiquitination-mediated NF-κB activation [73]. Therefore, the cryptic peptide targeting LUBAC and the HOIPINs may serve as drug seeds to treat ABC-DLBCL patients with a poor prognosis. Moreover, we showed the HOIPIN-1-mediated alleviation of imiquimod-induced psoriasis in model mice. To date, dysfunctions in skin barrier production, IL-23/IL17-mediated lymphocyte signaling, and the NF-κB pathway are reportedly involved in the pathogenesis of psoriasis [154,155]. Since LUBAC affects not only NF-κB but also the production of IL-17 [156], HOIPINs may be effective to treat psoriasis.

7. Conclusions and Perspectives

In this review, we have summarized the recent advances in identifying the regulators of LUBAC-mediated linear ubiquitination and its pathophysiological functions. We have highlighted the "linear ubiquitin code", and its "writer" (LUBAC), "erasers" (DUBs such as OTULIN and CYLD), and "decoders" (linear ubiquitin-binding proteins such as UBAN domain-containing proteins and A20). LUBAC principally activates the canonical NF-κB pathway and suppresses apoptosis. Therefore, the impaired LUBAC activity and the aberrant functions in linear ubiquitin decoders are associated with autoinflammatory and neurodegenerative diseases, and cancers. In particular, it is worthwhile to focus on the fact that linear ubiquitin is present in the protein aggregates of various neurodegenerative diseases, including ALS.

From these studies, we hypothesized that the aggregable proteins in wisp inclusions may be initially conjugated with K48-linked ubiquitin chains; however, they seem to be resistant to the proteasomal degradation due to misfolding (Figure 6). Concomitant with aging, maturation, and liquid–liquid phase separation (LLPS), LUBAC-mediated linear ubiquitin and/or K63-linked ubiquitin chains will be conjugated to NCI. We speculated that it may further generate complex-types of ubiquitin chains, such as K48/linear-branched chains, in the thick bundles. The linear polyubiquitin may serve as a scaffold to recruit the IKK complex and OPTN via the UBAN domain, thus inducing neuroinflammation followed by cell death and selective autophagy, such as mitophagy and aggrephagy (Figure 6). Interestingly, agitation experiments showed that linear polyubiquitin forms fibrillar aggregates more easily than K48-linked polyubiquitin [157]. Therefore, the linear ubiquitination may facilitate the formation of thick inclusions by LLPS. Thus, linear ubiquitination may have pleiotropic effects on neuroinflammation, protein folding, and proteostasis. Although we identified HOIPIN-8 as a potent inhibitor of LUBAC, further studies are necessary to generate potent and specific inhibitors of LUBAC for therapeutic treatments of neurodegenerative diseases.

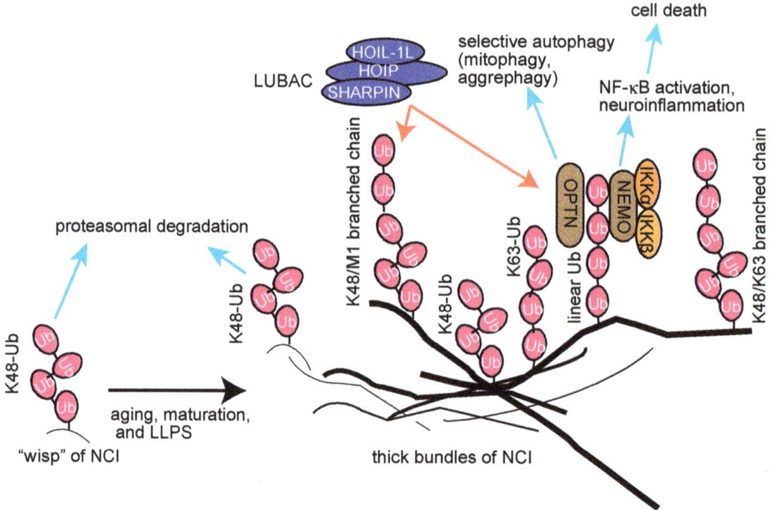

Figure 6. Proposed schema for the involvement of multiple ubiquitin chains, including linear ubiquitin, in neurodegenerative diseases, and their potential functions.

Author Contributions: D.O., Y.S., H.I., and F.T. performed literature research and wrote the paper. All authors have read and agreed to the published version of the manuscript.

Funding: This research was funded by MEXT/JSPS KAKENHI grants (Nos. JP16H06575, JP18H02619, and JP19K22541 to F.T., and JP18K06967 and JP19H05296 to D.O.), Osaka City University Strategic Research Grant 2017 for top priority research (F.T.), Takeda Science Foundation (F.T.), a Grant for Research Program on Hepatitis from the Japan Agency for Medical Research and Development (AMED – 19fk0210050h0001 to F.T.), GSK Japan Research Grant 2017 (D.O.), and a grant from the Nakatomi Foundation (D.O.).

Conflicts of Interest: The authors declare no conflict of interest. The funders had no role in the design of the study; in the collection, analyses, or interpretation of data; in the writing of the manuscript, or in the decision to publish the results.

Abbreviations

ABC-DLBCL	Activated B cell-like diffuse large B cell lymphoma
ABIN	A20-binding inhibitor of NF-κB
ALS	Amyotrophic lateral sclerosis
BCR	B cell receptor
DUB	Deubiquitinating enzyme
E1	Ubiquitin-activating enzyme
E2	Ubiquitin-conjugating enzyme
E3	Ubiquitin ligase
FTD	Frontotemporal dementia
GCB-DLBCL	Germinal center B cell-like diffuse large B cell lymphoma
HECT	Homologous to the E6AP carboxyl terminus
HOIL-1L	Heme-oxidized IRP2 ubiquitin ligase-1L
HOIP	HOIL-1L-interacting protein
IFN	Interferon
IKK	IκB kinase
IRF	Interferon regulatory factor
LDD	Linear ubiquitin chain determining domain
LLPS	Liquid-liquid phase separation
LSCC	Lung squamous cell carcinoma
LTM	LUBAC-tethering motif
LUBAC	Linear ubiquitin chain assembly complex
NCI	Neuronal cytoplasmic inclusion
NEMO	NF-κB-essential modulator
NF-κB	Nuclear factor-κB
NZF	Npl4-type zinc finger
OPTN	Optineurin
PAMPs	Pathogen-associated molecular patterns
PUB	PNGase/UBA or UBX
RBR	RING-in between-RING
RING	Really interesting new gene
SHARPIN	Shank-associated RH domain interactor
TBK1	TANK-binding kinase 1
TDP-43	TAR DNA-binding protein 43
TCR	T cell receptor
TLR	Toll-like receptor
TNF	Tumor necrosis factor
TNFR	TNF receptor
UBA	Ubiquitin-associated
UBAN	UBD in ABIN proteins and NEMO
UBL	Ubiquitin-like
ZF	Zinc finger

References

1. Hershko, A.; Ciechanover, A. The ubiquitin system. *Annu. Rev. Biochem.* **1998**, *67*, 425–479. [CrossRef] [PubMed]
2. McDowell, G.S.; Philpott, A. Non-canonical ubiquitylation: Mechanisms and consequences. *Int. J. Biochem. Cell Biol.* **2013**, *45*, 1833–1842. [CrossRef]
3. Bhogaraju, S.; Kalayil, S.; Liu, Y.; Bonn, F.; Colby, T.; Matic, I.; Dikic, I. Phosphoribosylation of Ubiquitin Promotes Serine Ubiquitination and Impairs Conventional Ubiquitination. *Cell* **2016**, *167*, 1636–1649.e13. [CrossRef] [PubMed]
4. Eisenhaber, B.; Chumak, N.; Eisenhaber, F.; Hauser, M.T. The ring between ring fingers (RBR) protein family. *Genome Biol.* **2007**, *8*, 209. [CrossRef] [PubMed]
5. Deshaies, R.J.; Joazeiro, C.A. RING domain E3 ubiquitin ligases. *Annu. Rev. Biochem.* **2009**, *78*, 399–434. [CrossRef] [PubMed]
6. Komander, D.; Rape, M. The ubiquitin code. *Annu. Rev. Biochem.* **2012**, *81*, 203–229. [CrossRef] [PubMed]
7. Tokunaga, F. Linear ubiquitination-mediated NF-κB regulation and its related disorders. *J. Biochem.* **2013**, *154*, 313–323. [CrossRef]
8. Asaoka, T.; Almagro, J.; Ehrhardt, C.; Tsai, I.; Schleiffer, A.; Deszcz, L.; Junttila, S.; Ringrose, L.; Mechtler, K.; Kavirayani, A.; et al. Linear ubiquitination by LUBEL has a role in Drosophila heat stress response. *EMBO Rep.* **2016**, *17*, 1624–1640. [CrossRef]
9. Yamanaka, K.; Ishikawa, H.; Megumi, Y.; Tokunaga, F.; Kanie, M.; Rouault, T.A.; Morishima, I.; Minato, N.; Ishimori, K.; Iwai, K. Identification of the ubiquitin-protein ligase that recognizes oxidized IRP2. *Nat. Cell Biol.* **2003**, *5*, 336–340. [CrossRef]
10. Kirisako, T.; Kamei, K.; Murata, S.; Kato, M.; Fukumoto, H.; Kanie, M.; Sano, S.; Tokunaga, F.; Tanaka, K.; Iwai, K. A ubiquitin ligase complex assembles linear polyubiquitin chains. *EMBO J.* **2006**, *25*, 4877–4887. [CrossRef]
11. Gerlach, B.; Cordier, S.M.; Schmukle, A.C.; Emmerich, C.H.; Rieser, E.; Haas, T.L.; Webb, A.I.; Rickard, J.A.; Anderton, H.; Wong, W.W.; et al. Linear ubiquitination prevents inflammation and regulates immune signalling. *Nature* **2011**, *471*, 591–596. [CrossRef] [PubMed]
12. Tokunaga, F.; Nakagawa, T.; Nakahara, M.; Saeki, Y.; Taniguchi, M.; Sakata, S.; Tanaka, K.; Nakano, H.; Iwai, K. SHARPIN is a component of the NF-κB-activating linear ubiquitin chain assembly complex. *Nature* **2011**, *471*, 633–636. [CrossRef] [PubMed]
13. Ikeda, F.; Deribe, Y.L.; Skanland, S.S.; Stieglitz, B.; Grabbe, C.; Franz-Wachtel, M.; van Wijk, S.J.; Goswami, P.; Nagy, V.; Terzic, J.; et al. SHARPIN forms a linear ubiquitin ligase complex regulating NF-κB activity and apoptosis. *Nature* **2011**, *471*, 637–641. [CrossRef] [PubMed]
14. Rittinger, K.; Ikeda, F. Linear ubiquitin chains: Enzymes, mechanisms and biology. *Open Biol.* **2017**, *7*, 170026. [CrossRef] [PubMed]
15. Iwai, K.; Fujita, H.; Sasaki, Y. Linear ubiquitin chains: NF-κB signalling, cell death and beyond. *Nat. Rev. Mol. Cell Biol.* **2014**, *15*, 503–508. [CrossRef]
16. Fujita, H.; Tokunaga, A.; Shimizu, S.; Whiting, A.L.; Aguilar-Alonso, F.; Takagi, K.; Walinda, E.; Sasaki, Y.; Shimokawa, T.; Mizushima, T.; et al. Cooperative Domain Formation by Homologous Motifs in HOIL-1L and SHARPIN Plays A Crucial Role in LUBAC Stabilization. *Cell Rep.* **2018**, *23*, 1192–1204. [CrossRef]
17. Wenzel, D.M.; Lissounov, A.; Brzovic, P.S.; Klevit, R.E. UBCH7 reactivity profile reveals parkin and HHARI to be RING/HECT hybrids. *Nature* **2011**, *474*, 105–108. [CrossRef]
18. Dove, K.K.; Stieglitz, B.; Duncan, E.D.; Rittinger, K.; Klevit, R.E. Molecular insights into RBR E3 ligase ubiquitin transfer mechanisms. *EMBO Rep.* **2016**, *17*, 1221–1235. [CrossRef]
19. Stieglitz, B.; Morris-Davies, A.C.; Koliopoulos, M.G.; Christodoulou, E.; Rittinger, K. LUBAC synthesizes linear ubiquitin chains via a thioester intermediate. *EMBO Rep.* **2012**, *13*, 840–846. [CrossRef]
20. Stieglitz, B.; Rana, R.R.; Koliopoulos, M.G.; Morris-Davies, A.C.; Schaeffer, V.; Christodoulou, E.; Howell, S.; Brown, N.R.; Dikic, I.; Rittinger, K. Structural basis for ligase-specific conjugation of linear ubiquitin chains by HOIP. *Nature* **2013**, *503*, 422–426. [CrossRef]
21. Smit, J.J.; Monteferrario, D.; Noordermeer, S.M.; van Dijk, W.J.; van der Reijden, B.A.; Sixma, T.K. The E3 ligase HOIP specifies linear ubiquitin chain assembly through its RING-IBR-RING domain and the unique LDD extension. *Embo J.* **2012**, *31*, 3833–3844. [CrossRef] [PubMed]

22. Lechtenberg, B.C.; Rajput, A.; Sanishvili, R.; Dobaczewska, M.K.; Ware, C.F.; Mace, P.D.; Riedl, S.J. Structure of a HOIP/E2~ubiquitin complex reveals RBR E3 ligase mechanism and regulation. *Nature* **2016**, *529*, 546–550. [CrossRef] [PubMed]

23. Kelsall, I.R.; Zhang, J.; Knebel, A.; Arthur, J.S.C.; Cohen, P. The E3 ligase HOIL-1 catalyses ester bond formation between ubiquitin and components of the Myddosome in mammalian cells. *Proc. Natl. Acad. Sci. USA* **2019**, *116*, 13293–13298. [CrossRef] [PubMed]

24. Allen, M.D.; Buchberger, A.; Bycroft, M. The PUB domain functions as a p97 binding module in human peptide N-glycanase. *J. Biol. Chem.* **2006**, *281*, 25502–25508. [CrossRef]

25. Schaeffer, V.; Akutsu, M.; Olma, M.H.; Gomes, L.C.; Kawasaki, M.; Dikic, I. Binding of OTULIN to the PUB domain of HOIP controls NF-κB signaling. *Mol. Cell* **2014**, *54*, 349–361. [CrossRef]

26. Elliott, P.R.; Leske, D.; Hrdinka, M.; Bagola, K.; Fiil, B.K.; McLaughlin, S.H.; Wagstaff, J.; Volkmar, N.; Christianson, J.C.; Kessler, B.M.; et al. SPATA2 Links CYLD to LUBAC, Activates CYLD, and Controls LUBAC Signaling. *Mol. Cell* **2016**, *63*, 990–1005. [CrossRef]

27. Kupka, S.; De Miguel, D.; Draber, P.; Martino, L.; Surinova, S.; Rittinger, K.; Walczak, H. SPATA2-Mediated Binding of CYLD to HOIP Enables CYLD Recruitment to Signaling Complexes. *Cell Rep.* **2016**, *16*, 2271–2280. [CrossRef]

28. Schlicher, L.; Wissler, M.; Preiss, F.; Brauns-Schubert, P.; Jakob, C.; Dumit, V.; Borner, C.; Dengjel, J.; Maurer, U. SPATA2 promotes CYLD activity and regulates TNF-induced NF-κB signaling and cell death. *EMBO Rep.* **2016**, *17*, 1485–1497. [CrossRef]

29. Wagner, S.A.; Satpathy, S.; Beli, P.; Choudhary, C. SPATA2 links CYLD to the TNF-alpha receptor signaling complex and modulates the receptor signaling outcomes. *EMBO J.* **2016**, *35*, 1868–1884. [CrossRef]

30. Sato, Y.; Fujita, H.; Yoshikawa, A.; Yamashita, M.; Yamagata, A.; Kaiser, S.E.; Iwai, K.; Fukai, S. Specific recognition of linear ubiquitin chains by the Npl4 zinc finger (NZF) domain of the HOIL-1L subunit of the linear ubiquitin chain assembly complex. *Proc. Natl. Acad. Sci. USA* **2011**, *108*, 20520–20525. [CrossRef]

31. Shimizu, S.; Fujita, H.; Sasaki, Y.; Tsuruyama, T.; Fukuda, K.; Iwai, K. Differential Involvement of the Npl4 Zinc Finger Domains of SHARPIN and HOIL-1L in Linear Ubiquitin Chain Assembly Complex-Mediated Cell Death Protection. *Mol. Cell. Biol.* **2016**, *36*, 1569–1583. [CrossRef] [PubMed]

32. Fujita, H.; Rahighi, S.; Akita, M.; Kato, R.; Sasaki, Y.; Wakatsuki, S.; Iwai, K. Mechanism underlying IkappaB kinase activation mediated by the linear ubiquitin chain assembly complex. *Mol. Cell. Biol.* **2014**, *34*, 1322–1335. [CrossRef] [PubMed]

33. Zhang, Q.; Lenardo, M.J.; Baltimore, D. 30 Years of NF-κB: A Blossoming of Relevance to Human Pathobiology. *Cell* **2017**, *168*, 37–57. [CrossRef] [PubMed]

34. Sasaki, Y.; Sano, S.; Nakahara, M.; Murata, S.; Kometani, K.; Aiba, Y.; Sakamoto, S.; Watanabe, Y.; Tanaka, K.; Kurosaki, T.; et al. Defective immune responses in mice lacking LUBAC-mediated linear ubiquitination in B cells. *Embo J.* **2013**, *32*, 2463–2476. [CrossRef]

35. de Almagro, M.C.; Goncharov, T.; Newton, K.; Vucic, D. Cellular IAP proteins and LUBAC differentially regulate necrosome-associated RIP1 ubiquitination. *Cell Death Dis.* **2015**, *6*, e1800. [CrossRef]

36. Ea, C.K.; Deng, L.; Xia, Z.P.; Pineda, G.; Chen, Z.J. Activation of IKK by TNFalpha requires site-specific ubiquitination of RIP1 and polyubiquitin binding by NEMO. *Mol. Cell* **2006**, *22*, 245–257. [CrossRef]

37. Haas, T.L.; Emmerich, C.H.; Gerlach, B.; Schmukle, A.C.; Cordier, S.M.; Rieser, E.; Feltham, R.; Vince, J.; Warnken, U.; Wenger, T.; et al. Recruitment of the linear ubiquitin chain assembly complex stabilizes the TNF-R1 signaling complex and is required for TNF-mediated gene induction. *Mol. Cell* **2009**, *36*, 831–844. [CrossRef]

38. Tokunaga, F.; Sakata, S.; Saeki, Y.; Satomi, Y.; Kirisako, T.; Kamei, K.; Nakagawa, T.; Kato, M.; Murata, S.; Yamaoka, S.; et al. Involvement of linear polyubiquitylation of NEMO in NF-κB activation. *Nat. Cell Biol.* **2009**, *11*, 123–132. [CrossRef]

39. Goto, E.; Tokunaga, F. Decreased linear ubiquitination of NEMO and FADD on apoptosis with caspase-mediated cleavage of HOIP. *Biochem. Biophys. Res. Commun.* **2017**, *485*, 152–159. [CrossRef]

40. Kensche, T.; Tokunaga, F.; Ikeda, F.; Goto, E.; Iwai, K.; Dikic, I. Analysis of nuclear factor-kappaB (NF-κB) essential modulator (NEMO) binding to linear and lysine-linked ubiquitin chains and its role in the activation of NF-κB. *J. Biol. Chem.* **2012**, *287*, 23626–23634. [CrossRef]

41. Lee, I.Y.; Lim, J.M.; Cho, H.; Kim, E.; Kim, Y.; Oh, H.K.; Yang, W.S.; Roh, K.H.; Park, H.W.; Mo, J.S.; et al. MST1 Negatively Regulates TNFalpha-Induced NF-κB Signaling through Modulating LUBAC Activity. *Mol. Cell* **2019**, *73*, 1138–1149. [CrossRef]

42. Meschede, J.; Sadic, M.; Furthmann, N.; Miedema, T.; Sehr, D.A.; Muller-Rischart, A.K.; Bader, V.; Berlemann, L.A.; Pilsl, A.; Schlierf, A.; et al. The parkin-coregulated gene product PACRG promotes TNF signaling by stabilizing LUBAC. *Sci. Signal.* **2020**, *13*, eaav1256. [CrossRef] [PubMed]

43. Tarantino, N.; Tinevez, J.Y.; Crowell, E.F.; Boisson, B.; Henriques, R.; Mhlanga, M.; Agou, F.; Israel, A.; Laplantine, E. TNF and IL-1 exhibit distinct ubiquitin requirements for inducing NEMO-IKK supramolecular structures. *J. Cell Biol.* **2014**, *204*, 231–245. [CrossRef] [PubMed]

44. Emmerich, C.H.; Ordureau, A.; Strickson, S.; Arthur, J.S.; Pedrioli, P.G.; Komander, D.; Cohen, P. Activation of the canonical IKK complex by K63/M1-linked hybrid ubiquitin chains. *Proc. Natl. Acad. Sci. USA* **2013**, *110*, 15247–15252. [CrossRef] [PubMed]

45. Meininger, I.; Krappmann, D. Lymphocyte signaling and activation by the CARMA1-BCL10-MALT1 signalosome. *Biol. Chem.* **2016**, *397*, 1315–1333. [CrossRef]

46. Teh, C.E.; Lalaoui, N.; Jain, R.; Policheni, A.N.; Heinlein, M.; Alvarez-Diaz, S.; Sheridan, J.M.; Rieser, E.; Deuser, S.; Darding, M.; et al. Linear ubiquitin chain assembly complex coordinates late thymic T-cell differentiation and regulatory T-cell homeostasis. *Nat. Commun.* **2016**, *7*, 13353. [CrossRef]

47. Klein, T.; Fung, S.Y.; Renner, F.; Blank, M.A.; Dufour, A.; Kang, S.; Bolger-Munro, M.; Scurll, J.M.; Priatel, J.J.; Schweigler, P.; et al. The paracaspase MALT1 cleaves HOIL1 reducing linear ubiquitination by LUBAC to dampen lymphocyte NF-κB signalling. *Nat. Commun.* **2015**, *6*, 8777. [CrossRef]

48. Yang, Y.K.; Yang, C.; Chan, W.; Wang, Z.; Deibel, K.E.; Pomerantz, J.L. Molecular Determinants of Scaffold-induced Linear Ubiquitinylation of B Cell Lymphoma/Leukemia 10 (Bcl10) during T Cell Receptor and Oncogenic Caspase Recruitment Domain-containing Protein 11 (CARD11) Signaling. *J. Biol. Chem.* **2016**, *291*, 25921–25936. [CrossRef]

49. Dubois, S.M.; Alexia, C.; Wu, Y.; Leclair, H.M.; Leveau, C.; Schol, E.; Fest, T.; Tarte, K.; Chen, Z.J.; Gavard, J.; et al. A catalytic-independent role for the LUBAC in NF-κB activation upon antigen receptor engagement and in lymphoma cells. *Blood* **2014**, *123*, 2199–2203. [CrossRef]

50. McCool, K.W.; Miyamoto, S. DNA damage-dependent NF-κB activation: NEMO turns nuclear signaling inside out. *Immunol. Rev.* **2012**, *246*, 311–326. [CrossRef]

51. Niu, J.; Shi, Y.; Iwai, K.; Wu, Z.H. LUBAC regulates NF-κB activation upon genotoxic stress by promoting linear ubiquitination of NEMO. *Embo J.* **2011**, *30*, 3741–3753. [CrossRef] [PubMed]

52. Damgaard, R.B.; Nachbur, U.; Yabal, M.; Wong, W.W.; Fiil, B.K.; Kastirr, M.; Rieser, E.; Rickard, J.A.; Bankovacki, A.; Peschel, C.; et al. The ubiquitin ligase XIAP recruits LUBAC for NOD2 signaling in inflammation and innate immunity. *Mol. Cell* **2012**, *46*, 746–758. [CrossRef]

53. Lopez-Castejon, G. Control of the inflammasome by the ubiquitin system. *FEBS J.* **2020**, *287*, 11–26. [CrossRef] [PubMed]

54. Rodgers, M.A.; Bowman, J.W.; Fujita, H.; Orazio, N.; Shi, M.; Liang, Q.; Amatya, R.; Kelly, T.J.; Iwai, K.; Ting, J.; et al. The linear ubiquitin assembly complex (LUBAC) is essential for NLRP3 inflammasome activation. *J. Exp. Med.* **2014**, *211*, 1333–1347. [CrossRef] [PubMed]

55. Walczak, H. TNF and ubiquitin at the crossroads of gene activation, cell death, inflammation, and cancer. *Immunol. Rev.* **2011**, *244*, 9–28. [CrossRef] [PubMed]

56. Joo, D.; Tang, Y.; Blonska, M.; Jin, J.; Zhao, X.; Lin, X. Regulation of Linear Ubiquitin Chain Assembly Complex by Caspase-Mediated Cleavage of RNF31. *Mol. Cell. Biol.* **2016**, *36*, 3010–3018. [CrossRef]

57. Peltzer, N.; Rieser, E.; Taraborrelli, L.; Draber, P.; Darding, M.; Pernaute, B.; Shimizu, Y.; Sarr, A.; Draberova, H.; Montinaro, A.; et al. HOIP deficiency causes embryonic lethality by aberrant TNFR1-mediated endothelial cell death. *Cell Rep.* **2014**, *9*, 153–165. [CrossRef]

58. Gijbels, M.J.; Zurcher, C.; Kraal, G.; Elliott, G.R.; HogenEsch, H.; Schijff, G.; Savelkoul, H.F.; Bruijnzeel, P.L. Pathogenesis of skin lesions in mice with chronic proliferative dermatitis (cpdm/cpdm). *Am. J. Pathol.* **1996**, *148*, 941–950.

59. HogenEsch, H.; Janke, S.; Boggess, D.; Sundberg, J.P. Absence of Peyer's patches and abnormal lymphoid architecture in chronic proliferative dermatitis (cpdm/cpdm) mice. *J. Immunol.* **1999**, *162*, 3890–3896.

60. Seymour, R.E.; Hasham, M.G.; Cox, G.A.; Shultz, L.D.; Hogenesch, H.; Roopenian, D.C.; Sundberg, J.P. Spontaneous mutations in the mouse Sharpin gene result in multiorgan inflammation, immune system dysregulation and dermatitis. *Genes Immun.* **2007**, *8*, 416–421. [CrossRef]

61. Kumari, S.; Redouane, Y.; Lopez-Mosqueda, J.; Shiraishi, R.; Romanowska, M.; Lutzmayer, S.; Kuiper, J.; Martinez, C.; Dikic, I.; Pasparakis, M.; et al. Sharpin prevents skin inflammation by inhibiting TNFR1-induced keratinocyte apoptosis. *eLife* **2014**, *3*, e03422. [CrossRef]

62. Rickard, J.A.; Anderton, H.; Etemadi, N.; Nachbur, U.; Darding, M.; Peltzer, N.; Lalaoui, N.; Lawlor, K.E.; Vanyai, H.; Hall, C.; et al. TNFR1-dependent cell death drives inflammation in Sharpin-deficient mice. *eLife* **2014**, *3*, e03464. [CrossRef] [PubMed]

63. Sharma, B.R.; Karki, R.; Lee, E.; Zhu, Q.; Gurung, P.; Kanneganti, T.D. Innate immune adaptor MyD88 deficiency prevents skin inflammation in SHARPIN-deficient mice. *Cell Death Differ.* **2019**, *26*, 741–750. [CrossRef]

64. Lee, M.S.; Kim, Y.J. Signaling pathways downstream of pattern-recognition receptors and their cross talk. *Annu. Rev. Biochem.* **2007**, *76*, 447–480. [CrossRef]

65. Davis, M.E.; Gack, M.U. Ubiquitination in the antiviral immune response. *Virology* **2015**, *479–480*, 52–65. [CrossRef] [PubMed]

66. Inn, K.S.; Gack, M.U.; Tokunaga, F.; Shi, M.; Wong, L.Y.; Iwai, K.; Jung, J.U. Linear ubiquitin assembly complex negatively regulates RIG-I- and TRIM25-mediated type I interferon induction. *Mol. Cell* **2011**, *41*, 354–365. [CrossRef] [PubMed]

67. Belgnaoui, S.M.; Paz, S.; Samuel, S.; Goulet, M.L.; Sun, Q.; Kikkert, M.; Iwai, K.; Dikic, I.; Hiscott, J.; Lin, R. Linear ubiquitination of NEMO negatively regulates the interferon antiviral response through disruption of the MAVS-TRAF3 complex. *Cell Host Microbe* **2012**, *12*, 211–222. [CrossRef]

68. Zuo, Y.; Feng, Q.; Jin, L.; Huang, F.; Miao, Y.; Liu, J.; Xu, Y.; Chen, X.; Zhang, H.; Guo, T.; et al. Regulation of the linear ubiquitination of STAT1 controls antiviral interferon signaling. *Nat. Commun.* **2020**, *11*, 1146. [CrossRef] [PubMed]

69. Zinngrebe, J.; Rieser, E.; Taraborrelli, L.; Peltzer, N.; Hartwig, T.; Ren, H.; Kovacs, I.; Endres, C.; Draber, P.; Darding, M.; et al. LUBAC deficiency perturbs TLR3 signaling to cause immunodeficiency and autoinflammation. *J. Exp. Med.* **2016**, *213*, 2671–2689. [CrossRef]

70. Khan, M.; Syed, G.H.; Kim, S.J.; Siddiqui, A. Hepatitis B Virus-Induced Parkin-Dependent Recruitment of Linear Ubiquitin Assembly Complex (LUBAC) to Mitochondria and Attenuation of Innate Immunity. *PLoS Pathog.* **2016**, *12*, e1005693. [CrossRef]

71. MacDuff, D.A.; Baldridge, M.T.; Qaqish, A.M.; Nice, T.J.; Darbandi, A.D.; Hartley, V.L.; Peterson, S.T.; Miner, J.J.; Iwai, K.; Virgin, H.W. HOIL1 Is Essential for the Induction of Type I and III Interferons by MDA5 and Regulates Persistent Murine Norovirus Infection. *J. Virol.* **2018**, *92*, e01368-18. [CrossRef] [PubMed]

72. Lafont, E.; Draber, P.; Rieser, E.; Reichert, M.; Kupka, S.; de Miguel, D.; Draberova, H.; von Massenhausen, A.; Bhamra, A.; Henderson, S.; et al. TBK1 and IKKepsilon prevent TNF-induced cell death by RIPK1 phosphorylation. *Nat. Cell Biol.* **2018**, *20*, 1389–1399. [CrossRef] [PubMed]

73. Oikawa, D.; Sato, Y.; Ohtake, F.; Komakura, K.; Hanada, K.; Sugawara, K.; Terawaki, S.; Mizukami, Y.; Phuong, H.T.; Iio, K.; et al. Molecular bases for HOIPINs-mediated inhibition of LUBAC and innate immune responses. *Commun. Biol.* **2020**, *3*, 163. [CrossRef] [PubMed]

74. Morishita, H.; Mizushima, N. Diverse Cellular Roles of Autophagy. *Annu. Rev. Cell Dev. Biol.* **2019**, *35*, 453–475. [CrossRef] [PubMed]

75. Fujita, N.; Morita, E.; Itoh, T.; Tanaka, A.; Nakaoka, M.; Osada, Y.; Umemoto, T.; Saitoh, T.; Nakatogawa, H.; Kobayashi, S.; et al. Recruitment of the autophagic machinery to endosomes during infection is mediated by ubiquitin. *J. Cell Biol.* **2013**, *203*, 115–128. [CrossRef] [PubMed]

76. Fiskin, E.; Bionda, T.; Dikic, I.; Behrends, C. Global Analysis of Host and Bacterial Ubiquitinome in Response to Salmonella Typhimurium Infection. *Mol. Cell* **2016**, *62*, 967–981. [CrossRef]

77. Noad, J.; von der Malsburg, A.; Pathe, C.; Michel, M.A.; Komander, D.; Randow, F. LUBAC-synthesized linear ubiquitin chains restrict cytosol-invading bacteria by activating autophagy and NF-κB. *Nat. Microbiol.* **2017**, *2*, 17063. [CrossRef]

78. Polajnar, M.; Dietz, M.S.; Heilemann, M.; Behrends, C. Expanding the host cell ubiquitylation machinery targeting cytosolic Salmonella. *EMBO Rep.* **2017**, *18*, 1572–1585. [CrossRef]

79. van Wijk, S.J.L.; Fricke, F.; Herhaus, L.; Gupta, J.; Hotte, K.; Pampaloni, F.; Grumati, P.; Kaulich, M.; Sou, Y.S.; Komatsu, M.; et al. Linear ubiquitination of cytosolic Salmonella Typhimurium activates NF-κB and restricts bacterial proliferation. *Nat. Microbiol.* **2017**, *2*, 17066. [CrossRef]

80. Komander, D.; Clague, M.J.; Urbe, S. Breaking the chains: Structure and function of the deubiquitinases. *Nat. Rev. Mol. Cell Biol.* **2009**, *10*, 550–563. [CrossRef]

81. Mevissen, T.E.T.; Komander, D. Mechanisms of Deubiquitinase Specificity and Regulation. *Annu. Rev. Biochem.* **2017**, *86*, 159–192. [CrossRef] [PubMed]

82. Abdul Rehman, S.A.; Kristariyanto, Y.A.; Choi, S.Y.; Nkosi, P.J.; Weidlich, S.; Labib, K.; Hofmann, K.; Kulathu, Y. MINDY-1 Is a Member of an Evolutionarily Conserved and Structurally Distinct New Family of Deubiquitinating Enzymes. *Mol. Cell* **2016**, *63*, 146–155. [CrossRef] [PubMed]

83. Haahr, P.; Borgermann, N.; Guo, X.; Typas, D.; Achuthankutty, D.; Hoffmann, S.; Shearer, R.; Sixma, T.K.; Mailand, N. ZUFSP Deubiquitylates K63-Linked Polyubiquitin Chains to Promote Genome Stability. *Mol. Cell* **2018**, *70*, 165–174. [CrossRef] [PubMed]

84. Lork, M.; Verhelst, K.; Beyaert, R. CYLD, A20 and OTULIN deubiquitinases in NF-κB signaling and cell death: So similar, yet so different. *Cell Death Differ.* **2017**, *24*, 1172–1183. [CrossRef] [PubMed]

85. Keusekotten, K.; Elliott, P.R.; Glockner, L.; Fiil, B.K.; Damgaard, R.B.; Kulathu, Y.; Wauer, T.; Hospenthal, M.K.; Gyrd-Hansen, M.; Krappmann, D.; et al. OTULIN antagonizes LUBAC signaling by specifically hydrolyzing Met1-linked polyubiquitin. *Cell* **2013**, *153*, 1312–1326. [CrossRef] [PubMed]

86. Rivkin, E.; Almeida, S.M.; Ceccarelli, D.F.; Juang, Y.C.; MacLean, T.A.; Srikumar, T.; Huang, H.; Dunham, W.H.; Fukumura, R.; Xie, G.; et al. The linear ubiquitin-specific deubiquitinase gumby regulates angiogenesis. *Nature* **2013**, *498*, 318–324. [CrossRef]

87. Fiil, B.K.; Damgaard, R.B.; Wagner, S.A.; Keusekotten, K.; Fritsch, M.; Bekker-Jensen, S.; Mailand, N.; Choudhary, C.; Komander, D.; Gyrd-Hansen, M. OTULIN restricts Met1-linked ubiquitination to control innate immune signaling. *Mol. Cell* **2013**, *50*, 818–830. [CrossRef]

88. Elliott, P.R.; Nielsen, S.V.; Marco-Casanova, P.; Fiil, B.K.; Keusekotten, K.; Mailand, N.; Freund, S.M.; Gyrd-Hansen, M.; Komander, D. Molecular basis and regulation of OTULIN-LUBAC interaction. *Mol. Cell* **2014**, *54*, 335–348. [CrossRef]

89. Takiuchi, T.; Nakagawa, T.; Tamiya, H.; Fujita, H.; Sasaki, Y.; Saeki, Y.; Takeda, H.; Sawasaki, T.; Buchberger, A.; Kimura, T.; et al. Suppression of LUBAC-mediated linear ubiquitination by a specific interaction between LUBAC and the deubiquitinases CYLD and OTULIN. *Genes Cells* **2014**, *19*, 254–272. [CrossRef]

90. Damgaard, R.B.; Walker, J.A.; Marco-Casanova, P.; Morgan, N.V.; Titheradge, H.L.; Elliott, P.R.; McHale, D.; Maher, E.R.; McKenzie, A.N.J.; Komander, D. The Deubiquitinase OTULIN Is an Essential Negative Regulator of Inflammation and Autoimmunity. *Cell* **2016**, *166*, 1215–1230. [CrossRef]

91. Zhou, Q.; Yu, X.; Demirkaya, E.; Deuitch, N.; Stone, D.; Tsai, W.L.; Kuehn, H.S.; Wang, H.; Yang, D.; Park, Y.H.; et al. Biallelic hypomorphic mutations in a linear deubiquitinase define otulipenia, an early-onset autoinflammatory disease. *Proc. Natl. Acad. Sci. USA* **2016**, *113*, 10127–10132. [CrossRef] [PubMed]

92. Nabavi, M.; Shahrooei, M.; Rokni-Zadeh, H.; Vrancken, J.; Changi-Ashtiani, M.; Darabi, K.; Manian, M.; Seif, F.; Meyts, I.; Voet, A.; et al. Auto-inflammation in a Patient with a Novel Homozygous OTULIN Mutation. *J. Clin. Immunol.* **2019**, *39*, 138–141. [CrossRef]

93. Damgaard, R.B.; Elliott, P.R.; Swatek, K.N.; Maher, E.R.; Stepensky, P.; Elpeleg, O.; Komander, D.; Berkun, Y. OTULIN deficiency in ORAS causes cell type-specific LUBAC degradation, dysregulated TNF signalling and cell death. *EMBO Mol. Med.* **2019**, *11*, e9324. [CrossRef] [PubMed]

94. Heger, K.; Wickliffe, K.E.; Ndoja, A.; Zhang, J.; Murthy, A.; Dugger, D.L.; Maltzman, A.; de Sousa, E.M.F.; Hung, J.; Zeng, Y.; et al. OTULIN limits cell death and inflammation by deubiquitinating LUBAC. *Nature* **2018**, *559*, 120–124. [CrossRef] [PubMed]

95. Douglas, T.; Saleh, M. Post-translational Modification of OTULIN Regulates Ubiquitin Dynamics and Cell Death. *Cell Rep.* **2019**, *29*, 3652–3663.e5. [CrossRef] [PubMed]

96. Verboom, L.; Martens, A.; Priem, D.; Hoste, E.; Sze, M.; Vikkula, H.; Van Hove, L.; Voet, S.; Roels, J.; Maelfait, J.; et al. OTULIN Prevents Liver Inflammation and Hepatocellular Carcinoma by Inhibiting FADD- and RIPK1 Kinase-Mediated Hepatocyte Apoptosis. *Cell Rep.* **2020**, *30*, 2237–2247. [CrossRef] [PubMed]

97. Zhao, M.; Song, K.; Hao, W.; Wang, L.; Patil, G.; Li, Q.; Xu, L.; Hua, F.; Fu, B.; Schwamborn, J.C.; et al. Non-proteolytic ubiquitination of OTULIN regulates NF-κB signaling pathway. *J. Mol. Cell Biol.* **2020**, *12*, 163–175. [CrossRef]

98. Stangl, A.; Elliott, P.R.; Pinto-Fernandez, A.; Bonham, S.; Harrison, L.; Schaub, A.; Kutzner, K.; Keusekotten, K.; Pfluger, P.T.; El Oualid, F.; et al. Regulation of the endosomal SNX27-retromer by OTULIN. *Nat. Commun.* **2019**, *10*, 4320. [CrossRef]

99. Bignell, G.R.; Warren, W.; Seal, S.; Takahashi, M.; Rapley, E.; Barfoot, R.; Green, H.; Brown, C.; Biggs, P.J.; Lakhani, S.R.; et al. Identification of the familial cylindromatosis tumour-suppressor gene. *Nat. Genet.* **2000**, *25*, 160–165. [CrossRef]

100. Regamey, A.; Hohl, D.; Liu, J.W.; Roger, T.; Kogerman, P.; Toftgard, R.; Huber, M. The tumor suppressor CYLD interacts with TRIP and regulates negatively nuclear factor kappaB activation by tumor necrosis factor. *J. Exp. Med.* **2003**, *198*, 1959–1964. [CrossRef]

101. Trompouki, E.; Hatzivassiliou, E.; Tsichritzis, T.; Farmer, H.; Ashworth, A.; Mosialos, G. CYLD is a deubiquitinating enzyme that negatively regulates NF-κB activation by TNFR family members. *Nature* **2003**, *424*, 793–796. [CrossRef] [PubMed]

102. Komander, D.; Reyes-Turcu, F.; Licchesi, J.D.; Odenwaelder, P.; Wilkinson, K.D.; Barford, D. Molecular discrimination of structurally equivalent Lys 63-linked and linear polyubiquitin chains. *EMBO Rep.* **2009**, *10*, 466–473. [CrossRef] [PubMed]

103. Hrdinka, M.; Fiil, B.K.; Zucca, M.; Leske, D.; Bagola, K.; Yabal, M.; Elliott, P.R.; Damgaard, R.B.; Komander, D.; Jost, P.J.; et al. CYLD Limits Lys63- and Met1-Linked Ubiquitin at Receptor Complexes to Regulate Innate Immune Signaling. *Cell Rep.* **2016**, *14*, 2846–2858. [CrossRef] [PubMed]

104. Tokunaga, F.; Nishimasu, H.; Ishitani, R.; Goto, E.; Noguchi, T.; Mio, K.; Kamei, K.; Ma, A.; Iwai, K.; Nureki, O. Specific recognition of linear polyubiquitin by A20 zinc finger 7 is involved in NF-κB regulation. *Embo J.* **2012**, *31*, 3856–3870. [CrossRef]

105. Draber, P.; Kupka, S.; Reichert, M.; Draberova, H.; Lafont, E.; de Miguel, D.; Spilgies, L.; Surinova, S.; Taraborrelli, L.; Hartwig, T.; et al. LUBAC-Recruited CYLD and A20 Regulate Gene Activation and Cell Death by Exerting Opposing Effects on Linear Ubiquitin in Signaling Complexes. *Cell Rep.* **2015**, *13*, 2258–2272. [CrossRef]

106. Sato, Y.; Goto, E.; Shibata, Y.; Kubota, Y.; Yamagata, A.; Goto-Ito, S.; Kubota, K.; Inoue, J.; Takekawa, M.; Tokunaga, F.; et al. Structures of CYLD USP with Met1- or Lys63-linked diubiquitin reveal mechanisms for dual specificity. *Nat. Struct. Mol. Biol.* **2015**, *22*, 222–229. [CrossRef]

107. Yamanaka, S.; Sato, Y.; Oikawa, D.; Goto, E.; Fukai, S.; Tokunaga, F.; Takahashi, H.; Sawasaki, T. Subquinocin, a small molecule inhibitor of CYLD and USP-family deubiquitinating enzymes, promotes NF-κB signaling. *Biochem. Biophys. Res. Commun.* **2020**, *524*, 1–7. [CrossRef]

108. Uematsu, A.; Kido, K.; Takahashi, H.; Takahashi, C.; Yanagihara, Y.; Saeki, N.; Yoshida, S.; Maekawa, M.; Honda, M.; Kai, T.; et al. The E3 ubiquitin ligase MIB2 enhances inflammation by degrading the deubiquitinating enzyme CYLD. *J. Biol. Chem.* **2019**, *294*, 14135–14148. [CrossRef]

109. Dobson-Stone, C.; Hallupp, M.; Shahheydari, H.; Ragagnin, A.M.G.; Chatterton, Z.; Carew-Jones, F.; Shepherd, C.E.; Stefen, H.; Paric, E.; Fath, T.; et al. CYLD is a causative gene for frontotemporal dementia—Amyotrophic lateral sclerosis. *Brain* **2020**, *143*, 783–799. [CrossRef]

110. Wagner, S.; Carpentier, I.; Rogov, V.; Kreike, M.; Ikeda, F.; Lohr, F.; Wu, C.J.; Ashwell, J.D.; Dotsch, V.; Dikic, I.; et al. Ubiquitin binding mediates the NF-κB inhibitory potential of ABIN proteins. *Oncogene* **2008**, *27*, 3739–3745. [CrossRef]

111. Verhelst, K.; Carpentier, I.; Kreike, M.; Meloni, L.; Verstrepen, L.; Kensche, T.; Dikic, I.; Beyaert, R. A20 inhibits LUBAC-mediated NF-κB activation by binding linear polyubiquitin chains via its zinc finger 7. *Embo J.* **2012**, *31*, 3845–3855. [CrossRef] [PubMed]

112. Fennell, L.M.; Rahighi, S.; Ikeda, F. Linear ubiquitin chain-binding domains. *FEBS J.* **2018**, *285*, 2746–2761. [CrossRef]

113. Rahighi, S.; Ikeda, F.; Kawasaki, M.; Akutsu, M.; Suzuki, N.; Kato, R.; Kensche, T.; Uejima, T.; Bloor, S.; Komander, D.; et al. Specific recognition of linear ubiquitin chains by NEMO is important for NF-κB activation. *Cell* **2009**, *136*, 1098–1109. [CrossRef] [PubMed]

114. Lo, Y.C.; Lin, S.C.; Rospigliosi, C.C.; Conze, D.B.; Wu, C.J.; Ashwell, J.D.; Eliezer, D.; Wu, H. Structural basis for recognition of diubiquitins by NEMO. *Mol. Cell* **2009**, *33*, 602–615. [CrossRef] [PubMed]

115. Doffinger, R.; Smahi, A.; Bessia, C.; Geissmann, F.; Feinberg, J.; Durandy, A.; Bodemer, C.; Kenwrick, S.; Dupuis-Girod, S.; Blanche, S.; et al. X-linked anhidrotic ectodermal dysplasia with immunodeficiency is caused by impaired NF-κB signaling. *Nat. Genet.* **2001**, *27*, 277–285. [CrossRef] [PubMed]

116. Hubeau, M.; Ngadjeua, F.; Puel, A.; Israel, L.; Feinberg, J.; Chrabieh, M.; Belani, K.; Bodemer, C.; Fabre, I.; Plebani, A.; et al. New mechanism of X-linked anhidrotic ectodermal dysplasia with immunodeficiency: Impairment of ubiquitin binding despite normal folding of NEMO protein. *Blood* **2011**, *118*, 926–935. [CrossRef]

117. Rezaie, T.; Child, A.; Hitchings, R.; Brice, G.; Miller, L.; Coca-Prados, M.; Heon, E.; Krupin, T.; Ritch, R.; Kreutzer, D.; et al. Adult-onset primary open-angle glaucoma caused by mutations in optineurin. *Science* **2002**, *295*, 1077–1079. [CrossRef]

118. Ying, H.; Yue, B.Y. Cellular and molecular biology of optineurin. *Int. Rev. Cell Mol. Biol.* **2012**, *294*, 223–258.

119. Maruyama, H.; Morino, H.; Ito, H.; Izumi, Y.; Kato, H.; Watanabe, Y.; Kinoshita, Y.; Kamada, M.; Nodera, H.; Suzuki, H.; et al. Mutations of optineurin in amyotrophic lateral sclerosis. *Nature* **2010**, *465*, 223–226. [CrossRef]

120. Ito, H.; Nakamura, M.; Komure, O.; Ayaki, T.; Wate, R.; Maruyama, H.; Nakamura, Y.; Fujita, K.; Kaneko, S.; Okamoto, Y.; et al. Clinicopathologic study on an ALS family with a heterozygous E478G optineurin mutation. *Acta Neuropathol.* **2011**, *122*, 223–229. [CrossRef]

121. Nakazawa, S.; Oikawa, D.; Ishii, R.; Ayaki, T.; Takahashi, H.; Takeda, H.; Ishitani, R.; Kamei, K.; Takeyoshi, I.; Kawakami, H.; et al. Linear ubiquitination is involved in the pathogenesis of optineurin-associated amyotrophic lateral sclerosis. *Nat. Commun.* **2016**, *7*, 12547. [CrossRef] [PubMed]

122. Verstrepen, L.; Carpentier, I.; Beyaert, R. The biology of A20-binding inhibitors of NF-κB activation (ABINs). *Adv. Exp. Med. Biol.* **2014**, *809*, 13–31. [CrossRef] [PubMed]

123. Lin, S.M.; Lin, S.C.; Hong, J.Y.; Su, T.W.; Kuo, B.J.; Chang, W.H.; Tu, Y.F.; Lo, Y.C. Structural Insights into Linear Tri-ubiquitin Recognition by A20-Binding Inhibitor of NF-κB, ABIN-2. *Structure* **2017**, *25*, 66–78. [CrossRef] [PubMed]

124. Herhaus, L.; van den Bedem, H.; Tang, S.; Maslennikov, I.; Wakatsuki, S.; Dikic, I.; Rahighi, S. Molecular Recognition of M1-Linked Ubiquitin Chains by Native and Phosphorylated UBAN Domains. *J. Mol. Biol.* **2019**, *431*, 3146–3156. [CrossRef]

125. Ma, A.; Malynn, B.A. A20: Linking a complex regulator of ubiquitylation to immunity and human disease. *Nat. Rev. Immunol.* **2012**, *12*, 774–785. [CrossRef]

126. Wertz, I.E.; O'Rourke, K.M.; Zhou, H.; Eby, M.; Aravind, L.; Seshagiri, S.; Wu, P.; Wiesmann, C.; Baker, R.; Boone, D.L.; et al. De-ubiquitination and ubiquitin ligase domains of A20 downregulate NF-κB signalling. *Nature* **2004**, *430*, 694–699. [CrossRef]

127. Garcia-Carbonell, R.; Wong, J.; Kim, J.Y.; Close, L.A.; Boland, B.S.; Wong, T.L.; Harris, P.A.; Ho, S.B.; Das, S.; Ernst, P.B.; et al. Elevated A20 promotes TNF-induced and RIPK1-dependent intestinal epithelial cell death. *Proc. Natl. Acad. Sci. USA* **2018**, *115*, E9192–E9200. [CrossRef]

128. Martens, A.; Priem, D.; Hoste, E.; Vetters, J.; Rennen, S.; Catrysse, L.; Voet, S.; Deelen, L.; Sze, M.; Vikkula, H.; et al. Two distinct ubiquitin-binding motifs in A20 mediate its anti-inflammatory and cell-protective activities. *Nat. Immunol.* **2020**, *21*, 381–387. [CrossRef]

129. Razani, B.; Whang, M.I.; Kim, F.S.; Nakamura, M.C.; Sun, X.; Advincula, R.; Turnbaugh, J.A.; Pendse, M.; Tanbun, P.; Achacoso, P.; et al. Non-catalytic ubiquitin binding by A20 prevents psoriatic arthritis-like disease and inflammation. *Nat. Immunol.* **2020**, *21*, 422–433. [CrossRef]

130. Aksentijevich, I.; Zhou, Q. NF-κB Pathway in Autoinflammatory Diseases: Dysregulation of Protein Modifications by Ubiquitin Defines a New Category of Autoinflammatory Diseases. *Front. Immunol.* **2017**, *8*, 399. [CrossRef]

131. Peltzer, N.; Darding, M.; Montinaro, A.; Draber, P.; Draberova, H.; Kupka, S.; Rieser, E.; Fisher, A.; Hutchinson, C.; Taraborrelli, L.; et al. LUBAC is essential for embryogenesis by preventing cell death and enabling haematopoiesis. *Nature* **2018**, *557*, 112–117. [CrossRef] [PubMed]

132. Boisson, B.; Laplantine, E.; Prando, C.; Giliani, S.; Israelsson, E.; Xu, Z.; Abhyankar, A.; Israel, L.; Trevejo-Nunez, G.; Bogunovic, D.; et al. Immunodeficiency, autoinflammation and amylopectinosis in humans with inherited HOIL-1 and LUBAC deficiency. *Nat. Immunol.* **2012**, *13*, 1178–1186. [CrossRef] [PubMed]

133. Wang, K.; Kim, C.; Bradfield, J.; Guo, Y.; Toskala, E.; Otieno, F.G.; Hou, C.; Thomas, K.; Cardinale, C.; Lyon, G.J.; et al. Whole-genome DNA/RNA sequencing identifies truncating mutations in RBCK1 in a novel Mendelian disease with neuromuscular and cardiac involvement. *Genome Med.* **2013**, *5*, 67. [CrossRef] [PubMed]

134. Nilsson, J.; Schoser, B.; Laforet, P.; Kalev, O.; Lindberg, C.; Romero, N.B.; Davila Lopez, M.; Akman, H.O.; Wahbi, K.; Iglseder, S.; et al. Polyglucosan body myopathy caused by defective ubiquitin ligase RBCK1. *Ann. Neurol.* **2013**, *74*, 914–919. [CrossRef]

135. Boisson, B.; Laplantine, E.; Dobbs, K.; Cobat, A.; Tarantino, N.; Hazen, M.; Lidov, H.G.; Hopkins, G.; Du, L.; Belkadi, A.; et al. Human HOIP and LUBAC deficiency underlies autoinflammation, immunodeficiency, amylopectinosis, and lymphangiectasia. *J. Exp. Med.* **2015**, *212*, 939–951. [CrossRef]

136. Oda, H.; Beck, D.B.; Kuehn, H.S.; Sampaio Moura, N.; Hoffmann, P.; Ibarra, M.; Stoddard, J.; Tsai, W.L.; Gutierrez-Cruz, G.; Gadina, M.; et al. Second Case of HOIP Deficiency Expands Clinical Features and Defines Inflammatory Transcriptome Regulated by LUBAC. *Front. Immunol.* **2019**, *10*, 479. [CrossRef]

137. Pasqualucci, L.; Zhang, B. Genetic drivers of NF-κB deregulation in diffuse large B-cell lymphoma. *Semin. Cancer Biol.* **2016**, *39*, 26–31. [CrossRef]

138. Yang, Y.; Schmitz, R.; Mitala, J.; Whiting, A.; Xiao, W.; Ceribelli, M.; Wright, G.W.; Zhao, H.; Yang, Y.; Xu, W.; et al. Essential role of the linear ubiquitin chain assembly complex in lymphoma revealed by rare germline polymorphisms. *Cancer Discov.* **2014**, *4*, 480–493. [CrossRef]

139. Yang, Y.; Kelly, P.; Shaffer, A.L., III; Schmitz, R.; Yoo, H.M.; Liu, X.; Huang, D.W.; Webster, D.; Young, R.M.; Nakagawa, M.; et al. Targeting Non-proteolytic Protein Ubiquitination for the Treatment of Diffuse Large B Cell Lymphoma. *Cancer Cell* **2016**, *29*, 494–507. [CrossRef]

140. Ruiz, E.J.; Diefenbacher, M.E.; Nelson, J.K.; Sancho, R.; Pucci, F.; Chakraborty, A.; Moreno, P.; Annibaldi, A.; Liccardi, G.; Encheva, V.; et al. LUBAC determines chemotherapy resistance in squamous cell lung cancer. *J. Exp. Med.* **2019**, *216*, 450–465. [CrossRef]

141. Boland, B.; Yu, W.H.; Corti, O.; Mollereau, B.; Henriques, A.; Bezard, E.; Pastores, G.M.; Rubinsztein, D.C.; Nixon, R.A.; Duchen, M.R.; et al. Promoting the clearance of neurotoxic proteins in neurodegenerative disorders of ageing. *Nat. Rev. Drug Discov.* **2018**, *17*, 660–688. [CrossRef] [PubMed]

142. Winklhofer, K.F.; Tatzelt, J.; Haass, C. The two faces of protein misfolding: Gain- and loss-of-function in neurodegenerative diseases. *EMBO J.* **2008**, *27*, 336–349. [CrossRef] [PubMed]

143. Cirulli, E.T.; Lasseigne, B.N.; Petrovski, S.; Sapp, P.C.; Dion, P.A.; Leblond, C.S.; Couthouis, J.; Lu, Y.F.; Wang, Q.; Krueger, B.J.; et al. Exome sequencing in amyotrophic lateral sclerosis identifies risk genes and pathways. *Science* **2015**, *347*, 1436–1441. [CrossRef] [PubMed]

144. Nakayama, Y.; Tsuji, K.; Ayaki, T.; Mori, M.; Tokunaga, F.; Ito, H. Linear Polyubiquitin Chain Modification of TDP-43-Positive Neuronal Cytoplasmic Inclusions in Amyotrophic Lateral Sclerosis. *J. Neuropathol. Exp. Neurol.* **2020**, *79*, 256–265. [CrossRef]

145. Nakayama, Y.; Sakamoto, S.; Tsuji, K.; Ayaki, T.; Tokunaga, F.; Ito, H. Identification of linear polyubiquitin chain immunoreactivity in tau pathology of Alzheimer's disease. *Neurosci. Lett.* **2019**, *703*, 53–57. [CrossRef]

146. van Well, E.M.; Bader, V.; Patra, M.; Sanchez-Vicente, A.; Meschede, J.; Furthmann, N.; Schnack, C.; Blusch, A.; Longworth, J.; Petrasch-Parwez, E.; et al. A protein quality control pathway regulated by linear ubiquitination. *EMBO J.* **2019**, *38*, e100730. [CrossRef]

147. Strickson, S.; Campbell, D.G.; Emmerich, C.H.; Knebel, A.; Plater, L.; Ritorto, M.S.; Shpiro, N.; Cohen, P. The anti-inflammatory drug BAY 11-7082 suppresses the MyD88-dependent signalling network by targeting the ubiquitin system. *Biochem. J.* **2013**, *451*, 427–437. [CrossRef]

148. Sakamoto, H.; Egashira, S.; Saito, N.; Kirisako, T.; Miller, S.; Sasaki, Y.; Matsumoto, T.; Shimonishi, M.; Komatsu, T.; Terai, T.; et al. Gliotoxin suppresses NF-κB activation by selectively inhibiting linear ubiquitin chain assembly complex (LUBAC). *ACS Chem. Biol.* **2015**, *10*, 675–681. [CrossRef]

149. De Cesare, V.; Johnson, C.; Barlow, V.; Hastie, J.; Knebel, A.; Trost, M. The MALDI-TOF E2/E3 Ligase Assay as Universal Tool for Drug Discovery in the Ubiquitin Pathway. *Cell Chem. Biol.* **2018**, *25*, 1117–1127. [CrossRef]

150. Johansson, H.; Isabella Tsai, Y.C.; Fantom, K.; Chung, C.W.; Kumper, S.; Martino, L.; Thomas, D.A.; Eberl, H.C.; Muelbaier, M.; House, D.; et al. Fragment-Based Covalent Ligand Screening Enables Rapid Discovery of Inhibitors for the RBR E3 Ubiquitin Ligase HOIP. *J. Am. Chem. Soc.* **2019**, *141*, 2703–2712. [CrossRef]

151. Tsai, Y.I.; Johansson, H.; Dixon, D.; Martin, S.; Chung, C.W.; Clarkson, J.; House, D.; Rittinger, K. Single-Domain Antibodies as Crystallization Chaperones to Enable Structure-Based Inhibitor Development for RBR E3 Ubiquitin Ligases. *Cell Chem. Biol.* **2020**, *27*, 83–93. [CrossRef] [PubMed]

152. Katsuya, K.; Hori, Y.; Oikawa, D.; Yamamoto, T.; Umetani, K.; Urashima, T.; Kinoshita, T.; Ayukawa, K.; Tokunaga, F.; Tamaru, M. High-Throughput Screening for Linear Ubiquitin Chain Assembly Complex (LUBAC) Selective Inhibitors Using Homogenous Time-Resolved Fluorescence (HTRF)-Based Assay System. *SLAS Discov.* **2018**, *23*, 1018–1029. [CrossRef] [PubMed]

153. Katsuya, K.; Oikawa, D.; Iio, K.; Obika, S.; Hori, Y.; Urashima, T.; Ayukawa, K.; Tokunaga, F. Small-molecule inhibitors of linear ubiquitin chain assembly complex (LUBAC), HOIPINs, suppress NF-κB signaling. *Biochem. Biophys. Res. Commun.* **2019**, *509*, 700–706. [CrossRef] [PubMed]

154. Harden, J.L.; Krueger, J.G.; Bowcock, A.M. The immunogenetics of Psoriasis: A comprehensive review. *J. Autoimmun.* **2015**, *64*, 66–73. [CrossRef] [PubMed]

155. Greb, J.E.; Goldminz, A.M.; Elder, J.T.; Lebwohl, M.G.; Gladman, D.D.; Wu, J.J.; Mehta, N.N.; Finlay, A.Y.; Gottlieb, A.B. Psoriasis. *Nat. Rev. Dis. Primers* **2016**, *2*, 16082. [CrossRef]

156. Yu, K.; Phu, L.; Varfolomeev, E.; Bustos, D.; Vucic, D.; Kirkpatrick, D.S. Immunoaffinity enrichment coupled to quantitative mass spectrometry reveals ubiquitin-mediated signaling events. *J. Mol. Biol.* **2015**, *427*, 2121–2134. [CrossRef]

157. Morimoto, D.; Walinda, E.; Fukada, H.; Sou, Y.S.; Kageyama, S.; Hoshino, M.; Fujii, T.; Tsuchiya, H.; Saeki, Y.; Arita, K.; et al. The unexpected role of polyubiquitin chains in the formation of fibrillar aggregates. *Nat. Commun.* **2015**, *6*, 6116. [CrossRef]

International Journal of
Molecular Sciences

Review

Cracking the Monoubiquitin Code of Genetic Diseases

Raj Nayan Sewduth [1,2,†], Maria Francesca Baietti [1,2,†] and Anna A. Sablina [1,2,*]

[1]   VIB-KU Leuven Center for Cancer Biology, VIB, Herestraat 49, 3000 Leuven, Belgium;
     rajnayan.sewduth@kuleuven.vib.be (R.N.S.); mariafrancesca.baietti@kuleuven.vib.be (M.F.B.)

[2]   Department of Oncology, KU Leuven, Herestraat 49, 3000 Leuven, Belgium

[*]   Correspondence: anna.sablina@kuleuven.vib.be

[†]   These authors contributed equally to the work.

Received: 7 April 2020; Accepted: 23 April 2020; Published: 25 April 2020

Abstract: Ubiquitination is a versatile and dynamic post-translational modification in which single ubiquitin molecules or polyubiquitin chains are attached to target proteins, giving rise to mono- or poly-ubiquitination, respectively. The majority of research in the ubiquitin field focused on degradative polyubiquitination, whereas more recent studies uncovered the role of single ubiquitin modification in important physiological processes. Monoubiquitination can modulate the stability, subcellular localization, binding properties, and activity of the target proteins. Understanding the function of monoubiquitination in normal physiology and pathology has important therapeutic implications, as alterations in the monoubiquitin pathway are found in a broad range of genetic diseases. This review highlights a link between monoubiquitin signaling and the pathogenesis of genetic disorders.

Keywords: ubiquitin system; genetic diseases; ubiquitin ligase; deubiquitinases; monoubiquitin signaling; vesicular trafficking; protein complex formation

---

1. Introduction

Ubiquitination is a reversible post-translational modification process during which the highly conserved 76-aminoacid protein ubiquitin is conjugated to target proteins. Ubiquitin can be conjugated to a protein substrate via distinct mechanisms. Monoubiquitination is the attachment of a single ubiquitin molecule to a single lysine residue on a substrate protein, whereas multi-monoubiquitination is the conjugation of a single ubiquitin molecule to multiple lysine residues. Polyubiquitination occurs when ubiquitin molecules are attached end-to-end to a lysine residue on a substrate protein to form a poly-ubiquitin chain. In this case, ubiquitin molecules are conjugated through one of the seven lysine residues present on the ubiquitin itself (K6, K11, K27, K29, K33, K48, and 63) or the N-terminal methionine (M1). While most of the studies have described the role of specific polyubiquitination, such as K48-linked polyubiquitination for proteasomal degradation [1,2] or K63-linked polyubiquitination for vesicular trafficking [3], emerging evidences implicate monoubiquitination and multi-monoubiquitination in controlling numerous aspects of protein function, such as degradation, subcellular localization, and protein–protein interaction. In this review, we focus on the role of monoubiquitin conjugation in normal physiology and genetic disease.

1.1. Monoubiquitination in Protein Function

Even though K48-linked polyubiquitination is the "canonical" signal for proteasomal degradation, a handful of proteins are degraded following monoubiquitination. Degradative monoubiquitination targets proteins smaller than 150 amino acids with low structural disorder, whereas polyubiquitination recognizes proteins with highly disorganized structures independently of their size. Monoubiquitination-dependent substrates of proteasomal degradation are enriched for genes associated with carbohydrate transport and oxidative stress response pathways [4]. Because carbohydrate transporters are plasma membrane proteins, this finding is consistent with previous studies linking down-regulation of membrane receptors to monoubiquitination-mediated endocytosis [5]. The stability of ribosomal and proteasomal subunits is also regulated by monoubiquitination [6]. Multi-monoubiquitination could also act as a degradation signal [7]. For instance, multi-monoubiquitination of Cyclin B1 mediated by the Anaphase-promoting complex (APC/C) is sufficient to promote Cyclin B1 proteasomal degradation and mitotic exit [7].

The attachment of a single ubiquitin can serve as a signal for specific subcellular localization. It is well established that monoubiquitination triggers endocytosis of receptor tyrosine kinases (RTKs) and the NOTCH receptor 1 (NOTCH) from the plasma membrane [8,9]. Monoubiquitination targets the TNF receptor associated factor 4 (TRAF4) to cell–cell junctions where it is required to promote cell migration [10]. Dissociation of lipidated proteins from the plasma membrane could also be induced by ubiquitin conjugation [11]. Specifically, ubiquitin conjugated to K170 of the small GTPase Harvey rat sarcoma viral oncogene homolog (HRAS) sequesters its farnesyl and palmitoyl groups, impairing association of HRAS with the membrane [12]. Monoubiquitination is also involved in the control of nuclear/cytoplasmic shuttling. Vascular Endothelial Growth Factor (VEGF)-induced monoubiquitination of the actin-binding protein filamin B (FLNB) was shown to inhibit the nuclear translocation of Histone Deacetylase 7 (HDAC7). The monoubiquitinated form of filamin B binds to the nuclear localization signal of HDAC7, thereby transiently preventing its re-entry into the nucleus and mitigating the transcriptional repressor activity of HDAC7 on the genes required for VEGF-mediated responses [13]. In contrast, ubiquitin conjugation induces the nuclear export of the neddylation regulator, Defective In Cullin Neddylation (DCNL1). It was suggested that ubiquitin conjugation might act as a nuclear export signal that promotes the interaction with nuclear export machinery [14]. Furthermore, the monoubiquitination of the cancer stem cells marker CD133 (prominin-1) promotes its secretion into extracellular vesicles by facilitating the interaction of CD133 with the vesicular sorting protein tumor susceptibility gene 101 (TSG101) [15].

Multiple studies highlight a key function of monoubiquitination in controlling protein–protein interactions, both in positive and negative ways. Ubiquitin conjugation facilitates the interaction either by creating additional interaction surface or by attracting proteins containing ubiquitin-binding domains. Ubiquitin-binding domains have been classified into nearly 25 subfamilies. These domains are typically independently folded modular domains of up to 150 amino acids showing remarkable structural heterogeneity, which utilize diverse surfaces to contact ubiquitin [16,17]. Ubiquitin-binding activities are hard to detect and predict, both structurally and bioinformatically; thus, many of such ubiquitin-binding proteins are yet to be identified. Especially taking into account that a subset of ubiquitin interactors has no obvious structurally defined ubiquitin-binding domains, such as the disordered Proteasome Complex Subunit SEM1/DSS1, in which ubiquitin-binding sites are characterized by acidic and hydrophobic residues [18]. The low-affinity interactions between ubiquitin-binding domains and ubiquitin are required for rapid, timely, and reversible cellular responses to a particular stimulus. To illustrate this, insulin-like growth factor 1(IGF-I)-induced monoubiquitination of the insulin receptor substrate (IRS-2) favors its interaction with the ubiquitin-binding protein EPSIN1, thereby enhancing IRS-2-mediated signaling and cell proliferation induced by IGF-I [19]. T-cell stimulation leads to monoubiquitination of the paracaspase MALT1 (mucosa-associated lymphoid tissue protein-1) that promotes the formation and stabilization of the catalytically active MALT1 dimer [20]. On top of this, auto-monoubiquitination of several E3 ligases, such as neural precursor cell-expressed developmentally down-regulated 4 (NEDD4), E3 ubiquitin-protein ligase Itchy (ITCH), and X-linked inhibitor of apoptosis (XIAP), is implicated in the recruitment of substrates and regulation of the ligase activity. For instance, the auto-monoubiquitination of NEDD4 promotes the recruitment of its substrate epidermal growth factor receptor substrate 15 (EPS15) [21].

Monoubiquitination can also block protein–protein interactions by either creating steric hindrance or inducing an autoinhibitory conformation for proteins containing ubiquitin-binding domains. As an example for the first mechanism, the attachment of ubiquitin within the effector-binding domain of the RAS-like GTPase, RAS-related protein B (RALB) sterically inhibits its binding to the downstream effector exocyst complex component EXO84 [22]. Another example is the inhibitory effect of monoubiquitination on the aggregation of amyloid proteins. Specifically, N-terminal monoubiquitination of presynaptic neuronal proteins, tau^{K18} and α-synuclein, changes their aggregation properties, resulting in structurally distinct aggregate structures that are cleared through proteasomal degradation instead of accumulating in cells [23]. On the other hand, the monoubiquitination of proteins that contain ubiquitin-binding domains could impose their autoinhibitory conformation, thus providing an intrinsic switch-off mechanism. This mechanism is used to regulate the degradative activity of the proteasome. Ubiquitin conjugation to the proteasome regulatory subunit RPN10 blocks the binding between its ubiquitin-interacting motif and interactors containing ubiquitin conjugates [24]. Monoubiquitination of the endocytic proteins, EPS15, hepatocyte growth factor-regulated tyrosine kinase substrate (HGS or HRS), ubiquitin-associated and SH3 domain-containing protein A and B (UBASH3A and UNASH3B, also called STS2 and STS1) leads to intramolecular interactions between ubiquitin and their ubiquitin-binding domains, thereby preventing their binding to ubiquitinated targets and inhibiting their trafficking [25].

By differentially modifying the protein–protein interaction network, monoubiquitination provides precise and timely transmission of biological information [26]. To illustrate this, ubiquitin conjugation of the RAS-like GTPase RALB modifies its affinity for downstream effectors, inhibiting the interaction with EXO84 and facilitating its binding to exocyst complex component 2 (EXOC2 or SEC5). Thus, monoubiquitination within the effector-binding domain provides the switch for the dual functions of RALB in autophagy and innate immune responses [22]. Another example is the monoubiquitination of t-SNARE syntaxin 5 (SYN5) in early mitosis that disrupts SNARE complex formation. Subsequently, ubiquitinated SYN5 recruits p97/VCP (valosin-containing protein) to the mitotic Golgi fragments and promotes post-mitotic Golgi reassembly; thus, ubiquitin conjugation regulates Golgi membrane dynamics during the cell cycle [27].

Importantly, the attachment of a single ubiquitin molecule to specific lysine residues can generate diverse substrate–ubiquitin structures, leading to different functional outcomes. This is well-described for the RAS-like small GTPases that undergo monoubiquitination at several lysines. While monoubiquitination of the GTPase Ras-related protein RAB5 at K165 affects its GTPase activity, attachment of ubiquitin at K140 alters the ability of RAB5 to bind and activate its downstream effectors. Ubiquitin conjugation to specific lysines also differentially affects the activity and subcellular localization of the Rat sarcoma (RAS) proto-oncogenes [28]. Specifically, the monoubiquitination of RAS at K117 promotes GTP loading, whereas monoubiquitination at K147 impairs GTP hydrolysis activity, both resulting in RAS activation. On the other hand, monoubiquitination at K170 impairs RAS association to the membrane and downstream signaling [12,29–31]. Altogether, these examples demonstrate a fundamental role of monoubiquitination in controlling protein function.

1.2. Enzymes Controlling Monoubiquitination

The ubiquitination reaction involves three enzymes: ubiquitin activating enzyme (E1), ubiquitin conjugating enzymes (E2), and ubiquitin ligase (E3) enzymes [32]. Although there are only 2 E1s and 30–50 E2s, approximately 600–700 E3 ubiquitin ligases are encoded in the human genome, conferring target specificity to the ubiquitination reaction. The process is reversible as ubiquitin hydrolases, also called deubiquitinating enzymes (DUBs), can hydrolyze the covalently bound ubiquitin peptides.

E2 conjugating enzymes attach ubiquitin molecules to lysine residues on protein substrate or ubiquitin molecules, prompting mono- or poly-ubiquitination. E2s share a highly conserved catalytic core domain called the ubiquitin conjugation domain (UBC). The UBC contains a specific catalytic cysteine that forms a thioester bond with ubiquitin. Most of the studies have been focused on ubiquitin-conjugating enzymes 2S and 3A (UBE2S, UBE3A) and ubiquitin-conjugating enzyme E2 variant 1 (UEV1A) which are responsible for pro-degradative K48-linked polyubiquitination and UBE2N, which is important for pro-trafficking K63-linked polyubiquitination. These E2-conjugating enzymes also have intrinsic ability to form free polyubiquitin chains. On the other hand, several E2 ubiquitin ligases, such as ubiquitin-conjugating enzymes 2A, 2K, 2T, and 2W (UBE2A, UBE2K, UBE2T, and UBE2W) are only able to generate monoubiquitin chains. When E2s are charged with ubiquitin molecules, some of them exhibit enhanced affinity for co-factors containing ubiquitin-binding domains that confer specificity of ubiquitin linkages to specific substrates [33].

The substrate specificity is given by the E3 ubiquitin ligases that control the transfer of ubiquitin to a lysine residue on the substrate. The E3 ligases identify their protein substrate within the cellular pool of proteins, by recognizing specific protein sequences or chemical motifs in the substrate. The E3 ligases can be classified into RING (really interesting new gene), HECT (homology to E6AP C-terminus), and RING-related types [34]. The activity of the RING E3s is specified by a RING domain, which promotes ubiquitin transfer from the E2 to the substrate, whereas the substrate-recruiting module is responsible for the substrate recognition. The RING domain and the substrate-recruiting module of the RING E3 ligase can be found in a single polypeptide, as in the case of CBL (Casitas B-lineage lymphoma), or in separate subunits of a multi-complex E3s, as for the heterodimer of BRCA1 (breast cancer 1) and BARD1 (BRCA1-associated RING domain 1), Cullin–RING ligases, and APC [34]. The HECT E3s have a catalytic cysteine residue that can form a thioester bond directly with ubiquitin [34]. While the HECT domain represents the catalytic domain, the substrate specificity of HECT E3s is determined by their respective N-terminal extensions. The RING-between-RING (RBR) E3s define a third class of ubiquitin ligases distinct from the RING and HECT types. The RBR E3s are characterized by RING1 and RING2 domains and a central in-between-RINGs (IBR) zinc-binding domain. Substrate ubiquitination by RBR ligases is a multistep process. It starts with the recognition of the ubiquitinated E2 by RING1, followed by the transfer of ubiquitin onto the catalytic cysteine in RING2 to form the thioester intermediate, and finally, the transfer onto the substrate [35].

Most E3 ligases are able to generate different types of ubiquitin chains, depending on the E2 conjugating enzyme with which they preferentially interact. It was shown that the monoubiquitinating E2 enzymes have stronger affinity to RING E3 ligases. For example, UBE2A binds the RING E3 ligase RAD18, whereas UBE2K and UBE2T have high affinity to the RING E3 ligase BRCA1 [36,37]. Furthermore, a process termed coupled monoubiquitination is responsible for the monoubiquitination of ubiquitin-binding domain-containing proteins. The proposed model suggests that monoubiquitinated substrates cannot be further polyubiquitinated because the ubiquitin-binding domain interacts intramolecularly with the attached ubiquitin, thus disrupting the association of the substrate with the monoubiquitinated E3 enzyme [21]. Monoubiquitination can also be achieved by engaging a DUB that trims the assembled poly-ubiquitin chain to produce monoubiquitinated proteins.

Given that around 5% of the human genome encodes ubiquitin system components [38], it is not surprising that alterations of the ubiquitination machinery have been observed in multiple disease conditions [39]. Here, we will discuss how dysregulation of monoubiquitin signaling due to germline mutations of E3 ligases or DUBs could contribute to the development of genetic disorders (Table 1).

Table 1. Genetic diseases associated with genes regulating monoubiquitination. Short list of substrates modified by the indicated E2 conjugating enzymes, E3 ligases, and ubiquitin hydrolases (DUBs) are shown, together with the indication of the modulated cellular functions and the type of mutations detected in patients.

Disease	Gene	Type of Enzyme	Monoubiquitinated Substrate	Cellular Function	Disease-Associated Mutations
X-linked syndromic mental retardation	UBE2A	Ubiquitin-conjugating enzyme E2 A	PCNA [40]; Histone H2B [41]	DNA damage tolerance pathway [42–44]; epigenetic regulation [41]	Loss of function: missense mutations, microdeletions, larger deletions [45,46]
Autosomal recessive juvenile parkinsonism	Parkin or PARK2	RBR E3 ubiquitin ligase	VDAC1 [47,48]	Mitophagy, apoptosis [49,50]	Loss of function: missense mutations, deletions [51]
Fanconi Anemia	UBE2T	Ubiquitin-conjugating enzyme E2 T	FANCD2/FANCI [52,53]	Cross-linked DNA repair [54,55]	Loss of function: missense mutations [56]
	FANCL	PHD FINGER E3 ubiquitin ligase			Loss of function: missense, frameshift mutations [57]
	BRCA1	RING E3 ubiquitin ligase	FANCD2/FANCI [58]		Loss of function: missense frameshift mutations, deletions [59]
Charcot-Marie-Tooth disease	LRSAM1	RING E3 ubiquitin ligase	TSG101 [60]	Endosomal sorting [61]	Loss of function: missense, frameshift mutations [62]
Cushing disease	USP8	Ubiquitin specific peptidase 8	EGFR [63,64]; CHMP1B [65]	Endosomal sorting [66,67]	Gain of function: missense mutations [68,69]
Noonan Syndrome	LZTR1	BTB-Kelch ubiquitin ligase adaptor	RAS [12,30]; CHMP1B [70]	RAS localization and signaling [12,30]; VEGFR trafficking and signaling [70]	Loss of function: missense, frameshift mutations [71,72]
	CBL	RING E3 ubiquitin ligase	SH3KBP1 [73]	EGFR trafficking and signaling [74]	Loss of function: missense mutations [75–77]
Autoimmune disorder associated to facial dysmorphism	ITCH	HECT E3 ubiquitin ligase	TIEG1 [78]; SMN [79]	Nuclear translocation of FOXP3 [78], translocation of SMN to Cajal body [79]	Loss of function: frameshift mutations [80]

2. Genetic Diseases Associated with Dysregulated Monoubiquitination

2.1. X-linked Syndromic Mental Retardation

Germline alterations of the ubiquitin-conjugating enzyme E2 A (*UBE2A*) gene coding for the ubiquitin-conjugating enzyme 2A [40] are associated with X-linked syndromic mental retardation, a disease characterized by abnormal intellectual development and dysmorphic features, such as large head, wide mouth, almond-shaped eyes, and onychodystrophy. Patients with this syndrome carry intragenic point mutations or microdeletions of *UBE2A* or larger Xq24 deletions encompassing the *UBE2A* region [45,46]. The *Ube2a* knockout mice present defects in spatial learning tasks, but no other severe phenotypes [81]. This indicates that the mouse model only partially recapitulates the phenotype observed in patients suffering from *UBE2A*-linked mental retardation.

Molecular mechanisms linking *UBE2A* mutations to neurodevelopmental disorders are not fully understood. A study focused on the fly *UBE2A* homolog *Rad6A* implicates defective mobilization of the E3 ligase PARKIN as a cause of abnormal vesicle trafficking and dysregulated clearance of dysfunctional mitochondria in neurons. These abnormalities are suspected to contribute to the neurodevelopmental phenotype in patients with *UBE2A* deficiency syndrome [82]. On the other hand, one of the best-described functions of UBE2A is to promote monoubiquitination of proliferating cell nuclear antigen (PCNA) in a complex with the RING-Type E3 ubiquitin transferase RAD18. PCNA monoubiquitination can be switched to polyubiquitination in the presence of helicase-like transcription factor (HLTF). Two distinct branches of the DNA damage tolerance pathways are activated by either mono-, or polyubiquitinated PCNA to rescue a stalled replication fork and ensure continuous DNA synthesis. Monoubiquitinated PCNA favors low-fidelity translesion DNA synthesis, whereas PCNA polyubiquitination induces high-fidelity homology-dependent DNA repair [42]. Defects in DNA damage response could explain some of the developmental aspects of X-linked mental retardation [43,44]. *PCNA* mutations in patients also cause ataxia-telangiectasia-like disorder-2, a disease showing development delay [83]. Moreover, the disease-associated G23R mutation of UBE2A disrupts the binding site for RAD18 [84]. This suggests that the UBE2A/RAD18/PCNA axis might be at least partially responsible for the pathogenesis in mental retardation (Figure 1A).

In complex with the E3 ligase ring finger protein 20 (RNF20), UBE2A also promotes the monoubiquitination of histone H2B [41]. Monoubiquitinated H2B not only regulates global transcriptional elongation [85,86] but also plays an essential role in the regulation of inducible genes involved in cell differentiation [87–89] and inflammation [90]. However, a potential role of UBE2A-mediated monoubiquitination of H2B in mental retardation is still to be elucidated.

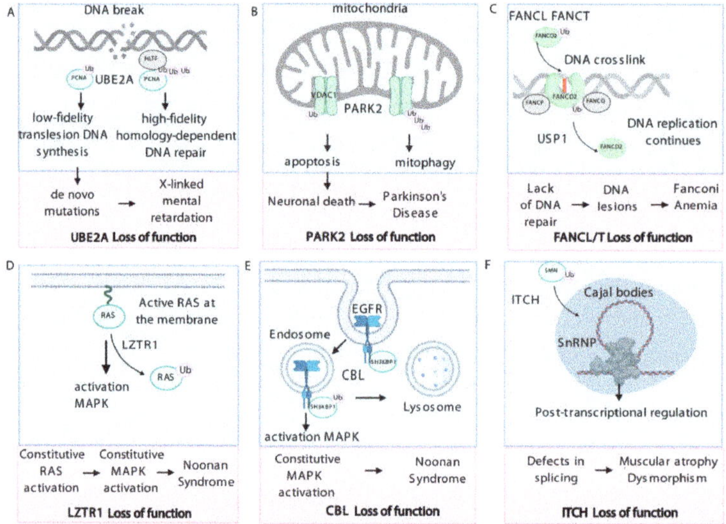

Figure 1. The role of monoubiquitination in human diseases. (**A**) Ubiquitin-conjugating enzyme E2 A (UBE2A) loss of function impairs proliferating cell nuclear antigen (PCNA)-mediated DNA repair that partially explains developmental aspects of X-linked mental retardation. (**B**) Parkinson Protein 2 (PARK2) regulates mitophagy and apoptosis by controlling poly- and monoubiquitination of voltage-dependent anion-selective channel 1 (VDAC1). Dysregulation of VDAC1 ubiquitination contributes to the development of Parkinson's disease. (**C**) Mutations in Fanconi Anemia complementation group L/T (*FANCL/T*) lead to DNA repair deficiency. Monoubiquitinated FA group D2 (FANCD2)/FA complementation group I (FANCI) heterodimer binds DNA, whereas deubiquitination of FANCD2 allows to re-start DNA replication. (**D**) The Rat sarcoma (RAS) GTPases are monoubiquitinated by the Leucine Zipper Like Transcription Regulator 1 (LZTR1)- Cullin 3 (CUL3) complex, inhibiting RAS association with the membrane and activating of RAS signaling. Hyperactivation of the mitogen activated protein kinase (MAPK) caused by LZTR1 loss of function leads to the Noonan syndrome phenotypes. (**E**) Casitas B-lineage lymphoma (CBL)-mediated monoubiquitination of SH3 domain-containing kinase-binding protein 1 (SH3KBP1) recruits active epidermal growth factor receptor (EGFR) for degradation. *CBL* mutations lead to up-regulation of the MAPK pathway that partially explains its contribution to the development of Noonan syndrome. (**F**) Mutations in E3 ubiquitin-protein ligase Itchy (*ITCH*) impair the monoubiquitination of survival motor neuron (SMN) that dysregulates translocation to Cajal bodies and affects post-transcriptional regulation of gene expression, linking ITCH loss of function to the development of spinal muscular atrophy.

2.2. Parkinson's Disease

PARKIN, or Parkinson Protein 2 (PARK2), is a RBR-type E3 ubiquitin ligase mutated in autosomal recessive juvenile parkinsonism [51], a form of familial Parkinson's disease, defined by an onset before 40 years of age and characterized by slow movement and tremor (Table 1). *PARK2* is also mutated in other neurological diseases such as retropulsion, dystonia, hyperreflexia, and sensory axonal neuropathy [91] causing olfactory impairment [92]. In these different pathologies, loss of PARK2 function causes death of selective neuron populations, such as the dopaminergic neurons [93]. Deletion of *Parkin* in mice leads to motor and cognitive deficits [94] caused by catecholaminergic neuronal death and the subsequent loss of norepinephrine in some regions of the brain [95]. The *Parkin* knockout mice also show enhanced hepatocyte proliferation, macroscopic hepatic tumors in aged mice, higher sensitivity to myocardial infarction, and a strong inflammatory phenotype [96].

PARKIN maintains mitochondrial health through mitochondrial quality control and generation of mitochondrial-derived vesicles, followed by whole-organellar degradation, a process called mitophagy [97]. Mitophagy is vital for the removal of damaged mitochondria and toxic mitochondrial proteins, protecting neuronal cells from apoptosis [49]. Dysregulation of these processes plays a key role in Parkinson's disease [50]. PARKIN was shown to mediate both polyubiquitination and monoubiquitination depending on the protein context [47]. This dual activity of PARKIN differentially affects function of its substrates such as voltage-dependent anion-selective channel 1 (VDAC1), which transports ions and small molecules at the mitochondrial outer membrane. Defect in VDAC1 polyubiquitination hinders PARKIN-mediated mitophagy, whereas dysregulation of VDAC1 monoubiquitination induces apoptosis. This suggests that the dual regulation of mitophagy and apoptosis by Parkin via VDAC1 poly- and monoubiquitination is critical in protecting cells from the pathogenesis of Parkinson's disease [48] (Figure 1B). PARKIN also mediates the multi-monoubiquitination of heat shock protein 70 (HSP70) and heat shock cognate 70 (HSC70), leading to their association to insoluble substrates, consistent with a degradation-independent role for this type of ubiquitin modification [98]. These data strongly implicate PARKIN-mediated monoubiquitination in the development of Parkinson's disease.

2.3. Fanconi Anemia

Fanconi anemia (FA) is a disorder caused by the genetic inactivation of crosslink repair. FA is characterized by abnormal development, bone marrow failure, hypogonadism, and marked cancer susceptibility. Autosomal recessive mutations in any one of 20 genes (*FANCA* to *FANCQ*) result in this genetic disorder, and collectively, the FANC gene products function in a FA−DNA crosslink repair pathway [99]. Several of the FANC genes form a large monoubiquitination complex (FA core complex). Ubiquitin conjugating enzyme E2 T (UBE2T, also called FANCT) and E3 ligase FA complementation group L (FANCL) are key enzymes in the FA core complex and are mutated in different subtypes of FA, causing FA-T and FA-L respectively [56,57] (Table 1).

Several animal models were generated to study the disease. *Ube2t* knockout in zebrafish leads to hypersensitivity to DNA damage and reversion of female-to-male sex [99], reflecting the hypogonadism phenotype occurring in FA patients. *Fancl* knockout in mice leads to decrease of fertility and defects in the proliferation of germ cells. Bone marrow cells isolated from the *Fancl* knockout mice were also hypersensitive to DNA crosslinking agent, mitomycin C [57].

The hallmark of FA is a high frequency of chromosomal aberrations caused by defects in DNA repair and hypersensitivity to DNA crosslinking agents in cells isolated from patients [100]. When DNA replication is stalled, the FA core complex is activated and monoubiquitinates the FA group D2 (FANCD2) and FA complementation group I (FANCI) heterodimer [52]. Monoubiquitinated FANCD2/FANCI heterodimer adopts a closed conformation, creating a channel that encloses double-stranded DNA [53]. The ubiquitin residue plays a key role in this process as it acts as a covalent molecular pin to trap the complex on DNA. Moreover, monoubiquitinated FANCD2 serves as a signal to recruit to the replication fork, DNA repair proteins that contain ubiquitin-binding motifs [101], such as FA complementation group P (FANCP, also called SLX4) and FA complementation group Q (FANCQ, also called ERCC), to remove the cross-linked DNA [54,55] (Figure 1C). DNA crosslinking repair is completed when ubiquitin-specific protease 1 (USP1) reverses the monoubiquitination of FANCD2 [102–104]. Interestingly, knockout of *Usp1* in mice also recapitulates FA phenotypes, including perinatal lethality, infertility, and crosslinker hypersensitivity. Finally, FANCD2 monoubiquitination could also be mediated by the ubiquitin ligase activity of the E3 ligase BRCA1 [58]. *BRCA1* is also mutated in a subset of patients suffering from Fanconi anemia D1 (FANCD1), identifying another key ubiquitin ligase in the pathogenesis of FA [59].

2.4. Charcot-Marie-Tooth Disease

Mutations of LRSAM1 (leucine-rich repeat and sterile alpha motif containing 1), an E3 ligase with RING ZINC finger domains and leucine-rich repeats, are associated with cases of early-onset Parkinson's disease [105] and to Charcot-Marie-Tooth disease [62] (Table 1). Charcot-Marie-Tooth disease affects the peripheral nervous system and is characterized by progressive muscular atrophy. Knockdown of *Lrsam1* in zebrafish leads to disturbed neurodevelopment with a less organized neural structure and affected tail formation and movement. *Lrsam1* mutant mice present a normal neuromuscular structure and only mild neuropathy phenotype in aged mice, but higher sensitivity to neurotoxic agents that cause axonal degeneration [106].

Mechanistically, LRSAM1 promotes monoubiquitination of TSG101, a component of the endosomal sorting complex required for transport (ESCRT)-1. LRSAM1-mediated monoubiquitination of TSG101 enables recycling of TSG101-containing sorting complexes and cargo reloading [60]. Several proteins directly regulating post-translational processing and intracellular trafficking, such as tripartite motif-containing 2 (TRIM2) and RAS-related small GTPase RAB7, are also found to be mutated in Charcot-Marie-Tooth disease, confirming the link between vesicle trafficking and Charcot-Marie-Tooth disorder [61].

2.5. Cushing Disease

Mutations of the ubiquitin-specific peptidase 8 (USP8) are associated with pituitary adenoma tumors, also called Cushing disease [68] (Table 1). In Cushing disease, the adrenocorticotropic hormone (ACTH)-producing pituitary adenoma tumors secrete cortisol in the blood, resulting in obesity, diabetes, hypertension as well as additional cerebrovascular, cardiac, and reproductive disorders [107]. Disease-associated *USP8* mutations are located within or adjacent to the 14-3-3 binding motif. The diminished ability of mutant UPS8 to interact with 14-3-3 proteins enhances the proteolytic cleavage of USP8, leading to the generation of an activated catalytic fragment [69].

Enhanced activity of mutant USP8 in Cushing disease impairs the down-regulation of the epidermal growth factor receptor (EGFR) pathway due to increased EGFR deubiquitination. Consequently, sustained EGFR signaling in pituitary adenoma leads to enhanced promoter activity of the gene encoding proopiomelanocortin (POMC), the precursor of ACTH [63,64]. However, a subset of *USP8* mutations are not associated with higher EGFR expression, and mutations in *USP8* rarely occur in other tumor types, suggesting that USP8-dependent mechanisms other than EGFR up-regulation cannot be ruled out to be responsible for the pathogenesis of Cushing disease [66].

The role of USP8 in endosomal sorting complexes required for transport (ESCRT)-mediated trafficking of RTKs remains controversial [108]. USP8 contains an N-terminal microtubule interacting and transport domain, which has unveiled its potential to interact with charged multivesicular body proteins (CHMP), components of the ESCRT-III complex [67]. One of the proposed mechanisms is the regulation of multi-monoubiquitination of charged multivesicular body protein 1b (CHMP1B) [65]. CHMP1B, a part of the endosomal ESCRT-III complex involved in endosomal budding, is multi-monoubiquitinated in response to growth factor stimulation [65]. Deubiquitination of CHMP1B favors its assembly into a membrane-associated ESCRT-III polymer complexes and modulates the dynamics of endosomal sorting [65]. Further studies on the USP8 targets may shed new light on our understanding of USP8 contribution to membrane receptor trafficking and chemotherapy resistance in Cushing's disease.

2.6. Noonan Syndrome

Noonan syndrome (NS) is a complex disease where patients present various clinical features including short stature, dysmorphia, and different cardiac defects [71,72]. Infants affected by the disease suffer from polyhydramnios, pleural effusions, or edema. Recent studies have linked NS to lung lymphangiectasis that is characterized by vessel dilatation and hemorrhages. NS is caused by germline mutations in genes that encode protein components of the RAS/mitogen activated protein kinase (MAPK) pathway. The vast majority of these mutations result in increased RAS/MAPK signaling [75].

Germline mutations of leucine zipper-like transcriptional regulator 1 (LZTR1) are associated with Noonan syndrome and familial schwannomatosis [71,109] (Table 1). Schwannomatosis is characterized by multiple 'schwannomas', benign tumors of the peripheral and spinal nerves [110]. *Lztr1* loss in mice results in perinatal lethality due to cardiovascular dysfunction [12,70]. *Lztr1* haploinsufficiency in mice partially recapitulates NS phenotypes, including heart malformation, craniofacial features, and bleeding abnormalities [70]. *LZTR1* loss in Schwann cells promotes their proliferation and dedifferentiation [12].

LZTR1 serves as a substrate adaptor for Cullin3 (CUL3) ubiquitin ligase complex that controls ubiquitination of the RAS GTPases [12,30,111]. LZTR1-mediated monoubiquitination of RAS at K170 attenuates RAS association with the membrane, inhibiting the RAS signaling pathway [12,30]. Disease-associated *LZTR1* mutations diminish either LZTR1-CUL3 complex formation or its interaction with RAS proteins, which impairs RAS ubiquitination. Dysregulation of RAS monoubiquitination results in the hyperactivation of the MAPK pathway and explains the role of LZTR1 in NS and schwannomatosis (Figure 1D). LZTR1 was also reported to regulate vesicular trafficking of VEGFR through CHMP1B ubiquitination. Loss of *Lztr1* reduces multi-monoubiquitination of CHMP1B, thus blocking the disassembly of the ESCRT complex and the trafficking of endosomal VEGFR. This ultimately leads to the activation of VEGF signaling and explains the bleeding abnormalities detected in NS patients [70].

The RING-type E3 ligase Casitas B-lineage lymphoma (CBL) is also found to be mutated in Noonan syndrome-like disorder, as well as early-onset juvenile myelomonocytic leukemia (JMML) (Table 1). JMML is a myeloproliferative disorder characterized by malignant transformation of the hematopoietic stem cell (HSC) causing oncogenic transformation of HSC-derived cells. Most common *CBL* mutations in patients affect the RING domains, such as the Y371 residue, as its phosphorylation is important for activation of the ligase function [76,77]. Deletion of *c-Cbl* in mice causes a decrease of B- and T-lymphocyte function causing lymphopenia [112], partially recapitulating JMML. *c-Cbl* knockout mice also present increased sensitivity to cardiac infarct [113], echoing the cardiovascular defects detected in NS patients.

CBL is a key regulator of internalization and turnover of different receptor tyrosine kinases, as it monoubiquitinates endosomal adaptors important for the recycling and/or degradation of the RTKs [114]. Loss of CBL results in endosomal accumulation of the RTK and uncontrolled activation of the downstream pathways [73]. One of the endosomal adaptors modified by CBL is SH3 domain-containing kinase-binding protein 1 (SH3KBP1, also called CIN85), which undergoes ubiquitination in response to EGF stimulation. After EGFR is internalized, monoubiquitinated SH3KBP1 recruits active EGFR to the multivesicular bodies (MVB) that targets EGFR for degradation. Loss of *CBL* leads to constitutive activation of EGF signaling [74]. The alterations in the trafficking and degradation of these signaling proteins lead to up-regulation of the MAPK pathway that could at least partially explain the mechanism by which CBL contributes to human disease (Figure 1E). Interestingly, deletion of the *SH3KBP1* gene in B-cells in mice results in impaired T-cell-independent antibody responses and abnormal B-cell receptor response, partially recapitulating JMML phenotypes [115].

2.7. Autoimmune Disorder

Mutations in the HECT-domain E3 ligase Itchy (ITCH) are found in a rare form of autoimmune disorder associated with facial dysmorphism, organomegaly, and developmental delay (Table 1). These patients also present severe autoimmune inflammatory cell infiltration of the lungs, liver, and gut [80]. Genetic analysis of the patients identified homozygosity for a frameshift mutation of *ITCH* caused by a 1-bp insertion, resulting in a truncated protein [80]. Mice lacking expression of the *Itch* gene develop a wide spectrum of immunologic phenotypes, such as lung and stomach inflammation, as well as hyperplasia of lymphoid cells and itching [116]. *Itch* deficiency in mice also renders mice resistant to TNFα-induced acute liver failure [117].

Different mechanisms have been proposed for the role of ITCH in human pathogenesis. ITCH was shown to regulate the stability and subcellular localization of multiple targets by mediating not only poly-ubiquitination, but also monoubiquitnation. ITCH-mediated monoubiquitination promotes degradation of the key regulators of T-cell anergy, phospholipase C-γ1 (PLC-γ1) and protein kinase C-theta (PKC-θ) [118]. ITCH can also suppress inflammation by controlling ubiquitination of tumor necrosis factor alpha-induced protein 3 (TNFAIP3, also called A20) [119]. Moreover, in T cells, ITCH mediates monoubiquitination of TGF-β inducible early gene-1 (TIEG1) [78]. Monoubiquitinated TIEG1 is translocated to the nucleus and triggers expression of Forkhead Box P3 (FOXP3), a master regulator of T cell function (Treg cells). *Tieg1* deficient mice are not able to suppress lung inflammation, indicating that deficiency in TIEG1 ubiquitination can explain ITCH-mediated pathogenesis in patients with autoimmune disorder.

Furthermore, ITCH regulates subcellular localization of survival motor neuron (SMN) by mediating its monoubiquitination [79]. Specifically, SMN is expressed mainly in the nucleus, where it accumulates in subnuclear structures such as the Cajal body. Dysregulation of SMN ubiquitination significantly impairs its co-localization with small nuclear ribonucleoprotein (snRNP) in Cajal body foci [79]. Importantly, SMN loss of function causes spinal muscular atrophy, a neuromuscular disease characterized by motor neurons degeneration, suggesting that the ITCH/SMN axis might be a major driver of muscular atrophy and dysmorphism (Figure 1E).

3. Discussion

Up to now, the role of non-degradative monoubiquitination in human disease has been relatively understudied, as focus was mostly falling on the role of polyubiquitination in proteasomal degradation. Recent emerging evidences have highlighted the key function of monoubiquitination in a wide range of cellular processes. The findings listed here only represent the most characterized enzymes controlling monoubiquitination. Nonetheless, it reflects the high prevalence of alterations of the monoubiquitin pathway in such a broad array of genetic disorders. This suggests that disruption of the monoubiquitin pathway may be a major force driving the pathogenic phenotypes of such diseases.

The abundance of the ubiquitin-related enzymes mutated in genetic disorders indicates that targeting the ubiquitin pathway might have a utility for a range of genetic diseases. However, at present, we lack detailed knowledge on how monoubiquitin signals are generated and how they are decoded by the cell. This is challenged by the diversity and complexity of the ubiquitin pathway. Moreover, monoubiquitinated proteins might not have been accurately identified, because polyubiquitinated conjugates are recognized more efficiently by anti-ubiquitin specific antibodies. This leads to underestimation of the pool of monoubiquitinated proteins present in the cell and challenges their characterization. The development of novel tools to purify monoubiquitinated proteins using high-affinity ubiquitin-binding domains and synthetic biology approaches to efficiently generate monoubiquitinated proteins overcoming these issues.

It is also worth noting that when looking at the few drugs that were developed to target the ubiquitin pathway, most are meant to inhibit its functioning. Several inhibitors targeting the ubiquitinating enzymes described in this review have been reported. Ubiquitin variants that block the E2-ubiquitin binding surface of the RING domain of CBL were shown to specifically inhibit the activity of phosphorylated CBL [120,121]. A high-throughput screening to identify ITCH inhibitors discovered that clomipramine, a common antidepressant drug, blocks ITCH autoubiquitination and affects the ability of ITCH to ubiquitinate its substrates [122]. Screening for the inhibitors of UBE2T/FANCL identified two compounds that sensitize cells to DNA crosslinking [123]. Pharmacological inhibition of USP8 was shown to effectively suppress ACTH synthesis in vitro without causing any significant cytotoxicity, indicating its potential for the management of ACTH hypersecretion in Cushing's disease [124]. However, considering that the disease-associated alterations of the ubiquitin ligases and DUBs are mostly loss of function, inhibitors targeting these enzymes would not be beneficial. This indicates that there is a need to develop novel strategies for targeted therapies of genetic diseases [125]. Several screens identified compounds activating PARKIN ubiquitin ligase activity [126] and enhancing mitophagy [127], such as the compound described in patent WO2018023029. While no in vivo validation is available for this compound yet, this demonstrates the feasibility of identification of E3 ligase activators, opening novel therapeutic options for patients with genetic disorders.

This review collected multiple evidences that monoubiquitination is a highly relevant process in the pathogenesis of a wide range of genetic diseases; however, further research is necessary to identify specific entry points for therapeutic intervention of monoubiquitination-dependent signaling pathways.

Funding: This work was supported by H2020 European Research Council (Ub-RASDisease); Research Foundation Flanders (FWO) postdoctoral fellowship (MFB).

Conflicts of Interest: The authors declare no conflict of interest.

Abbreviations

FA	Fanconi Anemia
NS	Noonan Syndrome
JMML	Juvenile myelomonocytic leukemia

References

1. Thrower, J.S.; Hoffman, L.; Rechsteiner, M.; Pickart, C.M. Recognition of the polyubiquitin proteolytic signal. *EMBO J.* **2000**, *19*, 94–102. [CrossRef] [PubMed]
2. Matyskiela, M.E.; Martin, A. Design principles of a universal protein degradation machine. *J. Mol. Biol.* **2013**, *425*, 199–213. [CrossRef] [PubMed]
3. Erpapazoglou, Z.; Walker, O.; Haguenauer-Tsapis, R. Versatile roles of K63-linked ubiquitin chains in trafficking. *Cells* **2014**, *3*, 1027–1088. [CrossRef] [PubMed]
4. Braten, O.; Livneh, I.; Ziv, T.; Admon, A.; Kehat, I.; Caspi, L.H.; Gonen, H.; Bercovich, B.; Godzik, A.; Jahandideh, S.; et al. Numerous proteins with unique characteristics are degraded by the 26S proteasome following monoubiquitination. *Proc. Natl. Acad. Sci. USA* **2016**, *113*, E4639–E4647. [CrossRef]
5. Hicke, L.; Dunn, R. Regulation of Membrane protein transport by ubiquitin and ubiquitin-binding proteins. *Annu. Rev. Cell Dev. Biol.* **2003**, *19*, 141–172. [CrossRef]
6. Jung, Y.; Kim, H.D.; Yang, H.W.; Kim, H.J.; Jang, C.Y.; Kim, J. Modulating cellular balance of Rps3 mono-ubiquitination by both Hel2 E3 ligase and Ubp3 deubiquitinase regulates protein quality control. *Exp. Mol. Med.* **2017**, *49*, e390. [CrossRef]
7. Dimova, N.V.; Hathaway, N.A.; Lee, B.H.; Kirkpatrick, D.S.; Berkowitz, M.L.; Gygi, S.P.; Finley, D.; King, R.W. APC/C-mediated multiple monoubiquitylation provides an alternative degradation signal for cyclin B1. *Nat. Cell Biol.* **2012**, *14*, 168–176. [CrossRef]
8. Gupta-Rossi, N.; Six, E.; LeBail, O.; Logeat, F.; Chastagner, P.; Olry, A.; Israël, A.; Brou, C. Monoubiquitination and endocytosis direct γ-secretase cleavage of activated Notch receptor. *J. Cell Biol.* **2004**, *166*, 73–83. [CrossRef]
9. Haglund, K.; Sigismund, S.; Polo, S.; Szymkiewicz, I.; Di Fiore, P.P.; Dikic, I. Multiple monoubiquitination of RTKs is sufficient for their endocytosis and degradation. *Nat. Cell Biol.* **2003**, *5*, 461–466. [CrossRef]

10. Wang, X.; Jin, C.; Tang, Y.; Tang, L.Y.; Zhang, Y.E. Ubiquitination of tumor necrosis factor receptor-associated factor 4 (TRAF4) by smad ubiquitination regulatory factor 1 (Smurf1) regulates motility of breast epithelial and cancer cells. *J. Biol. Chem.* **2013**, *288*, 21784–21792. [CrossRef]

11. Jura, N.; Scotto-Lavino, E.; Sobczyk, A.; Bar-Sagi, D. Differential modification of Ras proteins by ubiquitination. *Mol. Cell* **2006**, *21*, 679–687. [CrossRef]

12. Steklov, M.; Pandolfi, S.; Baietti, M.F.; Batiuk, A.; Carai, P.; Najm, P.; Zhang, M.; Jang, H.; Renzi, F.; Cai, Y.; et al. Mutations in LZTR1 drive human disease by dysregulating RAS ubiquitination. *Science* **2018**, *362*, 1177–1182. [CrossRef] [PubMed]

13. Su, Y.-T.; Gao, C.; Liu, Y.; Guo, S.; Wang, A.; Wang, B.; Erdjument-Bromage, H.; Miyagi, M.; Tempst, P.; Kao, H.-Y. Monoubiquitination of filamin B regulates vascular endothelial growth factor-mediated trafficking of histone deacetylase 7. *Mol. Cell. Biol.* **2013**, *33*, 1546–1560. [CrossRef] [PubMed]

14. Wu, K.; Yan, H.; Fang, L.; Wang, X.; Pfleger, C.; Jiang, X.; Huang, L.; Pan, Z.Q. Mono-ubiquitination drives nuclear export of the human DCN1-like protein hDCNL. *J. Biol. Chem.* **2011**, *286*, 34060–34070. [CrossRef] [PubMed]

15. Yang, F.; Xing, Y.; Li, Y.; Chen, X.; Jiang, J.; Ai, Z.; Wei, Y. Monoubiquitination of cancer stem cell marker CD133 at lysine 848 regulates its secretion and promotes cell migration. *Mol. Cell. Biol.* **2018**, *38*. [CrossRef] [PubMed]

16. Dikic, I.; Wakatsuki, S.; Walters, K.J. Ubiquitin-binding domains from structures to functions. *Nat. Rev. Mol. Cell Biol.* **2009**, *10*, 659–671. [CrossRef]

17. Garner, T.P.; Strachan, J.; Shedden, E.C.; Long, J.E.; Cavey, J.R.; Shaw, B.; Layfield, R.; Searle, M.S. Independent interactions of ubiquitin-binding domains in a ubiquitin-mediated ternary complex. *Biochemistry* **2011**, *50*, 9076–9087. [CrossRef]

18. Paraskevopoulos, K.; Kriegenburg, F.; Tatham, M.H.; Rösner, H.I.; Medina, B.; Larsen, I.B.; Brandstrup, R.; Hardwick, K.G.; Hay, R.T.; Kragelund, B.B.; et al. Dss1 is a 26S proteasome ubiquitin receptor. *Mol. Cell* **2014**, *56*, 453–461. [CrossRef]

19. Fukushima, T.; Yoshihara, H.; Furuta, H.; Kamei, H.; Hakuno, F.; Luan, J.; Duan, C.; Saeki, Y.; Tanaka, K.; Iemura, S.I.; et al. Nedd4-induced monoubiquitination of IRS-2 enhances IGF signalling and mitogenic activity. *Nat. Commun.* **2015**, *6*, 6780. [CrossRef]

20. Pelzer, C.; Cabalzar, K.; Wolf, A.; Gonzalez, M.; Lenz, G.; Thome, M. The protease activity of the paracaspase MALT1 is controlled by monoubiquitination. *Nat. Immunol.* **2013**, *14*, 337–345. [CrossRef]

21. Woelk, T.; Oldrini, B.; Maspero, E.; Confalonieri, S.; Cavallaro, E.; Di Fiore, P.P.; Polo, S. Molecular mechanisms of coupled monoubiquitination. *Nat. Cell Biol.* **2006**, *8*, 1246–1254. [CrossRef] [PubMed]

22. Simicek, M.; Lievens, S.; Laga, M.; Guzenko, D.; Aushev, V.N.; Kalev, P.; Baietti, M.F.; Strelkov, S.V.; Gevaert, K.; Tavernier, J.; et al. The deubiquitylase USP33 discriminates between RALB functions in autophagy and innate immune response. *Nat. Cell Biol.* **2013**, *15*, 1220–1230. [CrossRef]

23. Ye, Y.; Klenerman, D.; Finley, D. N-Terminal ubiquitination of amyloidogenic proteins triggers removal of their oligomers by the proteasome holoenzyme. *J. Mol. Biol.* **2020**, *432*, 585–596. [CrossRef] [PubMed]

24. Isasa, M.; Katz, E.J.; Kim, W.; Yugo, V.; González, S.; Kirkpatrick, D.S.; Thomson, T.M.; Finley, D.; Gygi, S.P.; Crosas, B. Monoubiquitination of RPN10 regulates substrate recruitment to the proteasome. *Mol. Cell* **2010**, *38*, 733–745. [CrossRef] [PubMed]

25. Fallon, L.; Bélanger, C.M.L.; Corera, A.T.; Kontogiannea, M.; Regan-Klapisz, E.; Moreau, F.; Voortman, J.; Haber, M.; Rouleau, G.; Thorarinsdottir, T.; et al. A regulated interaction with the UIM protein Eps15 implicates parkin in EGF receptor trafficking and PI(3)K-Akt signalling. *Nat. Cell Biol.* **2006**, *8*, 834–842. [CrossRef] [PubMed]

26. Hoeller, D.; Crosetto, N.; Blagoev, B.; Raiborg, C.; Tikkanen, R.; Wagner, S.; Kowanetz, K.; Breitling, R.; Mann, M.; Stenmark, H.; et al. Regulation of ubiquitin-binding proteins by monoubiquitination. *Nat. Cell Biol.* **2006**, *8*, 163–169. [CrossRef]

27. Huang, S.; Tang, D.; Wang, Y. Monoubiquitination of syntaxin 5 regulates golgi membrane dynamics during the cell cycle. *Dev. Cell* **2016**, *38*, 73–85. [CrossRef]

28. Shin, D.; Na, W.; Lee, J.-H.; Kim, G.; Baek, J.; Park, S.H.; Choi, C.Y.; Lee, S. Site-specific monoubiquitination downregulates Rab5 by disrupting effector binding and guanine nucleotide conversion. *eLife* **2017**, *6*, e29154. [CrossRef]

29. Sasaki, A.T.; Carracedo, A.; Locasale, J.W.; Anastasiou, D.; Takeuchi, K.; Kahoud, E.R.; Haviv, S.; Asara, J.M.; Pandolfi, P.P.; Cantley, L.C. Ubiquitination of K-Ras enhances activation and facilitates binding to select downstream effectors. *Sci. Signal.* **2011**, *4*, ra13. [CrossRef]

30. Bigenzahn, J.W.; Collu, G.M.; Kartnig, F.; Pieraks, M.; Vladimer, G.I.; Heinz, L.X.; Sedlyarov, V.; Schischlik, F.; Fauster, A.; Rebsamen, M.; et al. LZTR1 is a regulator of RAS ubiquitination and signaling. *Science* **2018**, *362*, 1171–1177. [CrossRef]

31. Baker, R.; Lewis, S.M.; Sasaki, A.T.; Wilkerson, E.M.; Locasale, J.W.; Cantley, L.C.; Kuhlman, B.; Dohlman, H.G.; Campbell, S.L. Site-specific monoubiquitination activates Ras by impeding GTPase-activating protein function. *Nat. Struct. Mol. Biol.* **2013**, *20*, 46–52. [CrossRef] [PubMed]

32. Hershko, A. Ubiquitin: Roles in protein modification and breakdown. *Cell* **1983**, *34*, 11–12. [CrossRef]

33. Ikeda, F.; Crosetto, N.; Dikic, I. What determines the specificity and outcomes of ubiquitin signaling? *Cell* **2010**, *143*, 677–681. [CrossRef] [PubMed]

34. Buetow, L.; Huang, D.T. Structural insights into the catalysis and regulation of E3 ubiquitin ligases. *Nat. Rev. Mol. Cell Biol.* **2016**, *17*, 626–642. [CrossRef] [PubMed]

35. Berndsen, C.E.; Wolberger, C. New insights into ubiquitin E3 ligase mechanism. *Nat. Struct. Mol. Biol.* **2014**, *21*, 301–307. [CrossRef] [PubMed]

36. de Oliveira, J.F.; do Prado, P.F.V.; da Costa, S.S.; Sforça, M.L.; Canateli, C.; Ranzani, A.T.; Maschietto, M.; de Oliveira, P.S.L.; Otto, P.A.; Klevit, R.E.; et al. Mechanistic insights revealed by a UBE2A mutation linked to intellectual disability. *Nat. Chem. Biol.* **2019**, *15*, 62–70. [CrossRef]

37. Stewart, M.D.; Duncan, E.D.; Coronado, E.; DaRosa, P.A.; Pruneda, J.N.; Brzovic, P.S.; Klevit, R.E. Tuning BRCA1 and BARD1 activity to investigate RING ubiquitin ligase mechanisms. *Protein Sci.* **2017**, *26*, 475–483. [CrossRef]

38. George, A.J.; Hoffiz, Y.C.; Charles, A.J.; Zhu, Y.; Mabb, A.M. A comprehensive atlas of E3 ubiquitin ligase mutations in neurological disorders. *Front. Genet.* **2018**, *9*, 29. [CrossRef]

39. Popovic, D.; Vucic, D.; Dikic, I. Ubiquitination in disease pathogenesis and treatment. *Nat. Med.* **2014**, *20*, 1242–1253. [CrossRef]

40. Zhang, H.; Yu, D.S. One stone, two birds: CDK9-directed activation of UBE2A regulates monoubiquitination of both H2B and PCNA. *Cell Cycle* **2012**, *11*, 2418. [CrossRef]

41. Kim, J.; Guermah, M.; McGinty, R.K.; Lee, J.S.; Tang, Z.; Milne, T.A.; Shilatifard, A.; Muir, T.W.; Roeder, R.G. RAD6-mediated transcription-coupled H2B ubiquitylation directly stimulates H3K4 methylation in human cells. *Cell* **2009**, *137*, 459–471. [CrossRef] [PubMed]

42. Zhao, L.; Todd Washington, M. Translesion synthesis: Insights into the selection and switching of DNA polymerases. *Genes* **2017**, *8*, 24. [CrossRef] [PubMed]

43. Alpatov, R.; Lesch, B.J.; Nakamoto-Kinoshita, M.; Blanco, A.; Chen, S.; Stützer, A.; Armache, K.J.; Simon, M.D.; Xu, C.; Ali, M.; et al. A chromatin-dependent role of the fragile X mental retardation protein FMRP in the DNA damage response. *Cell* **2014**, *157*, 869–881. [CrossRef] [PubMed]

44. Juhász, S.; Elbakry, A.; Mathes, A.; Löbrich, M. ATRX promotes DNA repair synthesis and sister chromatid exchange during homologous recombination. *Mol. Cell* **2018**, *71*, 11–24. [CrossRef]

45. Budny, B.; Badura-Stronka, M.; Materna-Kiryluk, A.; Tzschach, A.; Raynaud, M.; Latos-Bielenska, A.; Ropers, H.H. Novel missense mutations in the ubiquitination-related gene UBE2A cause a recognizable X-linked mental retardation syndrome. *Clin. Genet.* **2010**, *77*, 541–551. [CrossRef]

46. Ma, D.; Tan, J.; Zhou, J.; Zhang, J.; Cheng, J.; Luo, C.; Liu, G.; Wang, Y.; Xu, Z. A novel splice site mutation in the UBE2A gene leads to aberrant mRNA splicing in a Chinese patient with X-linked intellectual disability type Nascimento. *Mol. Genet. Genomic Med.* **2019**, *7*, e976. [CrossRef]

47. Hampe, C.; Ardila-Osorio, H.; Fournier, M.; Brice, A.; Corti, O. Biochemical analysis of Parkinson's disease-causing variants of Parkin, an E3 ubiquitin - Protein ligase with monoubiquitylation capacity. *Hum. Mol. Genet.* **2006**, *15*, 2059–2075. [CrossRef]

48. Ham, S.J.; Lee, D.; Yoo, H.; Jun, K.; Shin, H.; Chung, J. Decision between mitophagy and apoptosis by Parkin via VDAC1 ubiquitination. *Proc. Natl. Acad. Sci. USA* **2020**, *117*, 4281–4291. [CrossRef]

49. Wade Harper, J.; Ordureau, A.; Heo, J.M. Building and decoding ubiquitin chains for mitophagy. *Nat. Rev. Mol. Cell Biol.* **2018**, *19*, 93–108. [CrossRef]

50. Ryan, B.J.; Hoek, S.; Fon, E.A.; Wade-Martins, R. Mitochondrial dysfunction and mitophagy in Parkinson's: From familial to sporadic disease. *Trends Biochem. Sci.* **2015**, *40*, 200–210. [CrossRef]

51. Shimura, H.; Hattori, N.; Kubo, S.I.; Mizuno, Y.; Asakawa, S.; Minoshima, S.; Shimizu, N.; Iwai, K.; Chiba, T.; Tanaka, K.; et al. Familial Parkinson disease gene product, parkin, is a ubiquitin-protein ligase. *Nat. Genet.* **2000**, *25*, 302–305. [CrossRef] [PubMed]

52. Kottemann, M.C.; Smogorzewska, A. Fanconi anaemia and the repair of Watson and Crick DNA crosslinks. *Nature* **2013**, *493*, 356–363. [CrossRef] [PubMed]

53. Alcón, P.; Shakeel, S.; Chen, Z.A.; Rappsilber, J.; Patel, K.J.; Passmore, L.A. FANCD2–FANCI is a clamp stabilized on DNA by monoubiquitination of FANCD2 during DNA repair. *Nat. Struct. Mol. Biol.* **2020**, *27*, 240–248. [CrossRef] [PubMed]

54. Jasti, V.P.; Sharma, S.; Althaus, I.; Weinstein, I.; Ramani, S. Repairing interstrand DNA crosslinks (ICL): Characterization of an ICL incision/lesion bypass polymerase complex regulated by the fanconi anemia pathway. *FASEB J.* **2017**, *31*.

55. Jones, N.; Makki, F.; Davis, G.; Marriott, A.; Wilson, J.; Kupfer, G. The Fanconi anaemia pathway and the repair of DNA lesions resulting from endogenous aldehydes and chemotherapeutic agents. *Mutagenesis* **2014**, *29*, 80.

56. Hira, A.; Yoshida, K.; Sato, K.; Okuno, Y.; Shiraishi, Y.; Chiba, K.; Tanaka, H.; Miyano, S.; Shimamoto, A.; Tahara, H.; et al. Mutations in the gene encoding the E2 conjugating enzyme UBE2T cause fanconi anemia. *Am. J. Hum. Genet.* **2015**, *96*, 1001–1007. [CrossRef]

57. Meetei, A.R.; De Winter, J.P.; Medhurst, A.L.; Wallisch, M.; Waisfisz, Q.; Van de Vrugt, H.J.; Oostra, A.B.; Yan, Z.; Ling, C.; Bishop, C.E.; et al. A novel ubiquitin ligase is deficient in Fanconi anemia. *Nat. Genet.* **2003**, *35*, 165–170. [CrossRef]

58. Garcia-Higuera, I.; Taniguchi, T.; Ganesan, S.; Meyn, M.S.; Timmers, C.; Hejna, J.; Grompe, M.; D'Andrea, A.D. Interaction of the Fanconi anemia proteins and BRCA1 in a common pathway. *Mol. Cell* **2001**, *7*, 249–262. [CrossRef]

59. Hirsch, B.; Shimamura, A.; Moreau, L.; Baldinger, S.; Hag-Alshiekh, M.; Bostrom, B.; Sencer, S.; D'Andrea, A.D. Association of biallelic BRCA2/FANCD1 mutations with spontaneous chromosomal instability and solid tumors of childhood. *Blood* **2004**, *103*, 2554–2559. [CrossRef]

60. Amit, I.; Yakir, L.; Katz, M.; Zwang, Y.; Marmor, M.D.; Citri, A.; Shtiegman, K.; Alroy, I.; Tuvia, S.; Reiss, Y.; et al. Tal, a Tsg101-specific E3 ubiquitin ligase, regulates receptor endocytosis and retrovirus budding. *Genes Dev.* **2004**, *18*, 1737–1752. [CrossRef]

61. Rossor, A.M.; Polke, J.M.; Houlden, H.; Reilly, M.M. Clinical implications of genetic advances in charcot-marie-tooth disease. *Nat. Rev. Neurol.* **2013**, *9*, 562–571. [CrossRef] [PubMed]

62. Nicolaou, P.; Cianchetti, C.; Minaidou, A.; Marrosu, G.; Zamba-Papanicolaou, E.; Middleton, L.; Christodoulou, K. A novel LRSAM1 mutation is associated with autosomal dominant axonal Charcot-Marie-Tooth disease. *Eur. J. Hum. Genet.* **2013**, *21*, 190–194. [CrossRef] [PubMed]

63. Reincke, M.; Sbiera, S.; Hayakawa, A.; Komada, M. Mutations in the deubiquitinase gene USP8 cause Cushing's disease. *Nat. Genet.* **2015**, *47*, 31–38. [CrossRef] [PubMed]

64. Ma, Z.Y.; Song, Z.J.; Chen, J.H.; Wang, Y.F.; Li, S.Q.; Zhou, L.F.; Mao, Y.; Li, Y.M.; Hu, R.G.; Zhang, Z.Y.; et al. Recurrent gain-of-function USP8 mutations in Cushing's disease. *Cell Res.* **2015**, *3*, 306–317. [CrossRef] [PubMed]

65. Crespo-Yàñez, X.; Aguilar-Gurrieri, C.; Jacomin, A.C.; Journet, A.; Mortier, M.; Taillebourg, E.; Soleilhac, E.; Weissenhorn, W.; Fauvarque, M.O. CHMP1B is a target of USP8/UBPY regulated by ubiquitin during endocytosis. *PLoS Genet.* **2018**, *14*, 1–24. [CrossRef] [PubMed]

66. Ballmann, C.; Thiel, A.; Korah, H.E.; Reis, A.-C.; Saeger, W.; Stepanow, S.; Köhrer, K.; Reifenberger, G.; Knobbe-Thomsen, C.B.; Knappe, U.J.; et al. USP8 mutations in pituitary Cushing adenomas—Targeted analysis by next-generation sequencing. *J. Endocr. Soc.* **2018**, *2*, 266–278. [CrossRef] [PubMed]

67. Row, P.E.; Liu, H.; Hayes, S.; Welchman, R.; Charalabous, P.; Hofmann, K.; Clague, M.J.; Sanderson, C.M.; Urbé, S. The MIT domain of UBPY constitutes a CHMP binding and endosomal localization signal required for efficient epidermal growth factor receptor degradation. *J. Biol. Chem.* **2007**, *282*, 30929–30937. [CrossRef]

68. Theodoropoulou, M.; Reincke, M.; Fassnacht, M.; Komada, M. Decoding the genetic basis of cushing's disease: Usp8 in the spotlight. *Eur. J. Endocrinol.* **2015**, *173*, M73–M83. [CrossRef]

69. Dufner, A.; Knobeloch, K.P. Ubiquitin-specific protease 8 (USP8/UBPy): A prototypic multidomain deubiquitinating enzyme with pleiotropic functions. *Biochem. Soc. Trans.* **2019**, 1867–1879. [CrossRef]

70. Sewduth, R.N.; Pandolfi, S.; Steklov, M.; Sheryazdanova, A.; Zhao, P. The Noonan syndrome gene Lztr1 controls cardiovascular function by regulating vesicular trafficking. *Circ. Res.* **2020**. [CrossRef]

71. Johnston, J.J.; van der Smagt, J.J.; Rosenfeld, J.A.; Pagnamenta, A.T.; Alswaid, A.; Baker, E.H.; Blair, E.; Borck, G.; Brinkmann, J.; Craigen, W.; et al. Autosomal recessive Noonan syndrome associated with biallelic LZTR1 variants. *Genet. Med.* **2018**, *20*, 1175–1185. [CrossRef] [PubMed]

72. Jacquinet, A.; Bonnard, A.; Capri, Y.; Martin, D.; Sadzot, B.; Bianchi, E.; Servais, L.; Sacré, J.P.; Cavé, H.; Verloes, A. Oligo-astrocytoma in LZTR1-related Noonan syndrome. *Eur. J. Med. Genet.* **2020**, *63*, 103617. [CrossRef] [PubMed]

73. Swaminathan, G.; Tsygankov, A.Y. The Cbl family proteins: Ring leaders in regulation of cell signaling. *J. Cell. Physiol.* **2006**, *209*, 21–43. [CrossRef] [PubMed]

74. Haglund, K.; Shimokawa, N.; Szymkiewicz, I.; Dikic, I. Cbl-directed monoubiquitination of CIN85 is involved in regulation of ligand-induced degradation of EGF receptors. *Proc. Natl. Acad. Sci. USA* **2002**, *99*, 12191–12196. [CrossRef]

75. Roberts, A.E.; Allanson, J.E.; Tartaglia, M.; Gelb, B.D. Noonan syndrome. *Lancet* **2013**, *381*, 333–342. [CrossRef]

76. Niemeyer, C.M.; Kang, M.W.; Shin, D.H.; Furlan, I.; Erlacher, M.; Bunin, N.J.; Bunda, S.; Finklestein, J.Z.; Sakamoto, K.M.; Gorr, T.A.; et al. Germline CBL mutations cause developmental abnormalities and predispose to juvenile myelomonocytic Leukemia. *Nat. Genet.* **2010**, *42*, 641. [CrossRef]

77. Pathak, A.; Pemov, A.; McMaster, M.L.; Dewan, R.; Ravichandran, S.; Pak, E.; Dutra, A.; Lee, H.J.; Vogt, A.; Zhang, X.; et al. Juvenile myelomonocytic leukemia due to a germline CBL Y371C mutation: 35-year follow-up of a large family. *Hum. Genet.* **2015**, *134*, 775–787. [CrossRef]

78. Venuprasad, K.; Huang, H.; Harada, Y.; Elly, C.; Subramaniam, M.; Spelsberg, T.; Su, J.; Liu, Y.C. The E3 ubiquitin ligase Itch regulates expression of transcription factor Foxp3 and airway inflammation by enhancing the function of transcription factor TIEG1. *Nat. Immunol.* **2008**, *9*, 245–253. [CrossRef]

79. Han, K.J.; Foster, D.; Harhaj, E.W.; Dzieciatkowska, M.; Hansen, K.; Liu, C.W. Monoubiquitination of survival motor neuron regulates its cellular localization and Cajal body integrity. *Hum. Mol. Genet.* **2016**, *25*, 1392–1405. [CrossRef]

80. Lohr, N.J.; Molleston, J.P.; Strauss, K.A.; Torres-Martinez, W.; Sherman, E.A.; Squires, R.H.; Rider, N.L.; Chikwava, K.R.; Cummings, O.W.; Morton, D.H.; et al. Human ITCH E3 ubiquitin ligase deficiency causes syndromic multisystem autoimmune disease. *Am. J. Hum. Genet.* **2010**, *86*, 447–453. [CrossRef]

81. Bruinsma, C.F.; Savelberg, S.M.C.; Kool, M.J.; Jolfaei, M.A.; Van Woerden, G.M.; Baarends, W.M.; Elgersma, Y. An essential role for UBE2A/HR6A in learning and memory and mGLUR-dependent long-term depression. *Hum. Mol. Genet.* **2016**, *25*, 1–8. [CrossRef] [PubMed]

82. Haddad, D.M.; Vilain, S.; Vos, M.; Esposito, G.; Matta, S.; Kalscheuer, V.M.; Craessaerts, K.; Leyssen, M.; Nascimento, R.M.P.; Vianna-Morgante, A.M.; et al. Mutations in the intellectual disability gene Ube2a cause neuronal dysfunction and impair parkin-dependent mitophagy. *Mol. Cell* **2013**, *50*, 831–843. [CrossRef] [PubMed]

83. Baple, E.L.; Chambers, H.; Cross, H.E.; Fawcett, H.; Nakazawa, Y.; Chioza, B.A.; Harlalka, G.V.; Mansour, S.; Sreekantan-Nair, A.; Patton, M.A.; et al. Hypomorphic PCNA mutation underlies a human DNA repair disorder. *J. Clin. Invest.* **2014**, *124*, 3137–3146. [CrossRef] [PubMed]

84. Hibbert, R.G.; Huang, A.; Boelens, R.; Sixma, T.K. E3 ligase Rad18 promotes monoubiquitination rather than ubiquitin chain formation by E2 enzyme Rad6. *Proc. Natl. Acad. Sci. USA* **2011**, *108*, 5590–5595. [CrossRef]

85. Pavri, R.; Zhu, B.; Li, G.; Trojer, P.; Mandal, S.; Shilatifard, A.; Reinberg, D. Histone H2B monoubiquitination functions cooperatively with FACT to regulate elongation by RNA polymerase II. *Cell* **2006**, *125*, 703–717. [CrossRef]

86. Minsky, N.; Shema, E.; Field, Y.; Schuster, M.; Segal, E.; Oren, M. Monoubiquitinated H2B is associated with the transcribed region of highly expressed genes in human cells. *Nat. Cell Biol.* **2008**, *10*, 483–488. [CrossRef]

87. Karpiuk, O.; Najafova, Z.; Kramer, F.; Hennion, M.; Galonska, C.; König, A.; Snaidero, N.; Vogel, T.; Shchebet, A.; Begus-Nahrmann, Y.; et al. The histone H2B monoubiquitination regulatory pathway is required for differentiation of multipotent stem cells. *Mol. Cell* **2012**, *45*, 705–713. [CrossRef]

88. Materne, P.; Vázquez, E.; Sánchez, M.; Yague-Sanz, C.; Anandhakumar, J.; Migeot, V.; Antequera, F.; Hermand, D. Histone H2B ubiquitylation represses gametogenesis by opposing RSC-dependent chromatin remodeling at the ste11 master regulator locus. *eLife* **2016**, *5*, e13500. [CrossRef]

89. Fuchs, G.; Shema, E.; Vesterman, R.; Kotler, E.; Wolchinsky, Z.; Wilder, S.; Golomb, L.; Pribluda, A.; Zhang, F.; Haj-Yahya, M.; et al. RNF20 and USP44 regulate stem cell differentiation by modulating H2B monoubiquitylation. *Mol. Cell* **2012**, *46*, 662–673. [CrossRef]

90. Tarcic, O.; Pateras, I.S.; Cooks, T.; Shema, E.; Kanterman, J.; Ashkenazi, H.; Boocholez, H.; Hubert, A.; Rotkopf, R.; Baniyash, M.; et al. RNF20 links histone H2B ubiquitylation with inflammation and inflammation-associated cancer. *Cell Rep.* **2016**, *14*, 1462–1476. [CrossRef]

91. Khan, N.L.; Graham, E.; Critchley, P.; Schrag, A.E.; Wood, N.W.; Lees, A.J.; Bhatia, K.P.; Quinn, N. Parkin disease: A phenotypic study of a large case series. *Brain* **2003**, *126*, 1279–1292. [CrossRef] [PubMed]

92. Alcalay, R.N.; Siderowf, A.; Ottman, R.; Caccappolo, E.; Mejia-Santana, H.; Tang, M.X.; Rosado, L.; Louis, E.; Ruiz, D.; Waters, C.; et al. Olfaction in Parkin heterozygotes and compound heterozygotes: The CORE-PD study. *Neurology* **2011**, *76*, 319–326. [CrossRef] [PubMed]

93. Hattori, N.; Mizuno, Y. Twenty years since the discovery of the parkin gene. *J. Neural Transm.* **2017**, *124*, 1037–1054. [CrossRef] [PubMed]

94. Itier, J.M.; Ibáñez, P.; Mena, M.A.; Abbas, N.; Cohen-Salmon, C.; Bohme, G.A.; Laville, M.; Pratt, J.; Corti, O.; Pradier, L.; et al. Parkin gene inactivation alters behaviour and dopamine neurotransmission in the mouse. *Hum. Mol. Genet.* **2003**, *12*, 2277–2291. [CrossRef]

95. Von Coelln, R.; Thomas, B.; Savitt, J.M.; Lim, K.L.; Sasaki, M.; Hess, E.J.; Dawson, V.L.; Dawson, T.M. Loss of locus coeruleus neurons and reduced startle in parkin null mice. *Proc. Natl. Acad. Sci. USA* **2004**, *101*, 10744–10749. [CrossRef]

96. Kubli, D.A.; Zhang, X.; Lee, Y.; Hanna, R.A.; Quinsay, M.N.; Nguyen, C.K.; Jimenez, R.; Petrosyans, S.; Murphy, A.N.; Gustafsson, Å.B. Parkin protein deficiency exacerbates cardiac injury and reduces survival following myocardial infarction. *J. Biol. Chem.* **2013**, *288*, 915–926. [CrossRef]

97. Xiong, H.; Wang, D.; Chen, L.; Yeun, S.C.; Ma, H.; Tang, C.; Xia, K.; Jiang, W.; Ronai, Z.; Zhuang, X.; et al. Parkin, PINK1, and DJ-1 form a ubiquitin E3 ligase complex promoting unfolded protein degradation. *J. Clin. Invest.* **2009**, *119*, 650–660. [CrossRef]

98. Moore, D.J.; West, A.B.; Dikeman, D.A.; Dawson, V.L.; Dawson, T.M. Parkin mediates the degradation-independent ubiquitination of Hsp70. *J. Neurochem.* **2008**, *105*, 1806–1819. [CrossRef]

99. Ramanagoudr-Bhojappa, R.; Carrington, B.; Ramaswami, M.; Bishop, K.; Robbins, G.M.; Jones, M.P.; Harper, U.; Frederickson, S.C.; Kimble, D.C.; Sood, R.; et al. Multiplexed CRISPR/Cas9-mediated knockout of 19 Fanconi anemia pathway genes in zebrafish revealed their roles in growth, sexual development and fertility. *PLoS Genet.* **2018**, *14*, e1007821. [CrossRef]

100. Rickman, K.A.; Lach, F.P.; Abhyankar, A.; Donovan, F.X.; Sanborn, E.M.; Kennedy, J.A.; Sougnez, C.; Gabriel, S.B.; Elemento, O.; Chandrasekharappa, S.C.; et al. Deficiency of UBE2T, the E2Ubiquitin ligase necessary for FANCD2 and FANCI ubiquitination, causes FA-T subtype of Fanconi anemia. *Cell Rep.* **2015**, *12*, 35–41. [CrossRef]

101. Tan, W.; van Twest, S.; Leis, A.; Bythell-Douglas, R.; Murphy, V.J.; Sharp, M.; Parker, M.W.; Crismani, W.M.; Deans, A.J. Monoubiquitination by the human Fanconi Anemia core complex clamps FANCI:FANCD2 on DNA in filamentous arrays. *eLife* **2020**, *9*, e54128. [CrossRef] [PubMed]

102. Cohn, M.A.; Kowal, P.; Yang, K.; Haas, W.; Huang, T.T.; Gygi, S.P.; D'Andrea, A.D. A UAF1-Containing Multisubunit Protein Complex Regulates the Fanconi Anemia Pathway. *Mol. Cell* **2007**, *28*, 786–797. [CrossRef] [PubMed]

103. Nijman, S.M.B.; Huang, T.T.; Dirac, A.M.G.; Brummelkamp, T.R.; Kerkhoven, R.M.; D'Andrea, A.D.; Bernards, R. The deubiquitinating enzyme USP1 regulates the fanconi anemia pathway. *Mol. Cell* **2005**, *17*, 331–339. [CrossRef] [PubMed]

104. Oestergaard, V.H.; Langevin, F.; Kuiken, H.J.; Pace, P.; Niedzwiedz, W.; Simpson, L.J.; Ohzeki, M.; Takata, M.; Sale, J.E.; Patel, K.J. Deubiquitination of FANCD2 is required for DNA crosslink repair. *Mol. Cell* **2007**, *28*, 798–809. [CrossRef] [PubMed]

105. Aerts, M.B.; Weterman, M.A.J.; Quadri, M.; Schelhaas, H.J.; Bloem, B.R.; Esselink, R.A.; Baas, F.; Bonifati, V.; van de Warrenburg, B.P. A LRSAM1 mutation links Charcot-Marie-Tooth type 2 to Parkinson's disease. *Ann. Clin. Transl. Neurol.* **2016**, *3*, 146–149. [CrossRef]

106. Bogdanik, L.P.; Sleigh, J.N.; Tian, C.; Samuels, M.E.; Bedard, K.; Seburn, K.L.; Burgess, R.W. Loss of the E3 ubiquitin ligase LRSAM1 sensitizes peripheral axons to degeneration in a mouse model of Charcot-Marie-Tooth disease. *DMM Dis. Model. Mech.* **2013**, *6*, 780–792. [CrossRef]

107. Cohen, M.; Persky, R.; Stegemann, R.; Hernández-Ramírez, L.C.; Zeltser, D.; Lodish, M.B.; Chen, A.; Keil, M.F.; Tatsi, C.; Faucz, F.R.; et al. Germline USP8 mutation associated with pediatric Cushing disease and other clinical features: A new syndrome. *J. Clin. Endocrinol. Metab.* **2019**, *104*, 4676–4682. [CrossRef]

108. Madshus, I.H.; Stang, E. Internalization and intracellular sorting of the EGF receptor: A model for understanding the mechanisms of receptor trafficking. *J. Cell Sci.* **2009**, *122*, 3433–3439. [CrossRef]

109. Paganini, I.; Chang, V.Y.; Capone, G.L.; Vitte, J.; Benelli, M.; Barbetti, L.; Sestini, R.; Trevisson, E.; Hulsebos, T.J.; Giovannini, M.; et al. Expanding the mutational spectrum of LZTR1 in schwannomatosis. *Eur. J. Hum. Genet.* **2015**, *23*, 963–968. [CrossRef]

110. Ruggieri, M.; Praticò, A.D.; Evans, D.G. Diagnosis, management, and new therapeutic options in childhood neurofibromatosis Type 2 and related forms. *Semin. Pediatr. Neurol.* **2015**, *22*, 240–258. [CrossRef]

111. Castel, P.; Cheng, A.; Cuevas-Navarro, A.; Everman, D.B.; Papageorge, A.G.; Simanshu, D.K.; Tankka, A.; Galeas, J.; Urisman, A.; McCormick, F. RIT1 oncoproteins escape LZTR1-mediated proteolysis. *Science* **2019**, *363*, 1226–1230. [CrossRef] [PubMed]

112. Rathinam, C.; Flavella, R.A. c-Cbl deficiency leads to diminished lymphocyte development and functions in an age-dependent manner. *Proc. Natl. Acad. Sci. USA* **2010**, *107*, 8316–8321. [CrossRef] [PubMed]

113. Rafiq, K.; Kolpakov, M.A.; Seqqat, R.; Guo, J.; Guo, X.; Qi, Z.; Yu, D.; Mohapatra, B.; Zutshi, N.; An, W.; et al. C-Cbl inhibition improves cardiac function and survival in response to myocardial ischemia. *Circulation* **2014**, *129*, 2031–2043. [CrossRef] [PubMed]

114. Mosesson, Y.; Shtiegman, K.; Katz, M.; Zwang, Y.; Vereb, G.; Szollosi, J.; Yarden, Y. Endocytosis of receptor tyrosine kinases is driven by monoubiquitylation, not polyubiquitylation. *J. Biol. Chem.* **2003**, *278*, 21323–21326. [CrossRef] [PubMed]

115. Kometani, K.; Yamada, T.; Sasaki, Y.; Yokosuka, T.; Saito, T.; Rajewsky, K.; Ishiai, M.; Hikida, M.; Kurosaki, T. CIN85 drives B cell responses by linking BCR signals to the canonical NF-κB pathway. *J. Exp. Med.* **2011**, *208*, 1447–1457. [CrossRef]

116. Perry, W.L.; Hustad, C.M.; Swing, D.A.; Norene O'Sullivan, T.; Jenkins, N.A.; Copeland, N.G. The itchy locus encodes a novel ubiquitin protein ligase that is disrupted in a(18H) mice. *Nat. Genet.* **1998**, *18*, 143–146. [CrossRef]

117. Chang, L.; Kamata, H.; Solinas, G.; Luo, J.L.; Maeda, S.; Venuprasad, K.; Liu, Y.C.; Karin, M. The E3 ubiquitin ligase itch couples JNK activation to TNFα-induced cell death by inducing c-FLIPL turnover. *Cell* **2006**, *124*, 601–613. [CrossRef]

118. Heissmeyer, V.; Macián, F.; Im, S.H.; Varma, R.; Feske, S.; Venuprasad, K.; Gu, H.; Liu, Y.C.; Dustin, M.L.; Rao, A. Calcineurin imposes T cell unresponsiveness through targeted proteolysis of signaling proteins. *Nat. Immunol.* **2004**, *5*, 255–265. [CrossRef]

119. Shembade, N.; Harhaj, N.S.; Parvatiyar, K.; Copeland, N.G.; Jenkins, N.A.; Matesic, L.E.; Harhaj, E.W. The E3 ligase Itch negatively regulates inflammatory signaling pathways by controlling the function of the ubiquitin-editing enzyme A20. *Nat. Immunol.* **2008**, *9*, 254–262. [CrossRef]

120. Wang, J.; Li, L.; Jiang, H. C-Cbl inhibition: A novel therapeutic approach for attenuating myocardial ischemia and reperfusion injury. *Int. J. Cardiol.* **2015**, *186*, 50–51. [CrossRef]

121. Gabrielsen, M.; Buetow, L.; Nakasone, M.A.; Ahmed, S.F.; Sibbet, G.J.; Smith, B.O.; Zhang, W.; Sidhu, S.S.; Huang, D.T. A general strategy for discovery of inhibitors and activators of RING and U-box E3 ligases with ubiquitin variants. *Mol. Cell* **2017**, *68*, 456–470. [CrossRef] [PubMed]

122. Rossi, M.; Rotblat, B.; Ansell, K.; Amelio, I.; Caraglia, M.; Misso, G.; Bernassola, F.; Cavasotto, C.N.; Knight, R.A.; Ciechanover, A.; et al. High throughput screening for inhibitors of the HECT ubiquitin E3 ligase ITCH identifies antidepressant drugs as regulators of autophagy. *Cell Death Dis.* **2014**, *5*, e1203. [CrossRef] [PubMed]

123. Cornwell, M.J.; Thomson, G.J.; Coates, J.; Belotserkovskaya, R.; Waddell, I.D.; Jackson, S.P.; Galanty, Y. Small-Molecule Inhibition of UBE2T/FANCL-Mediated Ubiquitylation in the Fanconi Anemia Pathway. *ACS Chem. Biol.* **2019**, *14*, 2148–2154. [CrossRef] [PubMed]

124. Theodoropoulou, M.; Perez-Rivas, L.; Albani, A.; Stalla, G.; Buchfelder, M.; Flitsch, J.; Honegger, J.; Rachinger, W.; Reincke, M. Efficacy of pharmacological USP8 inhibition in human Cushing's disease tumours in vitro. *Endocr. Abstr.* **2018**, *56*. [CrossRef]

125. Humbert, O.; Davis, L.; Maizels, N. Targeted gene therapies: Tools, applications, optimization. *Crit. Rev. Biochem. Mol. Biol.* **2012**, *47*, 264–281. [CrossRef]

126. Scott, H.L.; Buckner, N.; Fernandez-Albert, F.; Pedone, E.; Postiglione, L.; Shi, G.; Allen, N.; Wong, L.F.; Magini, L.; Marucci, L.; et al. A dual druggable genome-wide siRNA and compound library screening approach identifies modulators of parkin recruitment to mitochondria. *J. Biol. Chem.* **2020**, *295*, 3285–3300. [CrossRef]
127. Miller, S.; Muqit, M.M.K. Therapeutic approaches to enhance PINK1/Parkin mediated mitophagy for the treatment of Parkinson's disease. *Neurosci. Lett.* **2019**, *705*, 7–13. [CrossRef]

International Journal of
Molecular Sciences

Review

Deubiquitinase MYSM1 in the Hematopoietic System and beyond: A Current Review

Amanda Fiore [1,2], **Yue Liang** [1,2], **Yun Hsiao Lin** [1,2], **Jacky Tung** [1,2], **HanChen Wang** [1,2,3], **David Langlais** [2,3,4] **and Anastasia Nijnik** [1,2,*]

1   Department of Physiology, McGill University, Montreal, QC 3655, Canada;
    amanda.fiore@mail.mcgill.ca (A.F.); yue.liang@mail.mcgill.ca (Y.L.); yun.h.lin@mail.mcgill.ca (Y.H.L.);
    lin.tung@mail.mcgill.ca (J.T.); han.c.wang@mail.mcgill.ca (H.W.)
2   Research Centre on Complex Traits, McGill University, Montreal, QC 3649, Canada; david.langlais@mcgill.ca
3   Department of Human Genetics, McGill University, Montreal, QC 3640, Canada
4   McGill University Genome Centre, Montreal, QC 740, Canada
*   Correspondence: anastasia.nijnik@mcgill.ca; Tel.: +1-514-398-5567

Received: 4 April 2020; Accepted: 20 April 2020; Published: 24 April 2020

Abstract: MYSM1 has emerged as an important regulator of hematopoietic stem cell function, blood cell production, immune response, and other aspects of mammalian physiology. It is a metalloprotease family protein with deubiquitinase catalytic activity, as well as SANT and SWIRM domains. MYSM1 normally localizes to the nucleus, where it can interact with chromatin and regulate gene expression, through deubiquitination of histone H2A and non-catalytic contacts with other transcriptional regulators. A cytosolic form of MYSM1 protein was also recently described and demonstrated to regulate signal transduction pathways of innate immunity, by promoting the deubiquitination of TRAF3, TRAF6, and RIP2. In this work we review the current knowledge on the molecular mechanisms of action of MYSM1 protein in transcriptional regulation, signal transduction, and potentially other cellular processes. The functions of MYSM1 in different cell types and aspects of mammalian physiology are also reviewed, highlighting the key checkpoints in hematopoiesis, immunity, and beyond regulated by MYSM1. Importantly, mutations in MYSM1 in human were recently linked to a rare hereditary disorder characterized by leukopenia, anemia, and other hematopoietic and developmental abnormalities. Our growing knowledge of MYSM1 functions and mechanisms of actions sheds important insights into its role in mammalian physiology and the etiology of the MYSM1-deficiency disorder in human.

Keywords: deubiquitinase; hematopoiesis; hematopoietic stem cells; immune response; regulation of gene expression

---

1. Introduction

1.1. Overview of MYSM1 Protein Structure and Catalytic Activity

In recent years, Myb-like, SWIRM, and MPN domains 1 (MYSM1) has emerged as an essential regulator of hematopoiesis, immunity, and other aspects of mammalian physiology. It is primarily a nuclear chromatin-interacting protein, with orthologues found only in the vertebrate species, indicating more recent evolutionary origins and suggesting specialized biological functions.

Structurally MYSM1 comprises SANT, SWIRM, and MPN domains (Figure 1). The SANT domain of MYSM1 is structurally similar to the DNA-binding domain of transcription factor cMYB [1,2]. It can bind to DNA in vitro [2] and is required for MYSM1 association with histones in vivo [3], however, whether the DNA-binding is sequence specific is not known. The SWIRM domain of MYSM1, in contrast, does not have direct DNA binding activity [4], and although it is a common domain-type for chromatin-associated proteins [4,5], it was shown to be dispensable for MYSM1 interactions with histones, at least in some systems [3]. Importantly, the MPN metalloprotease domain of MYSM1 is the catalytic domain, characterized by Zn2$^+$-binding, the JAMM motif with consensus sequence EXnHSHX$_7$SX$_2$D [6], and deubiquitinase catalytic activity (DUB) [6].

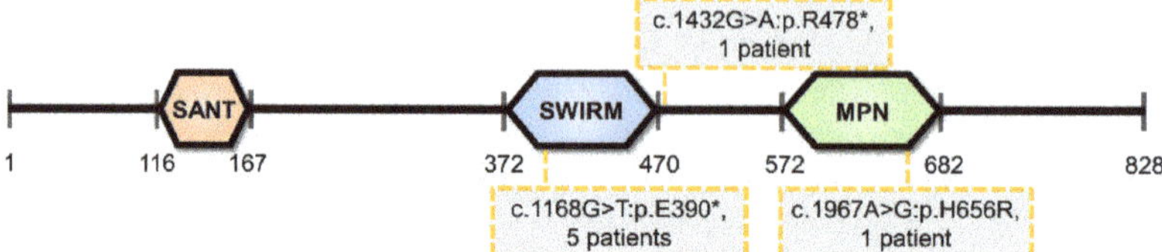

Figure 1. MYSM1 Protein: Domain Structure and Mutations in Human Patients. MYSM1 comprises SANT, SWIRM, and MPN domains. MYSM1 mutations reported in human MYSM1-deficiency syndrome patients include the p.H656R substitution within the critical JAMM-motif of the catalytic domain, and the nonsense variants p.E390* and p.Arg478* truncating MYSM1 protein upstream of the catalytic domain. All mutations in the patients are in a homozygous state.

Our knowledge of the physiological substrates of MYSM1 DUB-catalytic activity continues to expand. Histone H2A, monoubiquitinated at K119, was the first substrate of MYSM1 to be described (H2AK119ub) [7]. More recently MYSM1 was also demonstrated to cleave K63-polyubiquitin chains on cytosolic substrates TRAF3, TRAF6, and RIP2 [3,8]. MYSM1 catalytic activity against polyubiquitin chains of different geometries [9] was subsequently tested using in vitro assays, and MYSM1 was shown to cleave M1, K6, and K27 chains, but was catalytically inactive against K11, K29, K33, and K48 chains [8]. This suggests that our knowledge of MYSM1 substrates remains incomplete and further targets of its catalytic activity may be discovered in the future.

Despite some recent advances, the precise role of MYSM1 in shaping the epigenetic landscape of different mammalian cells remains poorly understood. Histone H2AK119ub is a repressive epigenetic mark deposited on chromatin primarily by polycomb repressive complex 1 (PRC1), for long-term gene silencing during cell differentiation and lineage specification [10–12]. In contrast, histone H2BK120ub is an epigenetic mark of transcriptionally active genes [13,14]. The catalytic activity of MYSM1 against histone H2AK119ub, but not H2BK120ub, therefore indicates its primary role as an activator of gene expression [7]. However, MYSM1 is one of many DUBs in mammalian cells that can catalyze H2AK119ub deubiquitination [15]. Other major DUBs with specificity for histone H2AK119ub over H2BK120ub are BAP1 and USP16, while USP3, USP12, USP22, and USP44 deubiquitinate both H2AK119ub and H2BK120ub, as well as different non-histone substrates [15]. How MYSM1 cooperates with the other DUBs to regulate the genome-wide landscape of histone H2A ubiquitination and the gene expression profiles of different mammalian cell types remains poorly understood. Furthermore, MYSM1 activities against polyubiquitinated forms of histone H2A at DNA damage foci [15,16] or against other less well-characterized ubiquitinated histones and histone variants [11,17,18] to our knowledge have not been investigated.

1.2. MYSM1-Deficiency in Human and Mouse: Mechanistic Insights and Biomedical Significance

Characterization of *Mysm1*-deficient mouse strains has been instrumental in revealing the essential functions of MYSM1 in different aspects of mammalian physiology [19,20]. *Mysm1$^{-/-}$* mice exhibit partial embryonic lethality, growth retardation, skeletal and coat pigmentation defects, and complex hematopoietic and immune phenotypes [19], while no phenotypic abnormalities are reported in *Mysm1$^{+/-}$* heterozygous mice. The International Mouse Phenotyping Consortium (IMPC) [21–23] provides extensive primary data on these phenotypes [24]. This and the far-reaching data from recent publications will be covered in depth in subsequent sections of this review.

MYSM1 mutations in patients with an inherited bone marrow failure syndrome (IBMFS) highlight the biomedical importance of understanding MYSM1 activities and functions. Five patients with homozygous *MYSM1* mutations were characterized in depth [25–27], carrying either a p.H656R substitution within the critical JAMM-motif of the catalytic domain, or a nonsense variant p.E390* truncating MYSM1 protein upstream of the catalytic domain [25–27] (Figure 1). The variants were therefore predicted to severely impact or fully abolish MYSM1 catalytic activity. All five patients exhibit anemia and leukopenia [25–27], in some cases associated with growth retardation [26,27], developmental malformations [27], and neurodevelopmental delay [27]. B cells were severely depleted in all patients [25–27], correlating with reduced serum IgM levels [25–27], although IgG levels were reduced in only one of the patients [26]. Most patients also showed a reduction in NK cell numbers, while neutropenia was noted in three patients, and T cell depletion was observed in two patients [25–27]. Skeletal and craniofacial abnormalities reported in two patients included limb shortening (rhizomelia) and midface hypoplasia [27]. Additionally, a sixth patient also carrying the p.E390* *MYSM1* mutation in a homozygous state was reported to have neutrophilic panniculitis, as well as reduced B cell count, anemia, and a mild growth retardation [28]. Finally, a novel homozygous mutation p.R478* in *MYSM1* was recently identified via whole-exome sequencing in a patient diagnosed with Diamond-Blackfan anemia, a disorder characterized by anemia and to a lesser extent other hematological and developmental abnormalities [29].

Multilineage defects in hematopoiesis in the *MYSM1*-deficient patients suggest impaired hematopoietic stem cell (HSC) function, and this was further supported by several lines of evidence. Thus, a reduction in the frequency of CD34$^+$ hematopoietic stem and progenitor cells was reported in the patient homozygous for the *MYSM1*:c.1967A>G:p.H656R mutation [26]. Remarkably, this patient experienced a spontaneous genetic reversion, restoring the normal sequence of one *MYSM1*-allele in hematopoietic cells and resulting in a correction of all immunohematological defects [26]. Assuming that the reversion mutation originated in a single HSC, we can conclude that restoration of MYSM1 function provided a very strong selective advantage and allowed one HSC clone to reconstitute normal wild-type hematopoiesis in competition with a large pool of MYSM1-deficient stem and progenitor cells. This attests to the true importance of MYSM1 for normal HSC function in human [26]. Consistent with this, immunohematological defects in several other *MYSM1*-syndrome patients were successfully cured via allogeneic hematopoietic stem cell transplantation (HSCT) [27].

The high MYSM1 protein homology between human and mouse (87%) and the similarities in the phenotypes of *MYSM1*-deficiency between the species indicate that studies of MYSM1 activities and functions in mouse models can provide important insights into the etiology of *MYSM1*-deficiency syndrome in human. The emerging data from many murine studies published over the past 8 years will be reviewed below.

2. Molecular Functions of MYSM1 Protein

2.1. MYSM1 Is a Transcriptional Regulator in Hematopoietic Lineage Specification

MYSM1 was originally characterized as an epigenetic regulator that promotes the expression of androgen receptor target genes in prostate cancer cell lines, through deubiquitination of histone H2AK119ub [7]. With the discovery of major hematopoietic dysfunction in *Mysm1*-deficient mouse models, most subsequent studies of MYSM1 in transcriptional regulation have focused on hematopoietic cells. These studies elucidated the role of MYSM1 in the de-repression of a range of genes essential for normal stem cell differentiation and lineage specification in hematopoiesis. The genes de-repressed by MYSM1 include: *Ebf1* in B cell progenitors [20], *Pax5* in naïve B cells [30], *miR150* in B1a cells [31], *Id2* in NK cell progenitors [32], *Flt3* in dendritic cell precursors [33], and *Gfi1* in hematopoietic stem and progenitor cells [34] (Figure 2). At many of these loci, loss of MYSM1 resulted in increased binding of PRC1 complex proteins and elevated levels of histone H2AK119ub [20,30,32–34], suggesting that MYSM1 antagonizes PRC1-mediated histone ubiquitination and transcriptional repression.

Increased levels of histone H3K27me3 were also observed at several loci [20,30,33,34], suggesting indirect cross-talk between MYSM1-deficiency and PRC2 complex activity. Regrettably, only a few studies specifically tested whether the catalytic activity of MYSM1 is essential for its transcriptional activity [7].

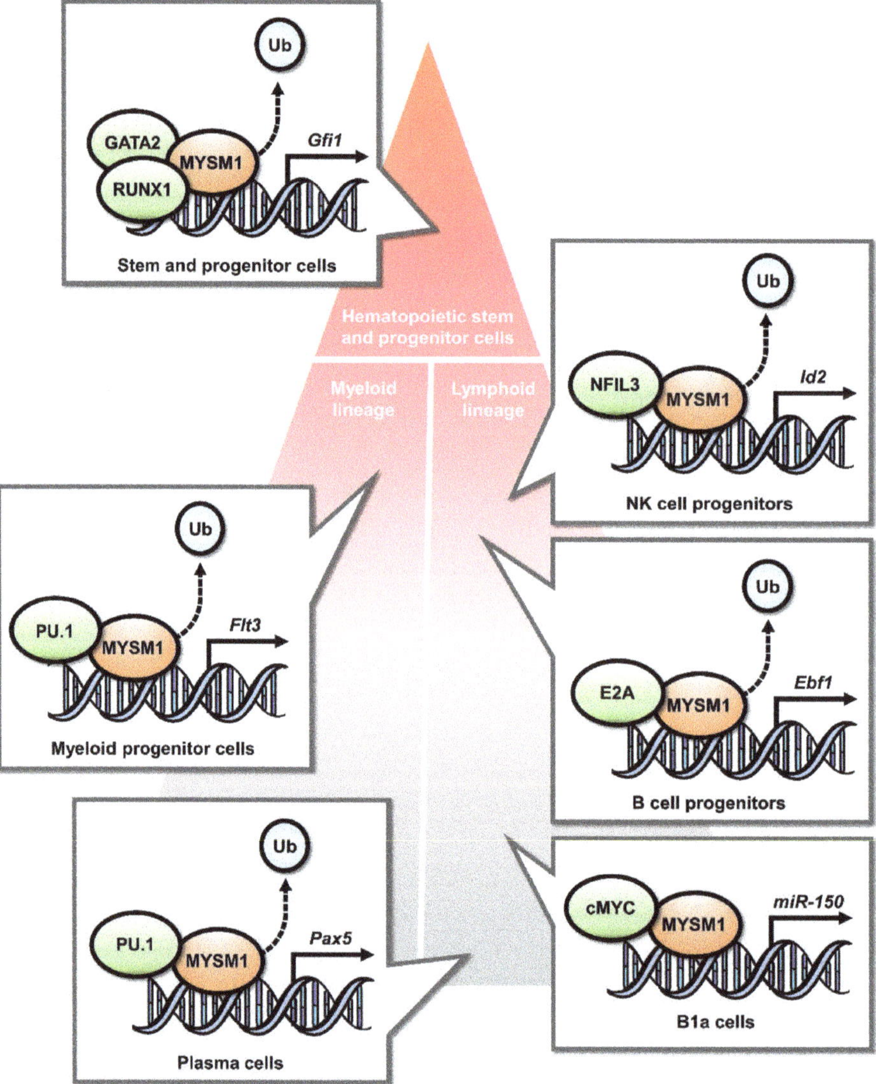

Figure 2. Overview of the reported roles of MYSM1 in the transcriptional regulation of hematopoiesis. MYSM1 was shown to de-repress the expression of *Ebf1* in B cell progenitors [20], *Pax5* in naïve B cells [30], *miR150* in B1a cells [31], *Id2* in NK cell progenitors [32], *Flt3* in dendritic cell precursors [33], and *Gfi1* in hematopoietic stem and progenitor cells [34], through deubiquitination of histone H2AK119ub, and interactions with hematopoietic transcription factors E2A [20], PU.1 [30,33], GATA2 [34], RUNX1 [34], cMYC [31], and NFIL3 [32].

MYSM1 was also shown to interact with many transcriptional regulators through co-immunoprecipitation assays, including essential hematopoietic transcription factors E2A [20], PU.1 [30,33], GATA2 [34], RUNX1 [34], cMYC [31], and NFIL3 [32], as well as the histone acetylase pCAF [7], and the BRG1 and BRM catalytic components of the chromatin remodeling complex SWI/SNF [20] (Figure 2). In these studies, the loss of MYSM1 was commonly associated with reduced recruitment of these transcription factors to specific gene promoters and repression of target gene expression [7,20,30–34]. In contrast, pCAF was suggested to act upstream of MYSM1, as MYSM1 catalytic activity was enhanced on hyperacetylated chromatin but inhibited with pCAF-knockdown [7]. However, many unanswered questions remain about the structural basis and functional significance of these MYSM1 protein interactions. Although the interactions were validated through co-immunoprecipitation [7,20,30–34], it is unclear whether the binding is direct and its structural basis remains unknown. It is also unclear whether the binding of MYSM1 to the transcription factors is needed for their enhanced recruitment to specific MYSM1-regulated genes, or whether MYSM1 enhances their recruitment indirectly through chromatin opening and remodeling. Finally, the above studies are restricted to a small number of MYSM1 regulated loci, and the relevance of the proposed mechanisms to MYSM1-transcriptional regulation on a genome-wide scale remains unknown.

2.2. MYSM1 Is a Regulator of Signal Transduction Pathways in Innate Immunity

Recent studies indicated an alternative function for MYSM1 as a regulator of the signal transduction pathways of innate immunity in the cytosol, independent of MYSM1-mediated regulation of gene expression at chromatin [3,8]. Thus, a cytosolic pool of MYSM1 protein is produced transiently in mouse macrophages, in response to inflammatory stimulation or infection, involving de novo protein synthesis and subsequent proteasomal degradation [3]. This pool of MYSM1 antagonizes K63-polyubiquitination of TRAF3 and TRAF6 in TLR signal transduction pathways, and K63/K27/M1-polyubiquitination of RIP2 downstream of NOD2 [3,8] (Figure 3). Interestingly, the SWIRM and MPN domains of MYSM1 are required for this activity, but the SANT domain is dispensable [3,8]. Loss of cytosolic MYSM1 function was implicated in the enhanced production of inflammatory cytokine and type-I interferons seen in *Mysm1*[-/-] macrophages. It also likely accounts for the increased susceptibility to septic shock and peritonitis [3,8], and enhanced clearance of viral infection in mice with systemic or myeloid lineage-restricted *Mysm1*-deletion [3].

Whether MYSM1 has cytosolic functions in other cell types and in regulation of other signaling pathways remains unknown and merits further investigation. Of note, *Mysm1*[-/-] adipose-derived stem cells were also recently shown to produce higher levels of inflammatory cytokines and to exacerbate the pathology of colitis in mouse models, although MYSM1 regulation of *miR150*-expression was proposed as the underlying mechanism [35,36]. The exact mechanisms mediating MYSM1 retention in the cytosol also remain to be further characterized. Finally, the role of MYSM1 in the regulation of signal transduction in human innate immune response requires further investigation, including the impact on the symptoms and etiology of the human *MYSM1*-deficiency syndrome.

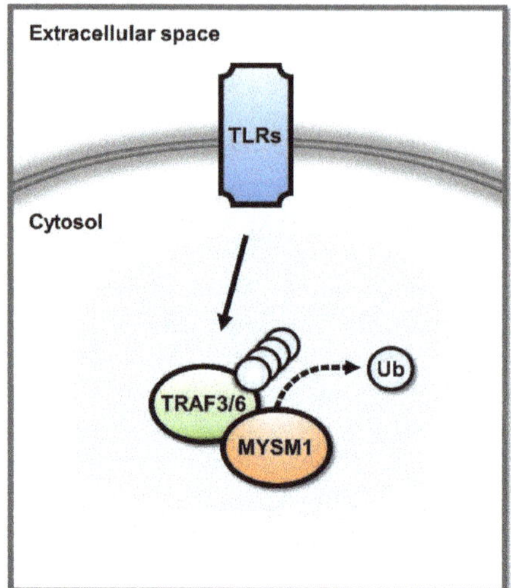
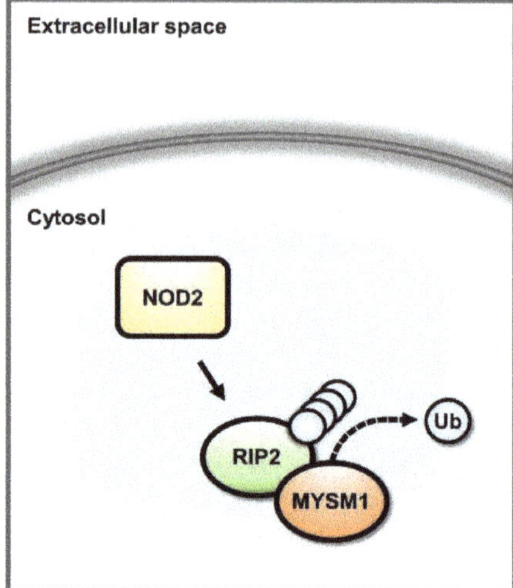

Figure 3. Overview of the cytosolic functions of the MYSM1 protein in the regulation of signal transduction in innate immunity and inflammatory responses in macrophages. MYSM1 was shown to promote deubiquitination of TRAF3, TRAF6, and RIP2 in the signal transduction pathways of TLR and NOD2 pattern recognition receptors, thus, repressing the innate immune and inflammatory responses of macrophages [3,8].

2.3. Putative Roles of MYSM1 in DNA Repair

Given the prominent roles of ubiquitination of histone H2A and other proteins in DNA repair [15,16], links between MYSM1 and DNA repair were suggested in several studies [37]. A proteomic screen identified MYSM1 among a large set of putative substrates phosphorylated by ATM and ATR in response to irradiation induced DNA damage [38]. A comprehensive screen of DUB proteins further demonstrated MYSM1 localization to DNA damage foci, and showed that *Mysm1*-knockdown in cell lines triggers spontaneous accumulation of DNA double strand breaks (DSB) and results in impaired repair of induced DNA damage [39]. Increased γH2AX levels were also seen in hematopoietic cells from *MYSM1*-deficiency syndrome patients [27], suggesting increased levels of DNA damage. Furthermore, *Mysm1*-deficient mice were hypersensitive to whole-body ionizing radiation [34] and accumulated increased levels of DNA damage in UV-treated skin [40]. We also observed spontaneous accrual of γH2AX in *Mysm1*-deficient mouse hematopoietic stem and progenitor cells [19], however, this was alleviated in *Mysm1$^{-/-}$p53$^{-/-}$* and *Mysm1$^{-/-}$Puma$^{-/-}$* double-knockout mouse strains [41], suggesting that it was a result of elevated cell apoptosis, rather than of a DNA repair deficiency. To our knowledge, MYSM1 effects on K63-polyubiquitination of histone H2A or other proteins involved in DNA repair or DNA damage response have not been reported. Although BRCC36 is known to be the major DUB with activity on polyubiquitinated histone H2A at DNA damage foci [15,16], a possible role for MYSM1 does merit further investigation.

2.4. MYSM1 and the p53 Stress Response Pathway

p53 is a transcription factor commonly called the "guardian of the genome" due to its central role in regulating cellular responses to stress, including cell cycle arrest, senescence, apoptosis, DNA repair, and autophagy [42,43]. Although p53 activation was classically studied in the context of DNA damage and oncogene activation, a variety of cellular stresses are now known to trigger its activity.

Several studies noted an increase in p53 protein levels and activation of p53-regulated stress response genes in *Mysm1*-deficient mouse hematopoietic stem and progenitor cells [19,44,45]. This was also observed after an inducible deletion of *Mysm1* in adult mice, indicating that MYSM1 is required constitutively to repress p53-activation [46].

Importantly, the silencing of p53 stress response in *Mysm1$^{-/-}$p53$^{-/-}$* double-knockout mice resulted in a visible rescue of *Mysm1$^{-/-}$* phenotype, restoring mouse body size, normal morphology of the hind-limbs, tail and skin [44,45,47,48], the cellularity of the blood and lymphoid organs, and the numbers of hematopoietic stem and progenitor cells in the mouse bone marrow [44,45]. *Mysm1$^{-/-}$p53$^{-/-}$* HSCs were fully capable of reconstituting hematopoiesis following transplantation, indicating restoration of normal HSC functions [44]. Altogether, these studies indicated that p53 activation is the common mechanism driving hematopoietic dysfunction and other phenotypic abnormalities in *Mysm1*-deficiency.

Further characterization of the mouse strains deficient for the important mediators of p53-dependent apoptosis *Bbc3*/PUMA and cell cycle arrest *Cdkn1a*/p21 elucidated the specific pathways downstream of p53 mediating the *Mysm1*-deficiency phenotype. In particular, these studies demonstrated a partial rescue of *Mysm1*-phenotypes in *Mysm1$^{-/-}$Bbc3$^{-/-}$* but not in *Mysm1$^{-/-}$Cdkn1a$^{-/-}$* animals [41]. The studies therefore concluded that the p53-driven over-expression of *Bbc3*/PUMA drives the apoptosis of hematopoietic multipotent progenitor cells (MPPs) and the depletion myeloid lineage cells, but not the arrest in lymphocyte development in *Mysm1*-deficiency [41].

The molecular mechanisms connecting *Mysm1*-deficiency and p53-activation remain somewhat controversial. Interaction between MYSM1 and p53 proteins was demonstrated in hematopoietic cell lines through co-immunoprecipitation (co-IP) [41], and MYSM1 binding to the promoters of several p53 stress response genes was also reported, suggesting possible mechanisms for p53 regulation by MYSM1. Thus, in thymocytes, MYSM1 binds at the *Cdkn2a* locus that encodes upstream regulators of p53 activation p16/INK4A and p19/ARF, and the expression and protein levels p19/ARF are elevated in *Mysm1*-deficient thymocytes [45]. In pro-B cell lines, MYSM1 binds at the promoters of p53-regulated genes *Bbc3*/PUMA and *Cdkn1a*/p21 that encode important mediators of p53-dependent apoptosis and cell cycle arrest, respectively [41]. MYSM1-binding sites at these promoters coincide with known p53-binding sites, and *Mysm1*-knockdown results in increased p53 binding, increased levels of activating histone marks H3K27ac and H3K4me3, and increased gene expression [41], suggesting that MYSM1 is a negative regulator of these loci. It is important to note however that these experiments were limited to a small set of putative MYSM1-regulated loci [41,45], and thus do not rule out existence of other mechanistic links between *Mysm1*-deficiency and p53 activation. The studies were also conducted primarily in lymphoid precursor cells [41,45], while a conditional deletion of *Mysm1* in these cells in vivo has minimal impact on lymphocyte development [49,50], stressing the importance of conducting further mechanistic studies on the cross-talk between MYSM1 and p53 in earlier hematopoietic precursors and HSCs.

3. MYSM1 Regulated Checkpoints in Mammalian Physiology

3.1. Essential and Cell Intrinsic Functions of MYSM1 in Hematopoietic Stem Cells

Many lines of evidence support the essential role of MYSM1 in hematopoiesis and HSCs. Thus, multilineage defects in hematopoiesis are a common feature of *Mysm1*-deficiency in human patients and mouse models, with *Mysm1$^{-/-}$* mice exhibiting a severe depletion of B cells and NK cells, and a milder reduction in T cells, myeloid leukocytes, and erythroid lineage cells [19]. Importantly, MYSM1 regulates hematopoiesis through cell intrinsic mechanisms, as demonstrated by its expression in HSCs and other hematopoietic cells [34], mouse-to-mouse bone marrow transplantation experiments [19,34], and the selective deletions of *Mysm1* within hematopoietic cells in Cre/loxP mouse models [51,52]. Further mechanistic studies in mice established that *Mysm1*-deficiency preserves HSC numbers, as defined by the cell surface markers Lin$^-$cKit$^+$Sca1$^+$CD150$^+$CD48$^-$CD34$^-$Flt3$^-$, however, these cells exhibit loss quiescence and fail to engraft hematopoiesis following transplantation, indicating loss of HSC function [19,34]. The loss of HSC function is observed in both *Mysm1$^{-/-}$* fetal liver and adult bone marrow [46], indicating the importance of MYSM1 in hematopoiesis through multiple stages of ontogenesis. Its role specifically in the emergence of HSCs in embryogenesis to our knowledge has not been specifically investigated. Importantly, data from human patients also supports the cell-intrinsic role of MYSM1 in HSC function, as shown by the correction of all immunohematological defects in one of the patients following a spontaneous genetic reversion of the *MYSM1*:p.H656R mutation in the patient's hematopoietic cells [26].

At the molecular level, MYSM1 was shown to regulate HSC function in part through the induction of the gene encoding transcription factor GFI1 [34]. Indeed, *Gfi1*-expression was reduced in *Mysm1*-/- HSCs, and MYSM1 was shown to bind to *Gfi1* promoter in lineage-negative bone marrow cells [34]. MYSM1 was required for normal recruitment of transcription factors GATA2 and RUNX1 to the *Gfi1* locus, while loss of MYSM1 resulted in increased recruitment of the PRC1 complex proteins RING1B and BMI1 and increased levels of histone H2AK119ub [34] (Figure 2). Furthermore, retroviral expression of *Gfi1* in *Mysm1*-/- cells could partly restore HSC quiescence and function [34]. Despite strong mechanistic evidence, it is important to note that these studies did not conduct genome-wide analyses of MYSM1-regulated genes in HSCs, and therefore cannot rule out possible alternative mechanisms for MYSM1 regulation of HSC function. Indeed a number of recent studies reported activation of p53 stress response in *Mysm1*-/- hematopoietic stem and progenitor cells (HSPCs) and a rescue of hematopoiesis and other *Mysm1*-deficiency phenotypes in *Mysm1*-/-p53-/- double-knockout mice [44,45]. Although a mechanistic link between *Gfi1*-loss and p53 activation was previously reported in other models [53,54], it is not clear at this point how p53 activation and the reduction in *Gfi1* expression are mechanistically connected in the *Mysm1*-/- mouse model.

3.2. MYSM1 in B Cell Development and Humoral Immune Response

Depletion of the B cell lineage is one of the major common features of *MYSM1*-deficiency in human patients and mouse models. The arrest in early B cell development in *Mysm1*-deficient mice was linked to the MYSM1 role in the induction of the expression of *Ebf1* gene, encoding an essential transcription factor for B cell lineage specification [20]. MYSM1 binding was detected at the *Ebf1*-locus promoter in B cells, *Ebf1*-expression was reduced in *Mysm1*-deficient pre-pro-B cells, and retroviral expression of *Ebf1* in *Mysm1*-deficient bone marrow could partly rescue B cell development in an in vitro assay [20]. Reduced expression of *Ebf1* in *Mysm1*-deficiency correlated with reduced recruitment of transcription factor E2A, increased recruitment of PRC1 complex proteins, and increased levels of histone H2AK119ub at the *Ebf1*-locus [20] (Figure 2). An interaction of MYSM1 with the BRM and BRG1 components of the SWI/SNF chromatin remodeling complex was also suggested to have a role in the regulation of *Ebf1* locus expression [20].

Surprisingly, despite the severe depletion of B cells, *Mysm1*-/- mice had normal antibody levels, could mount normal antigen-specific antibody responses to immunization, and had increased frequencies of antigen-specific plasma cells in their lymphoid organs, although the absolute numbers of plasma cells were nevertheless reduced [30]. Furthermore, *Mysm1*-/- plasma cells showed enhanced levels of antibody secretion in vitro and had altered gene expression profiles with downregulation of B cell lineage transcription factors *Pax5* and *Bach2*, and increased expression of plasma cell transcription factors *Blimp1* and *Xbp1* [30]. Altogether, MYSM1 was proposed to inhibit plasma cell differentiation by maintaining *Pax5* expression, via histone H2AK119-deubiquitination and recruitment of transcription factor PU.1 to the *Pax5* locus [30] (Figure 2). In another study, MYSM1 was also reported to act as a negative regulator of B1a cell development or expansion, by maintaining the expression of microRNA miR-150, working in concert with transcription factor cMYC [31] (Figure 2). This microRNA is a known regulator of hematopoiesis and lymphocyte development, modulating the expression of multiple targets through post-transcriptional mechanisms, such as repression of translation and promoting the degradation of specific mRNAs [36]. It is important to note however that although B1a cells represented a larger fraction of the overall B cell population in *Mysm1*-/- mice, they were nevertheless strongly reduced in absolute numbers in this animal model [31].

To characterize the checkpoints in lymphocyte development and adaptive immune response regulated by MYSM1, independently of MYSM1 functions at the earlier stages of hematopoiesis, we performed a deletion of *Mysm1* at different stages of B and T cell development in Cre/loxP mouse models [49,50]. Surprisingly, *Mysm1*-deletion from the pro-B cell stage (mb1-Cre) resulted in a mild depletion of B cell numbers, while *Mysm1*-deletion at the pre-B cell stage (CD19-cre) or in mature follicular B cells (CD21-cre) allowed the normal numbers of B cells to be maintained [50]. These B cells also retained normal levels of *Ebf1* and *Pax5* gene expression, despite a full loss of *Mysm1* transcript [50]. This indicates that the severe loss of B cells in *Mysm1*$^{-/-}$ mice is primarily due to upstream defects in lymphoid lineage specification, and not due to the cell-intrinsic functions of MYSM1 in B cells. Intriguingly, mature splenic B cells from *Mysm1*$^{fl/fl}$ mb1-cre mice had altered responses to in vitro stimulation, including increased expression of activation markers, impaired survival, and reduced proliferation, while *Mysm1*$^{fl/fl}$ CD21-cre B cells showed no such defects. This suggests that MYSM1 activity at early stages of B cell development may be required for epigenetic programing to allow normal responses to stimulation and engagement in immune response in mature B cells. Additionally, *Mysm1*$^{fl/fl}$ mb1-cre mice had normal serum antibody levels, normal antigen-specific antibody titres in response to immunization, and showed a trend toward an increase in plasma cell numbers, despite a reduction in overall B cell numbers [50], providing some support for the negative role of MYSM1 in the regulation of plasma cell differentiation [30].

Overall, the above studies indicated that MYSM1 plays a critical role in the early stages of B cell lineage specification and possibly also to a lesser extent in the regulation of plasma cell development in B-cell-mediated immune response. The exact mechanisms for B cell lineage depletion discussed above remain to be reconciled with the apparent role of p53 and the rescue of B cell development seen in *Mysm1*$^{-/-}$*p53*$^{-/-}$ mice [19,44,45]. Furthermore, the striking differences in the B cell phenotypes of mice with systemic versus conditional *Mysm1*-deletion highlight the importance of understanding the specific functions of MYSM1 at different stages in B cell lineage development, and the possible mechanisms of epigenetic memory, through which MYSM1 activity at early stages of development may affect B cell functions later in their maturation history.

3.3. MYSM1 in T Cell and NK Cell Development

Several studies explored the defects in T cell and NK cell differentiation in *Mysm1*-deficient mouse models. Thus *Mysm1*$^{-/-}$ mice have a severe reduction in mature NK cell numbers in the bone marrow, blood, and lymphoid organs [32]. This defect in NK cell maturation is attributed to the cell intrinsic role of MYSM1 in the induction of the gene encoding transcription factor ID2, known to be important for the NK cell lineage [32]. MYSM1 binding was detected at the *Id2*-locus promoter and loss of MYSM1 was associated with reduced recruitment of transcription factor NFIL3, increased recruitment of PRC1 complex proteins, increased levels of histone H2AK119ub, and reduced gene expression [32] (Figure 2). To our knowledge, the role of p53 activation in *Mysm1*-deficient NK cell dysfunction has not been addressed [19,44,45].

T cells are also significantly reduced in absolute numbers in the peripheral lymphoid organs of *Mysm1*$^{-/-}$ mice [19,45,55], and this is associated with a severe reduction in the cellularity of the thymus and a depletion of all thymocyte subsets [19,45,55], including the early thymic progenitors (ETPs) [45]. The underlying cellular mechanisms likely include the upstream defects in lymphoid lineage specification within the bone marrow [19,45,55], as well as elevated levels of cell apoptosis within the thymus [45]. At the molecular level, activation of the p53-stress response was extensively characterized and shown to be functionally important, based on the rescue of T cell development in *Mysm1*$^{-/-}$*p53*$^{-/-}$ mice [44,45]. The suggested triggers for p53 activation in *Mysm1*$^{-/-}$ include the putative roles of MYSM1 in the maintenance of p19ARF expression within the thymus [45], and IRF2 and IRF8 expression in bone marrow progenitor cells [55].

To characterize MYSM1 functions within the T cell lineage, independently of its roles at earlier stages of hematopoiesis, we performed a conditional deletion of *Mysm1* in Cre/loxP mouse models [49].

Deletion of *Mysm1* in the thymus from the double-negative 3 (DN3, Lck-Cre) or the double-positive (DP, CD4-Cre) stage in T cell lineage development resulted in normal thymocyte numbers and normal progression of thymic T cell development [49]. In peripheral lymphoid organs, mild depletion of CD8 T cells and elevated expression of activation markers on both CD8 and CD4 T cells were observed [49]. T cell responses to in vitro stimulation were also altered, with CD8 T cells showing elevated cytokine production, reduced induction of granzyme B, impaired proliferation, and increased apoptosis, while CD4 T cells showed only impaired proliferation [49]. Altogether, this indicated that the defects in T cell development in *Mysm1*-deficiency are primarily the result of upstream defects in lymphoid lineage specification within the bone marrow, but suggested some role for MYSM1 in the regulation of mature CD8 T cell activation, proliferation, and survival. Importantly, increased levels of p53 were observed in CD8 T cells from the *Mysm1*$^{fl/fl}$ CD4-cre mice [49], suggesting that the mechanisms for CD8 T cell dysfunction in this model might also be p53-dependent.

3.4. MYSM1 in Myeloid Lineage Cell Development and Innate Immune Response

The development of myeloid lineage leukocytes is also significantly impaired in *Mysm1*$^{-/-}$ mice, as demonstrated by the reduction in the absolute numbers of granulocytes, monocytes, and macrophages in peripheral lymphoid organs, and their progenitors in the bone marrow [33]. A profound requirement for MYSM1 in the development of dendritic cells (DCs), including both conventional and plasmacytoid lineages, was also reported [33]. Expression of receptor tyrosine kinase *Flt3*, known to be essential for DC development, was reduced in *Mysm1*$^{-/-}$ hematopoietic progenitors, and exogenous expression of *Flt3* through retroviral transduction in these cells could partially rescue DC development in vitro [33]. MYSM1 bound to the *Flt3* gene promoter, and loss of MYSM1 resulted in reduced recruitment of the major hematopoietic transcription factor PU.1 and an increase in repressive histone marks H2AK119ub and H3K27me3 at the *Flt3* locus [33] (Figure 2).

MYSM1 was also shown to have a direct role in the regulation of macrophage activation, in response to inflammatory stimuli and infection. Importantly, these MYSM1 activities were linked not to MYSM1 functions as a transcriptional regulator at chromatin, but relied on a cytosolic pool of MYSM1 protein that was transiently produced in activated macrophages [3,8]. This pool of MYSM1 worked to remove polyubiquitin chains from TRAF3, TRAF6, and RIP2, thereby, silencing TLR and NOD2 signal transduction pathways [3,8] (Figure 3). *Mysm1*$^{-/-}$ macrophages therefore showed enhanced production of inflammatory cytokines and type-I interferons, and mice with either systemic or myeloid lineage-restricted *Mysm1*-deletion were more susceptible to septic shock and peritonitis [3,8], but showed enhanced clearance of viral infections [3]. In a separate study, *Mysm1*$^{-/-}$ macrophages were confirmed to have elevated production of pro-inflammatory cytokines in response to stimulation, as well as increased proliferation, increased apoptosis, and enhanced capacity to control melanoma tumor growth in vivo in mouse models [56].

3.5. MYSM1 Functions beyond the Hematopoietic System

MYSM1 functions beyond the hematopoietic and immune system are apparent from visual examination of *Mysm1*-deficient mouse strains, which exhibit growth retardation, dysmorphology of hind-limbs and tail, and abnormal coat pigmentation (white belly patch) [19]. The most comprehensive source of information on the phenotypes of *Mysm1*-deficiency is provided by the International Mouse Phenotyping Consortium (IMPC) [21–23], and this catalogues other skeletal, metabolic, neurological, and behavioral phenotypes. Despite the comprehensive characterization of *Mysm1*$^{-/-}$ phenotypes by IMPC, only a few systems beyond hematopoiesis and immunity have been explored at a mechanistic level [24]. For example, skin phenotypes of *Mysm1*$^{-/-}$ mice were further characterized in several studies, reporting reduced cellularity and altered morphology of interfollicular epidermis, abnormal patterning of hair follicles and sebaceous glands, and disruptions of the hair follicle cycle [47,57,58]. This was associated with depletion of epidermal stem cells, reduced colony formation by epidermal progenitors, and activation of the transcriptional signatures of p53 stress response in the skin samples [47].

Abnormalities in the in vitro differentiation of *Mysm1^-/-^* mesenchymal stem cells (MSCs), such as enhanced adipogenesis, were also reported, and could be functionally linked to osteopenia and skeletal abnormalities in *Mysm1*-deficiency [48].

3.6. Possible Roles of MYSM1 in Cancer

Several recent studies explored the role of MYSM1 in various cancer models and suggested both pro- and anti-carcinogenic functions of MYSM1 protein. *Mysm1*-deficiency is known to result in p53-activation in hematopoietic cells [19,44,45], indicating that MYSM1 normally represses p53 activation, and suggesting it as a possible target for therapeutic p53-activation in hematological malignancies that retain wild type p53 function. At the same time, *Mysm1*-deficient mice are known to develop spontaneous thymic lymphomas at 6–9 months of age [44]. These tumors have not been extensively characterized, may carry genetic changes that silence the p53 stress response pathway, and may arise through mechanisms common with other mouse strains with impaired production of thymic progenitors [59]. Several other studies also suggested a role for MYSM1 as a positive regulator of cancer progression. Thus, MYSM1 was originally characterized as a positive regulator of androgen receptor target gene expression in human prostate cancer cell lines [7]. Furthermore, elevated MYSM1 protein levels were observed in human melanomas compared to normal melanocytes, and *Mysm1*-knockdown impaired the proliferation and survival of melanoma cell lines [40]. Moreover, increased MYSM1 protein levels were also seen in human colorectal tumors, compared to adjacent normal mucosa [60]. No malignancies have been reported to date in *MYSM1*-deficiency syndrome in human [25–27], however, due to the low patient numbers and their young age, our understanding of cancer susceptibility in human *MYSM1*-deficiency may be incomplete.

4. Conclusions

The discovery of *MYSM1* mutations in patients with a hereditary developmental and bone marrow failure syndrome raises strong interest in understanding the MYSM1-regulated checkpoints in mammalian physiology and the underlying molecular mechanisms of MYSM1 activity in these systems. Over the past decade, research in mouse models has provided numerous important insights, although many unanswered questions and opportunities for discoveries remain. The role of MYSM1 as a transcriptional regulator of hematopoiesis and immune cell development has been most extensively studied, although the diverse findings remain to be consolidated and reconciled. In particular, genome-wide characterization of MYSM1-regulated genes in the relevant hematopoietic cell types promise to provide many new insights into MYSM1 functions. The recently discovered role of MYSM1 in the regulation of signal transduction in the cytosol is highly interesting, however, research to date has focused only on the signaling pathways of innate immunity and primarily in macrophages. Furthermore, except for several recent insightful studies, MYSM1 functions and mechanisms of action beyond hematopoiesis and immunity remain poorly characterized. We expect that many new studies in coming years will address these and other outstanding questions, providing insights into the basic molecular mechanisms regulating mammalian development and physiology and further understanding into the disease pathology in *MYSM1*-deficiency in humans.

Author Contributions: Y.L., Y.H.L., J.T., H.W., D.L., A.N. contributed to manuscript preparation. A.F. contributed to manuscript preparation and the Figures. All authors have read and agreed to the published version of the manuscript.

Funding: This work was funded by the Canadian Institutes of Health Research (CIHR, Project Grant PJT-153016, Operating Grant MOP-123403), and supported by the Canadian Foundation for Innovation (CFI). A.N. is a Canada Research Chair Tier II in Hematopoiesis and Lymphocyte Differentiation. A.F. was a recipient of the Frederick Banting Tri-Council Scholarship, and was previously supported by FRQS Masters Training Studentship and the Max & Jane Childress Entrance Fellowship from the Department of Physiology of McGill University. Y.L. was supported by the Richard Birks Fellowship from the Department of Physiology of McGill University. Y.H.L. is a recipient of the Cole Foundation Studentship, and was previously supported by the Frederick Banting Tri-Council Graduate Scholarship. H.C.W. is a recipient of the Fonds de Recherche du Québec Santé (FRQS) Masters Training Studentship. D.L. was supported by the Fonds de Recherche du Québec Santé (FRQS).

Conflicts of Interest: The authors declare no conflicts of interest.

References

1. Ogata, K.; Morikawa, S.; Nakamura, H.; Sekikawa, A.; Inoue, T.; Kanai, H.; Sarai, A.; Ishii, S.; Nishimura, Y. Solution structure of a specific DNA complex of the Myb DNA-binding domain with cooperative recognition helices. *Cell* **1994**, *79*, 639–648. [CrossRef]
2. Yoneyama, M.; Tochio, N.; Umehara, T.; Koshiba, S.; Inoue, M.; Yabuki, T.; Aoki, M.; Seki, E.; Matsuda, T.; Watanabe, S.; et al. Structural and functional differences of SWIRM domain subtypes. *J. Mol. Biol.* **2007**, *369*, 222–238. [CrossRef] [PubMed]
3. Panda, S.; Nilsson, J.A.; Gekara, N.O. Deubiquitinase MYSM1 Regulates Innate Immunity through Inactivation of TRAF3 and TRAF6 Complexes. *Immunity* **2015**, *43*, 647–659. [CrossRef] [PubMed]
4. Qian, C.; Zhang, Q.; Li, S.; Zeng, L.; Walsh, M.J.; Zhou, M.M. Structure and chromosomal DNA binding of the SWIRM domain. *Nat. Struct. Mol. Biol.* **2005**, *12*, 1078–1085. [CrossRef]
5. Aravind, L.; Iyer, L.M. The SWIRM domain: A conserved module found in chromosomal proteins points to novel chromatin-modifying activities. *Genome Biol.* **2002**, *3*, RESEARCH0039. [CrossRef] [PubMed]
6. Birol, M.; Echalier, A. Structure and Function of MPN (Mpr1/Pad1 N-terminal) Domain-Containing Proteins. *Curr. Protein Pept. Sc.* **2014**, *15*, 504–517. [CrossRef]
7. Zhu, P.; Zhou, W.; Wang, J.; Puc, J.; Ohgi, K.A.; Erdjument-Bromage, H.; Tempst, P.; Glass, C.K.; Rosenfeld, M.G. A histone H2A deubiquitinase complex coordinating histone acetylation and H1 dissociation in transcriptional regulation. *Mol. Cell* **2007**, *27*, 609–621. [CrossRef]
8. Panda, S.; Gekara, N.O. The deubiquitinase MYSM1 dampens NOD2-mediated inflammation and tissue damage by inactivating the RIP2 complex. *Nat. Commun.* **2018**, *9*, 4654. [CrossRef]
9. Swatek, K.N.; Komander, D. Ubiquitin modifications. *Cell Res.* **2016**, *26*, 399–422. [CrossRef]
10. Steffen, P.A.; Ringrose, L. What are memories made of? How Polycomb and Trithorax proteins mediate epigenetic memory. *Nat. Rev. Mol. Cell Biol.* **2014**, *15*, 340–356. [CrossRef]
11. Vidal, M.; Starowicz, K. Polycomb complexes PRC1 and their function in hematopoiesis. *Exp. Hematol.* **2017**, *48*, 12–31. [CrossRef] [PubMed]
12. Wang, H.; Wang, L.; Erdjument-Bromage, H.; Vidal, M.; Tempst, P.; Jones, R.S.; Zhang, Y. Role of histone H2A ubiquitination in Polycomb silencing. *Nature* **2004**, *431*, 873–878. [CrossRef]
13. Pavri, R.; Zhu, B.; Li, G.; Trojer, P.; Mandal, S.; Shilatifard, A.; Reinberg, D. Histone H2B monoubiquitination functions cooperatively with FACT to regulate elongation by RNA polymerase II. *Cell* **2006**, *125*, 703–717. [CrossRef] [PubMed]
14. Chandrasekharan, M.B.; Huang, F.; Sun, Z.W. Histone H2B ubiquitination and beyond: Regulation of nucleosome stability, chromatin dynamics and the trans-histone H3 methylation. *Epigenetics* **2010**, *5*, 460–468. [CrossRef] [PubMed]
15. Belle, J.I.; Nijnik, A. H2A-DUBbing the mammalian epigenome: Expanding frontiers for histone H2A deubiquitinating enzymes in cell biology and physiology. *Int. J. Biochem. Cell Biol.* **2014**, *50*, 161–174. [CrossRef] [PubMed]
16. Jackson, S.P.; Durocher, D. Regulation of DNA damage responses by ubiquitin and SUMO. *Mol. Cell* **2013**, *49*, 795–807. [CrossRef] [PubMed]
17. Han, J.; Zhang, H.; Zhang, H.; Wang, Z.; Zhou, H.; Zhang, Z. A Cul4 E3 ubiquitin ligase regulates histone hand-off during nucleosome assembly. *Cell* **2013**, *155*, 817–829. [CrossRef]
18. Sarcinella, E.; Zuzarte, P.C.; Lau, P.N.; Draker, R.; Cheung, P. Monoubiquitylation of H2A.Z distinguishes its association with euchromatin or facultative heterochromatin. *Mol. Cell Biol.* **2007**, *27*, 6457–6468. [CrossRef]
19. Nijnik, A.; Clare, S.; Hale, C.; Raisen, C.; McIntyre, R.E.; Yusa, K.; Everitt, A.R.; Mottram, L.; Podrini, C.; Lucas, M.; et al. The critical role of histone H2A-deubiquitinase Mysm1 in hematopoiesis and lymphocyte differentiation. *Blood* **2012**, *119*, 1370–1379. [CrossRef]
20. Jiang, X.X.; Nguyen, Q.; Chou, Y.; Wang, T.; Nandakumar, V.; Yates, P.; Jones, L.; Wang, L.; Won, H.; Lee, H.R.; et al. Control of B cell development by the histone H2A deubiquitinase MYSM1. *Immunity* **2011**, *35*, 883–896. [CrossRef]

21. Skarnes, W.; Rosen, B.; West, A.; Koutsourakis, M.; Bushell, W.; Iyer, V.; Cox, T.; Jackson, D.; Severin, J.; Biggs, P.; et al. A conditional knockout resource for genome-wide analysis of mouse gene function. *Nature* **2011**, *474*, 337–342. [CrossRef] [PubMed]

22. White, J.K.; Gerdin, A.K.; Karp, N.A.; Ryder, E.; Buljan, M.; Bussell, J.N.; Salisbury, J.; Clare, S.; Ingham, N.J.; Podrini, C.; et al. Genome-wide Generation and Systematic Phenotyping of Knockout Mice Reveals New Roles for Many Genes. *Cell* **2013**, *154*, 452–464. [CrossRef] [PubMed]

23. Dickinson, M.E.; Flenniken, A.M.; Ji, X.; Teboul, L.; Wong, M.D.; White, J.K.; Meehan, T.F.; Weninger, W.J.; Westerberg, H.; Adissu, H.; et al. High-throughput discovery of novel developmental phenotypes. *Nature* **2016**, *537*, 508–514. [CrossRef] [PubMed]

24. International Mouse Phenotyping Consortium. Available online: https://www.mousephenotype.org/data/genes/MGI:2444584 (accessed on 28 March 2020).

25. Alsultan, A.; Shamseldin, H.E.; Osman, M.E.; Aljabri, M.; Alkuraya, F.S. MYSM1 is mutated in a family with transient transfusion-dependent anemia, mild thrombocytopenia, and low NK- and B-cell counts. *Blood* **2013**, *122*, 3844–3845. [CrossRef]

26. Le Guen, T.; Touzot, F.; Andre-Schmutz, I.; Lagresle-Peyrou, C.; France, B.; Kermasson, L.; Lambert, N.; Picard, C.; Nitschke, P.; Carpentier, W.; et al. An in vivo genetic reversion highlights the crucial role of Myb-Like, SWIRM, and MPN domains 1 (MYSM1) in human hematopoiesis and lymphocyte differentiation. *J. Allergy Clin. Immunol.* **2015**, *136*, 1619–1626 e1615. [CrossRef]

27. Bahrami, E.; Witzel, M.; Racek, T.; Puchalka, J.; Hollizeck, S.; Greif-Kohistani, N.; Kotlarz, D.; Horny, H.P.; Feederle, R.; Schmidt, H.; et al. Myb-like, SWIRM, and MPN domains 1 (MYSM1) deficiency: Genotoxic stress-associated bone marrow failure and developmental aberrations. *J. Allergy Clin. Immunol.* **2017**, *140*, 1112–1119. [CrossRef] [PubMed]

28. Nanda, A.; Al-Abboh, H.; Zahra, A.; Al-Sabah, H.; Gupta, A.; Adekile, A.D. Neutrophilic Panniculitis in a child with MYSM1 deficiency. *Pediatr. Dermatol.* **2019**, *36*, 258–259. [CrossRef] [PubMed]

29. Ulirsch, J.C.; Verboon, J.M.; Kazerounian, S.; Guo, M.H.; Yuan, D.; Ludwig, L.S.; Handsaker, R.E.; Abdulhay, N.J.; Fiorini, C.; Genovese, G.; et al. The Genetic Landscape of Diamond-Blackfan Anemia. *Am. J. Hum. Genet.* **2018**, *103*, 930–947. [CrossRef] [PubMed]

30. Jiang, X.X.; Chou, Y.; Jones, L.; Wang, T.; Sanchez, S.; Huang, X.F.; Zhang, L.; Wang, C.; Chen, S.Y. Epigenetic Regulation of Antibody Responses by the Histone H2A Deubiquitinase MYSM1. *Sci. Rep.* **2015**, *5*, 13755. [CrossRef]

31. Jiang, X.X.; Liu, Y.; Li, H.; Gao, Y.; Mu, R.; Guo, J.; Zhang, J.; Yang, Y.M.; Xiao, F.; Liu, B.; et al. MYSM1/miR-150/FLT3 inhibits B1a cell proliferation. *Oncotarget* **2016**, *7*, 68086–68096. [CrossRef]

32. Nandakumar, V.; Chou, Y.; Zang, L.; Huang, X.F.; Chen, S.Y. Epigenetic control of natural killer cell maturation by histone H2A deubiquitinase, MYSM1. *Proc. Natl. Acad. Sci. USA* **2013**, *110*, E3927–E3936. [CrossRef] [PubMed]

33. Won, H.; Nandakumar, V.; Yates, P.; Sanchez, S.; Jones, L.; Huang, X.F.; Chen, S.Y. Epigenetic control of dendritic cell development and fate determination of common myeloid progenitor by Mysm1. *Blood* **2014**, *124*, 2647–2656. [CrossRef]

34. Wang, T.; Nandakumar, V.; Jiang, X.X.; Jones, L.; Yang, A.G.; Huang, X.F.; Chen, S.Y. The control of hematopoietic stem cell maintenance, self-renewal, and differentiation by Mysm1-mediated epigenetic regulation. *Blood* **2013**, *122*, 2812–2822. [CrossRef]

35. Wang, Y.H.; Huang, X.H.; Yang, Y.M.; He, Y.; Dong, X.H.; Yang, H.X.; Zhang, L.; Wang, Y.; Zhou, J.; Wang, C.; et al. Mysm1 epigenetically regulates the immunomodulatory function of adipose-derived stem cells in part by targeting miR-150. *J. Cell Mol. Med.* **2019**, *23*, 3737–3746. [CrossRef] [PubMed]

36. He, Y.; Jiang, X.; Chen, J. The role of miR-150 in normal and malignant hematopoiesis. *Oncogene* **2014**, *33*, 3887–3893. [CrossRef] [PubMed]

37. Jacq, X.; Kemp, M.; Martin, N.M.B.; Jackson, S.P. Deubiquitylating Enzymes and DNA Damage Response Pathways. *Cell Biochem. Biophys.* **2013**, *67*, 25–43. [CrossRef] [PubMed]

38. Matsuoka, S.; Ballif, B.A.; Smogorzewska, A.; McDonald, E.R.; Hurov, K.E.; Luo, J.; Bakalarski, C.E.; Zhao, Z.; Solimini, N.; Lerenthal, Y.; et al. ATM and ATR substrate analysis reveals extensive protein networks responsive to DNA damage. *Science* **2007**, *316*, 1160–1166. [CrossRef]

39. Nishi, R.; Wijnhoven, P.; le Sage, C.; Tjeertes, J.; Galanty, Y.; Forment, J.V.; Clague, M.J.; Urbe, S.; Jackson, S.P. Systematic characterization of deubiquitylating enzymes for roles in maintaining genome integrity. *Nat. Cell Biol.* **2014**, *16*, 1016–1026. [CrossRef]

40. Wilms, C.; Kroeger, C.M.; Hainzl, A.V.; Banik, I.; Bruno, C.; Krikki, I.; Farsam, V.; Wlaschek, M.; Gatzka, M.V. MYSM1/2A-DUB is an epigenetic regulator in human melanoma and contributes to tumor cell growth. *Oncotarget* **2017**, *8*, 67287–67299. [CrossRef]

41. Belle, J.I.; Petrov, J.C.; Langlais, D.; Robert, F.; Cencic, R.; Shen, S.; Pelletier, J.; Gros, P.; Nijnik, A. Repression of p53-target gene Bbc3/PUMA by MYSM1 is essential for the survival of hematopoietic multipotent progenitors and contributes to stem cell maintenance. *Cell Death Differ.* **2016**, *23*, 759–775. [CrossRef]

42. Bieging, K.T.; Mello, S.S.; Attardi, L.D. Unravelling mechanisms of p53-mediated tumour suppression. *Nat. Rev. Cancer* **2014**, *14*, 359–370. [CrossRef] [PubMed]

43. Levine, A.J.; Oren, M. The first 30 years of p53: Growing ever more complex. *Nat. Rev. Cancer* **2009**, *9*, 749–758. [CrossRef] [PubMed]

44. Belle, J.I.; Langlais, D.; Petrov, J.C.; Pardo, M.; Jones, R.G.; Gros, P.; Nijnik, A. p53 mediates loss of hematopoietic stem cell function and lymphopenia in Mysm1 deficiency. *Blood* **2015**, *125*, 2344–2348. [CrossRef] [PubMed]

45. Gatzka, M.; Tasdogan, A.; Hainzl, A.; Allies, G.; Maity, P.; Wilms, C.; Wlaschek, M.; Scharffetter-Kochanek, K. Interplay of H2A deubiquitinase 2A-DUB/Mysm1 and the p19/p53 axis in hematopoiesis, early T-cell development and tissue differentiation. *Cell Death Differ.* **2015**, *22*, 1451–1462. [CrossRef] [PubMed]

46. Forster, M.; Belle, J.I.; Petrov, J.C.; Ryder, E.J.; Clare, S.; Nijnik, A. Deubiquitinase MYSM1 Is Essential for Normal Fetal Liver Hematopoiesis and for the Maintenance of Hematopoietic Stem Cells in Adult Bone Marrow. *Stem Cells Dev.* **2015**, *24*, 1865–1877. [CrossRef]

47. Wilms, C.; Krikki, I.; Hainzl, A.; Kilo, S.; Alupei, M.; Makrantonaki, E.; Wagner, M.; Kroeger, C.M.; Brinker, T.J.; Gatzka, M. 2A-DUB/Mysm1 Regulates Epidermal Development in Part by Suppressing p53-Mediated Programs. *Int. J. Mol. Sci.* **2018**, *19*, 687. [CrossRef]

48. Haffner-Luntzer, M.; Kovtun, A.; Fischer, V.; Prystaz, K.; Hainzl, A.; Kroeger, C.M.; Krikki, I.; Brinker, T.J.; Ignatius, A.; Gatzka, M. Loss of p53 compensates osteopenia in murine Mysm1 deficiency. *FASEB J.* **2018**, *32*, 1957–1968. [CrossRef]

49. Forster, M.; Boora, R.K.; Petrov, J.C.; Fodil, N.; Albanese, I.; Kim, J.; Gros, P.; Nijnik, A. A role for the histone H2A deubiquitinase MYSM1 in maintenance of CD8(+) T cells. *Immunology* **2017**, *151*, 110–121. [CrossRef]

50. Forster, M.; Farrington, K.; Petrov, J.C.; Belle, J.I.; Mindt, B.C.; Witalis, M.; Duerr, C.U.; Fritz, J.H.; Nijnik, A. MYSM1-dependent checkpoints in B cell lineage differentiation and B cell-mediated immune response. *J. Leukoc. Biol.* **2017**, *101*, 643–654. [CrossRef]

51. Petrov, J.C.; Nijnik, A. Mysm1 expression in the bone marrow niche is not essential for hematopoietic maintenance. *Exp. Hematol.* **2017**, *47*, 76–82 e73. [CrossRef]

52. Huo, Y.; Li, B.Y.; Lin, Z.F.; Wang, W.; Jiang, X.X.; Chen, X.; Xi, W.J.; Yang, A.G.; Chen, S.Y.; Wang, T. MYSM1 Is Essential for Maintaining Hematopoietic Stem Cell (HSC) Quiescence and Survival. *Med. Sci. Monit.* **2018**, *24*, 2541–2549. [CrossRef] [PubMed]

53. Khandanpour, C.; Phelan, J.D.; Vassen, L.; Schutte, J.; Chen, R.; Horman, S.R.; Gaudreau, M.C.; Krongold, J.; Zhu, J.; Paul, W.E.; et al. Growth factor independence 1 antagonizes a p53-induced DNA damage response pathway in lymphoblastic leukemia. *Cancer Cell* **2013**, *23*, 200–214. [CrossRef] [PubMed]

54. Vadnais, C.; Chen, R.; Fraszczak, J.; Hamard, P.J.; Manfredi, J.J.; Moroy, T. A novel regulatory circuit between p53 and GFI1 controls induction of apoptosis in T cells. *Sci. Rep.* **2019**, *9*, 6304. [CrossRef] [PubMed]

55. Huang, X.F.; Nandakumar, V.; Tumurkhuu, G.; Wang, T.; Jiang, X.; Hong, B.; Jones, L.; Won, H.; Yoshii, H.; Ozato, K.; et al. Mysm1 is required for interferon regulatory factor expression in maintaining HSC quiescence and thymocyte development. *Cell Death Dis.* **2016**, *7*, e2260. [CrossRef] [PubMed]

56. Zhao, X.; Huang, X.H.; Dong, X.H.; Wang, Y.H.; Yang, H.X.; Wang, Y.; He, Y.; Liu, S.; Zhou, J.; Wang, C.; et al. Deubiquitinase Mysm1 regulates macrophage survival and polarization. *Mol. Biol. Rep.* **2018**, *45*, 2393–2401. [CrossRef] [PubMed]

57. Liakath-Ali, K.; Vancollie, V.E.; Heath, E.; Smedley, D.P.; Estabel, J.; Sunter, D.; Ditommaso, T.; White, J.K.; Ramirez-Solis, R.; Smyth, I.; et al. Novel skin phenotypes revealed by a genome-wide mouse reverse genetic screen. *Nat. Commun.* **2014**, *5*, 3540. [CrossRef] [PubMed]

58. DiTommaso, T.; Jones, L.K.; Cottle, D.L.; Program, W.M.G.; Gerdin, A.K.; Vancollie, V.E.; Watt, F.M.; Ramirez-Solis, R.; Bradley, A.; Steel, K.P.; et al. Identification of genes important for cutaneous function revealed by a large scale reverse genetic screen in the mouse. *PLoS Genet.* **2014**, *10*, e1004705. [CrossRef]

59. Martins, V.C.; Busch, K.; Juraeva, D.; Blum, C.; Ludwig, C.; Rasche, V.; Lasitschka, F.; Mastitsky, S.E.; Brors, B.; Hielscher, T.; et al. Cell competition is a tumour suppressor mechanism in the thymus. *Nature* **2014**, *509*, 465–470. [CrossRef]

60. Li, Y.; Li, J.; Liu, H.; Liu, Y.; Cui, B. Expression of MYSM1 is associated with tumor progression in colorectal cancer. *PLoS ONE* **2017**, *12*, e0177235. [CrossRef]

MDPI
St. Alban-Anlage 66
4052 Basel
Switzerland
Tel. +41 61 683 77 34
Fax +41 61 302 89 18
www.mdpi.com

International Journal of Molecular Sciences Editorial Office
E-mail: ijms@mdpi.com
www.mdpi.com/journal/ijms